Studies in Systems, Decision and Control

Volume 251

The series "Studies in Systems, Decision and Control" (SSDC) covers both new developments and advances, as well as the state of the art, in the various areas of broadly perceived systems, decision making and control–quickly, up to date and with a high quality. The intent is to cover the theory, applications, and perspectives on the state of the art and future developments relevant to systems, decision making, control, complex processes and related areas, as embedded in the fields of engineering, computer science, physics, economics, social and life sciences, as well as the paradigms and methodologies behind them. The series contains monographs, textbooks, lecture notes and edited volumes in systems, decision making and control spanning the areas of Cyber-Physical Systems, Autonomous Systems, Sensor Networks, Control Systems, Energy Systems, Automotive Systems, Biological Systems, Vehicular Networking and Connected Vehicles, Aerospace Systems, Automation, Manufacturing, Smart Grids, Nonlinear Systems, Power Systems, Robotics, Social Systems, Economic Systems and other. Of particular value to both the contributors and the readership are the short publication timeframe and the world-wide distribution and exposure which enable both a wide and rapid dissemination of research output.

** Indexing: The books of this series are submitted to ISI, SCOPUS, DBLP, Ulrichs, MathSciNet, Current Mathematical Publications, Mathematical Reviews, Zentralblatt Math: MetaPress and Springerlink.

More information about this series at http://www.springer.com/series/13304

Jan Treur

Network-Oriented Modeling for Adaptive Networks: Designing Higher-Order Adaptive Biological, Mental and Social Network Models

Springer

Jan Treur
Social AI Group, Department
of Computer Science
Vrije Universiteit Amsterdam
Amsterdam, Noord-Holland, The Netherlands

ISSN 2198-4182 ISSN 2198-4190 (electronic)
Studies in Systems, Decision and Control
ISBN 978-3-030-31447-7 ISBN 978-3-030-31445-3 (eBook)
https://doi.org/10.1007/978-3-030-31445-3

This Springer imprint is published by the registered company Springer Nature Switzerland AG
The registered company address is: Gewerbestrasse 11, 6330 Cham, Switzerland

Preface

This book addresses the challenging topic of modeling adaptive networks, which often have inherently complex behavior. Networks by themselves usually can be modeled using a neat, declarative and conceptually transparent Network-Oriented Modeling approach. For adaptive networks changing the network's structure, it is different; often separate procedural specifications are added for the adaptation process. This leaves you with a less transparent, hybrid specification, part of which often is more at a programming level than at a modeling level. This book presents an overall Network-Oriented Modeling approach by which designing adaptive network models becomes much easier, as also the adaptation process is modeled in a neat, declarative and conceptually transparent Network-Oriented Modeling manner, like the network itself. Due to this dedicated overall Network-Oriented Modeling approach, no procedural, algorithmic or programming skills are needed to design complex adaptive network models. A dedicated software environment is available to run these adaptive network models from their high-level specifications. Moreover, as adaptive networks are described in a network format as well, the approach can simply be applied iteratively, so that higher order adaptive networks in which network adaptation itself is adaptive too, can be modeled just as easily; for example, this can be applied to model metaplasticity from Cognitive Neuroscience. The usefulness of this approach is illustrated in the book by many examples of complex (higher order) adaptive network models for a wide variety of biological, mental and social processes.

The book has been written with multidisciplinary Master and Ph.D. students in mind without assuming much prior knowledge, although also some elementary mathematical analysis is not completely avoided. The detailed presentation makes that it can be used as an introduction in Network-Oriented Modeling for adaptive networks. Sometimes overlap between chapters can be found in order to make it easier to read each chapter separately. In each of the chapters, in the Discussion section specific publications and authors are indicated that relate to the material presented in the chapter. The specific mathematical details concerning difference and differential equations have been concentrated in Chaps. 10–15 in Part IV and Part V, which easily can be skipped, if desired. For a modeler who just wants to use this modeling approach, Chaps. 1–9 provide a good introduction.

The material in this book is being used in teaching undergraduate and graduate students with multidisciplinary background or interests. Lecturers can contact me for additional material such as slides, assignments and software. The content of the book has benefited much from cooperation with students and (past and current) members of the Social AI Group (formerly the Behavioural Informatics Group) at the Vrije Universiteit Amsterdam. In particular, I am grateful to Fakhra Jabeen, Nimat Ullah and Joey van den Heuvel who gave useful feedback on earlier versions of most of the chapters.

Amsterdam, The Netherlands Jan Treur
July 2019

Contents

Part I
Introduction

Chapter 1
On Adaptive Networks and Network Reification

Abstract This chapter is a brief preview of what can be expected in this book, with some pointers to various chapters and sections. First, it is discussed how networks can be adaptive in different ways and according to different orders. A variety of examples of first and second-order adaptation are summarized, and the possibility of adaptation of order higher than two is discussed. After this, the notion of network reification is briefly summarized and how it can be used to model adaptive networks in a transparent and network-oriented manner. It is pointed out how repeated application of network reification can be used to model adaptive networks with the adaptation of multiple orders. Finally, it is discussed how mathematical analysis of emerging behavior of a network not only can be applied to non-adaptive base networks, but also to reified adaptive networks.

1.1 Introduction

To model dynamics in real-world processes, different dynamic modeling approaches have been developed based on some type of dynamical system architecture; e.g., Ashby (1960), Port and van Gelder (1995). Within the dynamic modeling area in general, adaptive behaviour is a nontrivial and interesting challenge. For network-oriented dynamic modeling approaches in particular, the considered model structure is a network structure. It turns out that for many real-world domains, network models often show some form of network adaptation by which some of the network structure characteristics change over time. This can be described by *network adaptation principles* specifying how exactly certain characteristics of a network structure change over time. A well-known example of such an adaptation principle concerns adaptation of the connection weights in Mental Networks by *Hebbian learning* (Hebb 1949), which within Cognitive Neuroscience is considered a form of *plasticity* (see Sect. 1.4.2 below). Another example of an adaptation principle is adaptation of the connection weights in Social Networks by *bonding based on homophily*; e.g., Byrne (1986), McPherson et al. (2001), Pearson et al. (2006), Sharpanskykh and Treur (2014).

J. Treur, *Network-Oriented Modeling for Adaptive Networks: Designing Higher-Order Adaptive Biological, Mental and Social Network Models*, Studies in Systems, Decision and Control 251, https://doi.org/10.1007/978-3-030-31445-3_1

The adaptive behavior itself can also be adaptive, which leads to adaptation of different orders; for example, within Cognitive Neuroscience *metaplasticity* determines under which circumstances and to which extent plasticity occurs in Mental Networks; e.g., Abraham and Bear (1996), Magerl et al. (2018), Sehgal et al. (2013), Schmidt et al. (2013), Zelcer et al. (2006). To model adaptive networks in a neat and easily manageable manner is a nontrivial challenge, and even more so when they are adaptive of higher-order.

The notion of *network reification* as introduced in this book is a means to model adaptive networks in a more transparent manner within a Network-Oriented Modelling perspective. Reification literally means representing something abstract as a material or more real concrete thing (Merriam-Webster and Oxford dictionaries). This concept is used in different scientific areas in which it has been shown to provide substantial advantages in expressivity and transparency of models, and, in particular, within AI; e.g., Davis and Buchanan (1977), Davis (1980), Bowen and Kowalski (1982), Demers and Malenfant (1995), Galton (2006), Hofstadter (1979), Sterling and Shapiro (1996), Sterling and Beer (1989), Weyhrauch (1980). Specific cases of reification from a linguistic or logical perspective are representing relations between objects by objects themselves, or representing more complex statements by objects or numbers. For network models, reification can be applied by reifying the network structure characteristics in the form of additional network states (called reification states) within an extended network. Multilevel reified networks can be used to model networks which are adaptive of different orders. It is also discussed how mathematical analysis can be applied to reified networks.

In this chapter, in Sect. 1.2 various examples of first and second-order adaptation are summarized, and in Sect. 1.3 the possibility of adaptation of order higher than 2 is discussed. In Sect. 1.4 the notion of network reification is briefly discussed and how it can be used to model adaptive networks in a more transparent manner; in Sect. 1.5 it is discussed how repeated application of network reification can be used to model adaptive networks with adaptation of multiple orders. In Sect. 1.6 it is discussed how mathematical analysis of emerging behavior of a network not only can be applied to base networks, but also to reification states in reified networks. Finally, Sect. 1.7 is a discussion.

1.2 First- and Second-Order Adaptation

In this section, a brief overview is given of a variety of known adaptation principles of different orders. It shows the wide range of potential applications for a Network-Oriented Modeling approach based on network reification. First, first-order adaptation is addressed in Sect. 1.2.1; next, adaptation of second-order is addressed in Sect. 1.2.2.

1.2.1 First-Order Adaptation

There are many well-known examples of first-order adaptive networks, for example, related to or inspired by adaptation principles from Cognitive Neuroscience, Cognitive Science or Social Science. Just a few examples are listed below. Although the majority of the first-order network adaptation principles known in the literature consider adaptations of connection weights over time, also other characteristics of the network structure can be considered to be adaptive, for example, the way in which incoming impact is aggregated or the speed of processing, as the last two bullets point out:

- Mental or neural networks equipped with a Hebbian learning adaptation principle (Hebb 1949) to adapt connection weights over time ('neurons that fire together, wire together'); see Sects. 1.4 and 1.6 below, Chap. 3, Sect. 3.6.1, and Chap. 4 in this book.
- Mental networks in which an adaptation principle describes how stress affects the connections ('state-connection modulation'); e.g., (Sousa et al. 2012; Treur and Mohammadi Ziabari 2018), see also and Chap. 3, Sect. 3.6.4, and Chap. 5 in this book.
- Mental networks in which an adaptation principle describes how context factors can affect the excitability of states; e.g., (Chandra and Barkai 2018); see also and Chap. 3, Sect. 3.6.5, and Chap. 4 in this book.
- Social networks equipped with an adaptation principle for bonding based on homophily (Byrne 1986; McPherson et al. 2001; Pearson et al. 2006; Sharpanskykh and Treur 2014; Beukel et al. 2019; Blankendaal et al. 2016; Boomgaard et al. 2018) to adapt connection weights over time ('birds of a feather flock together'); see also Chap. 3, Sect. 3.6.1, and Chap. 6 in this book.
- Social networks equipped with a triadic closure adaptation principle to adapt connection weights over time ('friends of my friends will become my friends'); e.g., Rapoport (1953), Banks and Carley (1996), see also Chap. 3, Sect. 3.6.2 in this book.
- Social networks equipped with a preferential attachment adaptation principle expressing that connections are strengthened preferably to nodes that have more and/or stronger connections ('more becomes more'); e.g., Barabasi and Albert (1999); see also Chap. 3, Sect. 3.6.3 in this book.
- Adaptive social network models and analysis of these networks for a variety of application domains can be found in the work around the toolkit for dynamic network analysis and visualization ORA (Carley 2017; Carley et al. 2013b). Among the many applications are (Carley et al. 2013a; Carley and Pfeffer 2012; Merrill et al. 2015).
- Neural networks equipped with (machine) learning mechanisms such as back-propagation or deep learning to adapt connection weights over time; e.g., LeCun et al. (2015).
- Adaptive functions for aggregation of incoming impact and activation of nodes. For example, as mentioned above, their threshold values to model adaptive

intrinsic properties of neurons such as their excitability; e.g., Chandra and Barkai (2018). As another example, the mechanism for the formation of an opinion based on multiple incoming opinions may change over time from selecting the maximal value of them to using the average instead; e.g., Chap. 3, Sects. 3.6.7 and 3.7 in this book. Yet another example describes how for multicriteria decision-making criteria weight factors are changed over time.
- Adaptive speed of states to model adaptive processing speed, for example, the response time of a person depending on workload, or intake of certain chemicals that affect response time; e.g., Chap. 3, Sects. 3.6.6 and 3.7 in this book.

As several real-world examples show, adaptation principles may be adaptive themselves too, according to certain second-order adaptation principles. This will be discussed next.

1.2.2 Second-Order Adaptation

Second-order adaptation can occur in different forms. From recent literature it is apparent that in real world domains characteristics representing adaptation principles often can still change over time, depending on circumstances. The notion of metaplasticity or second-order adaptation has become an important topic within Cognitive Neuroscience and Social Sciences. Some examples are:

- In literature such as Abraham and Bear (1996), Chandra and Barkai (2018), Daimon et al. (2017), Magerl et al. (2018), Parsons (2018), Robinson et al. (2016), Sehgal et al. (2013), Schmidt et al. (2013), Zelcer et al. (2006) various studies are reported which show how adaptation of synapses as described, for example, by first-order adaptation principles based on Hebbian learning can be modulated by a second-order adaptation principle suppressing the first-order adaptation process or amplifying it, thus some form of *metaplasticity* is described. Factors affecting synaptic plasticity as reported are presynaptic or postsynaptic activation, previous (learning) experiences, stress, or intake of certain chemicals or medicine; e.g., (Robinson et al. 2016): 'Adaptation accelerates with increasing stimulus exposure' (p. 2). This is addressed in Chap. 4, Sects. 4.4 and 4.5 in this book.
- From the Social Science area, in an adaptive social network based on a first-order adaptation principle for bonding based on homophily (McPherson et al. 2001) the similarity measure determining how similar two persons are may change over time by a second-order adaptation principle, for example, due to age or other varying circumstances. As an example, for somebody who is very busy or already has a lot of connections the requirements for being similar might become more strict; e.g., see Treur (2018b, 2019b) and Chap. 6 in this book.
- Also in the Social Science area the second-order adaptation concept called 'inhibiting adaptation' can be found Carley (2001, 2002, 2006). The idea is that

networked organisations need to be adaptive in order to survive in a dynamic world. However, some types of circumstances affect this first-order adaptivity in a negative manner, for example, frequent changes of persons or (other) resources. Such circumstances can be considered as inhibiting the adaptation capabilities of the organisation. Especially in Carley (2006) it is described in some detail how such a second-order adaptation principle based on inhibiting the first-order adaptation can be exploited as a strategy to attack organisations that are considered harmful or dangerous such as terrorist networks, by creating circumstances that indeed achieve inhibiting adaptation.

The second item above on adaptive adaptation principles for bonding based on homophily is illustrated in more detail in Chap. 6 in this book; see also (Treur 2018b). For the first item, adaptive adaptation principles for Hebbian learning have been considered in which the adaptation speed (learning rate) and the persistence factor for the first-order Hebbian learning adaptation principle are changing based on a second-order adaptation principle; see Chap. 4, Sects. 4.4 and 4.5 in this book.

In Fessler et al. (2015) some interesting ideas are put forward on first and second-order adaptation for the area of evolutionary adaptive processes.

> For example, the S-curve in the human spine reflects the determinative influence of the original function of the spine as a suspensory beam in a quadrupedal mammal, in contrast to its current function as a load-bearing pillar: whereas the original design functioned efficiently in a horizontal position, the transition to bipedality required the introduction of bends in the spine to position weight over the pelvis (Lovejoy 2005). The resulting configuration makes humans prone to lower-back injury, illustrating how path dependence can both set the stage for kludgy designs and constrain their optimality. Moreover, the combination of bipedality and pressures favoring large brain size in humans exacerbates a conflict between the biomechanics of locomotion (favoring a narrow pelvis) and the need to accommodate a large infant skull during parturition. This increases the importance of higher-order adaptations such as relaxin, a hormone that loosens ligaments during pregnancy, allowing the pelvic bones to separate. (Fessler et al. 2015)

According to this the following types of adaptation can be considered for the human spine:

- First-order adaptation:
  for quadrupedal mammals, a straight horizontal spine is an advantage.
- Second-order adaptation:
  transition to bipedality requires the introduction of bends in the spine to position weight over the pelvis; this makes humans prone to lower-back injury.

Similarly, the following types of adaptation can be considered for the human pelvis:

- First-order adaptation:
  bipedality favors a narrow pelvis.
- Second-order adaptation:
  larger brain size needs a wider pelvis: using relaxin allowing the pelvic bones to separate during giving birth.

1.3 Higher-Order Adaptation

A next question is whether also relevant examples of third- or even higher-order adaptation can be found. First, it will be discussed what orders of adaptation are addressed in the literature. After, a few examples from an evolutionary context are pointed out.

1.3.1 What Orders of Adaptation Are Addressed?

A Google Scholar search in this direction on February 3, 2019 resulted in the following outcomes:

"adaptation"	4 million hits
"second-order adaptation"	360 hits
"third-order adaptation"	14 hits
"fourth-order adaptation"	3 hits.

This shows a very fast decreasing pattern. Further inspection of the left graph in Fig. 1.1 depicting the logarithm of the number of hits against the order clearly shows that the pattern is not just (negatively) exponential. Instead, the right graph in Fig. 1.1 depicting the double logarithm of the number of hits fits much better. The dotted line in that graph is a linear trendline with linear formula in x and y as indicated: the double logarithm seems to allow an almost linear approximation, so the number of hits is in the order of a double (negative) exponential pattern

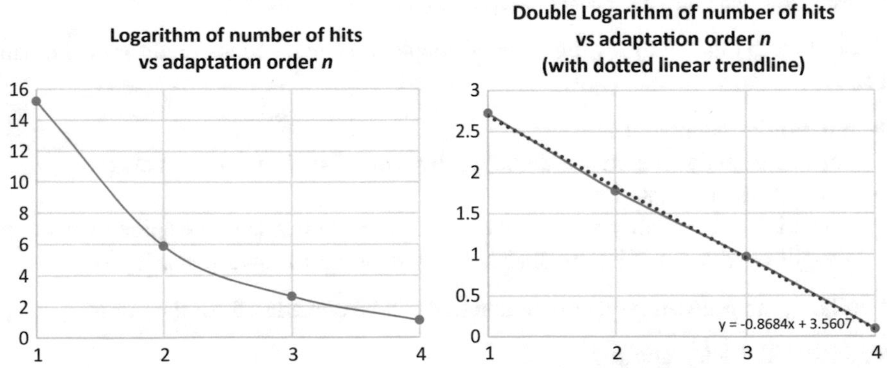

Fig. 1.1 Logarithm (left graph) and double logarithm (right graph) of the number of hits of adaptivity of different orders (vertical axis) in Google Scholar versus the order of adaptation (horizontal axis). The dotted linear trendline shows that the double logarithm of the number of hits has an almost linear dependence of the order of adaptivity

$e^{35.18782e-0.8684n}$ as a function of the order n. This very strongly decreasing pattern may suggest that within a research context, an adaptation of order higher than two is not often considered a very useful or applicable notion.

1.3.2 *Examples of Adaptation of Order Higher Than Two from an Evolutionary Context*

As one of the hits found for third-order adaptation, in (Fessler et al. 2015) the following is put forward referring to second and third-order adaptation:

> Also of relevance here, one form of disgust, pathogen disgust, functions in part as a third-order adaptation, as disease-avoidance responses are up-regulated in a manner that compensates for the increases in vulnerability to pathogens that accompany pregnancy and preparation for implantation – changes that are themselves a second-order adaptation addressing the conflict between maternal immune defenses and the parasitic behavior of the half-foreign conceptus (Fessler et al. 2005; Jones et al. 2005; Fleischman and Fessler 2011). (Fessler et al. 2015)

It could be argued that this domain of evolutionary development is not exactly comparable to the types of application domains considered in the current chapter and book. For example, in evolutionary processes not organisms in their daily life are considered but species on an evolutionary relevant long term time scale. However, at least in a metaphorical sense, this evolutionary domain might provide an interesting source of inspiration.

In Chaps. 7 and 8 the question on adaptive networks of order higher than 2 comes back. Then two application contexts will be addressed for higher-order (higher than 2) adaptive network models: in Chap. 7 one for evolutionary processes as described in the above quote, and in Chap. 8 one for the notion of Strange Loop as put forward by Hofstadter (1979, 2007).

1.4 Using Network Reification to Model Adaptive Networks

In this section, first the often used hybrid approach to adaptive networks is discussed and next the approach based on network reification presented in the current book.

1.4.1 *The Hybrid Approach to Model Adaptive Networks*

Adaptive networks are often modeled in a hybrid manner by considering two types of models that interact with each other (see Fig. 1.2):

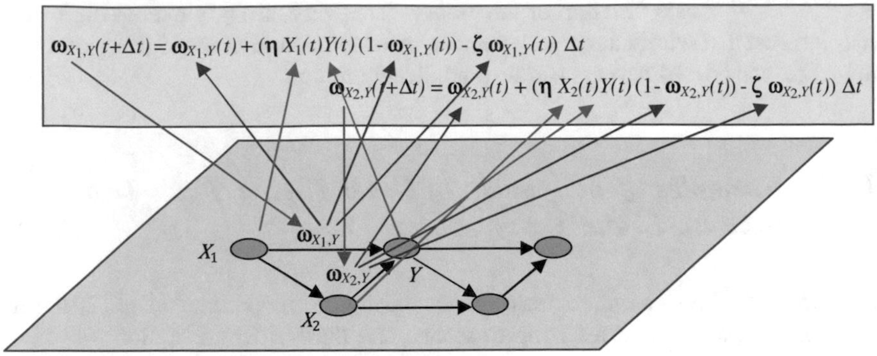

Fig. 1.2 Adaptive network model as a combination of and interaction between a network model and a non-network model

(1) a *network model* for the dynamics within the base network
(2) an *adaptation model* for the dynamics of the network structure characteristics of the base network.

The latter dynamic model is usually specified in a format outside the context of the Network-Oriented Modeling perspective as used for the base network itself. It is most often specified in the form of some procedural or algorithmic programming specification used to run the difference or differential equations underlying the network adaptation process. This non-network dynamic model interacts intensively with the dynamic model for the internal network dynamics of the base network; in Social Network context sometimes this interaction is termed co-evolution; e.g., Holme and Newman (2006), Treur (2019b). An example from the neurocognitive area is modeling *Hebbian learning* (Hebb 1949); this is also what is addressed in Fig. 1.2 in a hybrid manner. Hebbian learning is based on the principle that strengthening of a connection between neurons over time may take place when both states are often active simultaneously: 'neurons that fire together, wire together'. As an illustration, consider the connection from states X_i to state Y. At each point in time t, within the adaptation model (top level rectangle in Fig. 1.2), the change of the weights $\boldsymbol{\omega}_{X_i,Y}(t)$ depends both on the states $X_i(t)$ and $Y(t)$ representing the activation values of the states within the base network (indicated by the blue upward arrows in Fig. 1.2). Note that the blue rectangle only gives a part of the specification needed for the adaptation process in the hybrid situation. Also procedural code is needed to run these equations, which for the sake of simplicity is left out of this picture.

In the picture $\boldsymbol{\eta}$ is the learning rate and $\boldsymbol{\zeta}$ the extinction rate; see also Treur (2016) Chap. 2, p. 93. For this Hebbian learning example, the adaptation principle strengthens a connection when the learning part $\boldsymbol{\eta} X_i(t)Y(t)\ (\mathbf{1}-\boldsymbol{\omega}_{X_i,Y}(t))$ is higher than the extinction part $\boldsymbol{\zeta}\,\boldsymbol{\omega}_{X_i,Y}(t)$ and weakens it when this is opposite. Within the base network, the values for the connection weights as determined by the adaptation model are used all the time (indicated in Fig. 1.2 by the red downward arrows), in order to determine the specific dynamics of each base state. This leaves us with a hybrid model consisting of one network model and one non-network model (see Fig. 1.2), each with their own software components to run them, with interactions between these two different types of models [upward from $\boldsymbol{\omega}_{X_i,Y}(t)$, $X_i(t)$ and $Y(t)$, and downward from $\boldsymbol{\omega}_{X_i,Y}(t + \Delta t)$]. The hybrid approach for adaptive networks was also followed in (Treur 2016) and (Treur and Mohammadi Ziabari 2018). For each new adaptation principle, a new piece of software had to be added. This experience led to the motivation to develop the alternative approach described in this book; for a preview see Sect. 1.4.2.

1.4.2 Modeling Adaptive Networks Based on Network Reification

One class of dynamic modeling approaches is referred to by Network-Oriented Modeling; for this class, some form of network structure is used as basic architecture. In particular, for the Network-Oriented Modeling approach as addressed in Treur (2016, 2019a) the basic architecture chosen is a temporal-causal network architecture defined by three network structure characteristics (for more details, see Chap. 2, or the above references):

(a) **Connectivity of the network**

- *connection weights* $\boldsymbol{\omega}_{X,Y}$ for each connection from a state (or node) X to a state Y.

(b) **Aggregation of multiple connections in the network**

- a *combination function* $\mathbf{c}_Y(..)$ for each state Y to determining the aggregation of incoming causal impacts.

(c) **Timing in the network**

- a *speed factor* $\boldsymbol{\eta}_Y$ for each state Y.

Such a network architecture can be used to model in a dynamic manner a wide variety of natural processes and human mental and social processes, based on causal relations that are identified in various empirical scientific disciplines, as has been shown in Treur (2016, 2017).

Recently it has been found out how adaptive networks can be modeled differently from the hybrid approach discussed in Sect. 1.4.1, thereby modeling the whole process in a more transparent Network-Oriented Modeling manner by one overall network extending the base network (Treur 2018a); this will be discussed in more detail in Chap. 3. This process of extending the base network by reification states has been called *network reification*. Reification literally means representing something abstract as a material or concrete thing, or making something abstract more concrete or real (Merriam-Webster and Oxford dictionaries). It is used in different scientific areas in which it has been shown to provide substantial advantages in expressivity and transparency of models; e.g., Davis and Buchanan (1977), Davis (1980), Bowen and Kowalski (1982), Demers and Malenfant (1995), Galton (2006), Hofstadter (1979), Sterling and Shapiro (1996), Sterling and Beer (1989), Weyhrauch (1980). For example, strongly enhanced expressive power and more support for modeling adaptivity in a transparent manner are achieved.

Network reification provides similar advantages; it will be shown in Chap. 3 how network reification can be used to explicitly represent all kinds of well-known (e.g., from Cognitive Neuroscience and Social Science) adaptation principles for networks in a more declarative, transparent and unified manner. Examples of such adaptation principles include, among others, principles for Hebbian learning (to model plasticity in the brain), as already mentioned above, and for bonding based on homophily (to model adaptive social networks). Writing procedural or algorithmic specifications and programming code as usually applied for network adaptation in the hybrid approach is not needed anymore. Both the dynamics of the states within the base network and the dynamics of the network structure are run, not by two different interacting software components as in the hybrid case, but by one and the same generic computational reified network engine based on one universal difference equation as described in detail in Chaps. 9 and 10.

Basically, network reification for a temporal-causal network means that for the adaptive network structure characteristics $\boldsymbol{\omega}_{X,Y}$, $\mathbf{c}_Y(..)$, $\boldsymbol{\eta}_Y$ for each state Y of the base network, additional network states $\mathbf{W}_{X,Y}$, $\mathbf{C}_Y$, $\mathbf{H}_Y$ (called *reification states*) are introduced respectively; see the blue upper plane in Fig. 1.3. Here for practical reasons the combination function reification $\mathbf{C}_Y$ actually is a vector ($\mathbf{C}_{1,Y}$, $\mathbf{C}_{2,Y}$, $\mathbf{C}_{3,Y}$,..) of a number of reification states representing weights for a weighted average of basic combination functions from the *combination function library* (see Chap. 9 for more details). Moreover, combination functions usually have some parameters that also can be reified by reification states $\mathbf{P}_{i,j,Y}$, but for the current chapter for the sake of simplicity, these will be left out of consideration. Including reification states for (some of) the characteristics of the base network structure in an extended network is one step. As a next step, the dynamics of the reification states themselves and their impact on base

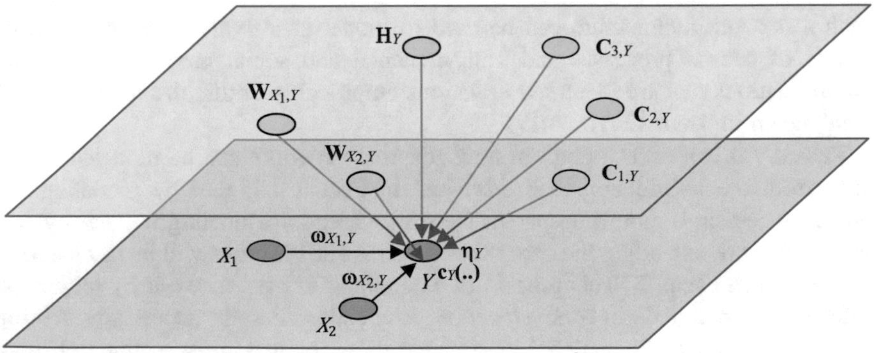

Fig. 1.3 Extending the base network (pink plane) by reification states (blue plane)

state Y are described by additional network structure of the extended network; this additional network structure is also in temporal-causal network format and replaces the blue non-network-like part of the hybrid model and the upward and downward arrows from Fig. 1.2. As a result, an extended network (called a *reified network*) is obtained that embeds the base network in it. This extended network is still a temporal-causal network, as will be pointed out in Chap. 3, Sect. 3.5, and illustrated and proven in more detail in Chap. 10.

Note that the combination function of base state Y has to incorporate these downward causal connections; for details about this, see Chap. 3 or, in more detail, Chaps. 9 and 10. Then a reified network structure is obtained that explicitly represents the characteristics of the (adaptive) base network structure by some of its states, and, moreover, it represents how exactly this base network evolves over time based on adaptation principles that change the base network structure. This construction as described in more detail in Chap. 3 provides an extended temporal-causal network that is called a *reified network architecture*. Like any other state, reification states are defined by three general network structure characteristics connectivity (a), aggregation (b), and timing (c), mentioned above:

(a) For the reification states their *connectivity* in terms of their incoming and outgoing connections has different functions:

- The *outgoing downward causal connections* (the pink downward arrows in Figs. 1.3 or 1.4) from the reification states $\mathbf{W}_{X,Y}$, $\mathbf{C}_Y$, $\mathbf{H}_Y$ to state Y represent the specific causal impact (their special effect) each of these reification states have on Y. These downward causal impacts are standard per type of reification state, and make that the original network characteristics $\boldsymbol{\omega}_{X,Y}$, $\mathbf{c}_Y(..)$, $\boldsymbol{\eta}_Y$ need not be used anymore, as $\mathbf{W}_{X,Y}$, $\mathbf{C}_Y$, $\mathbf{H}_Y$ are used in their place.
- The *upward (or leveled) causal connections* (blue arrows) to the reification states give them the dynamics as desired. They are used to specify, together with the combination function that is chosen and the downward connection, the particular adaptation principle that is addressed.

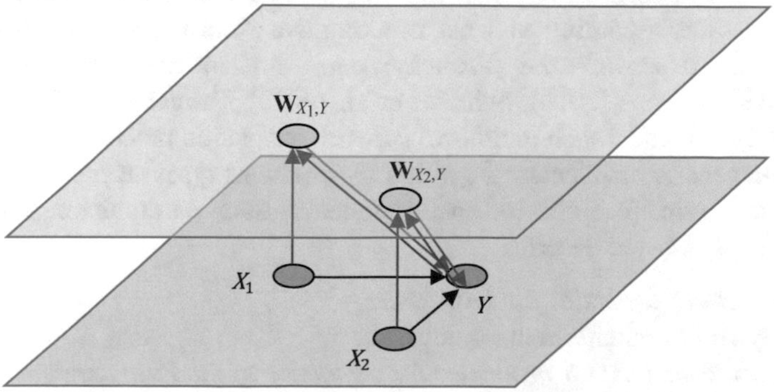

Fig. 1.4 Graphical conceptual representation of a reified network for Hebbian learning in mental networks

This is illustrated in more detail in Fig. 1.4 where network reification for Hebbian learning is modeled. The upward causal connections (blue arrows) to the reification states $\mathbf{W}_{X_1,Y}$ and $\mathbf{W}_{X_2,Y}$ are from the states X_1, X_2, and Y from the base network. Hebbian learning can be described in a simple manner by 'neurons that fire together, wire together'. To incorporate the 'firing together' part, the upward causal connections to $\mathbf{W}_{X_i,Y}$ from the states X_i and Y are needed to express a Hebbian learning adaptation principle. The upward connections are usually assumed to have weight 1. Note that usually also connections from each reification state to itself with weight 1 are applied, but in pictures as in Fig. 1.4 they are often left out.

(b) Concerning the *aggregation* of their incoming causal impacts, in Sect. 1.6 below a combination function that can be used for the reification states $\mathbf{W}_{X_i,Y}$ for Hebbian Learning will be discussed.

(c) Finally, like any other state in a temporal-causal network, reification states have their own *timing* in terms of speed factors. In this case these speed factors represent the *adaptation speed*.

The network reification idea is illustrated for one specific adaptation principle, namely Hebbian learning. In Chap. 3 it is shown how this works for many other well-known adaptation principles.

1.5 Modeling Higher-Order Adaptive Networks by Multilevel Network Reification

What is discussed in Sect. 1.4 is not the end of the story. Reified networks form again a basic temporal-causal network structure defined by certain network structure characteristics. Adaptation principles represented by that reified network structure may themselves be adaptive too, according to certain *second-order adaptation principles*. For example, for real-world processes *plasticity* in Mental Networks as described by Hebbian learning is not a constant feature, but usually varies over time, according to what in Cognitive Neuroscience has been called *metaplasticity* (or *second-order plasticity*); e.g., Abraham and Bear (1996), Magerl et al. (2018), Parsons (2018), Schmidt et al. (2013), Sehgal et al. (2013), Zelcer et al. (2006). To model such multilevel network adaptation processes, it is useful to have some generic architecture in which the different types of adaptation can be modeled in a principled and transparent manner. Such an architecture should be able to distinguish and describe:

(1) *Base network dynamics for base states*
The dynamics within the base network.

(2) *First-order adaptation principles for dynamics of the base network structure*
The dynamics of the base network structure by first-order network adaptation principles.

(3) *Second-order adaptation principles for dynamics of the first-order adaptation principles*
The adaptation of these first-order adaptation principles by second-order adaptation principles.
(4) *Higher-order adaptation principles for dynamics of the second-order adaptation principles*
Maybe still more levels of (higher-order) adaptation.
(5) *Interlevel interactions*
The interactions between these levels.

Such distinctions indeed can be made within a Network-Oriented Modeling framework using the notion of *multilevel reified network architecture*. This type of architecture is obtained by subsequently applying network reification as pointed out in Sect. 1.4 on the reified network structures as well. Repeating multiple times this construction of a reified network architecture provides a multilevel reified network architecture, which will be discussed and illustrated in more detail in Chap. 4.

This multilevel reified (temporal-causal) network architecture is the basis of the implementation of a *dedicated software environment* developed by the author in Matlab, which is discussed in Chap. 9. This environment takes as input so called *role matrices* specifying the network structure characteristics for the different types of states in a designed network model and can just run the model based on them, using a generic *computational reified network engine* included in the environment.

For a (nonadaptive) base network, role matrices are nothing more than a neatly structured way to show in table format (e.g., in Word or in Excel) all values for the characteristics of the model, for example as used in a given simulation scenario; see Chap. 2, Sect. 2.4 and Box 2.1. Each role matrix groups together (for all states), the data of a specific type. In this way, there are five different types of role matrices:

mb for the base connectivity of the network
mcw for the specific connection weights
ms for the speed factors
mcfw for the combination functions used with their weights
mcfp for the parameters of these combination functions.

Again, in case of a nonadaptive network, these five matrices just contain all relevant data (the values) in a standardly structured way as shown in Box 2.1 in Chap. 2. In a reified network, the role matrices are only slightly different, as some of the entries of the role matrices are no values anymore (as now they represent adaptive characteristics), but instead just specify the name of the reification state for this (dynamic) value. This specifies the standard downward causal connection for that reification state: the pink downward arrow in a picture such as Fig. 1.4. The notion of role matrix will be introduced for a base network in detail in Chap. 2 and it will be shown how to apply them in reified networks in Chap. 3.

1.6 Mathematical Analysis of Reified Networks

In this section it will be briefly discussed how the specific structure of reified temporal-causal network models supports mathematical analysis and verification of them.

1.6.1 Mathematical Analysis of a Base Network

As indicated above in Sect. 1.4.2, the choice for temporal-causal networks as basic network architecture comes with three basic concepts for the network structure used for modeling and analysis of the dynamics within a network. It uses state values *X* *(t)* (usually within the interval [0, 1]) over time for each node *X* and is based on the following *network structure characteristics*:

(a) **Connectivity**

- a *connection weight* $\boldsymbol{\omega}_{X,Y}$ for each connection from a state *X* to a state *Y*, together with state value *X(t)* defining the single impact $\boldsymbol{\omega}_{X,Y}X(t)$ of *X* on *Y*.

(b) **Aggregation**

- a *combination function* $\mathbf{c}_Y(..)$ for each state *Y* to determine the aggregated impact $\mathbf{c}_Y(\boldsymbol{\omega}_{X_1,Y}X(t), \ldots, \boldsymbol{\omega}_{X_k,Y}X(t)$ for the incoming single impacts $\boldsymbol{\omega}_{X_i,Y}X(t)$ of the states $X_1,\ldots, X_k$ with outgoing connections to *Y*

(c) **Timing**

- a *speed factor* $\boldsymbol{\eta}_Y$ for each state *Y* by which the effect of the aggregated impact on state *Y* is given a proper timing.

These notions are very helpful in designing a dynamic model of a network as they allow the modeler to concentrate on the choices for these values and functions (which are just declarative mathematical objects), and make that he or she will not feel any need to consider procedural specifications or even to do programming. This is enabled by dedicated software environments developed in Matlab that take these basic concepts as input and just run simulations based on them: (Mohammadi Ziabari and Treur 2019) for temporal-causal networks and Treur (2019c) for reified temporal-causal networks, the latter of which is described in Chap. 9. To understand how this can happen, within these software environments based on the network structure characteristics input $\boldsymbol{\omega}_{X,Y}$, $\mathbf{c}_Y(..)$, $\boldsymbol{\eta}_Y$, for each state *Y* the following difference equation is formed and executed; for a more detailed explanation and motivation, see Chap. 2 or Treur (2016), Chap. 2:

$$Y(t+\Delta t) = Y(t) + \boldsymbol{\eta}_Y[\mathbf{c}_Y(\boldsymbol{\omega}_{X_1,Y}X_1(t), \ldots, \boldsymbol{\omega}_{X_k,Y}X_k(t)) - Y(t)]\Delta t \tag{1.1}$$

As reified temporal-causal networks are themselves also temporal-causal networks, these three concepts and the above difference equation format basically also are applied to the states in a reified network. The only difference is that in case of adaptive network structure characteristics $\boldsymbol{\omega}_{X_i,Y}$, $\boldsymbol{\eta}_Y$, $\mathbf{c}_Y(..)$ the values of the reification states for them are incorporated, in the sense that their dynamic values are used in Eq. (1.1) instead of static values.

For mathematical analysis, based on the above equation, a simple criterion in terms of the network structure characteristics $\boldsymbol{\omega}_{X,Y}$, $\mathbf{c}_Y(..)$, $\boldsymbol{\eta}_Y$ can be formulated for the network reaching an *equilibrium* which is a situation in which no state changes anymore (Chap. 2, Sect. 2.5):

Criterion for an equilibrium of a temporal-causal network model

For all Y it holds

$$\boldsymbol{\eta}_Y = 0 \quad \text{or} \quad \mathbf{c}_Y(\boldsymbol{\omega}_{X_1,Y}X_1(t), \ldots, \boldsymbol{\omega}_{X_k,Y}X_k(t)) = Y(t) \tag{1.2}$$

The equation is also called an *equilibrium equation*. This criterion can easily be used for analysis of the equilibria of the network model with a central role of the combination functions together with connection weights and speed factors. It will be used for analysis of many of the examples in this book, in order to find out what emerging behaviour can be expected when performing simulations. Such results of mathematical analysis can be used for verification of an implemented network model. If the outcomes of a simulation contradict some of these analysis results, then there is an error that has to be addressed.

Following this line one step further, in Chaps. 11 and 12 a more extensive analysis is presented of possible combination functions together with the network connectivity to model convergence of social contagion to one common value. It is shown that such analysis often results in theorems of the form that if the combination function used and the network connectivity have certain properties (for example, strictly monotonous and scalar-free combination functions, and a strongly connected network), then the network behaviour has certain properties (e.g., when the network ends up in an equilibrium state, all state values are equal).

As reified networks inherit the structure of a temporal-causal network, it turns out that a similar line can be followed for the reification states in a reified network. This is addressed in Chap. 13 (for bonding by homophily) and Chap. 14 (for Hebbian learning). A brief preview (for a Hebbian learning adaptation principle) of this can be found in Sect. 1.6.2 below, after the following preparations.

For each reification state a specific combination function is needed; for the Hebbian learning case considered here, this combination function $\mathbf{c}_{\mathbf{W}_{X_i,Y}}(..)$ for reification state $\mathbf{W}_{X_i,Y}$ is called $\mathbf{hebb}_{\boldsymbol{\mu}}(..)$ and can simply be specified by a formula of not even one full line; see also Chap. 3, Sect. 3.6.1:

$$\mathbf{hebb}_{\boldsymbol{\mu}}(V_1, V_2, W) = V_1 V_2 (1 - W) + \boldsymbol{\mu} W \quad (1.3)$$

Here V_1, V_2 are variable used for the activation levels $X_i(t)$ and $Y(t)$ of two connected states (with connection weight 1) X_i and Y, and W for the value $\mathbf{W}_{X_i,Y}(t)$ of connection weight reification state $\mathbf{W}_{X_i,Y}$ (also connected to itself with weight 1) $\boldsymbol{\mu}$ is a parameter for the persistence factor. Such a simple declarative one line specification as shown in (1.3) which just specifies the core of the adaptation principle, is in strong contrast with a larger number of lines of procedural programming code as usually needed in a hybrid approach as discussed in Sect. 1.4.1. Note that the extinction parameter $\boldsymbol{\zeta}$ used in Fig. 1.2, and also in (Treur 2016) relates to the persistence parameter $\boldsymbol{\mu}$ as follows:

$$\boldsymbol{\mu} = 1 - \boldsymbol{\zeta}/\boldsymbol{\eta} \quad \text{or} \quad \boldsymbol{\zeta} = (1 - \boldsymbol{\mu})\boldsymbol{\eta} \quad (1.4)$$

In Chap. 15, Sect. 15.2 it is shown that based on this, the two ways of modeling are actually mathematically equivalent. However, the latter way based on (1.1) and (1.3) is more transparent and more uniform with the rest of the model, so that no hybrid form of modeling is needed anymore.

1.6.2 Mathematical Analysis Applied to Reification States

The criterion for an equilibrium formulated above can not only be applied to base states, but also to reification states. This enables to find out what emerging behaviour is possible for a reified adaptive network, in the sense of equilibria. For example, as discussed in Sects. 1.4 and 1.6.1 for Hebbian learning (see Fig. 1.4), combination function (1.3) can be used for the reification state $\mathbf{W}_{X,Y}$ for an adaptive connection weight $\boldsymbol{\omega}_{X,Y}$. The connections to the reification state $\mathbf{W}_{X,Y}$ are given weight 1. Assuming nonzero learning speed, the above criterion for an equilibrium applied to the reification state $\mathbf{W}_{X,Y}$ provides the *equilibrium equation* (see also Chap. 3, Sect. 3.6.1, and Chap. 14)

$$V_1 V_2 (1 - W) + \boldsymbol{\mu} W = W \quad (1.5)$$

where W is the value of the reification state $\mathbf{W}_{X,Y}$ and V_1 and V_2 of the connected base states X and Y. In Box 1.1 it is shown how this equation can be rewritten by elementary mathematical rules into an equation of the form $W = \ldots V_1 \ldots V_2 \ldots \boldsymbol{\mu} \ldots$ (an expression in terms of V_1, V_2, and $\boldsymbol{\mu}$).

Box 1.1 Rewriting the equilibrium equation for Hebbian learning for a reified connection weight W

$V_1V_2\,(1-W) + \mu W = W$	*equilibrium equation*
$V_1V_2 - V_1V_2\,W + \mu W = W$	*distribution for* V_1V_2
$V_1V_2 = W + V_1V_2\,W - \mu W$	*rearranging sides*
$V_1V_2 = (1 + V_1V_2 - \mu)\,W$	*(anti)distribution for* W
$V_1V_2 = (1 - \mu + V_1V_2)\,W$	*commutation of* $-\mu$ *and* V_1V_2
$W = \dfrac{V_1V_2}{(1-\mu) + V_1V_2}$	*dividing by* $(1 - \mu + V_1V_2)$

So, the equation

$$W = \frac{V_1 V_2}{(1-\mu) + V_1 V_2} \tag{1.6}$$

for W is obtained. In case $V_1V_2 = 1$, the outcome is

$$W = \frac{1}{2-\mu} \tag{1.7}$$

This is the maximal value that can be achieved for the connection weight. In case of full persistence $\mu = 1$, the equilibrium Eq. (1.7) results in

$$V_1 V_2 = 0 \,\text{or}\, W = 1 \tag{1.8}$$

In that case, the connection weight will always reach the value 1 unless one of the states gets value 0.

This illustrates how the central role of combination functions together with connection weights and speed factors does not only support design and simulation, but also mathematical analysis of a reified network model. In this way, in Chap. 14 a more extensive analysis is made of possible functions (and their properties) to model Hebbian learning, and in Chap. 13 a similar analysis for different functions (and their properties) that can be used for bonding by homophily. It can be seen in these two chapters, similar to what is found in Chaps. 11 and 12, that such an analysis can result in theorems of the form that if the combination function used has certain properties (and the network has a certain type of connectivity), then the network behaviour has certain properties.

1.7 Discussion

In this chapter, it was briefly previewed how networks can be adaptive in different ways and according to different orders. A variety of examples of first- and second-order adaptation were summarized, and pointers were given to chapters in this book where they are addressed in more detail. Also, the possibility of adaptation of order higher than 2 was discussed, which for now can be considered an open question to which Chaps. 7 and 8 will return. The modeling approach presented in this book allows treatment of such adaptation of order higher than two, but real-world examples of it cannot be found easily, as was illustrated by numbers of hits via Google Scholar. The notion of network reification was briefly introduced (as a preview of Chap. 3) and it was shown how it can be used to model adaptive networks in a more transparent and Network-Oriented manner than the usual hybrid approaches. It was discussed how repeated application of network reification can be used to model adaptive networks with adaptation of multiple orders (as a preview of Chap. 4). For this multilevel reified temporal-causal network architecture a dedicated software environment was developed by the author in Matlab. This can be used for modeling and simulation of adaptive networks of any order; it is described in Chap. 9, and more background on the format used in Chap. 10. Finally, it was discussed here how mathematical analysis of emerging behavior of a network not only can be applied to base networks (as addressed in more detail in Chaps. 11 and 12), but also to reified networks (as addressed in more detail in Chaps. 13 and 14).

The basic elements used in Network-Oriented Modeling based on reified temporal-causal networks are declarative: connection weights, combination functions, speed factors are declarative mathematical objects. Together these elements assemble in a standard manner a set of first-order difference or differential equations (1.1), which are declarative temporal specifications. The model's behavior is fully determined by these declarative specifications, given some initial values. The modeling process is strongly supported by using these declarative building blocks. Very complex adaptive patterns can be modeled easily, and in (temporal) declarative form. As mentioned, a dedicated software environment including a generic computational reified network engine (see Chap. 9) takes care of running these high level specifications.

Therefore in many chapters, especially Chaps. 1–9, mathematical and procedural details are kept at a minimum thus obtaining optimal readability for a wide group of readers with diverse multidisciplinary backgrounds. As the Network-Oriented Modeling approach based on reified temporal-causal networks presented in this book abstracts from specific implementation details, making use of the dedicated software environment, modeling can be done without having to design procedural or algorithmic specifications. Moreover, a modeler does not even need to explicitly specify difference or differential equations to get a simulation done, as these are already taken care for by the software environment, based on the modeler's input in the form of the conceptual representation of the network model (the so-called role matrices). Therefore, in Chaps. 1–9 all underlying specific procedural elements and difference or differential equations will usually not be discussed although sometimes the underlying universal difference and differential equation is briefly mentioned; it will be discussed more extensively in Chap. 10. The only mathematical details that will be addressed in Chaps. 1–9 for design of a network model concern the combination functions used, most of which are already given in the available combination function library. For analysis of the emerging behaviour of a network model also these combination functions are central, as the equilibrium equations are based on them, as shown in Sect. 1.6 above and, for example, Eq. (1.5). However, for those readers who still want to see more mathematical details that are covered in the software environment, Chap. 15 presents these in more depth in different sections, as a kind of appendices to many of the chapters.

The causal modeling area has a long history in AI; e.g., Kuipers (1984), Kuipers and Kassirer (1983). Reified temporal-causal network models are part of a newer branch in this causal modeling area. In this branch dynamics is added to causal models, making them temporal as in Treur (2016), but the main contribution in the current book is that a way to specify (multi-order) adaptivity in causal models is added, thereby conceptually using ideas on meta-level architectures that also have a long history in AI; e.g., Davis and Buchanan (1977), Davis (1980), Bowen and Kowalski (1982), Demers and Malenfant (1995), Galton (2006), Hofstadter (1979), Hofstadter (2007), Sterling and Shapiro (1996), Sterling and Beer (1989), Weyhrauch (1980). So, this new Reified Network-Oriented Modeling approach connects two different areas with a long tradition in AI, thereby strongly extending the applicability of causal modeling to dynamic and adaptive notions such as plasticity and metaplasticity of any order, which otherwise would be out of reach of causal modeling.

References

Abraham, W.C., Bear, M.F.: Metaplasticity: the plasticity of synaptic plasticity. Trends Neurosci. **19**(4), 126–130 (1996)

Ashby, W.R.: Design for a Brain. Chapman and Hall, London (second extended edition). First edition, 1952 (1960)

Banks, D.L., Carley, K.M.: Models for network evolution. J. Math. Soc. **21**, 173–196 (1996)

Barabasi, A.-L., Albert, R.: Emergence of scaling in random networks. Science **286**, 509–512 (1999)

Beukel, S.V.D., Goos, S.H., Treur, J.: An adaptive temporal-causal network model for social networks based on the homophily and more- becomes-more principle. Neurocomputing **338**, 361–371 (2019)

Blankendaal, R., Parinussa, S., Treur, J.: A temporal-causal modelling approach to integrated contagion and network change in social networks. In: Proceedings of the 22nd European Conference on Artificial Intelligence, ECAI'16. Frontiers in Artificial Intelligence and Applications, vol. 285, pp. 1388–1396. IOS Press (2016)

Boomgaard, G., Lavitt, F., Treur, J.: Computational analysis of social contagion and homophily based on an adaptive social network model. In: Koltsova, O., Ignatov, D.I., Staab, S. (eds.) Social Informatics: Proceedings of the 10th International Conference on Social Informatics,

SocInfo'18, vol. 1. Lecture Notes in Computer Science vol. 11185, pp. 86–101. Springer Publishers (2018)

Bowen, K.A., Kowalski, R.: Amalgamating language and meta-language in logic programming. In: Clark, K., Tarnlund, S. (eds.) Logic Programming, pp. 153–172. Academic Press, New York (1982)

Byrne, D.: The attraction hypothesis: do similar attitudes affect anything? J. Pers. Soc. Psychol. **51** (6), 1167–1170 (1986)

Carley, K.M.: Inhibiting adaptation. In: Proceedings of the 2002 Command and Control Research and Technology Symposium, pp. 1–10. Naval Postgraduate School, Monterey, CA (2002)

Carley, K.M.: Destabilization of covert networks. Comput. Math. Organ. Theor. **12**, 51–66 (2006)

Carley, K.M.: ORA: a toolkit for dynamic network analysis and visualization. In: Alhajj, R., Rokne, J. (eds.) Encyclopedia of Social Network Analysis and Mining. Springer (2017). https://doi.org/10.1007/978-1-4614-7163-9_309-1

Carley, K.M., Lee, J.-S., Krackhardt, D.: Destabilizing networks. Connections **24**(3), 31–34 (2001)

Carley, K.M., Pfeffer, J.: Dynamic network analysis (DNA) and ORA. In: Schmorrow, D.D., Nicholson, D.M. (eds.) Advances in Design for Cross-Cultural Activities Part I, pp. 265–274. CRC, Boca Raton (2012)

Carley, K.M., Pfeffer, J., Liu, H., Morstatter, F., Goolsby, R.: Near real time assessment of social media using geo temporal network analytics. In: Proceedings of 2013 IEEE/ACM International Conference on Advances in Social Networks Analysis and Mining (ASONAM), Niagara Falls, 25–28 Aug 2013 (2013a)

Carley, K.M., Reminga, J., Storrick, J., Pfeffer, J., Columbus, D.: ORA user's guide 2013. Carnegie Mellon University, School of Computer Science, Institute for Software Research. Technical report, CMU-ISR-13-108 (2013b)

Chandra, N., Barkai, E.: A non-synaptic mechanism of complex learning: modulation of intrinsic neuronal excitability. Neurobiol. Learn. Mem. **154**, 30–36 (2018)

Daimon, K., Arnold, S., Suzuki, R., Arita, T.: The emergence of executive functions by the evolution of second–order learning. Artif. Life Robot. **22**, 483–489 (2017)

Davis, R.: Meta-rules: reasoning about control. Artif. Intell. **15**, 179–222 (1980)

Davis, R., Buchanan, B.G.: Meta-level knowledge: overview and applications. In: Proceedings of the 5th International Joint Conference on AI, IJCAI'77, pp. 920–927 (1977)

Demers, F.N., Malenfant, J.: Reflection in logic, functional and object oriented programming: a short comparative study. In: IJCAI'95 Workshop on Reflection and Meta-Level Architecture and their Application in AI, pp. 29–38 (1995)

Fessler, D.M.T., Clark, J.A., Clint, E.K.: Evolutionary psychology and evolutionary anthropology. In: Buss, D.M. (ed.) The Handbook of Evolutionary Psychology, pp. 1029–1046. Wiley, New York (2015)

Fessler, D.M.T., Eng, S.J., Navarrete, C.D.: Elevated disgust sensitivity in the first trimester of pregnancy: evidence supporting the compensatory prophylaxis hypothesis. Evol. Hum. Behav. **26**(4), 344–351 (2005)

Fleischman, D.S., Fessler, D.M.T.: Progesterone's effects on the psychology of disease avoidance: support for the compensatory behavioral prophylaxis hypothesis. Horm. Behav. **59**(2), 271–275 (2011)

Galton, A.: Operators vs. arguments: the ins and outs of reification. Synthese **150**, 415–441 (2006)

Hebb, D.O.: The Organization of Behavior: A Neuropsychological Theory (1949)

Hofstadter, D.R.: Gödel, Escher, Bach. Basic Books, New York (1979)

Hofstadter, D.R.: I Am a Strange Loop. Basic Books, New York (2007)

Holme, P., Newman, M.E.J.: Nonequilibrium phase transition in the coevolution of networks and opinions. Phys. Rev. E **74**(5), 056108 (2006)

Jones, B.C., Perrett, D.I., Little, A.C., Boothroyd, L., Cornwell, R.E., Feinberg, D.R., Tiddeman, B.P., Whiten, S., Pitman, R.M., Hillier, S.G., Burt, D.M., Stirrat, M.R., Law Smith, M.J., Moore, F.R.: Menstrual cycle, pregnancy and oral contraceptive use alter attraction to apparent health in faces. Proc. R. Soc. B **5**(272), 347–354 (2005)

Kuipers, B.J.: Commonsense reasoning about causality: deriving behavior from structure. Artif. Intell. **24**, 169–203 (1984)

Kuipers, B.J., Kassirer, J.P.: How to discover a knowledge representation for causal reasoning by studying an expert physician. In: Proceedings of the Eighth International Joint Conference on Artificial Intelligence, IJCAI'83. William Kaufman, Los Altos, CA (1983)

LeCun, Y., Bengio, Y., Hinton, G.: Deep learning. Nature **521**, 436–444 (2015)

Lovejoy, C.O.: The natural history of human gait and posture. Part 2. Hip and thigh. Gait Posture **21**(1), 113–124 (2005)

Magerl, W., Hansen, N., Treede, R.D., Klein, T.: The human pain system exhibits higher-order plasticity (metaplasticity). Neurobiol. Learn. Mem. **154**, 112–120 (2018)

McPherson, M., Smith-Lovin, L., Cook, J.M.: Birds of a feather: homophily in social networks. Annu. Rev. Sociol. **27**, 415–444 (2001)

Merrill, J.A., Sheehan, B., Carley, K.M., Stetson, P.D.: Transition networks in a cohort of patients with congestive heart failure. a novel application of informatics methods to inform care coordination. Appl. Clin. Inform. **6**(3), 548–564 (2015). https://doi.org/10.4338/aci-2015-02-ra-0021

Mohammadi Ziabari, S.S., Treur, J.: A modeling environment for dynamic and adaptive network models implemented in Matlab. In: Proceedings of the 4th International Congress on Information and Communication Technology (ICICT2019). Springer Publishers (2019)

Parsons, R.G.: Behavioral and neural mechanisms by which prior experience impacts subsequent learning. Neurobiol. Learn. Mem. **154**, 22–29 (2018)

Pearson, M., Steglich, C., Snijders, T.: Homophily and assimilation among sport-active adolescent substance users. Connections **27**(1), 47–63 (2006)

Port, R.F., van Gelder, T.: Mind as motion: explorations in the dynamics of cognition. MIT Press, Cambridge, MA (1995)

Rapoport, A.: Spread of Information through a Population with Socio-structural Bias: I. Assumption of transitivity. Bull. Math. Biophys. **15**, 523–533 (1953)

Robinson, B.L., Harper, N.S., McAlpine, D.: Meta-adaptation in the auditory midbrain under cortical influence. Nat. Commun. **7**, 13442 (2016)

Sehgal, M., Song, C., Ehlers, V.L., Moyer Jr., J.R.: Learning to learn—intrinsic plasticity as a metaplasticity mechanism for memory formation. Neurobiol. Learn. Mem. **105**, 186–199 (2013)

Schmidt, M.V., Abraham, W.C., Maroun, M., Stork, O., Richter-Levin, G.: Stress-Induced metaplasticity: from synapses to behavior. Neuroscience **250**, 112–120 (2013)

Sharpanskykh, A., Treur, J.: Modelling and analysis of social contagion in dynamic networks. Neurocomputing **146**, 140–150 (2014)

Sousa, N., Almeida, O.F.X.: Disconnection and reconnection: the morphological basis of (mal) adaptation to stress. Trends Neurosci. **35**(12), 742–751 (2012). https://doi.org/10.1016/j.tins.2012.08.006. Epub 2012 Sep 21 (2012)

Sterling, L., Shapiro, E.: The Art of Prolog, Chap. 17, pp. 319–356. MIT Press (1996)

Sterling, L., Beer, R.: Metainterpreters for expert system construction. J. Logic Program. **6**, 163–178 (1989)

Treur, J.: Network-Oriented Modeling: Addressing Complexity of Cognitive, Affective and Social Interactions. Springer Publishers (2016)

Treur, J.: On the applicability of network-oriented modeling based on temporal-causal networks: why network models do not just model networks. J. Inf. Telecommun. **1**(1), 23–40 (2017)

Treur, J.: Network reification as a unified approach to represent network adaptation principles within a network. In: Proceeding of the 7th International Conference on Theory and Practice of Natural Computing, TPNC'18. Lecture Notes in Computer Science, vol. 11324, pp. 344–358. Springer Publishers (2018a)

Treur, J.: Multilevel network reification: representing higher order adaptivity in a network. In: Proceeding of the 7th International Conference on Complex Networks and their Applications, ComplexNetworks'18, vol. 1. Studies in Computational Intelligence, vol. 812, pp. 635–651. Springer (2018b)

Treur, J.: The ins and outs of network-oriented modeling: from biological networks and mental networks to social networks and beyond. In: Transactions on Computational Collective Intelligence, vol. 32, pp. 120–139. Springer Publishers. Contents of Keynote Lecture at ICCCI'18. (2019a)

Treur, J.: Mathematical analysis of the emergence of communities based on coevolution of social contagion and bonding by homophily. In: Applied Network Science, vol. 4, p. 39. https://doi-org.vu-nl.idm.oclc.org/10.1007/s41109-019-0130-7 (2019b)

Treur, J.: Design of a Software Architecture for Multilevel Reified Temporal-Causal Networks. https://doi.org/10.13140/rg.2.2.23492.07045. URL: https://www.researchgate.net/publication/333662169 (2019c)

Treur, J., Mohammadi Ziabari, S.S.: An adaptive temporal-causal network model for decision making under acute stress. In: Nguyen, N.T., Trawinski, B., Pimenidis, E., Khan, Z. (eds.) Computational Collective Intelligence: Proceeding of the 10th International Conference, ICCCI 2018, vol. 2. Lecture Notes in Computer Science, vol. 11056, pp. 13–25. Springer Publishers (2018)

Weyhrauch, R.W.: Prolegomena to a theory of mechanized formal reasoning. Artif. Intell. **13**, 133–170 (1980)

Zelcer, I., Cohen, H., Richter-Levin, G., Lebiosn, T., Grossberger, T., Barkai, E.: A cellular correlate of learning-induced metaplasticity in the hippocampus. Cereb. Cortex **16**, 460–468 (2006)

Chapter 2
Ins and Outs of Network-Oriented Modeling

Abstract Network-Oriented Modeling has successfully been applied to obtain network models for a wide range of phenomena, including Biological Networks, Mental Networks, and Social Networks. In this chapter, it is discussed how the interpretation of a network as a causal network and taking into account dynamics in the form of temporal-causal networks, brings more depth. Thus main characteristics for a network structure are obtained: Connectivity in terms of the connections and their weights, Aggregation of multiple incoming connections in terms of combination functions, and Timing in terms of speed factors. The basics and the scope of applicability of such a Network-Oriented Modelling approach are discussed and illustrated. This covers, for example, Social Network models for social contagion or information diffusion, and Mental Network models for cognitive and affective processes. From the more fundamental side, it will be discussed how emerging network behavior can be related to network structure.

Keywords Network-Oriented Modeling · Temporal-causal network

2.1 Introduction

Network-Oriented Modeling is a relatively new way of modeling that is especially useful to model intensively interconnected and interactive processes. It has been applied to model networks for biological, mental, and social processes, and still more. The aim of this chapter is to discuss the ins and outs of this modeling perspective in more detail, without considering network reification yet, as that will be the subject of Chap. 3. It is discussed how the interpretation of a network as a causal network and taking into account dynamics brings more depth in the Network-Oriented Modeling perspective, leading to the notion of temporal-causal network as introduced in (Treur 2016). In a temporal-causal network, nodes represent states with values that vary over time, and connections represent causal relations describing how states affect each other.

J. Treur, *Network-Oriented Modeling for Adaptive Networks: Designing Higher-Order Adaptive Biological, Mental and Social Network Models*, Studies in Systems, Decision and Control 251, https://doi.org/10.1007/978-3-030-31445-3_2

The wide scope of applicability (Treur 2016, 2017) of such a Network-Oriented Modelling approach will be discussed and illustrated. This covers, for example, network models for principles of social contagion or information diffusion, and network models for mental processes. When network reification as introduced in more detail in Chap. 3 is also taken into account, many kinds of adaptive network models are covered, for example for principles of evolving social networks, such as the homophily principle, or for Hebbian learning in Mental Networks. From the methodological side, it will be discussed how mathematical analysis can be used to identify the relation between emerging behaviour of the network and network structure.

In this chapter, in Sect. 2.2 first the conceptual background of Network-Oriented Modeling is discussed, leading to a conceptual representation of a temporal-causal network, which defines such a network. Next, in Sect. 2.3 the numerical foundation is discussed, including a precise definition of a numerical representation by which a temporal-causal network model gets its intended dynamic semantics, and which can be used for simulation and analysis. Section 2.4 introduces role matrices as a useful specification format for temporal-causal networks. In Sect. 2.5 the interesting challenge to determine how emerging network behaviour relates to network structure and some results on this relation are briefly discussed. In Sect. 2.6 the scope of applicability is discussed. Finally, Sect. 2.7 is a discussion.

2.2 Network-Oriented Modeling: Conceptual Background

Network-Oriented Modeling is applied in a wide variety of areas. The general pattern is that some type of process in some domain *X* is described by a network structure, and this type of network is called an *X* Network or *X* Network model. Note that such a network is considered as a modelling concept, not as reality. Some examples are:

- Modeling the dynamics of propagation of chemical activity in cells based on the concentration levels of chemicals by Biological Network models
- Modeling the dynamics of propagation of neural activity based on activation levels of neurons by Neural Network models
- Modeling the dynamics of propagation of mental activity based on engaging mental states by Mental Network models
- Modeling the dynamics of propagation of individual activity based on activation of personal states by Social Network models; e.g.,
  - Information diffusion; e.g., in social media
  - Opinion spread; e.g., in political campaigns
  - Emotion contagion; e.g., one smile triggering the other
  - Activity contagion; e.g., following each other

These are just four types of domains *X* where processes, in reality, are modelled by network models, which then can be called *X* Networks with *X* = Biological, Neural, Mental, or Social.

2.2.1 The Unifying Potential of Networks

As an illustration, consider the following two examples, one for a Biological Network, and one for a Mental Network. The example of a Biological Network shown in Fig. 2.1 describes how bacteria generate and regulate their behaviour on the one hand based on their genetical background as encoded in their DNA, and on the other hand based on the situational context of the environment; see also Jonker et al. (2008). For the general perspective on modelling the cell's metabolic and life processes as biochemical networks ('the dynamic biochemical networks of life'), see also Westerhoff et al. (2014a, b). For example:

> Living organisms persist by virtue of complex interactions among many components organized into dynamic, environment-responsive networks that span multiple scales and dimensions. Biological networks constitute a type of information and communication technology (ICT): they receive information from the outside and inside of cells, integrate and interpret this information, and then activate a response. Biological networks enable molecules within cells, and even cells themselves, to communicate with each other and their environment. (Westerhoff et al. 2014b, p. 1)

As a second example, the Mental Network shown in Fig. 2.2 describes how human behaviour is generated and regulated by desires and intentions, and beliefs

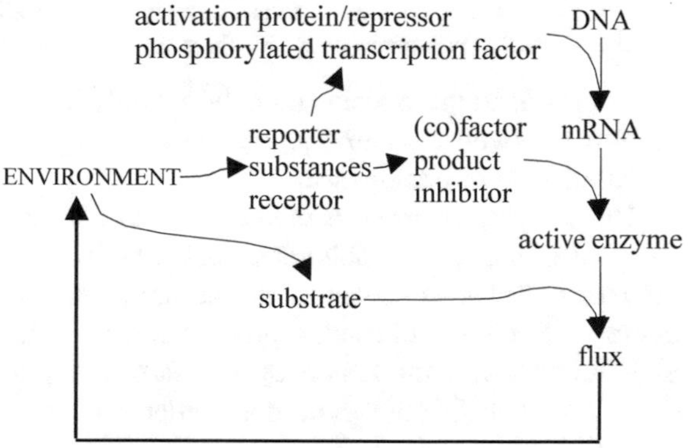

Fig. 2.1 Example of a Biological Network for bacterial behaviour based on its biochemistry; adapted picture from Jonker et al. (2008), Fig. 1 left hand side, p. 3

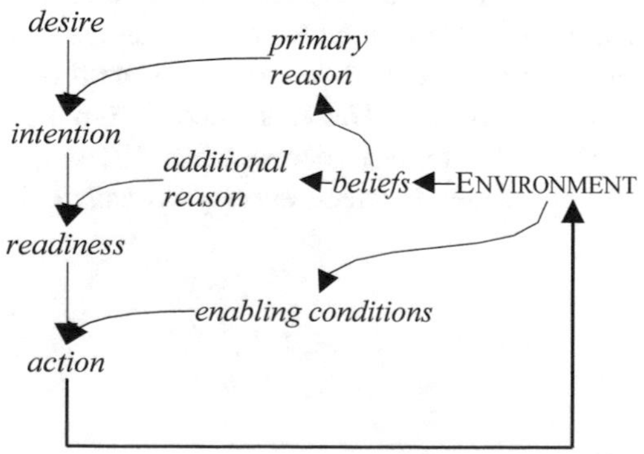

Fig. 2.2 Example of a Mental Network for behaviour based on Beliefs, Desires and Intentions (BDI); adapted picture from Jonker et al. (2008), Fig. 1 right hand side, p. 3

about the environment. Within Philosophy of Mind, Kim (1996) describes Mental Networks based on causal relations as follows:

> Mental events are conceived as nodes in a complex causal network that engages in causal transactions with the outside world by receiving sensory inputs and emitting behavioral outputs. (Kim 1996, p. 104)

As can be noted similar network structures may describe different types of processes; see the isomorphic structures in Figs. 2.1 and 2.2, where actually the latter is a mirror image of the former. The Network-Oriented perspective provides a form of unification so that different types of processes become comparable, and we can, for example, compare the processes underlying human intelligence and behaviour to the processes underlying bacterial behaviour, as described in more detail in Jonker et al. (2002, 2008), Westerhoff et al. (2014b). For example:

> We have become accustomed to associating brain activity – particularly activity of the human brain – with a phenomenon we call "intelligence." Yet, four billion years of evolution could have selected networks with topologies and dynamics that confer traits analogous to this intelligence, even though they were outside the intercellular networks of the brain. Here, we explore how macromolecular networks in microbes confer intelligent characteristics, such as memory, anticipation, adaptation and reflection and we review current understanding of how network organization reflects the type of intelligence required for the environments in which they were selected. We propose that, if we were to leave terms such as "human" and "brain" out of the defining features of "intelligence," all forms of life – from microbes to humans – exhibit some or all characteristics consistent with "intelligence". (Westerhoff et al. 2014b, p. 1)

The emphasis in this quote is on how not only in the brain, but even in the smallest life forms network structure, organisation, and dynamics are used to realise many if not all aspects of intelligence.

This unifying perspective of networks for different domains can be seen in many cases. For example, a politician such as Boris Johnson can be seen as a big influencer for the population in the UK, for example, concerning the Brexit dilemma. Such an influencing process can be described by social contagion in a Social Network; in the same way a flock of sheep following a leader sheep can be described by social contagion in a similar network, where the leading sheep is the big influencer.

As network structures for different domains may look similar, this suggests that there is a high potential for unification and exchange across different domains. For example, can we learn more about Mental Networks by studying Social Networks? Or can we develop Network Theory from a unified perspective that can be applied in both areas, or even in more areas? These questions indicate some of the promises and challenges in what nowadays is called Network Science.

2.2.2 *On the Meaning of the Basic Elements in a Network*

There are, however, some issues that may have to be addressed to enable the further development of this perspective of a unified Network Science. A main issue is that not every network may have the same form concerning definition and semantics. Then unification may be not so easy. What actually is a network? What does a node mean? How should we interpret what a connection is or does? Are all connections considered equal? And what if there are multiple connections to one node? Should we interpret this as a kind of conjunction (AND), or disjunction (OR), or maybe something in between, like some average; then what kind of average?

Is a network just an abstract graph structure with nodes and connections and nothing more, and in particular no further semantics? Then in fact Network Science = Graph Theory, which is an already existing area within Mathematics, and Network-Oriented Modeling could also be called Graph-Oriented Modeling. This perspective may provide a relevant stream, but will not be sufficient to further develop Network Science. For many applications, just a graph structure with only nodes and connections seems seriously underspecifying what is intended.

In many examples of applications of networks, such as those mentioned above, a notion of dynamics plays an important role. Shouldn't such dynamics be part of the definition or semantics of a network? These dynamics can concern dynamics of states (dynamics *within* a network: for example, diffusion or contagion of opinions or emotions in a network), but also dynamics of the network structure itself (dynamics *of* a network: for example, adaptive or evolving networks describing changing relationships between persons). Dynamics has a direct relation to causal relations describing how one state affects the other. The notions of dynamics and causality are fundamental for practically all scientific disciplines; these notions play an important unifying role in science and can be found in most of the scientific literature. Causal relations vary from how hitting a ball causes movement of the ball to how certain beliefs cause certain behaviour or how joining forces in a social movement causes a change in society, to name just a few cases.

2.2.3 *Meaning as Defined by the Notion of Temporal-Causal Network*

For the perspective on Network Science addressed in the current chapter these notions of causality and dynamics have been incorporated and are part of a more refined structure and semantics of the considered networks. More specifically, the nodes in a network are interpreted here as states (or state variables) that vary over time, and the connections are interpreted as causal relations that define how each state can affect other states over time. To acknowledge this perspective of dynamics and causality on networks, this type of network has been called a *temporal-causal network* (Treur 2016). Many examples of applications have demonstrated that all types of domains as listed above can be covered in this way; e.g., Treur (2016). In Sect. 2.6 below this wide applicability is briefly discussed; see also Treur (2017).

So, is there still some relevant graph perspective? A conceptual representation of a temporal-causal network model by a *labeled* graph still provides a fundamental basis. More specifically, a conceptual representation of a temporal-causal network model in the first place still involves representing in a declarative manner states and connections between them that, as discussed earlier, represent (causal) impacts of states on each other, as assumed to hold for the application domain addressed. This part of a conceptual representation is often depicted in a *conceptual picture* by a graph with nodes and directed connections. However, a *full conceptual representation* of a temporal-causal network model also includes a number of labels for such a graph. First, in reality, not all causal relations are equally strong, so some notion of strength of a connection is used as a label for connections. Second, when more than one causal relation affects a state, some way to aggregate multiple causal impacts on a state is used as a label for states. Third, a notion 'speed of change' is used for timing of the processes for a state. These three notions define the characteristics of the network structure; they are summarized as

(a) **Connectivity**

- connection weights from a state X to a state Y, denoted by $\boldsymbol{\omega}_{X,Y}$

(b) **Aggregation**

- a combination function for each state Y, denoted by $\mathbf{c}_Y(..)$

(c) **Timing**

- a speed factor for each state Y, denoted by $\boldsymbol{\eta}_Y$

They make the graph of states and connections a labeled graph (see Fig. 2.3), form the defining structure of a temporal-causal network model in the form of a

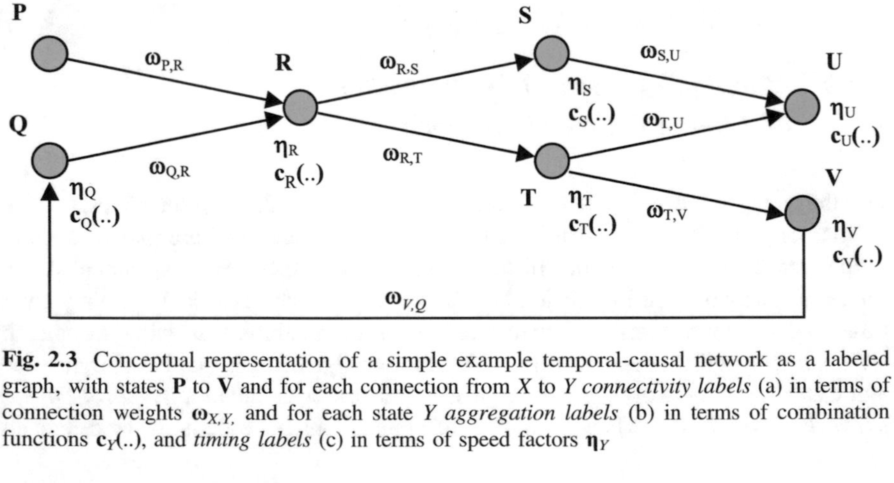

Fig. 2.3 Conceptual representation of a simple example temporal-causal network as a labeled graph, with states **P** to **V** and for each connection from X to Y *connectivity labels* (a) in terms of connection weights $\boldsymbol{\omega}_{X,Y}$ and for each state Y *aggregation labels* (b) in terms of combination functions $\mathbf{c}_Y(..)$, and *timing labels* (c) in terms of speed factors $\boldsymbol{\eta}_Y$

Table 2.1 Conceptual representation of a temporal-causal network model

Concepts	Notation	Explanation
States and connections	X, Y, $X \rightarrow Y$	Describes the nodes and links of a network structure (e.g., in graphical or matrix format)
Connection weight	$\boldsymbol{\omega}_{X,Y}$	The *connection weight* $\boldsymbol{\omega}_{X,Y} \in [-1, 1]$ represents the strength of the causal impact of state X on state Y through connection $X \rightarrow Y$
Aggregating multiple impacts on a state	$\mathbf{c}_Y(..)$	For each state Y (a reference to) a *combination function* $\mathbf{c}_Y(..)$ is chosen to combine the causal impacts of other states on state Y
Timing of the effect of causal impact	$\boldsymbol{\eta}_Y$	For each state Y a *speed factor* $\boldsymbol{\eta}_Y \geq 0$ is used to represent how fast a state is changing upon causal impact

conceptual representation; see also Table 2.1. Note that also connections from a state to itself are allowed, although often they are not depicted in a conceptual representation as shown in Fig. 2.3. Such connections can be used to give the state a more persistent character, as the old values are reused all the time. This may be relevant, in particular, for learning or adaptation.

Combination functions, in general, are similar to the functions used in a static manner in the (deterministic) Structural Causal Model perspective described, for example, in Mooij et al. (2013), Pearl (2000), Wright (1921), but in the Network-Oriented Modelling approach described here they are used in a dynamic manner. For example, Pearl (2000), p. 203, denotes nodes by V_i and combination functions by f_i; he also points at the issue of underspecification for aggregation of multiple connections mentioned in Sect. 2.2 above, as in the often used graph representations the role of combination functions f_i for nodes V_i, is lacking:

> Every causal model M can be associated with a directed graph, $G(M)$ (…) This graph merely identifies the endogeneous and background variables that have a direct influence on each V_i; it does not specify the functional form of f_i. (Pearl 2000, p. 203)

Therefore, if a graph representation is used, at least aggregation in terms of combination functions should be incorporated as labels, as indeed is done for temporal-causal networks, in order to avoid this problem of underspecification. That is the reason why aggregation in terms of combination functions is part of the definition of the network structure for temporal-causal networks, in addition to connectivity in terms of connections and their weights and timing in terms of speed factors.

Combination functions can have different forms, as there are many different approaches possible to address the issue of aggregating multiple impacts. For this aggregation, a library is available with a number of standard combination functions as options, but also own-defined functions can be added.

2.2.4 *Biological, Mental and Social Domains Ask for Networks*

In Sect. 2.1 it already was discussed how 'the dynamic biochemical networks of life' (Westerhoff et al. 2014a) are fundamental to describe life forms in the biological domain. For the mental domain, the mechanisms found within the area of Cognitive and Social Neuroscience also show how many parts in the brain have connections which are adaptive and often form cyclic pathways; such cycles are assumed to play an important role in many mental processes; see also Bell (1999), Potter (2007). It has been pointed out that to address such cyclic effects, a dynamic and adaptive perspective on causality is needed; e.g., Scherer (2009). Also, by Kim (1996) it is claimed that Mental Networks display cyclic network structures:

> (…) to explain what a given mental state is, we need to refer to other mental states, and explaining these can only be expected to require reference to further mental states, on so on – a process that can go on in an unending regress, or loop back in a circle. (Kim 1996, pp. 104–105)

For the social domain, intense interaction between persons also takes place based on mutual and usually cyclic relationships, by which they affect each other. Just one example from the context of modelling social systems or societies can be found in Naudé et al. (2008), where it is claimed that '*relational, network-oriented modelling approaches* are needed' to address human social complexity.

So, from the areas of biological processes (Westerhoff et al. 2014a, b), mental processes (Bell 1999; Kim 1996; Potter 2007; Scherer 2009) and social processes (Naudé et al. 2008), a notion of network is suggested as a basis of modeling, where connections between states or persons describe how they affect each other, thereby strongly suggesting causality and dynamics as crucial notions.

2.3 Numerical Representation of a Temporal-Causal Network

In this section, the numerical-mathematical foundations of temporal-causal networks are discussed in more detail. In Sect. 2.2 the choice made on how networks are interpreted conceptually was discussed based on the notions of temporality and causality, thus indicating semantics for networks based on the notion of temporal-causal network.

2.3.1 Numerical-Mathematical Formalisation

In the current section the interpretation based on temporality and causality is expressed in a formal-numerical way, thus associating semantics to any conceptual temporal-causal network specification in a detailed numerical-mathematically defined manner. This is done by showing how a conceptual representation as discussed in Sect. 2.2, based on states and connections enriched with labels for (a) connectivity (by connection weights), (b) aggregation (by combination functions), and (c) timing (by speed factors), defines a numerical representation (Treur 2016, Chap. 2). This is shown in Table 2.2, where Y is any state in the network and $X_1, \ldots, X_k$ are the states with outgoing connections to Y.

The difference equations in the last row in Table 2.2 form the numerical representation of a temporal-causal network model and can be used for simulation and mathematical analysis. They can also be written in differential equation format and are called the *basic difference or differential equations*:

$$\begin{aligned} Y(t+\Delta t) &= Y(t) + \boldsymbol{\eta}_Y[\mathbf{c}_Y(\boldsymbol{\omega}_{X_1,Y}X_1(t), \ldots, \boldsymbol{\omega}_{X_k,Y}X_k(t)) - Y(t)]\Delta t \\ \mathbf{d}Y(t)/\mathbf{d}t &= \boldsymbol{\eta}_Y[\mathbf{c}_Y(\boldsymbol{\omega}_{X_1,Y}X_1(t), \ldots, \boldsymbol{\omega}_{X_k,Y}X_k(t)) - Y(t)] \end{aligned} \tag{2.1}$$

Table 2.2 Numerical representation of a temporal-causal network model

concept	Representation	Explanation
State values over time t	$Y(t)$	At each time point t each state Y in the model has a real number value in [0, 1]
Single causal impact	$\mathbf{impact}_{X,Y}(t) = \boldsymbol{\omega}_{X,Y}\, X(t)$	At t state X with connection to state Y has an impact on Y, using connection weight $\boldsymbol{\omega}_{X,Y}$
Aggregating multiple impacts	$\mathbf{aggimpact}_Y(t)$ $= \mathbf{c}_Y(\mathbf{impact}_{X_1,Y}(t), \ldots, \mathbf{impact}_{X_k,Y}(t))$ $= \mathbf{c}_Y(\boldsymbol{\omega}_{X_1,Y}X_1(t), \ldots, \boldsymbol{\omega}_{X_k,Y}X_k(t))$	The aggregated causal impact of multiple states X_i on Y at t, is determined using a combination function $\mathbf{c}_Y(V_1, \ldots, V_k)$ and apply it to the k single causal impacts
Timing of the causal effect	$Y(t+\Delta t) = Y(t) + \boldsymbol{\eta}_Y[\mathbf{aggimpact}_Y(t) - Y(t)]\Delta t$ $=$ $Y(t) + \boldsymbol{\eta}_Y[\mathbf{c}_Y(\boldsymbol{\omega}_{X_1,Y}X_1(t), \ldots, \boldsymbol{\omega}_{X_k,Y}X_k(t)) - Y(t)]\Delta t$	The causal impact on Y is exerted over time gradually, using speed factor $\boldsymbol{\eta}_Y$; here the X_i are all states from which state Y has incoming connections

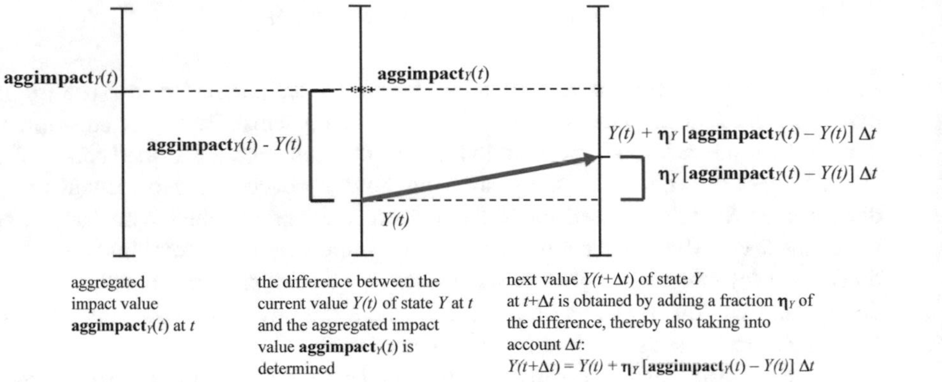

Fig. 2.4 How $\mathbf{aggimpact}_Y(t)$ makes a difference for state $Y(t)$ in the time step from t to $t + \Delta t$, using speed factor $\boldsymbol{\eta}_Y$ and taking into account step size Δt

This can be considered an interpretation of a network based on causality and dynamics as expressed in a formal-numerical way, thus associating semantics to any conceptual temporal-causal network representation in a detailed numerical-mathematically defined manner. Table 2.2 shows how a conceptual representation based on states and connections enriched with labels for connection weights, combination functions, and speed factors, can be transformed into a numerical representation (Treur 2016, Chap. 2). A more detailed explanation of this difference equation format, taken from Treur (2016), Chap. 2, pp. 60–61, is as follows; see also Fig. 2.4. The aggregated impact value $\mathbf{aggimpact}_Y(t)$ at time t pushes the value of Y up or down, depending on how it compares to the current value of Y. So, $\mathbf{aggimpact}_Y(t)$ is compared to the current value $Y(t)$ of Y at t by taking the difference between them (also see Fig. 2.4):

$$\mathbf{aggimpact}_Y(t) - Y(t)$$

If this difference is positive, which means that $\mathbf{aggimpact}_Y(t)$ at time t is higher than the current value of Y at t, in the time step from t to $t + \Delta t$ (for some small Δt) the value $Y(t)$ will increase in the direction of the higher value $\mathbf{aggimpact}_Y(t)$. This increase is done proportional to the difference, with proportion factor $\boldsymbol{\eta}_Y\ \Delta t$: the increase is (see Fig. 2.4):

$$\boldsymbol{\eta}_Y [\mathbf{aggimpact}_Y(t) - Y(t)] \Delta t$$

By this format, the network structure characteristic $\boldsymbol{\eta}_Y$ indeed acts as a speed factor by which it can be specified how fast state Y should change upon causal impact.

2.3.2 Combination Functions as Building Block for Aggregation

Often used examples of combination functions are the ones listed below: the *identity* **id**(.) for states with impact from only one other state, the *scaled maximum and minimum* $\mathbf{smax}_\lambda$(.) and $\mathbf{smin}_\lambda$(.), the *scaled sum* $\mathbf{ssum}_\lambda$(.) with scaling factor λ, the *advanced logistic sum* combination function $\mathbf{alogistic}_{\sigma,\tau}$(..) with steepness σ and threshold τ, and the Euclidean combination function $\mathbf{eucl}_{n,\lambda}$(.) where n is the order (which can be any nonzero natural number, but also any positive real number), and with scaling factor λ:

- *the identity* function for states with impact from only one other state

$$\mathbf{id}(V) = V$$

- the *scaled maximum and minimum* with scaling factor λ

$$\mathbf{smax}_\lambda(V_1, \ldots, V_k) = \mathbf{max}(V_1, \ldots, V_k)/\lambda$$
$$\mathbf{smin}_\lambda(V_1, \ldots, V_k) = \mathbf{min}(V_1, \ldots, V_k)/\lambda$$

- *scaled sum* with scaling factor λ

$$\mathbf{ssum}_\lambda(V_1, \ldots, V_k) = \frac{V_1 + \cdots + V_k}{\lambda}$$

- the *advanced logistic sum* combination function with steepness σ and threshold τ

$$\mathbf{alogistic}_{\sigma,\tau}(V_1, \ldots, V_k) = \left[\frac{1}{1 + \mathrm{e}^{-\sigma(V_1 + \ldots + V_k - \tau)}} - \frac{1}{1 + \mathrm{e}^{\sigma\tau})}\right](1 + \mathrm{e}^{-\sigma\tau})$$

- the Euclidean combination function $\mathbf{eucl}_{n,\lambda}$(.) where n is the order (which can be any nonzero natural number, but also any positive real number), and with scaling factor λ:

$$\mathbf{eucl}_{n,\lambda}(V_1, \ldots, V_k) = \sqrt[n]{\frac{V_1^n + \cdots + V_k^n}{\lambda}}$$

Scaling factors λ are used to normalise the values so that they fit in the intended interval for their values (usually the [0, 1] interval).

Note that for $\lambda = 1$, the scaled sum function is just the sum function **sum**(..), and this sum function can also be used as identity function in case of just one incoming connection. Furthermore, note that for $n = 1$ (first-order Euclidean combination function) we get the scaled sum function:

$$\mathbf{eucl}_{1,\lambda}(V_1, \ldots, V_k) = \mathbf{ssum}_{\lambda}(V_1, \ldots, V_k)$$

For $\boldsymbol{n}$ = 2 it is the second-order Euclidean combination function $\mathbf{eucl}_{2,\lambda}(..)$ defined by:

$$\mathbf{eucl}_{2,\lambda}(V_1, \ldots, V_k) = \sqrt{\frac{V_1^2 + \ldots + V_k^2}{\boldsymbol{\lambda}}}$$

This second-order Euclidean combination function is also often applied in aggregating the error value in optimisation and in parameter tuning using the root-mean-square deviation (RMSD), based on the Sum of Squared Residuals (SSR).

Combination functions as shown above are called *basic combination functions*. There is a *combination function library* containing these basic combination functions. Up till now the library contains 35 basic combination functions. However, it can easily be extended if the designer needs another combination function. For any network model some number m of them can be selected (usually just one or two, or at most a handful); they are represented in a standard format as $\text{bcf}_1(..)$, $\text{bcf}_2(..)$, …, $\text{bcf}_m(..)$. In principle, they use parameters $\boldsymbol{\pi}_{1,i,Y}, \boldsymbol{\pi}_{2,i,Y}$ such as the $\boldsymbol{\lambda}$, $\boldsymbol{\sigma}$, and $\boldsymbol{\tau}$ in the examples above. Including these parameters, the standard format used for basic combination functions is (with V_1, …, V_k the single causal impacts):

$$\text{bcf}_i\left(\boldsymbol{\pi}_{1,i,Y}, \boldsymbol{\pi}_{2,i,Y}, V_1, \ldots, V_k\right)$$

For each state Y just one basic combination function can be selected, but also a weighted average of them can be selected according to the following format

$$\mathbf{c}_Y(\boldsymbol{\pi}_{1,1,Y}, \boldsymbol{\pi}_{2,1,Y}, \ldots, \boldsymbol{\pi}_{1,m,Y}, \boldsymbol{\pi}_{2,m,Y}, \ldots, V_1, \ldots, V_k) = \frac{\boldsymbol{\gamma}_{1,Y}\text{bcf}_1\left(\boldsymbol{\pi}_{1,1,Y}, \boldsymbol{\pi}_{2,1,Y}, V_1, \ldots, V_k\right) + \cdots + \boldsymbol{\gamma}_{m,Y}\text{bcf}_m\left(\boldsymbol{\pi}_{1,m,Y}, \boldsymbol{\pi}_{2,m,Y}, V_1, \ldots, V_k\right)}{\boldsymbol{\gamma}_{1,Y} + \cdots + \boldsymbol{\gamma}_{m,Y}} \tag{2.2}$$

with *combination function weights* $\boldsymbol{\gamma}_{i,Y}$. Selecting only one of them for state Y, for example, $\text{bcf}_i(..)$, is done by putting weight $\boldsymbol{\gamma}_{i,Y} = 1$ and the other weights 0. This is a convenient way to indicate combination functions for a specific network model. The function $\mathbf{c}_Y(..)$ can just be indicated by the weight factors $\boldsymbol{\gamma}_{i,Y}$ and the parameters $\boldsymbol{\pi}_{i,j,Y}$. Note that in (2.2) the different basic combination functions are assumed to share the same variables V_1, …, V_k. If that is not intended, some functions may have to be adapted by adding auxiliary variables to get this right. An example of this can be found in Chap. 5.

So, the concepts $\boldsymbol{\omega}_{X,Y}$, $\boldsymbol{\eta}_Y$, $\boldsymbol{\gamma}_{i,Y}$, $\boldsymbol{\pi}_{i,j,Y}$ (all denoted by bold small Greek letters) represent the different characteristics of a network's structure. Together they fully define the network structure. They are summarised in Table 2.3. In Sect. 2.4 it is

Table 2.3 The characteristics defining the structure of a temporal-causal network model

Concept	Notation	Explanation
Connection weight	$\omega_{X,Y}$	Specifies the strength of the connection from state X to state Y
Speed factor	η_Y	Describes speed of change of state Y upon received causal impact
Combination function weight	$\gamma_{i,Y}$	Determines which combination function(s) are used for state Y
Combination function parameter	$\pi_{i,j,Y}$	The value of the jth parameter of the ith combination function for Y

Table 2.4 Connection matrix of the example of Fig. 2.3

Connection matrix		X_1	X_2	X_3	X_4	X_5	X_6	X_7
		P	Q	R	S	T	U	V
X_1	P			0.8				
X_2	Q			1				
X_3	R				0.9	1		
X_4	S						1	
X_5	T						0.7	1
X_6	U							
X_7	V		1					

shown how the format of role matrices can be used to specify these characteristics for a network model's structure.

For proper functioning of Euclidean combination functions, some constraints are used. First, in general, this function is only applied when all connection weights are positive, except in the specific case that $\boldsymbol{n}$ is an odd natural number. Moreover, also a constraint on the scaling factor λ is used. When no weights are negative, the maximal value of the outcome is achieved when for each X_i it holds $X_i(t) = 1$; then the maximal outcome is $\left((\Sigma_i \omega_{X_i,Y}^{\boldsymbol{n}})/\lambda\right)^{1/\boldsymbol{n}}$. To keep the outcomes within the [0, 1] interval 1, the scaling factor λ should be equal to or at least the sum of the $\boldsymbol{n}$th powers of all weights: $\lambda \geq \Sigma_i \omega_{X_i,Y}^{\boldsymbol{n}}$. In such cases the standard value $\lambda \geq \Sigma_i \omega_{X_i,Y}^{\boldsymbol{n}}$ is often used as a form of *normalisation*. All this also applies to scaled sum functions, as this is the case $\boldsymbol{n} = 1$.

2.4 Role Matrices to Specify a Network Model

To specify a network model being designed, often matrices are a useful means. The first type of matrix sometimes used is a *connection matrix*. This is a square matrix with on each of the two dimensions all states of the network, say $X_1, \ldots, X_n$. In cell (i, j) of the matrix in row i and column j (also denoted by $\mathbf{m}(i, j)$ with $\mathbf{m}$ the name of the matrix) it is indicated whether or not there is a connection from state X_i to state X_j (1 or 0) or the value of the weight of this connection is $(\boldsymbol{\omega}_{X_i,X_j})$. To get the idea, first the example shown in Fig. 2.3 is considered; see Table 2.4 for the connection matrix. For example, the 0.9 in cell (3, 4) indicates that there is a connection from X_3 to X_4 with weight 0.9.

Next, an example of a Social Network addressing social contagion (e.g., of opinions or emotions) is used as illustration: the (fully connected) network shown Fig. 2.5 with connection weights as shown in the square matrix Table 2.5. For example, in cell (4, 2) it is indicated that there is a connection from X_4 to X_2 with weight 0.15.

However, as described by the connection matrix above, the connection weights form only one of the network characteristics to specify a temporal-causal network model. The other ones, speed factors, and weight factors and parameters for the combination functions, are still missing in this matrix, and they are essential too; for example, see Sect. 2.2.3 and the quote from Pearl (2000) there. So, additional information on speed factors and on combination functions and their parameters is needed as well. Moreover, in many cases connection matrices are not very efficient, as usually each state in a network has only a limited number of connections, and then a connection matrix consists mainly of empty cells, which makes the space versus information ratio rather inefficient.

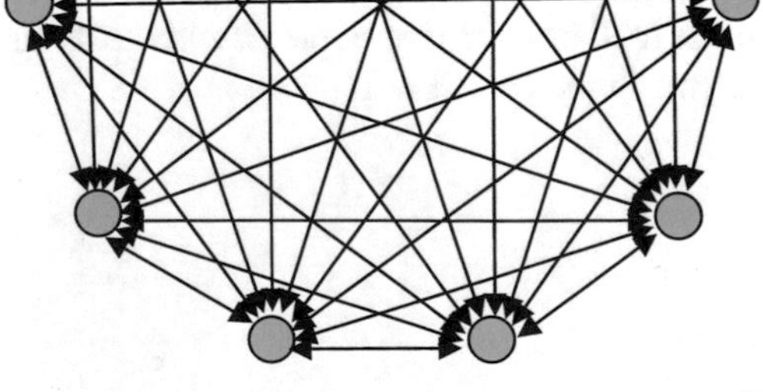

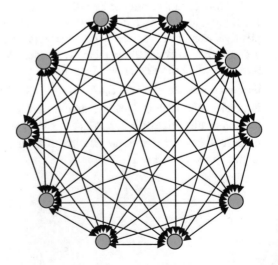

Fig. 2.5 The example Social Network

Table 2.5 Connection matrix of the example Social Network

Connection matrix	X_1	X_2	X_3	X_4	X_5	X_6	X_7	X_8	X_9	X_{10}
X_1		0.1	0.2	0.1	0.2	0.15	0.1	0.25	0.25	0.1
X_2	0.25		0.25	0.2	0.1	0.2	0.15	0.25	0.25	0.25
X_3	0.1	0.25		0.1	0.2	0.15	0.1	0.25	0.1	0.15
X_4	0.25	0.15	0.25		0.15	0.8	0.25	0.15	0.25	0.25
X_5	0.25	0.2	0.1	0.2		0.25	0.2	0.1	0.2	0.15
X_6	0.25	0.1	0.25	0.25	0.25		0.1	0.25	0.25	0.1
X_7	0.2	0.1	0.2	0.15	0.2	0.2		0.2	0.15	0.25
X_8	0.1	0.25	0.1	0.25	0.05	0.15	0.25		0.1	0.25
X_9	0.25	0.15	0.25	0.15	0.2	0.1	0.2	0.15		0.15
X_{10}	0.2	0.25	0.2	0.2	0.1	0.2	0.15	0.8	0.2	

2.4.1 Role Matrices as a Specification Format

To get a more complete and uniform, and more compact specification format, as an alternative to connection matrices, *role matrices* are introduced, according to the role played by the specified information. For example, as can be seen in Eq. (2.1) in Sect. 2.3.1, the numbers for the $\boldsymbol{\omega}_{X,Y}$, $\boldsymbol{\eta}_Y$ and the function $\mathbf{c}_{i,Y}(..)$ and parameters $\boldsymbol{\pi}_{j,i,Y}$ of it play completely different roles. These roles are made more explicit and neatly grouped below by the different role matrices in which they are specified. They cover the main elements of network structure (a) connectivity, (b) aggregation, and (c) timing as indicated in Sect. 2.2.3 above.

Note that the role matrices indeed have a more compact format than connection matrices, and also specify an ordering, which is important as combination functions used to aggregate the impact from multiple connections are not always symmetric in their arguments. Five roles are distinguished and there are five role matrices accordingly (for a first example, see Box 2.1). Here **mb** and **mcw** cover connectivity, **mcfw** and **mcfp** cover aggregation, and **ms** covers timing.

Note that for all role matrices, the first dimension, displayed as the vertical axis, is for the states of the network. In the row of a given state, the other states or values are listed that according to the role specified by that matrix affect this given state:

- **mb** *for the base network connectivity role*

Role matrix **mb** specifies on each row for a given state from which states it has incoming connections. The first (vertical) dimension is for states X_j and the second (horizontal) dimension for the list of states X_i from which the considered state X_j gets incoming connections: the names of these states X_i are indicated in the cells in the row of X_j. This information plays the role of the *base connectivity*. This matrix contains the information in graphical form specified by the arrows in a network picture.

- **mcw** *for the connection weights role*

Role matrix **mcw** specifies on each row for a given state X_j which are the connection weights for the states indicated in the corresponding cells in the base connectivity matrix **mb**. This information plays the role of the *connection weights.*

- **ms** *for the speed factors role*

Role matrix **ms** specifies for each state X_j its speed factor. This matrix has only one column. The first (vertical) dimension is for states and the second (horizontal) dimension for the column with speed values for each state. This information plays the role of the *speed factor.*

- **mcfw** *for the combination function weights role*

Role matrix **mcfw** specifies for each state X_j which basic combination functions $\text{bcf}_i(..)$ are used for it and with which weights $\gamma_{i,Y}$. The first (vertical) dimension is for states X_j and the second (horizontal) dimension for combination functions. This information $\gamma_{i,Y}$ plays the role of the *combination function weights.* A nonzero weight implies that the indicated combination function is used for that state. It is possible that for a given state there are nonzero weights for more than one combination function. This expresses that a weighted sum of multiple combination functions $\text{bcf}_i(\ldots)$ is used as combination function $\mathbf{c}_Y(\ldots)$ for that state:

$$\mathbf{c}_Y(\pi_{1,1,Y}, \pi_{2,1,Y}, \ldots, \pi_{1,m,Y}, \pi_{2,m,Y}, V_1, \ldots, V_k) \\ = \frac{\gamma_{1,Y}\text{bcf}_1(\pi_{1,1,Y}, \pi_{2,1,Y}, V_1, \ldots, V_k) + \cdots + \gamma_{m,Y}\text{bcf}_m(\pi_{1,m,Y}, \pi_{2,m,Y}, V_1, \ldots, V_k)}{\gamma_{1,Y} + \cdots + \gamma_{m,Y}} \tag{2.3}$$

For example, if $m = 2$, $\text{bcf}_1(\pi_{1,1,Y}, \pi_{2,1,Y}, V_1, \ldots, V_k)$ is the function **eucl(..)** and $\text{bcf}_2(\pi_{1,1,Y}, \pi_{2,1,Y}, V_1, \ldots, V_k)$ the function **alogistic(..)**, and $\gamma_{1,Y} = 3$, $\gamma_{1,Y} = 1$, then the outcome is

$$\frac{\gamma_{1,Y}\text{bcf}_1(\pi_{1,1,Y}, \pi_{2,1,Y}, V_1, \ldots, V_k) + \gamma_{2,Y}\text{bcf}_2(\pi_{1,m,Y}, \pi_{2,m,Y}, V_1, \ldots, V_k)}{\gamma_{1,Y} + \gamma_{2,Y}} \\ = \frac{3\,\mathbf{eucl}(\pi_{1,1,Y}, \pi_{2,1,Y}, V_1, \ldots, V_k) + \mathbf{alogistic}(\pi_{1,m,Y}, \pi_{2,m,Y}, V_1, \ldots, V_k)}{4} \\ = 0.75\,\mathbf{eucl}(\pi_{1,1,Y}, \pi_{2,1,Y}, V_1, \ldots, V_k) + 0.25\,\mathbf{alogistic}(\pi_{1,m,Y}, \pi_{2,m,Y}, V_1, \ldots, V_k)$$

- **mcfp** *for the combination function parameters role*

Role matrix **mcfp** specifies for each state X_j and each combination function, the parameters $\pi_{i,j,Y}$ of this combination function for state X_j. Note that this is a 3D matrix with as usual the first (vertical) dimension for the states, the second dimension for the parameters of the combination function, the third dimension for the combination function. This information plays the role of the *combination function parameter values.*

The first two role matrices **mb** and **mcw** can be considered a kind of more compact reformulation of the square matrices for connectivity and connection weights. However, there are two important differences. First, in the connectivity role matrix **mb** the row for a given state X_j displays the names of the states from which X_j gets incoming connections (not the outgoing connections); in a square connection matrix the elements of such a row are in the column for X_j. Second, the connection weights have a separate role matrix **mcw** with numbers for the weights in exactly the cells indicated in the base matrix **mb** for that connection.

Box 2.1 shows the complete specification by role matrices of the conceptual representation of the example network model from Fig. 2.3. Here it can be seen that the role matrices (7 × 2 = 14) are much more compact than connection matrices (7 × 7 = 49), so they are much more efficient as representation. In Box 2.2 this is shown for the example Social Network model of Fig. 2.5. For a fully connected network this condensation is just a modest improvement in efficiency of representation, but usually, networks have many more nodes than the ones connected to a given node.

Box 2.1 Conceptual representation of the example of Fig. 2.3 by role matrices

mb base connectivity		1	2
X_1	P		
X_2	Q	X_7	
X_3	R	X_1	X_2
X_4	S	X_3	
X_5	T	X_3	
X_6	U	X_4	X_5
X_7	V	X_5	

mcw connection weights		1	2
X_1	P		
X_2	Q	1	
X_3	R	0.8	1
X_4	S	0.9	
X_5	T	1	
X_6	U	1	0.7
X_7	V	1	

ms speed factors		1
X_1	P	0
X_2	Q	1
X_3	R	0.4
X_4	S	0.3
X_5	T	0.5
X_6	U	1
X_7	V	0.3

mcfw combination function weights		1 **eucl**	2 **alogis-tic**
X_1	P	1	
X_2	Q	1	
X_3	R	1	
X_4	S	1	
X_5	T	1	
X_6	U	1	
X_7	V	1	

mcfp	**function**	1 **eucl**		2 **alogistic**	
parameter		1 n	2 λ	1 σ	2 τ
X_1	P	1	1		
X_2	Q	1	1		
X_3	R	1	1.8		
X_4	S	1	0.9		
X_5	T	1	1		
X_6	U	1	1.7		
X_7	V	1	1		

In Box 2.1 it is shown in matrix **mcfw** for the combination function weights that for all states the Euclidean combination function **eucl** is chosen. In matrix **mcfp** for combination function parameters it is shown that order $\boldsymbol{n}$ = 1 is selected (which makes it the scaled sum function), and the scaling factors $\boldsymbol{\lambda}$ are indicated; note that they are chosen here as the sums of the weights of the incoming connections as shown in the rows of matrix **mcw** for the connection weights. Finally, matrix **ms** for speed factors just shows all speed factors that were chosen.

Box 2.2 Conceptual representation of the example Social Network model by role matrices (used in the second and third scenario in Fig. 2.5).

mb base connectivity	1	2	3	4	5	6	7	8	9
X_1	X_2	X_3	X_4	X_5	X_6	X_7	X_8	X_9	X_{10}
X_2	X_1	X_3	X_4	X_5	X_6	X_7	X_8	X_9	X_{10}
X_3	X_1	X_2	X_4	X_5	X_6	X_7	X_8	X_9	X_{10}
X_4	X_1	X_2	X_3	X_5	X_6	X_7	X_8	X_9	X_{10}
X_5	X_1	X_2	X_3	X_4	X_6	X_7	X_8	X_9	X_{10}
X_6	X_1	X_2	X_3	X_4	X_5	X_7	X_8	X_9	X_{10}
X_7	X_1	X_2	X_3	X_4	X_5	X_6	X_8	X_9	X_{10}
X_8	X_1	X_2	X_3	X_4	X_5	X_6	X_7	X_9	X_{10}
X_9	X_1	X_2	X_3	X_4	X_5	X_6	X_7	X_8	X_{10}
X_{10}	X_1	X_2	X_3	X_4	X_5	X_6	X_7	X_8	X_9

mcw connection weights	1	2	3	4	5	6	7	8	9
X_1	0.25	0.1	0.25	0.25	0.25	0.2	0.1	0.25	0.2
X_2	0.1	0.25	0.15	0.2	0.1	0.1	0.25	0.15	0.25
X_3	0.2	0.25	0.25	0.1	0.25	0.2	0.1	0.25	0.2
X_4	0.1	0.2	0.1	0.2	0.25	0.15	0.25	0.15	0.2
X_5	0.2	0.1	0.2	0.15	0.25	0.2	0.05	0.2	0.1
X_6	0.15	0.2	0.15	0.8	0.25	0.2	0.15	0.1	0.2
X_7	0.1	0.15	0.1	0.25	0.2	0.1	0.25	0.2	0.15
X_8	0.25	0.25	0.25	0.15	0.1	0.25	0.2	0.15	0.8
X_9	0.25	0.25	0.1	0.25	0.2	0.25	0.15	0.1	0.2
X_{10}	0.1	0.25	0.15	0.25	0.15	0.1	0.25	0.25	0.15

mcfw combination function weights	1 eucl	2 alogistic
X_1	1	
X_2	1	
X_3	1	
X_4	1	
X_5	1	
X_6	1	
X_7	1	
X_8	1	
X_9	1	
X_{10}	1	

mcfp function	1 eucl		2 alogistic	
parameter	1 n	2 λ	1 σ	2 τ
X_1	1	1.85		
X_2	1	1.55		
X_3	1	1.8		
X_4	1	1.6		
X_5	1	1.45		
X_6	1	2.2		
X_7	1	1.5		
X_8	1	2.4		
X_9	1	1.75		
X_{10}	1	1.65		

ms speed factors	1
X_1	0.8
X_2	0.5
X_3	0.8
X_4	0.5
X_5	0.5
X_6	0.5
X_7	0.8
X_8	0.5
X_9	0.5
X_{10}	0.5

In the base connectivity role matrix **mb** shown in Box 2.2 for the fully connected Social Network it is seen that each state has incoming connections from all other states. In the connection weight role matrix **mcw** the connection weights for these connections as indicated in **mb** are shown. For example in the row for state X_2 in the cell (2, 4) for the fourth incoming connection of X_2, the number 0.2 indicates that the state X_5 indicated in the corresponding cell (2, 4) in matrix **mb** has a connection to X_2 with weight 0.2.

In the combination function weight role matrix **mcfw** it is indicated that for all states the Euclidean combination function is chosen, with weight 1. In addition, in the combination function parameter role matrix **mcfp** the two parameters $\boldsymbol{n}$ (the order) and $\boldsymbol{\lambda}$ (the scaling factor) of this combination function are indicated for each

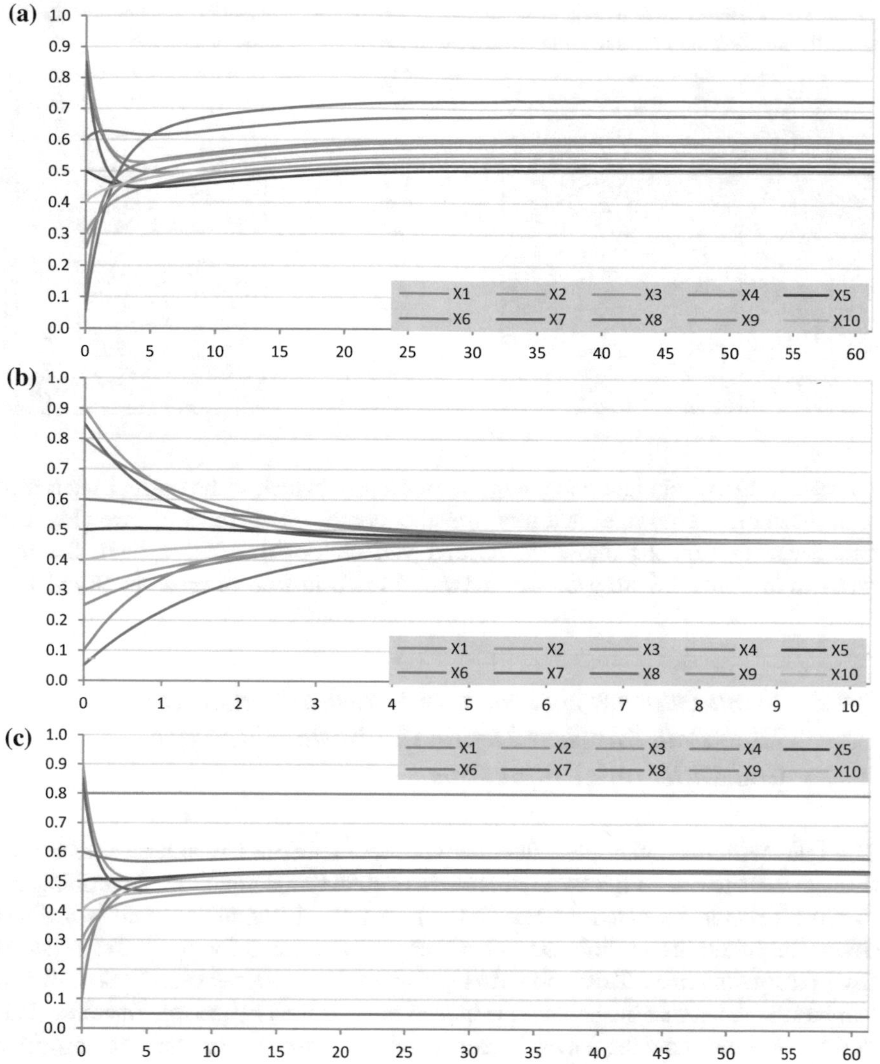

Fig. 2.6 Simulations for the example Social Network of Fig. 2.5 with **a** upper graph: advanced logistic sum combination functions with steepness $\sigma = 1.5$, threshold $\tau = 0.3$, as shown in Table 2.5, right hand side (no convergence to one common value), **b** middle graph: normalised scaled sum functions as shown in Box 2.1, matrix **mcfw** and **mcfp** (convergence to one common value), **c** normalised scaled sum functions with constant X_4 (at 0.8) and X_8 (at 0.05) (no convergence to one common value)

of the states. Finally, the speed factor role matrix **ms** indicates the speed factors of the different states.

In Fig. 2.6 three different simulations are shown for this network model. The first one uses the combination function $\mathbf{alogistic}_{\sigma,\tau}(..)$ and the other two the

Table 2.6 Another variant of example role matrices **mcfw** for combination function weights and **mcfp** for combination function parameters, used in the first simulation in Fig. 2.6

mcfw combination function weights	1 eucl	2 alogistic
X_1		1
X_2		1
X_3		1
X_4		1
X_5		1
X_6		1
X_7		1
X_8		1
X_9		1
X_{10}		1

mcfp function	1 eucl		2 alogistic	
parameter	1	2	1 σ	2 τ
X_1			1.5	0.3
X_2			1.5	0.3
X_3			1.5	0.3
X_4			1.5	0.3
X_5			1.5	0.3
X_6			1.5	0.3
X_7			1.5	0.3
X_8			1.5	0.3
X_9			1.5	0.3
X_{10}			1.5	0.3

normalised Euclidean function $\mathbf{eucl}_{1,\lambda}(..)$ of order 1 (which is the normalised scaled sum function). The role matrices for the combination function weights and parameters in Box 2.1 show the second variant. For the first variant, the role matrices in Table 2.6 are used for **mcfw** and **mcfp** instead of those in Box 2.1.

2.4.2 From Network Structure to Network Dynamics: How Role Matrices Define the Basic Difference and Differential Equations

The role matrices contain the values for all of the network structure characteristics that are used to define the basic difference or differential equations describing the network's dynamics based on Eq. (2.1) in Sect. 2.3.1. Actually, there is a direct derivation of the basic difference or differential equations for the different states from the role matrices. This is found as shown in Chap. 10, Sect. 10.6, Box 10.6. It can be seen there, that the equation is indeed fully defined by the role matrices. This derivation shows how the network structure characteristics determine the network's dynamics; see also Fig. 2.7. This is a relation between the network's structure and its dynamics at a basic level. In Chaps. 11–14 this relation between structure and dynamics will be analysed in more detail and at the higher level of properties of structure that entail properties of emerging behaviour (e.g., see Fig. 11.1 in Chap. 11).

Note that, although it may look a bit theoretical, from a practical perspective this is a very relevant derivation. It makes that the design of a network model can fully concentrate on the conceptual representation of the network's structure. As the numerical representation for the network's dynamics fully depends on that, the software environment developed (described in Chap. 9) just takes the role matrices as input and runs the model based on the implied difference equations generated internally by the software, without having to write or even see these

Table 2.7 Specification of initial values

Initial values

X_1	X_2	X_3	X_4	X_5	X_6	X_7	X_8	X_9	X_{10}
0.1	0.3	0.9	0.8	0.5	0.6	0.85	0.05	0.25	0.4

equations. So, for modeling networks in practice and exploring their behaviour no programming is needed, and even no difference equations need to be specified. Also for mathematical analysis of the network behaviour, usually the difference equations need not to be analysed, as a very simple criterion is available in terms of the network characteristics, that can be used; see also Sect. 2.5 below, and in Chaps. 11–14.

2.4.3 *Simulations for the Example Social Network Model*

For the above social network model, which models social contagion, for example, of opinions or emotions, simulations have been performed for different combination functions. Initial values were used as shown in Table 2.7.

In Fig. 2.7 thc threc differcnt simulations are shown, all with step size $\Delta t = 0.25$. For the upper graph, advanced logistic sum combination functions were used, for the middle graph normalized scaled sum functions, and in the lower graph scaled sum functions while two states remain constant (they have no incoming connections this time, so the cells in the columns for X_4 and X_8 in **mcw** all are 0 now). How can we explain these differences in emerging behavior from the structure of the networks? In Sect. 2.5 these results and their comparison are discussed and analysed in somc morc dctail.

2.5 Relating Emerging Network Behavior to Network Structure

The Network-Oriented Modeling approach based on temporal-causal networks does not only provide opportunities for simulation but also for mathematical analysis and to derive general theoretical results that predict or reflect behavior that is observed in specific cases of simulations. A general question for dynamic models is what patterns of behaviour will emerge, and how their emergence depends on the chosen

Conceptual Representation:
Network Structure
Numerical Representation:
Network Dynamics

Fig. 2.7 Network structure determines network behaviour

structure. Whether or not, in general, such relations between structure and emerging behavior can be found is sometimes a topic for discussion. However, in the context of the Network-Oriented Modeling approach based on temporal-causal networks considered here at least some results on this relation have been obtained.

Usually the structure of a network is described by a number of characteristics. For temporal-causal networks, in particular, such network structure characteristics are connectivity (in terms of connection weights), aggregation (in terms of combination functions) and timing (in terms of speed factors). So, the challenge is to find out how properties of connection weights, combination functions, and speed factors relate to emerging behavior.

2.5.1 Emerging Network Behaviour and Network Structure

Emerging network behaviour can be of different types. Three types are often distinguished:

- **Reaching an equilibrium**

A so called *equilibrium state* is reached, in which for all states the values do not change anymore. This often happens; for example, all three graphs in Fig. 2.6 show examples of this type.

- **Ending up in a limit cycle**

The behaviour ends up in a regular repeating pattern of values (a periodic pattern) for the states; this is called a *limit cycle*. In Fig. 2.8 an example of this is shown, taken from Treur (2016), Chap. 12.

- **Chaotic behaviour**

The behaviour is usually (loosely) called *chaotic* if there is no observed regularity in it. This means that at least no equilibrium is reached and also no periodic pattern as a limit cycle. Lorenz (1963) used as title for his paper on chaotic behaviour 'Deterministic Nonperiodic Flow'. In Mathematics, the area of Chaos Theory has developed more specific definitions for chaotic behaviour, usually involving that the outcome is very sensitive for the values of the initial settings; e.g., Lorenz (1963): the present determines the future but the approximate present does not approximately determine the future. An often cited example or metaphor is that a butterfly at one place in the world can cause a tornado somewhere else (the butterfly effect).

When all state values are in a bounded interval, for example, the [0, 1] interval, most often the first type of emerging behaviour is observed, but sometimes also the other two types can occur. An example (seemingly) showing the third type of emerging behaviour may be found in Chap. 6. Note that a pattern can initially look like this last type, but later on may still turn out to be one of the other two types.

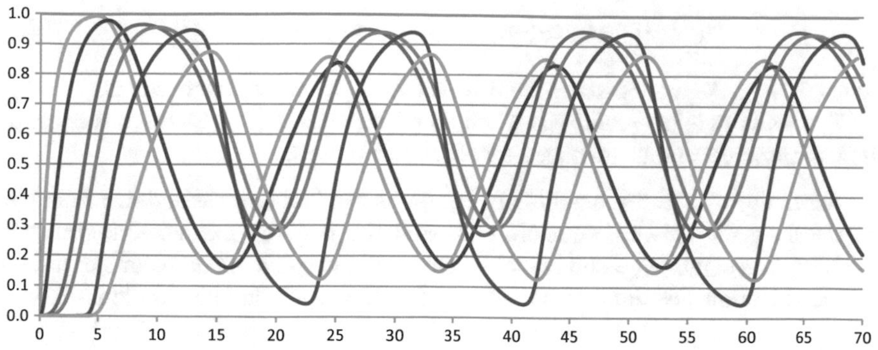

Fig. 2.8 Example simulation ending up in a limit cycle; adopted from Treur (2016), Chap. 12, Sect. 12.7, Fig. 12.7, p. 344

By mathematical analysis of the relation between network structure and network behaviour, the first type of emerging network behaviour (reaching an equilibrium) is relatively easy to explore; see also Treur (2016), Chap. 12. Below some of the basics for that are summarised. In particular, it can be addressed what values eventually will emerge; for example:

- Will the values of each state in the network separately in the end become constant?
- Will the values of different states eventually converge to a common value?
- Under which conditions on the structure of the network will this happen?

Such behaviour relates to what are called stationary points and equilibrium states, defined as follows:

Definition (stationary and equilibrium)
State Y is *stationary* or *has a stationary point* at time t if $\mathbf{d}Y(t)/\mathbf{d}t = 0$.
The network is in an *equilibrium state* at t if all states are stationary at t.

Note that a state may have a stationary point at some time point t, but later on still change its value; in particular, this happens when other states do not have a stationary point at that same time point t. For example, all peaks and dips in Fig. 2.8 indicate stationary points for the specific states, but no equilibrium occurs.

For a temporal-causal network, in particular, there is a simple *criterion* in terms of the network structure characteristics (speed factors $\boldsymbol{\eta}_Y$, connection weights $\boldsymbol{\omega}_{X_i,Y}$, and combination functions $\mathbf{c}_Y(..)$); this immediately follows from Eq. (2.1) in Sect. 2.3.1:

Criterion for stationary point and equilibrium

In a temporal-causal network model, state Y is stationary at t if and only if

$$\boldsymbol{\eta}_Y = 0 \quad \text{or} \quad \mathbf{c}_Y(\boldsymbol{\omega}_{X_1,Y} X_1(t), \ldots, \boldsymbol{\omega}_{X_k,Y} X_k(t)) = Y(t) \tag{2.4}$$

where X_1, …, X_k are the states from which Y has incoming connections.

The network is in *equilibrium* when for *all* states Y of the network this equation holds. These equations are called *equilibrium equations*.

Assuming $\boldsymbol{\eta}_Y > 0$, the equilibrium equations can just be written down from two of the three the network structure characteristics defining the network structure: connection weights $\boldsymbol{\omega}_{X,Y}$ and combination functions $\mathbf{c}_Y(..)$. As an example, in case the combination function is a scaled sum function, such an equation looks like

$$\frac{\boldsymbol{\omega}_{X_1,Y} X_1(t) + \cdots + \boldsymbol{\omega}_{X_k,Y} X_k(t)}{\boldsymbol{\lambda}} = Y(t) \tag{2.5}$$

Often the t is left out as it is a relation between the (constant) values; so the equilibrium equation becomes just an equation for these values:

$$\frac{\boldsymbol{\omega}_{X_1,Y} X_1 + \cdots + \boldsymbol{\omega}_{X_k,Y} X_k}{\boldsymbol{\lambda}} = Y$$

As an example, from the role matrices **mb** and **mcw** for base connectivity and connection weights, and **mcfp** for combination function parameters in Box 2.1 above it can be found that when using the scaled sum combination function in the above example network the equilibrium equation for X_3 is

$$\frac{0.8\,X_1 + X_2}{1.8} = X_3$$

which can be rewritten as

$$0.8\,X_1 + X_2 = 1.8\,X_3 \tag{2.6}$$

So, if in a simulation an equilibrium is reached, then the state values found should satisfy this relation (and also the other equilibrium equations). If not, then something is wrong and has to be resolved.

As another example, from the role matrices **mb** and **mcw** for base connectivity and connection weights, and **mcfp** for combination function parameters in Box 2.2 above it can be found that when using the scaled sum combination function in the above example Social Network the equilibrium equation for X_2 is

$$X_2 = \frac{0.1\,X_1 + 0.25\,X_3 + 0.15\,X_4 + 0.2\,X_5 + 0.1\,X_6 + 0.1\,X_7 + 0.25\,X_8 + 0.15\,X_9 + 0.25\,X_{10}}{1.55} \tag{2.7}$$

or

$$1.55\,X_2 = 0.1\,X_1 + 0.25\,X_3 + 0.15\,X_4 + 0.2\,X_5 + 0.1\,X_6 + 0.1\,X_7 + 0.25\,X_8 + 0.15\,X_9 + 0.25\,X_{10} \tag{2.8}$$

The equilibrium equations are a useful means to *verify the correctness* of the (implemented) network model. This can be done in two ways, as also described in Treur (2016), Chap. 12. The first is by taking the state values as observed for a stationary point or equilibrium in a simulation example, and substitute them in the equilibrium equations. If a serious deviation is found, that should be a reason to investigate the implemented model further to find and resolve some error. Another way, for an equilibrium, is to solve the equilibrium equations and compare the values found with the values observed in a simulation. Whether or not this can be done in an algebraic manner depends on the specific combination functions. These are very practical ways of using the relation of the network structure as specified by the role matrices (based on which the equilibrium equations are formulated) with emergent behaviour as generated by an implemented network model.

Also in a more general sense the relation between network structure and network behavior can be explored. A number of general properties of network structure have been identified such that they relate to similar emergent behavior. These network structure properties concern a connectivity property about how many states of the network are reachable from a given state, and some properties of combination functions. This will be briefly discussed in Sect. 2.5.2.

2.5.2 Network Structure Properties Relevant for Emerging Network Behaviour

It has been found out that some properties of network structure (in particular concerning aggregation and connectivity) underly the differences in emerging behaviour shown in Fig. 2.6. Chapters 11 and 12 address this in much more detail. In the current section just a brief introduction and summary is presented. First the relevant properties of aggregation, as specified by combination functions, that have been identified.

Definition (properties of combination functions)

Let $c(V_1, \ldots, V_k)$ be a function of values $V_1, \ldots, V_k$

(a) c(..) is *nonnegative* if $c(V_1, \ldots, V_k) \geq 0$
(b) c(..) *respects* 0 if $V_1, \ldots, V_k \geq 0 \Rightarrow [c(V_1, \ldots, V_k) = 0 \Leftrightarrow V_1 = \cdots = V_k = 0]$
(c) c(..) is *monotonically increasing* if

$$U_i \leq V_i \text{ for all } i \Rightarrow \mathrm{c}(U_1, \ldots, U_k) \leq \mathrm{c}(V_1, \ldots, V_k)$$

(d) c(..) is *strictly monotonically increasing* if

$$U_i \leq V_i \text{ for all } i, \text{and } U_j < V_j \text{ for at least one } j \Rightarrow \mathrm{c}(U_1, \ldots, U_k) < \mathrm{c}(V_1, \ldots, V_k)$$

(e) c(..) is *scalar-free* if $\mathrm{c}(\alpha V_1, \ldots, \alpha V_k) = \alpha\, \mathrm{c}(V_1, \ldots, V_k)$ for all $\alpha > 0$

The properties (a–c) are basic properties expected from most if not all combination functions. Properties (d) and (e) define a specific class of combination functions; this class includes all Euclidean combination functions, but logistic combination functions do not belong to this class as they are not scalar-free. In Chaps. 11 and 12 some theoretical results on emergent behaviour will be presented for this class, where also some other network properties concerning the network's connectivity and normalisation play a role.

Definition (normalised network)
A network is *normalised* or uses normalised combination functions if for each state Y it holds $\mathrm{c}_Y(\boldsymbol{\omega}_{X_1,Y}, \ldots, \boldsymbol{\omega}_{X_k,Y}) = 1$, where $X_1, \ldots, X_k$ are the states with outgoing connections to Y.

Note that $\mathrm{c}_Y(\boldsymbol{\omega}_{X_1,Y}, \ldots, \boldsymbol{\omega}_{X_k,Y})$ is an expression in terms of the parameter(s) of the combination function and $\boldsymbol{\omega}_{X_1,Y}, \ldots, \boldsymbol{\omega}_{X_k,Y}$. To require this expression to be equal to 1 provides a constraint on these parameters: an equation relating the parameter value(s) of the combination functions to the network structure characteristics $\boldsymbol{\omega}_{X_1,Y}, \ldots, \boldsymbol{\omega}_{X_k,Y}$. To satisfy this property, often the parameter(s) can be given suitable values. For example, for a Euclidean combination function, scaling factor $\boldsymbol{\lambda}_Y = \boldsymbol{\omega}^n_{X_1,Y} + \cdots + \boldsymbol{\omega}^n_{X_k,Y}$ will provide a normalised network. This can be done in general:

(1) **normalisation by adjusting the combination functions**

If any combination function $\mathbf{c}_Y(..)$ is replaced by $\mathbf{c}'_Y(..)$ defined as

$$\mathbf{c}'_Y(V_1, \ldots, V_k) = \mathbf{c}_Y(V_1, \ldots, V_k)/\mathbf{c}_Y(\boldsymbol{\omega}_{X_1,Y}, .., \boldsymbol{\omega}_{X_k,Y})$$

then the network becomes normalised: indeed $\mathbf{c}'_A(\boldsymbol{\omega}_{X_1,Y}, .., \boldsymbol{\omega}_{X_k,Y}) = 1$

(2) **normalisation by adjusting the connection weights (for scalar-free combination functions)**

For scalar-free combination functions also normalisation is possible by adapting the connection weights; define:

$$\boldsymbol{\omega}'_{X_i,Y} = \boldsymbol{\omega}_{X_i,Y}/\mathbf{c}_Y(\boldsymbol{\omega}_{X_1,Y}, .., \boldsymbol{\omega}_{X_k,Y})$$

Then the network becomes normalised; indeed it holds:

$$\mathbf{c}_Y(\omega'_{X_1,Y},..,\omega'_{X_k,Y}) = \mathbf{c}(\omega_{X_1,Y}/\mathbf{c}_Y(\omega_{X_1,Y},..,\omega_{X_k,Y}),..,\omega_{X_k,Y}/\mathbf{c}(\omega_{X_1,Y},..,\omega_{X_k,Y})) = 1$$

Another important determinant for emerging behaviour is connectivity: in how far the network has paths connecting any two states; for this, the following definition is used:

Definition State Y is *reachable* from state X if there is a directed path from X to Y

This property makes a difference between the third example simulation and the other two: from no state X_4 or X_8 is reachable in that third case as these states have no incoming connections.

2.5.3 *Relating Network Structure Properties to Emerging Network Behaviour*

Part of the mathematical analysis performed is summarised by the following theorem that has been derived.

Theorem 1 (equal equilibrium state values)
Suppose a network with nonnegative connections is based on normalised, strictly monotonically increasing and scalar-free combination functions.

(a) Suppose any state Y except at most one state, is reachable from all other states X. Then in an equilibrium state, all states have the same state value.
(b) Under the conditions of (a), the equilibrium state is attracting, and the common equilibrium state value lies in between the highest and lowest previous or initial state values.

Theorem 1 can be used to prove for many cases that in an equilibrium state all states have the same value. This includes cases in which the only combination functions used are Euclidean combination functions. Returning to the example simulations shown in Fig. 2.6, it turns out that in one case convergence to one common equilibrium value takes place, but in the other two cases that does not happen and instead some form of clustering seems to take place. How can we explain these differences in emerging behavior from the structure of the networks? This question can be answered based on the above properties. They show why for the second simulation in Fig. 2.6 convergence to one common value takes place, but not for the first and third case. The first case does not satisfy the scalar-free condition, and the third case does not satisfy the condition on reachability in Theorem 1; one exception is allowed but not two, as occurs in the third example in Fig. 2.6. In case of only one of X_4 and X_8 as exception, say X_4, there would be convergence to one common value: to the value of the one state that remains constant all the time. Note that these differences in emerging behaviour have no relation to linear or nonlinear equations, as Theorem 1 applies to, for example, all Euclidean combination functions, both to linear and nonlinear ones. This type of analysis will be addressed in much more detail and considering more types of functions and network connectivity in Chaps. 11 and 12.

2.6 The Wide Applicability of Network-Oriented Modeling

Many applications of Network-Oriented Modeling exist: Biological Networks, Neural Networks, Mental Networks, and Social Networks. It sometimes is a silent assumption that a Network-Oriented Modeling approach can only work for such specific application domains, where networks are felt as more or less already given or perceived in the real world. It has turned out that the applicability of Network-Oriented Modeling goes far beyond such domains as will be discussed below.

2.6.1 Network-Oriented Modeling Applies Beyond Perceived Networks

In Treur (2017) it is shown that the above-mentioned silent assumption is not a correct assumption. It has been shown that the applicability of the Network-Oriented Modeling approach based on temporal-causal networks is much wider. For example, it has been proven that modeling by temporal-causal networks subsumes modelling approaches based on the dynamical system perspective (Ashby 1960; Port and van Gelder 1995) or systems of first-order differential equations; see Treur (2017), Sect. 2.3. The dynamical system approach is not only often used to obtain dynamical cognitive models, but also to model processes in many other scientific domains, including biological and physical domains. Moreover, modeling by temporal-causal networks subsumes modelling approaches based on discrete (event) and agent simulation (Sarjoughian and Cellier 2001; Uhrmacher and Schattenberg 1998), including very basic computational notions such as finite state machines and transition systems; see Treur (2017), Sect. 2.4.

This shows that temporal-causal network models do not just model networks considered as given in the real world, but can be applied to model practically any type of process. Therefore, indeed the modelling approach is not limited only to Biological Networks, Neural Networks, Mental Networks, and Social Networks, but applies far beyond those types of domains. It shows that the specific temporal-causal interpretation and structure added to networks on top of a basic graph structure, as discussed in Sect. 2.2, does not introduce limitations compared to other dynamic modeling approaches that are based on difference or differential equations.

2.6.2 *Network-Oriented Modeling Applies to Network Adaptation*

In Chap. 1 it already has been pointed out how adaptive networks can be modelled too, using the notion of network reification. In this book, starting in Chap. 3, in different chapters, this is illustrated for many examples varying from adaptive Mental Networks based on a Hebbian learning principle to adaptive Social Networks based on a homophily principle, and more. Note that this illustrates once more how the presented Network-Oriented Modeling approach provides a unifying perspective across different domains, in this case, the mental domain for Hebbian learning and the social domain for bonding by homophily. Both can be described in a unified manner by a picture of the type as shown in Fig. 2.3, and by some combination function specifying the specific adaptation principle: see Chap. 3, Sect. 3.6.1 and Fig. 3.4.

2.7 Discussion

In this chapter, the ins and outs of the Network-Oriented Modeling perspective were discussed in some detail. Part of the material for this chapter is based on Treur (2019).

By committing to an interpretation of networks based on the notion of a temporal-causal network, more structure and more depth is obtained, and more dedicated support is possible. At first sight, it may suggest that it introduces a limitation to commit to a specific interpretation and structure of networks, but the proven wide scope of applicability of this Network-Oriented Modelling approach shows otherwise, as causality and temporality are very general concepts; e.g., see also Treur (2016, 2017). On the contrary, the specific network structure characteristics connection weights, combination functions, and speed factors allow for a quite sensitive and unifying way of modeling realistic processes. These characteristics also allow more theoretical depth, which was illustrated by presenting some mathematical results on how emerging network behaviour relates to specific properties of the network structure.

In the rest of the book, a number of fundamental themes are being developed in more depth. A major theme is *network reification* as already pointed out in Chap. 1. This provides a substantial enhancement of expressive power of the modelling format, in particular where it concerns adaptive networks where the network structure characteristics change over time. By considering dynamics not only for states but also for characteristics of the network structure such as connection weights, also adaptive processes are covered in the form of reified adaptive networks. Examples of this illustrate the unifying role that such reified temporal-causal network models can play, in particular by revealing similar structures in adaptive Mental Networks based on a Hebbian learning principle (Hebb 1949; Gerstner and

Kistler 2002), and adaptive Social Networks based on a homophily principle (McPherson et al. 2001). The structure provided by the notion of temporal-causal network in conjunction with the notion of network reification introduced in more depth in Chap. 3 provides the machinery to express such adaptive processes in a unified manner. More specifically, in Chap. 3 it is shown how the network reification construction can be defined in general, and it is illustrated by several examples for Mental and Social Networks how any network adaptation principle can be defined within the reified network. In Chap. 4 it is shown how this reification construction can be repeated, thus obtaining *multilevel network reification* in which, for example, adaptive adaptation principles can be represented explicitly.

Another fundamental theme being developed further in more depth is the *relation between network structure and emerging network behaviour*. Keeping in mind that network structure is defined by network characteristics Connectivity, Aggregation and Timing in terms of connection weights, speed factors, and combination functions, in this theme it is analysed how certain properties of these network structure characteristics relate to certain emerging behaviour (mainly focusing on equilibria). For example, can the values of the states for $t \to \infty$ be predicted from these characteristics? And in which cases will all states end up with the same common value? This is addressed in Chaps. 11 and 12, where in the latter chapter the network is analysed based on its *strongly connected components* (Harary et al. 1965) and *stratification* (Chen 2009) of the abstracted acyclic *condensation graph*. For reified networks for bonding based on homophily these questions are addressed in Chap. 13, and for Hebbian learning, this is addressed in Chap. 14.

A third area in which much development takes place is in the area of applications to certain biological, mental and social domains. The temporal-causal format makes it easy to represent causal domain knowledge in an understandable and executable manner. Several examples of applications in these domains illustrate this. In addition, it may be interesting to further investigate applications to the area of business economics, organisation modeling and management; e.g., Naudé et al. (2008).

References

Ashby, W.R.: Design for a Brain, 2nd edn. Wiley, New York (1960)

Bell, A.: Levels and loops: the future of artificial intelligence and neuroscience. Phil. Trans. R. Soc. Lond. B **354**, 2013–2020 (1999)

Chen, Y.: General spanning trees and reachability query evaluation. In: Desai, B.C. (ed.), Proceedings of the 2nd Canadian Conference on Computer Science and Software Engineering, C3S2E'09, pp. 243–252. ACM Press (2009)

Gerstner, W., Kistler, W.M.: Mathematical formulations of Hebbian learning. Biol. Cybern. **87**, 404–415 (2002)

Harary, F., Norman, R.Z., Cartwright, D.: Structural Models: An Introduction to the Theory of Directed Graphs. Wiley, New York (1965)

Hebb, D.: The Organisation of Behavior. Wiley, New York (1949)

Jonker, C.M., Snoep, J.L., Treur, J., Westerhoff, H.V., Wijngaards, W.C.A.: Putting intentions into cell biochemistry: an artificial intelligence perspective. J. Theor. Biol. **214**(2002), 105–134 (2002)

Jonker, C.M., Snoep, J.L., Treur, J., Westerhoff, H.V., Wijngaards, W.C.A.: BDI-modelling of complex intracellular dynamics. J. Theor. Biol. **251**, 1–23 (2008)

Kim, J.: Philosophy of Mind. Westview Press (1996)

Lorenz, E.N.: Deterministic nonperiodic flow. J. Atmos. Sci. **20**(2), 130–141 (1963)

McPherson, M., Smith-Lovin, L., Cook, J.M.: Birds of a feather: homophily in social networks. Annu. Rev. Sociol. **27**, 415–444 (2001)

Mooij, J.M., Janzing, D., Schölkopf, B.: From differential equations to structural causal models: the deterministic case. In: Nicholson, A., Smyth, P. (eds.), Proceedings of the 29th Annual Conference on Uncertainty in Artificial Intelligence (UAI-13), pp. 440–448. AUAI Press (2013). URL: http://auai.org/uai2013/prints/papers/24.pdf

Naudé, A., Le Maitre, D., de Jong, T., Mans, G.F.G., Hugo, W.: Modelling of spatially complex human-ecosystem, rural-urban and rich-poor interactions (2008). URL: https://www.researchgate.net/profile/Tom_De_jong/publication/30511313_Modelling_of_spatially_complex_human-ecosystem_rural-urban_and_rich-poor_interactions/links/02e7e534d3e9a47836000000.pdf

Pearl, J.: Causality. Cambridge University Press, Cambridge (2000)

Port, R.F., van Gelder, T.: Mind as motion: explorations in the dynamics of cognition. MIT Press, Cambridge, MA (1995)

Potter, S.M.: What can artificial intelligence get from neuroscience? In: Lungarella, M., Bongard, J., Pfeifer, R. (eds.) Artificial Intelligence Festschrift: The next 50 years. Springer, Berlin (2007)

Sarjoughian, H., Cellier, F.E. (eds.): Discrete Event Modeling and Simulation Technologies: A Tapestry of Systems and AI-Based Theories and Methodologies. Springer, Berlin (2001)

Scherer, K.R.: Emotions are emergent processes: they require a dynamic computational architecture. Phil. Trans. R. Soc. B **364**, 3459–3474 (2009)

Treur, J.: Network-Oriented Modeling: Addressing Complexity of Cognitive, Affective and Social Interactions. Springer Publishers, Berlin (2016) Downloadable at URL: https://link-springer-com.vu-nl.idm.oclc.org/book/10.1007/978-3-319-45213-5

Treur, J.: On the applicability of network-oriented modeling based on temporal-causal networks: why network models do not just model networks. J. Inf. Telecommun. **1**, 23–40 (2017)

Treur, J.: The ins and outs of network-oriented modeling: from biological networks and mental networks to social networks and beyond. In: Transactions on Computational Collective Intelligence vol. 32, pp. 120–139. Springer Publishers, Berlin. Contents of Keynote Lecture at ICCCI'18 (2019)

Uhrmacher, A., Schattenberg, B.: Agents in discrete event simulation. In: Proceedings of the European Symposium on Simulation (ESS '98, Nottingham, England). Society for Computer Simulation, San Diego, CA (1998)

Westerhoff, H.V., He, F., Murabito, E., Crémazy, F., Barberis, M.: Understanding principles of the dynamic biochemical networks of life through systems biology. In: Kriete, A., Eils, R. (eds.) Computational Systems Biology, 2nd edn, pp. 21–44. Academic Press, Oxford (2014a)

Westerhoff, H.V., Brooks, A.N., Simeonidis, E., García-Contreras, R., He, F., Boogerd, F.C., Jackson, V.J., Goncharuk, V., Kolodkin, A.: Macromolecular networks and intelligence in microorganisms. Front. Microbiol. **5**, Article 379 (2014b)

Wright, S.: Correlation and causation. J. Agric. Res. **20**, 557–585 (1921)

Part II
Modeling Adaptive Networks by Network Reification

Chapter 3
A Unified Approach to Represent Network Adaptation Principles by Network Reification

Abstract In this chapter, the notion of network reification is introduced: a construction by which a given (base) network is extended by adding explicit states representing the characteristics defining the base network's structure. This is explained for temporal-causal networks where connection weights, combination functions, and speed factors represent the characteristics for Connectivity, Aggregation, and Timing describing the network structure. Having the network structure represented in an explicit manner within the extended network enables to model the adaptation of the base network by dynamics within the reified network: an adaptive network is represented by a non-adaptive network. It is shown how the approach provides a unified modeling perspective on representing network adaptation principles across different domains. This is illustrated for a number of well-known network adaptation principles such as for Hebbian learning in Mental Networks and for network evolution based on homophily in Social Networks.

3.1 Introduction

Reification is a notion that is known from different scientific areas. Literally, it means representing something abstract as a material or concrete thing (Merriam-Webster dictionary), or making something abstract more concrete or real (Oxford dictionaries). Well known examples in linguistics, logic and knowledge representation domains are representing relations between objects as objects themselves (reified relations); this enables to introduce variables and relations over these reified relations. In this way, the expressivity of a language can be extended substantially. In such a way in logic, statements can be represented by term expressions over which predicates can be defined. This idea of reification has been applied in particular to many modeling and programming languages, for example, logical, functional, and object-oriented languages (e.g., Weyhrauch 1980; Bowen and Kowalski 1982; Bowen 1985; Sterling and Shapiro 1986; Sterling and Beer 1989; Demers and Malenfant 1995; Galton 2006). Also in fundamental research, the notion of reification plays an important role. For example, Gödel's

J. Treur, *Network-Oriented Modeling for Adaptive Networks: Designing Higher-Order Adaptive Biological, Mental and Social Network Models*, Studies in Systems, Decision and Control 251, https://doi.org/10.1007/978-3-030-31445-3_3

famous incompleteness theorems in Mathematical Logic depend on reification of logical statements by representing them by natural numbers over which predicates are used to express, for example, (non)provability of such statements (e.g., Smorynski 1977; Hofstadter 1979).

In this chapter, the general notion of reification is applied to networks in particular, and illustrated for a Network-Oriented Modeling approach based on temporal-causal networks (Treur 2016, 2019). A network (the base network) is extended by adding explicit network states representing characteristics of the network structure. In a temporal-causal network, the network structure is defined by three types of characteristics: connection weights (for Connectivity), combination functions (for Aggregation), and speed factors (for Timing). By reifying these characteristics of the base network as states in the extended network, and defining proper causal relations for them and with the other states, an extended, reified network is obtained which explicitly represents the structure of the base network, and how this network structure evolves over time. This enables to model dynamics *of* the base network by dynamics *within* the reified network: thus an adaptive network is represented as a non-adaptive network.

By the introduced concept of network reification it becomes possible to analyse network adaptation principles from an inherent network modeling perspective. Applying this, a unified framework is obtained to represent and compare network adaptation principles across different domains. To illustrate this, a number of well-known network adaptation principles are analysed and compared, including, for example, adaptation principles for Hebbian learning for Mental Networks, and for bonding based on homophily for Social Networks.

In Sect. 3.2 the Network-Oriented Modeling approach based on temporal-causal networks is briefly summarized. Next, in Sect. 3.3 the idea of reifying the network structure characteristics by additional reification states representing them is introduced. In Sect. 3.4 it is discussed how causal relations for these reified states can be defined by which they contribute to an aggregated causal effect on the states in the base network. In Sect. 3.5 the universal combination function and difference equation for the base states' dynamics is briefly presented, which generalises to reified networks what in Chap. 2 are called basic difference or differential equations. Section 3.6 shows how the obtained reification approach can be applied to analyse and unify many well-known network adaptation principles from a Network-Oriented Modeling perspective. In Sect. 3.7, as an illustration an example simulation within a developed software environment for network reification shows how an adaptive speed factor and an adaptive combination function can be used to model a scenario of a manager who adapts to an organisation. This example illustrates how the role matrices format to specify a non-reified network's structure as introduced in Chap. 2, can be generalised relatively easily to obtain a useful means to specify a reified network's structure. In Sect. 3.8 the (im)possibility of joint reification states for multiple base states or roles is briefly discussed. Section 3.9 presents an analysis of the added complexity of the reification construction, and Sect. 3.10 is a final discussion.

3.2 Temporal-Causal Networks: Structure and Dynamics

In general, a network structure is considered to be defined by nodes (or states) and connections between them. However, this only covers very general aspects of a network structure in which no distinctions can be made, for example, between different strengths of connections, and different ways in which multiple connections to the same node interact and work together. In this sense, in many cases a plain graph structure provides underspecification of a network structure. Also, Pearl (2000) points out this problem of underspecification in the context of causal networks from the (deterministic) Structural Causal Model perspective. In that context functions f_i for nodes V_i are used to specify how multiple impacts on the same node V_i should be combined, but this concept is lacking in a plain graph representation:

> Every causal model *M* can be associated with a directed graph, *G(M)* (…) This graph merely identifies the endogenous and background variables that have a direct influence on each V_i; it does not specify the functional form of f_i. (Pearl 2000), p. 203

3.2.1 Conceptual Representation of a Temporal-Causal Network Model

A conceptual representation of the network structure of a temporal-causal network model does involve representing in a declarative manner states and connections between them that represent (causal) impacts of states on each other. This part of the conceptual representation is often depicted in a conceptual picture by a graph with nodes and directed connections. However, a *full conceptual representation* of a temporal-causal network structure also includes a number of labels for such a graph. First, in reality, not all connections are equally strong, so some notion of *strength of a connection* $\boldsymbol{\omega}_{X,Y}$ is used as a label for connections (Connectivity). Second, a combination function $\mathbf{c}_Y(..)$ to *aggregate multiple impacts* on a state is used as a label for states (Aggregation). Third, for each state a notion of *speed factor* $\boldsymbol{\eta}_Y$ of a state is used as a label for timing of the state's processes (Timing). These three notions, called connection weight, combination function, and speed factor, make the graph of states and connections a labeled graph. This labeled graph forms the *defining network structure* of a temporal-causal network model in the form of a conceptual representation; see Table 3.1, adopted from (Treur 2019), and see Fig. 3.1 for an example of a basic fragment of a network with states X_1, X_2 and Y, and labels $\boldsymbol{\omega}_{X_1,Y}$, $\boldsymbol{\omega}_{X_2,Y}$ for connection weights, $\mathbf{c}_Y(..)$ for combination function, and $\boldsymbol{\eta}_Y$ for speed factor.

Combination functions can have different forms, as there are many different approaches possible to address the issue of combining multiple impacts. Combination functions provide a way to specify how multiple causal impacts on this state are aggregated. For this aggregation, pre-defined combination functions from a library can be used, or modified according to a pre-designed template.

Table 3.1 Conceptual representation of a temporal-causal network model: the network structure

Concepts	Notation	Explanation
States and connections	X, Y, $X \rightarrow Y$	Describes the nodes and links of a network structure (e.g., in graphical or matrix format)
Connection weight	$\boldsymbol{\omega}_{X,Y}$	The *connection weight* $\boldsymbol{\omega}_{X,Y} \in [-1, 1]$ represents the strength of the causal impact of state X on state Y through connection $X \rightarrow Y$
Aggregating multiple impacts on a state	$\mathbf{c}_Y(..)$	For each state Y (a reference to) a *combination function* $\mathbf{c}_Y(..)$ is chosen to combine the causal impacts of other states on state Y
Timing of the effect of impact	$\boldsymbol{\eta}_Y$	For each state Y a *speed factor* $\boldsymbol{\eta}_Y \geq 0$ is used to represent how fast a state is changing upon causal impact

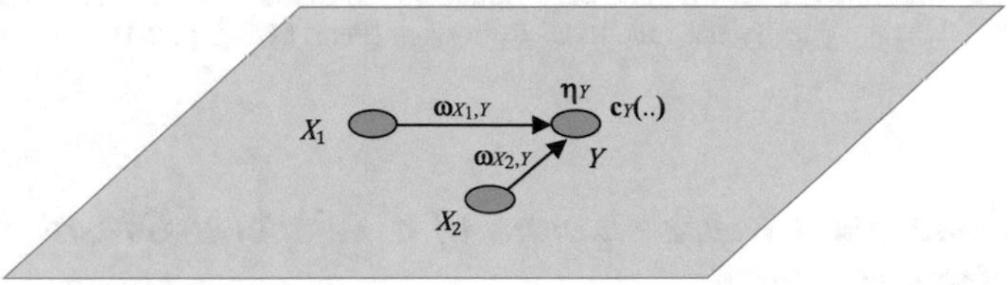

Fig. 3.1 A fragment of a temporal-causal network structure in a conceptual labeled graph representation

3.2.2 Numerical Representation of a Temporal-Causal Network Model

Next it is shown how a conceptual representation (based on states and connections enriched with labels for connection weights, combination functions, and speed factors), determines a numerical representation defining the network's intended dynamic semantics (Treur 2016), Chap. 2; see Table 3.2, adopted from (Treur 2019). Note that here $X_1, \ldots, X_k$ are the states from which state Y gets its incoming connections.

The difference equations in the last row in Table 3.2 form the numerical representation of the dynamics of a temporal-causal network model. They can be used for simulation and mathematical analysis, and also be written in differential equation format:

$$\begin{aligned} Y(t+\Delta t) &= Y(t) + \boldsymbol{\eta}_Y[\mathbf{c}_Y(\boldsymbol{\omega}_{X_1,Y}X_1(t), \ldots, \boldsymbol{\omega}_{X_k,Y}X_k(t)) - Y(t)]\Delta t \\ \mathbf{d}Y(t)/\mathbf{d}t &= \boldsymbol{\eta}_Y[\mathbf{c}_Y(\boldsymbol{\omega}_{X_1,Y}X_1(t), \ldots, \boldsymbol{\omega}_{X_k,Y}X_k(t)) - Y(t)] \end{aligned} \quad (3.1)$$

where the X_i are all states from which state Y gets its incoming connections.

Table 3.2 Numerical representation of a temporal-causal network model: the network dynamics

Concept	Representation	Explanation
State values over time *t*	$Y(t)$	At each time point *t* each state *Y* in the model has a real number value, usually in the [0, 1] interval
Single causal impact	$\mathbf{impact}_{X,Y}(t) = \boldsymbol{\omega}_{X,Y}X(t)$	At *t* state *X* with connection to state *Y* has an impact on *Y*, using connection weight $\boldsymbol{\omega}_{X,Y}$
Aggregating multiple causal impacts	$\mathbf{aggimpact}_Y(t)$ $= \mathbf{c}_Y(\mathbf{impact}_{X_1,Y}(t), \ldots, \mathbf{impact}_{X_k,Y}(t))$ $= \mathbf{c}_Y(\boldsymbol{\omega}_{X_1,Y}X_1(t), \ldots, \boldsymbol{\omega}_{X_k,Y}X_k(t))$	The aggregated causal impact of multiple states X_i on *Y* at *t*, is determined using combination function $\mathbf{c}_Y(..)$
Timing of the causal effect	$Y(t+\Delta t) = Y(t) + \boldsymbol{\eta}_Y[\mathbf{aggimpact}_Y(t) - Y(t)]\Delta t$ $= Y(t) + \boldsymbol{\eta}_Y[\mathbf{c}_Y(\boldsymbol{\omega}_{X_1,Y}X_1(t), \ldots, \boldsymbol{\omega}_{X_k,Y}X_k(t)) - Y(t)]\Delta t$	The causal impact on *Y* is exerted over time gradually, using speed factor $\boldsymbol{\eta}_Y$; here the X_i are all states from which state *Y* gets its incoming connections

3.2.3 Basic Combination Functions, Their Parameters and Combining Them

Often used examples of combination functions are shown in Table 3.3. As shown in Table 3.2 these functions are used by applying them on the single causal impacts for V_1, …, V_k for the states X_1, …, X_k from which state *Y* gets its incoming connections. They are the *identity* **id(.)** for states with impact from only one other state, the *scaled sum* combination function $\mathbf{ssum}_\lambda(..)$ with scaling factor $\boldsymbol{\lambda}$, and the *simple logistic sum* combination function $\mathbf{slogistic}_{\sigma,\tau}(..)$ and *advanced logistic sum* combination function $\mathbf{alogistic}_{\sigma,\tau}(..)$, both with steepness $\boldsymbol{\sigma}$ and threshold $\boldsymbol{\tau}$; see also (Treur 2016), Chap. 2, Table 2.10.

Other options for combination functions are the *scaled minimum* combination function $\mathbf{smin}_\lambda(..)$, *scaled maximum* combination function $\mathbf{smax}_\lambda(..)$, the *Euclidean* combination function of ***n***th-order with scaling factor $\boldsymbol{\lambda}$ (with *n* any number > 0,

Table 3.3 Often used combination functions

Name	Formula
Identity	$\mathbf{id}(V) = V$
Scaled sum	$\mathbf{ssum}_\lambda(V_1, \ldots, V_k) = \frac{V_1 + \ldots + V_k}{\lambda}$
Simple logistic	$\mathbf{slogistic}_{\sigma,\tau}(V_1, \ldots, V_k) = \frac{1}{1+e^{-\sigma(V_1+\ldots+V_k-\tau)}}$
Advanced logistic	$\mathbf{alogistic}_{\sigma,\tau}(V_1, \ldots, V_k) = \left[\frac{1}{1+e^{-\sigma(V_1+\ldots+V_k-\tau)}} - \frac{1}{1+e^{\sigma\tau}}\right](1+e^{-\sigma\tau})$
Scaled minimum	$\mathbf{smin}_\lambda(V_1, \ldots, V_k) = \frac{\min(V_1,\ldots,V_k)}{\lambda}$
Scaled maximum	$\mathbf{smax}_\lambda(V_1, \ldots, V_k) = \frac{\max(V_1,\ldots,V_k)}{\lambda}$
Euclidean function	$\mathbf{eucl}_{n,\lambda}(V_1, \ldots, V_k) = \sqrt[n]{\frac{V_1^n + \ldots + V_k^n}{\lambda}}$
Scaled geometric mean	$\mathbf{sgeomean}_\lambda(V_1, \ldots, V_k) = \sqrt[k]{\frac{V_1 * \ldots * V_k^n}{\lambda}}$

generalising the scaled sum $\mathbf{ssum}_\lambda$**(..)** for n= 1), and the *scaled geometric mean* combination function $\mathbf{sgeomean}_\lambda$**(..)**.

The above examples of combination functions are called *basic combination functions* and in a general format indicated by $\text{bcf}_i(..)$. As also discussed in Chap. 2, Sect. 2.3.2 they can be combined to form more complex combination functions by forming weighted averages of them with *combination function weight* factors $\gamma_1, \ldots, \gamma_m$ as follows

$$\mathbf{c}_Y(V_1, \ldots, V_k) = \frac{\gamma_{1,Y}\, \text{bcf}_1(V_1, \ldots, V_k) + \ldots + \gamma_{m,Y}\, \text{bcf}_m(V_1, \ldots, V_k)}{\gamma_{1,Y} + \ldots + \gamma_{m,Y}} \tag{3.2}$$

This type of representation (with the $\gamma_{j,Y}$ depending on time) for combination functions will also be used for combination function reification in Sect. 3.5. Usually, combination functions have parameters, for example, a scaling factor λ, or steepness σ and threshold τ for logistic functions. These *combination function parameters* can also be used as arguments in the notation $\text{bcf}_i(..)$, and denoted by $\pi_{i,j}$, so that it becomes $\text{bcf}_i(\pi_{1,1}, \pi_{1,2}, V_1, \ldots, V_k)$ and

$$\begin{aligned} &\mathbf{c}_Y(\pi_{1,1}, \pi_{1,2}, \ldots, \pi_{1,m}, \pi_{1,m}, V_1, \ldots, V_k) \\ &= \frac{\gamma_{1,Y}\text{bcf}_1(\pi_{1,1,Y}, \pi_{2,1,Y}, V_1, \ldots, V_k) + \ldots + \gamma_{m,Y}\text{bcf}_m(\pi_{1,m,Y}, \pi_{2,m,Y}, V_1, \ldots, V_k)}{\gamma_{1,Y} + \ldots + \gamma_{m,Y}} \end{aligned} \tag{3.3}$$

These characteristics γ will also be used in an adaptive manner for combination function reification in the example reified network models described in Sects. 3.6.7 and 3.7.

3.2.4 *Normalisation, Stationary Points and Equilibria for Temporal-Causal Network Models*

Often a combination function is assumed to be normalised by setting a proper scaling factor value. If the scaling factor is too low, an undesirable artificial upward bias may occur, and when the scaling factor is too high an artificial downward bias. Therefore, normalization of the combination functions is important to get a realistic simulation. The notion of normalisation is defined as follows.

Definition 1 (normalised) A network is *normalised* if for each state Y it holds $c_Y(\omega_{X_1,Y}, \ldots, \omega_{X_k,Y}) = 1$, where X_1, …, X_k are the states from which Y gets its incoming connections.

As an example, for a Euclidean combination function of ***n***th-order the scaling factor value choice

$$\lambda_Y = \omega^n_{X_1,Y} + \ldots + \omega^n_{X_k,Y}$$

will provide a normalised network. This can be done in general as follows:

Normalising a combination function
If any combination function $\mathbf{c}_Y(..)$ is replaced by $\mathbf{c}'_Y(..)$ defined as

$$\mathbf{c}'_Y(V_1, \ldots, V_k) = \mathbf{c}_Y(V_1, \ldots, V_k)/\mathbf{c}_Y(\omega_{X_1,Y}, .., \omega_{X_k,Y}) \tag{3.4}$$

where X_1, …, X_k are the states with outgoing connections to Y and assuming $\mathbf{c}_Y(\omega_{X_1,Y}, \ldots, \omega_{X_k,Y}) > 0$ for $\omega_{X_i,Y} > 0$, then the network becomes normalised.

For different example functions, following the normalisation step above, their normalised variants are given by Table 3.4.

Next, this section focuses on some tools that allow to analyse emerging behaviour and how it relates to the structure properties. The basic definition is as follows.

Definition 2 (stationary point and equilibrium) A state Y has a *stationary point* at t if $\mathbf{d}Y(t)/\mathbf{d}t = 0$. The network is in *equilibrium* at t if every state Y of the model has a stationary point at t.

Applying this definition to the specific differential equation format for a temporal-causal network model, the very simple criterion expressed in Lemma 1 can be formulated in terms of the temporal-causal network structure characteristics $\omega_{X,Y}$, $\mathbf{c}_Y(..)$, η_Y:

Lemma 1 (Criterion for a stationary point in a temporal-causal network) Let Y be a state and $X_1, \ldots, X_k$ the states with outgoing connections to state Y.

Table 3.4 Normalisation of the different examples of combination functions

Combination function	Notation	Normalising scaling factor λ	Normalised combination function
Identity function	$\mathbf{id}(.)$	$\omega_{X,Y}$	$V/\omega_{X,Y}$
Scaled sum	$\mathbf{ssum}_{\lambda}(V_1, \ldots, V_k)$	$\omega_{X_1,Y} + .. + \omega_{X_k,Y}$	$(V_1 + \ldots + V_k)/(\omega_{X_1,Y} + .. + \omega_{X_k,Y})$
Simple logistic	$\mathbf{slogistic}_{\sigma,\tau}(V_1, \ldots, V_k)$	$\mathbf{slogistic}_{\sigma,\tau}(\omega_{X_1,Y}, .., \omega_{X_k,Y})$	$\frac{1+e^{-\sigma(\omega_{X_1,Y}+\ldots+\omega_{X_k,Y}-\tau)}}{1+e^{-\sigma(V_1+\ldots+V_k-\tau)}}$
Advanced logistic	$\mathbf{alogistic}_{\sigma,\tau}(V_1, \ldots, V_k)$	$\mathbf{alogistic}_{\sigma,\tau}(\omega_{X_1,Y}, .., \omega_{X_k,Y})$	$\frac{\frac{1}{1+e^{-\sigma(V_1+\ldots+V_k-\tau)}} - \frac{1}{1+e^{\sigma\tau}}}{\frac{1}{1+e^{-\sigma(\omega_{X_1,Y}+\ldots+\omega_{X_k,Y}-\tau)}} - \frac{1}{1+e^{\sigma\tau}}}$
Scaled maximum	$\mathbf{smax}_{\lambda}(V_1, \ldots, V_k)$	$\max(\omega_{X_1,Y}, .., \omega_{X_k,Y})$	$\max(V_1, \ldots, V_k)/\max(\omega_{X_1,Y}, .., \omega_{X_k,Y})$
Scaled minimum	$\mathbf{smin}_{\lambda}(V_1, \ldots, V_k)$	$\min(\omega_{X_1,Y}, .., \omega_{X_k,Y})$	$\min(V_1, \ldots, V_k)/\min(\omega_{X_1,Y}, .., \omega_{X_k,Y})$
Euclidean	$\mathbf{eucl}_{n,\lambda}(V_1, \ldots, V_k)$	$\omega^n_{X_1,Y} + \ldots + \omega^n_{X_k,Y}$	$\sqrt[n]{\frac{V_1^n+\ldots+V_k^n}{\omega^n_{X_1,Y}+\ldots+\omega^n_{X_k,Y}}}$
Scaled geometric mean	$\mathbf{sgeomean}_{\lambda}(V_1, \ldots, V_k)$	$\omega_{X_1,Y} * .. * \omega_{X_k,Y}$	$\sqrt[k]{\frac{V_1*\ldots*V_k}{\omega_{X_1,Y}*\ldots*\omega_{X_k,Y}}}$

Then Y has a stationary point at t if and only if

$$\boldsymbol{\eta}_Y = 0 \quad \text{or} \quad \mathbf{c}_Y(\boldsymbol{\omega}_{X_1,Y}X_1(t), \ldots, \boldsymbol{\omega}_{X_k,Y}X_k(t)) = Y(t) \tag{3.5}$$

The latter equation is called a stationary point or equilibrium equation. This criterion will be used in Sects. 3.6 and 3.7 to determine in a straightforward manner the equilibrium equations for states for the different adaptation principles addressed.

3.3 Modeling Adaptive Networks by Network Reification

In general, the structure of a network is described by certain characteristics, such as connection weights. Usually, these network characteristics are considered static: they are assumed not to change over a period of time. This stands in the way of addressing network evolution, where the network structure does change.

Network evolution is studied usually in a hybrid manner by considering a separate dynamic model for additional variables representing network structure characteristics. Such a dynamic model is, for example, specified by a numerical mathematical form of difference or differential equations and a procedural description to simulate these equations. Such a description is different from and outside the context of the Network-Oriented Modeling perspective on dynamics used within the base network itself. In specific applications, still, this extra-network dynamical model has to interact intensively with the internal network dynamics of the base network. For example, see Chap. 1, Sect. 1.4, and in particular Fig. 1.2.

Network reification provides a way to address this in a more unified manner, staying more genuinely within the Network-Oriented Modeling perspective. Using network reification, the base network is extended by extra network states that represent the characteristics of the base network structure (Connectivity, Aggregation, and Timing). In this way, the whole model is specified by one network, a network extension of the base network. Thus the modeling stays within the network context. The new additional states representing the values for the network structure characteristics are called *reification states* for these characteristics. The network characteristics are *reified* by these states. The reification states are depicted in the upper plane in Fig. 3.2, together with the dashed lines indicating the representation relations with the network characteristics of the base network in the lower plane. What can be reified in temporal-causal networks, in particular, are the following characteristics of the network structure: the connection weights, combination functions, combination function parameters, and speed factors. For connection weights $\boldsymbol{\omega}_{X_i,Y}$ and speed factors $\boldsymbol{\eta}_Y$ their added reification states $\mathbf{W}_{X_i,Y}$ and $\mathbf{H}_Y$ represent the value of them.

For combination functions $\mathbf{c}_Y(..)$ the general idea is that from a theoretical perspective a coding is needed for all options for such functions by numbers; for example, assuming there is a countable number, the set of all of them is numbered by natural numbers n = 1, 2, …, and the reified state $\mathbf{C}_Y$ representing them actually represents that number. This is the general idea for addressing reification of

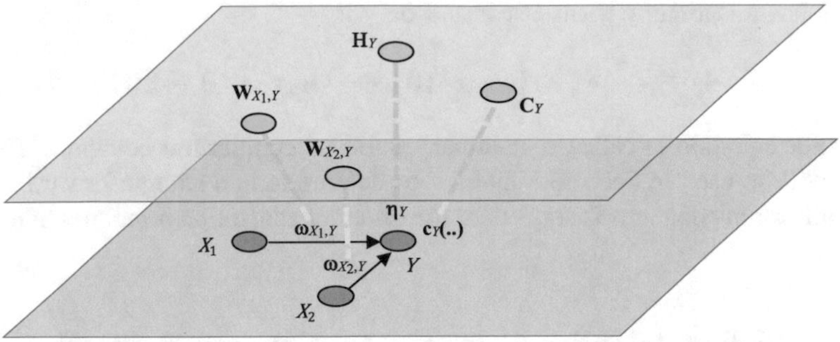

Fig. 3.2 Representation relations (the dashed lines) for connection weight reification states $\mathbf{W}_{X,Y}$, combination function reification states $\mathbf{C}_Y$ and speed factor reification states $\mathbf{H}_Y$: network state $\mathbf{W}_{X,Y}$ represents network characteristic $\boldsymbol{\omega}_{X,Y}$, network state $\mathbf{H}_Y$ represents network characteristic $\boldsymbol{\eta}_Y$, and network state $\mathbf{C}_Y$ represents network characteristic $\mathbf{c}_Y(..)$

combination functions; however, below a more refined approach is shown that is easier to use in practice.

By adding proper causal connections to the reification states (incoming arrows) within the extended network, these states are affected and therefore become adaptive. For many examples of this, see Sects. 3.6 and 3.7. Outward causal connections from reification states (outgoing downward arrows to the related base network states) make their intended special effect happen. This will be addressed in Sect. 3.4; see the pink downward arrows in Fig. 3.3.

3.4 Incorporating the Impact of Downward Causal Connections for Reification States

The added reification states need connections to obtain a well-connected overall network. As always, connections of a state are of two types: (1) outgoing connections, and (2) incoming connections. In the first place outward connections (1) from the reification states to the states in the base network are needed, in order to model how they have their special effect on the adaptive dynamics of the base network. More specifically, it has to be defined how the reification states contribute causally to an aggregated impact on the base network state. In addition to a downward connection, also the combination function has to be (re)defined for the aggregated impact on that base state. Both these downward causal relations and the combination functions are defined in a generic manner, related to the role of a specific network characteristic in the overall dynamics of a state in a temporal-causal network. That will be discussed in the current section and in Sect. 3.5.

In addition, incoming connections (2) of the reification states are added in order to model specific network adaptation principles. These may concern upward connections from the states of the base network to the reification states, or horizontal

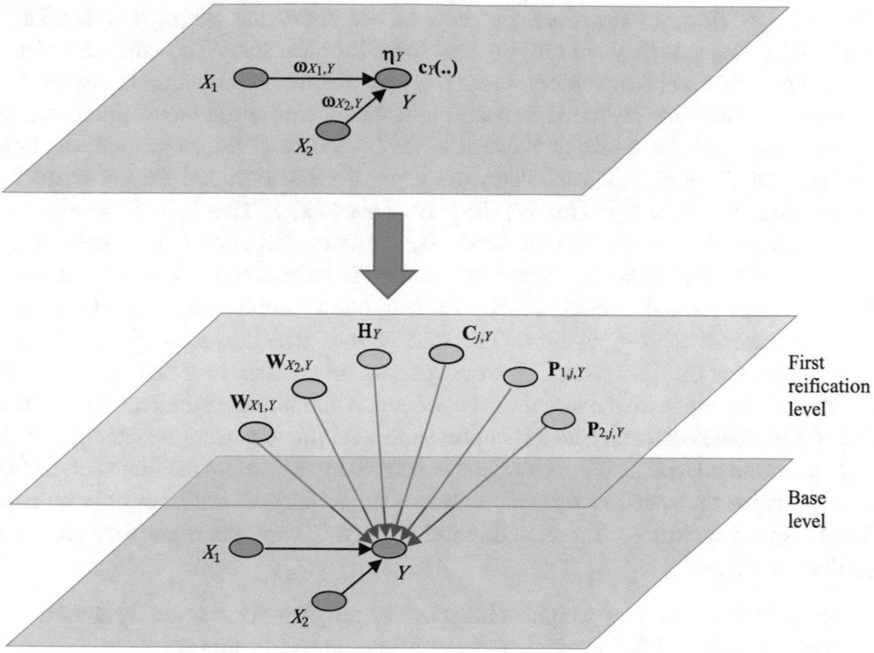

Fig. 3.3 Network reification for temporal-causal networks: downward connections from reification states to base network states

mutual connections between states within the upper plain, or both, depending on the specific network adaptation principles addressed. These connections are not generic but depend on the specific adaptation principle addressed; they will be discussed and illustrated for many cases in Sects. 3.6 and 3.7.

For the downward connections (1), the general pattern is that each of the reification states $\mathbf{W}_{X_i,Y}$, $\mathbf{H}_Y$ and $\mathbf{C}_Y$ for the reified network characteristics (connection weights, speed factors, and combination functions), has a specific causal connection to state *Y* in the base network according to its own special role. These are the (pink) downward arrows from the reification plane to the base plane in Fig. 3.3. Actually, $\mathbf{C}_Y$ is a vector of states ($\mathbf{C}_{1,Y}$, $\mathbf{C}_{2,Y}$, ...) with a (small) number of different components $\mathbf{C}_{1,Y}$, $\mathbf{C}_{2,Y}$, ... for different basic combination functions that will be explained below. Note that combination functions may contain some parameters, for example, for the scaled sum combination function the scaling factor $\boldsymbol{\lambda}$, and for the advanced logistic function the steepness $\boldsymbol{\sigma}$ and the threshold $\boldsymbol{\tau}$. For these parameters also reification states $\mathbf{P}_{i,j,Y}$ can be added, with the possibility to make them adaptive as well. More specifically, for each basic combination function represented by $\mathbf{C}_{j,Y}$ there are two parameters $\pi_{1,i}$ and $\pi_{2,i}$ that are reified by parameter reification states $\mathbf{P}_{1,j,Y}$ and $\mathbf{P}_{2,j,Y}$.

Note that the 3D layout of these figures and the depicted planes are just for understanding; in a mathematical or computational sense, they are not part of the network specification. However, for each of the reification states, it is crucial to know what it is that they are reifying and for what base state. Therefore the names of the reification states are chosen in such a way that this information is visible. For example, in the name $\mathbf{H}_Y$ the **H** indicates that it concerns speed factor (indicated by $\boldsymbol{\eta}$) reification and the subscript *Y* that it is for base state *Y*. So, in general, the bold capital letter **R** in $\mathbf{R}_{subscript}$ indicates the type of reification and the subscript the concerning base state *Y*, or (for **W**) the pair of states *X*, *Y*. This **R** indicates the *role* that is played by this reification state. This role corresponds one to one to the characteristics of the base network structure that is reified: connection weight $\boldsymbol{\omega}$ for Connectivity, speed factor $\boldsymbol{\eta}$ for Timing, basic combination function **c(..)**, and parameter $\boldsymbol{\pi}$ for Aggregation. The role defines which special effect the reification state has on base state *Y*. By the specific role matrix in which a downward connection is indicated, the downward connection and the special effect for that role is completely determined. The reification state with this connection cannot occur in any other role matrix then. For example, if such a downward link is indicated in role matrix **ms** for the speed factors, the value of the reification state can only be used for the speed factor, not for something else. In this way, there are four roles for reification states:

- the role of connection weight reification **W** reifying connection weights $\boldsymbol{\omega}$
- the role of speed factor reification **H** reifying speed factors $\boldsymbol{\eta}$
- the role of combination function reification $\mathbf{C}_j$ reifying combination functions **c(..)**
- the role of parameter reification $\mathbf{P}_{i,j}$ reifying combination function parameters $\boldsymbol{\pi}_{i,j}$

In accordance with this indicated role information, each reification state has exactly one downward causal connection, which goes to the specified base state *Y*, and in the reified network this downward connection has its special effect according to its role **R** in the aggregation of the causal impacts on *Y* by a new, dedicated combination function. Note that to keep a transparent one-to-one relation between a reification state representing one of the base network characteristics, and the actual value used for that characteristic in the dynamics of state *Y*, the (pink) downward links get automatically standard weight value 1; this cannot be changed.

The general picture is that the base states have more incoming connections now, some of which have specific roles, with special effects according to their role. Therefore, in the reified network new combination functions for the base states are needed to aggregate these special effects. These new combination functions can be expressed in a universal manner based on the original combination functions, and the different reification states, but to define them some work is needed. That will be done in Sect. 3.5, but also in a more extensive manner in Chap. 10.

3.5 The Universal Combination Function and Difference Equation for Reified Networks

In this section, the universal combination function and universal difference (or differential) equation for base states Y within a reified network are introduced. The universal difference (or differential) equation generalises what in Chap. 2, Sects. 2.3.1 and 2.4.2 are called the basic difference (or differential) equations. Recall from Sect. 3.2, that based on the basic combination functions $\text{bcf}_j(..)$ from the library, the general combination function format is expressed in terms of the network structure characteristics and the single impacts from the base states indicated by $V_1, \ldots, V_k$ as follows:

$$\mathbf{c}_Y(\pi_{1,1}, \pi_{1,2}, \ldots, \pi_{1,m}, \pi_{1,m}, V_1, \ldots, V_k) = \frac{\gamma_{1,Y}\,\text{bcf}_1(\pi_{1,1,Y}, \pi_{2,1,Y}, V_1, \ldots, V_k) + \ldots + \gamma_{m,Y}\text{bcf}_m(\pi_{1,m,Y}, \pi_{2,m,Y}, V_1, \ldots, V_k)}{\gamma_{1,Y} + \ldots + \gamma_{m,Y}} \quad (3.6)$$

To enable reification, the idea is that all network structure characteristics can become dynamic. This is the main step made here, compared to Chap. 2. In particular, within the combination function this holds for the γ and π characteristics, so that these characteristics can get an argument t for time:

$$\mathbf{c}_Y(t, \pi_{1,1}(t), \pi_{1,2}(t), \ldots, \pi_{1,m}(t), \pi_{1,m}(t), V_1, \ldots, V_k) = \frac{\gamma_{1,Y}(t)\text{bcf}_1(\pi_{1,1,Y}(t), \pi_{2,1,Y}(t), V_1, \ldots, V_k) + \ldots + \gamma_{m,Y}(t)\text{bcf}_m(\pi_{1,m,Y}(t)\pi_{2,m,Y}(t), V_1, \ldots, V_k)}{\gamma_{1,Y}(t) + \ldots + \gamma_{m,Y}(t)} \quad (3.7)$$

These combination functions become adaptive if for these dynamic characteristics γ and π, reification states **C** and **P** are introduced for their role, as shown in Fig. 3.3. Within the difference equation also the speed factor $\boldsymbol{\eta}_Y$ and the connection weights $\boldsymbol{\omega}_{X_i,Y}$ occur. Also, these network structure characteristics can be made dynamic by adding the argument t, and reification states **H** and **W** can be added for their role (see Fig. 3.3).

$$Y(t + \Delta t) = Y(t) + \boldsymbol{\eta}_Y(t)[\mathbf{c}_Y(t, \boldsymbol{\omega}_{X_1,Y}(t)X_1(t), \ldots, \boldsymbol{\omega}_{X_k,Y}(t)X_k(t)) - Y(t)]\Delta t \quad (3.8)$$

Using the above expressions (3.7) and (3.8) the difference and differential equation for state Y can be found based on the appropriate choice of the *universal combination function* $\mathbf{c}^*_Y(..)$ for Y in the reified network. This combination function in the reified network needs arguments for the reification states for all network structure characteristics as they are dynamic now and have an impact on state Y. So, in addition to the impacts within the original base network the new function $\mathbf{c}^*_Y(..)$ needs additional arguments indicated by variables H, C_1, …., C_m, $P_{1,1}$, $P_{2,1}$, …, $P_{1,m}$, $P_{2,m}$, W_1, …, W_k for the special effects of the different types of reification states on state Y, where:

- H is used for the speed factor reification $\mathbf{H}_Y(t)$ representing $\boldsymbol{\eta}_Y(t)$
- C_j for the combination function weight reification $\mathbf{C}_{j,Y}(t)$ representing $\gamma_{j,Y}(t)$
- $P_{i,j}$ for the combination function parameter reification $\mathbf{P}_{i,j,Y}(t)$ representing $\pi_{i,j,Y}(t)$
- W_i for the connection weight reification $\mathbf{W}_{X_i,Y}(t)$ representing $\boldsymbol{\omega}_{X_i,Y}(t)$.

It has been found out (for more details, see also Chap. 10) that the function $\mathbf{c}^*_Y(H, C_1, \ldots, C_m, P_{1,1}, P_{2,1}, \ldots, P_{1,m}, P_{2,m}, W_1, \ldots, W_k, V_1, \ldots, V_k, V)$ that is needed here can be defined as follows:

$$\begin{aligned}&\mathbf{c}^*_Y(H, C_1, \ldots, C_m, P_{1,1}, P_{2,1,\ldots,} P_{1,m}, P_{2,m}, W_1, \ldots, W_k, V_1, \ldots, V_k, V)\\ &= H\frac{C_1 \mathrm{bcf}_1(P_{1,1}, P_{2,1}, W_1V_1, .., W_kV_k) + \ldots + C_m \mathrm{bcf}_m(P_{1,m}, P_{2,m}, W_1V_1, .., W_kV_k)}{C_1 + \ldots + C_m} + (1-H)V\end{aligned} \tag{3.9}$$

where

- V_i for the state values $X_i(t)$ of base states X_i, which are the base states from which Y gets its incoming connections
- V for the state value $Y(t)$ of base state Y

This combination function shows the way in which the special impacts of the downward causal connections from the reification states (their special effects) according to their role are aggregated together with the other impacts within the base level. Then based on this combination function (and using speed factor with default value $\boldsymbol{\eta}^*_Y = 1$ and weights 1 for the incoming connections), within the reified network the *universal difference equation* for base state Y is

$$\begin{aligned}Y(t+\Delta t) &= Y(t)\\ &+ [\mathbf{c}^*_Y(\mathbf{H}_Y(t), \mathbf{C}_{1,Y}(t), \ldots, \mathbf{C}_{m,Y}(t), \mathbf{P}_{1,1,Y}(t), \mathbf{P}_{2,1,Y}(t), \ldots, \mathbf{P}_{1,m,Y}(t),\\ &\mathbf{P}_{2,m,Y}(t), \mathbf{W}_{X_1,Y}(t), \ldots, \mathbf{W}_{X_k,Y}(t), X_1(t), \ldots, X_k(t), Y(t)) - Y(t)]\Delta t\end{aligned} \tag{3.10}$$

In case of full reification this difference equation does not have any parameter for the network characteristics, it only has variables; therefore it has a universal form for every base state. So this is the way in which the special impact of the downward causal connections from the reification states is incorporated within the temporal-causal network format.

It can be verified using (3.9) by rewriting that this universal difference equation in (3.10) indeed is equivalent to the above difference equation in (3.8). For more explanation and background on this, see Chap. 10. The *universal differential equation* variant is as follows (leaving out the reference to t):

$$\mathbf{d}Y/\mathbf{d}t = \mathbf{H}_Y\left[\frac{\mathbf{C}_{1,Y}\text{bcf}_1(\mathbf{P}_{1,1,Y},\mathbf{P}_{2,1,Y},\mathbf{W}_{X_1,Y}X_1,..,\mathbf{W}_{X_k,Y}X_k)+\ldots+\mathbf{C}_{m,Y}\text{bcf}_m(\mathbf{P}_{1,m,Y},\mathbf{P}_{2,m,Y},\mathbf{W}_{X_1,Y}X_1,..,\mathbf{W}_{X_k,Y}X_k)}{\mathbf{C}_{1,Y}+\ldots+\mathbf{C}_{m,Y}} - Y\right] \tag{3.11}$$

The universal combination function was introduced above in (3.9) out of the blue, it may seem. But at least now it was shown that it fulfills what is required. In Chap. 10, it is shown in more detail how this universal combination function can be derived and it is illustrated for some cases. From an abstract point of view now it has been found that the class of temporal-causal network models is closed under the operation of reification.

Note that there are two important advantages in keeping the reified network in the form of a temporal-causal network model according to the standard format from Sect. 3.2. First is that now the reified network itself can also be reified again, in order to model second-order adaptation principles (as will be described in Chap. 4). Iteration of the reification step can only be done in a standard manner if every reification step makes a new temporal-causal network model and not an arbitrary complex dynamical system.

A second advantage is that mathematical analysis of equilibria can also be applied in a uniform manner, based on combination functions and the criterion on stationary points and equilibria formulated using them in Lemma 1 in Sect. 3.2.4. This allows such a mathematical analysis to relate emerging behaviour to properties of these combination functions. Chapters 11 to 14 are based on this, where Chaps. 11 and 12 address combination functions for base networks and Chaps. 13 and 14 address combination functions for reification states (for Hebbian learning and for bonding by homophily, respectively) at a first reification level. In all of these chapters, it is explored in general how certain relevant properties of the network structure entail certain properties of the emerging network behaviour. Aggregation characteristics as represented by combination functions are an essential element of the network structure; it turns out that such relevant properties of the network structure most often involve specific properties of the combination functions such as monotonicity and being scalar-free, sometimes in conjunction with some Connectivity properties such as the network being strongly connected.

More specifically, in all of these cases in Chaps. 11–14, analysis results were obtained of the uniform format that certain specific properties of the Aggregation characteristics as expressed by combination functions (sometimes together with certain properties of the network's Connectivity) entail certain properties of the network's emerging behaviour, mostly concerning equilibria that are reached. Typical properties of combination functions that are relevant for a base network are monotonicity and being scalar-free (see Chaps. 11 and 12). For combination functions describing bonding by homophily, a typical relevant property is having a tipping point for similarity where 'being alike' turns into 'not being alike' or conversely (see Chap. 13). Also for combination functions for Hebbian learning, some monotonicity properties turn out relevant for the emerging behaviour (see Chap. 14).

3.6 Using Network Reification for Unified Modeling of Network Adaptation Principles

In Sects. 3.4 and 3.5, it has been explained how the special effects of the downward causal connections for the different roles can be defined in a reified network, and how their contribution to a joint aggregated causal impact on base network states is specified by a generically defined universal combination function. This makes the reified network already work when each of the reification states has a constant value; it will then work just like a nonadaptive base network. However, availability of the reification states for the base network structure as explicit network states, which in principle can change over time, opens the possibility to define network adaptation principles in a Network-Oriented manner. This can be done by specifying

(a) The **connectivity** for reification states:
   - proper causal connections to the reification states

(b) The **aggregation** for reification states:
   - proper combination functions for them

(c) The **timing** for the reification states:
   - proper speed factors for the adaptation of them

This is not just specification by an arbitrary separate set of difference or differential equations or procedural description as for the traditional hybrid approach discussed in Chap. 1, Sect. 1.4.1 and shown in Fig. 1.2. The reification perspective offers a framework to specify network adaptation principles in a more transparent, unified and standardized Network-Oriented manner, and compare them to each other.

This will be illustrated below for a number of examples of well-known network adaptation principles: Hebbian learning in Mental Networks and homophily in Social Networks, triadic closure in Mental and Social Networks, and preferential attachment in Mental and Social Networks. These examples of network adaptation principles in Sects. 3.6.1–3.6.4 all focus on adaptive connection weights. By far most of the network adaptation principles described in the literature only concern Connectivity: they address adaptive connections as adaptive network characteristics. However, in Sects. 3.6.5, 3.6.6 and 3.6.7 it will be shown how other adaptive network characteristics concerning Aggregation and Timing such as adaptive excitability, adaptive speed factors and adaptive combination functions can have interesting applications as well. For example, in a Social Network, response time may depend on external factors such as workload, which varies over time; this can be modeled by a speed factor that all the time adapts to this work load. Also in a Social Network, the way in which someone aggregates opinions from others may also change over time. For example, due to circumstances such as bad experiences or new higher management, a manager may change in how inputs from different employees are incorporated in his or her own opinions and decision making. These applications of speed factor reification and combination function reification will be illustrated in Sect. 3.7 more extensively by an example reified network model and example simulations.

So, next a number of adaptation principles known from the literature are addressed. It is shown how they can be modeled by network reification, and, in particular, their *connectivity* (in terms of their connections) and their *aggregation* (in terms of their combination functions), and the entailed *emerging equilibria* are described and analysed.

3.6.1 Network Reification for Adaptation Principles for Hebbian Learning and Bonding by Homophily

Hebbian learning (Hebb 1949) is based on the principle that strengthening of a connection between neurons over a period of time may take place when both states are often active simultaneously: 'neurons that fire together, wire together'. The principle itself refers to Hebb (1949), and over time has gained more interest in the area of computational modeling due to more extensive empirical support (e.g., Bi and Poo 2001), and more advanced mathematical formulations (e.g., Gerstner and Kistler 2002).

A different principle in a different domain, namely the *bonding based on homophily principle* in Social Science, has exactly the same graphical representation. This principle states that within a social network the more similar two persons are, the stronger their connection will become: 'birds of a feather flock together' (e.g., McPherson et al. 2001).

Connectivity of the reification states for the Hebbian learning and bonding by homophily principles
In Fig. 3.4 it is shown how the Hebbian learning principle can be modeled conceptually in a reified network by upward arrows to the reification states for the connection weights: each connection weight reification state is affected by the two connected states, for the sake of simplicity with connection weights 1. Moreover, a connection of $\mathbf{W}_{X_i,Y}$ to itself is assumed, also with weight 1. For bonding by homophily, the social network connection is similarly affected by the connected states as well with weights 1, like for Hebbian learning, so also in that case Fig. 3.4 applies.

So, both cases share the same connectivity. However, this does not hold for their aggregation: to model these two adaptation principles, the combination functions still are not the same, as the state values have different effects on the connection weights; this is addressed next.

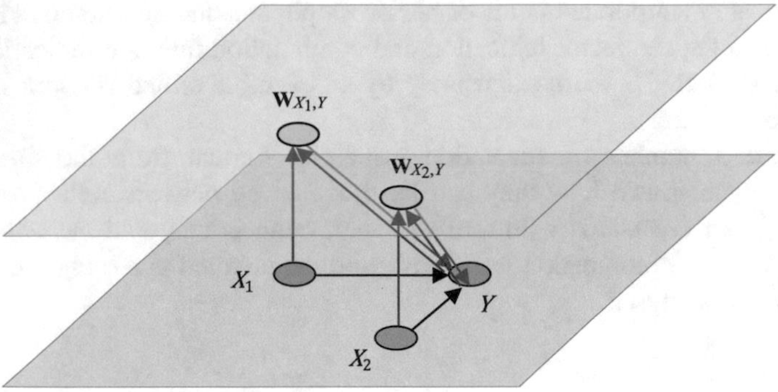

Fig. 3.4 Reified conceptual modeling of Hebbian learning in Mental Networks or Homophily in Social Networks

Aggregation by combination functions for the Hebbian learning and bonding by homophily principles

For Hebbian learning the combination function **hebb(..)** can be chosen, which is defined by

$$\mathbf{hebb}_{\boldsymbol{\mu}}(V_1, V_2, W) = V_1 V_2 (1 - W) + \boldsymbol{\mu} W \quad (3.12)$$

with $\boldsymbol{\mu}$ being the persistence parameter, where V_1 stands for $X_i(t)$, V_2 for $Y(t)$ and W for $\mathbf{W}_{X_i,Y}(t)$. This parameter describes in how far a learnt connection persists over time. Full persistence is indicated by $\boldsymbol{\mu} = 1$. If $\boldsymbol{\mu} < 1$, then some extent of extinction takes place; full extinction takes place for $\boldsymbol{\mu} = 0$. In the first part of the formula, the expression $V_1 V_2$ models the condition 'neurons that fire together', and the factor $(1 - W)$ takes care that the connection weight stays in the [0, 1] interval. For more options for Hebbian learning functions, see Chap. 14. Note that the function uses as third argument the current value W of the connection weight; this assumes that there is a connection from the reification state to itself, although in conceptual pictures such as in Fig. 3.4 such connections usually are not depicted; but they are specified in the role matrices. The same applies to many other reification states and adaptation principles.

For the bonding by homophily principle by (Blankendaal et al. 2016) or (Treur 2016), Chap. 11, Sect. 11.7, an option for the combination function is the *simple linear homophily function* $\mathbf{slhomo}_{\boldsymbol{\sigma},\boldsymbol{\tau}}(..)$:

$$\mathbf{slhomo}_{\boldsymbol{\sigma},\boldsymbol{\tau}}(V_1, V_2, W) = W + \boldsymbol{\alpha}(\boldsymbol{\tau} - |V_1 - V_2|)(1 - W)W \quad (3.13)$$

Here $\boldsymbol{\alpha}$ is the homophily modulation factor, and $\boldsymbol{\tau}$ the tipping point. Here the part $(\boldsymbol{\tau} - |V_1 - V_2|)$ models the condition 'birds of a feather': this part is positive if the difference between V_1 and V_2 is less than the tipping point $\boldsymbol{\tau}$ ('birds of a feather' is true) and negative when this difference is more than $\boldsymbol{\tau}$ ('birds of a feather' is false).

The factor $(1 - W)W$ takes care that W stays in the [0, 1] interval. As long as W is not 0 or 1, in the first case by the combination function a positive term is added to W, the combination function provides a value higher than W, so the connection weight will increase; in the second case a negative term is added, so the combination function provides a value lower than W, and the connection weight will decrease. In Chap. 13 more options for homophily functions are discussed.

Box 3.1 and 3.2 show mathematical analysis of emerging behaviour for these two adaptation principles.

Box 3.1 Mathematical analysis of emerging behaviour for the Hebbian learning principle

For a stationary point, applying the criterion (3.5) of Lemma 1 provides the following equilibrium equation for the above Hebbian learning combination function:

$$\begin{aligned}
&W = \mathbf{hebb}_{\mu}(V_1, V_2, W) = V_1V_2(1 - W) + \mu W && \Leftrightarrow\\
&W = V_1V_2 - V_1V_2W + \mu W && \Leftrightarrow\\
&W(1 + V_1V_2 - \mu) = V_1V_2 && \Leftrightarrow\\
&W = \frac{V_1V_2}{1-\mu+V_1V_2}
\end{aligned}$$

For example, when in an equilibrium both V_1 and V_2 have value 1, then $W = \frac{1}{2-\mu}$.

Box 3.2 Mathematical analysis of emerging behaviour for the bonding by homophily principle

For the above combination function for bonding based on homophily case, for a stationary point applying the criterion (3.5) of Lemma 1 provides the equilibrium equation:

$$\begin{aligned}
&W = \mathbf{slhomo}_{\sigma,\tau}(V_1, V_2, W) = W + \alpha(\tau - |V_1 - V_2|)(1 - W)W && \Leftrightarrow\\
&\alpha(\tau - |V_1 - V_2|)(1 - W)W = 0 && \Leftrightarrow\\
&W = 0 \quad \text{or} \quad W = 1 \quad \text{or} \quad |V_1 - V_2| = \tau
\end{aligned}$$

As in simulations, for example, with the scaled sum combination function for the base states, the third option here often turns out to be not attracting, this indicates that in an equilibrium a form of clustering is achieved with connection weights 1 between states within one cluster and connection weights 0 between states in different clusters; see also Chap. 13.

3.6.2 Network Reification for the Triadic Closure Adaptation Principle

Another adaptivity principle is the *triadic closure principle* from Social Science (Rapoport 1953; Granovetter 1973; Banks and Carley 1996): If two persons in a social network have a common friend, then there is a higher chance that they will become friends themselves.

Connectivity of the reification states for the triadic closure adaptation principle
The connectivity for this adaptation principle is modeled conceptually in graph form as shown in Fig. 3.5.
Here horizontal arrows between the reification states describe the effect of triadic closure. The weights of these connections from $\mathbf{W}_{X,Y}$ and $\mathbf{W}_{Y,Z}$ to $\mathbf{W}_{X,Z}$ may be 1 for the sake of simplicity, but may also have different values, for example, to express that one of the two has more influence than the other one.

Aggregation by a combination function for the triadic closure principle
The combination function for $\mathbf{W}_{XZ}(t)$ can, for example, be a scaled sum:

$$\mathbf{ssum}_\lambda(W_1, W_2) = \frac{W_1 + W_2}{\lambda} \tag{3.14}$$

where W_1 indicates $\mathbf{W}_{X,Y}(t)$ and W_2 indicates $\mathbf{W}_{Y,Z}(t)$ and when the black horizontal arrows from $\mathbf{W}_{X,Y}$ and $\mathbf{W}_{Y,Z}$ to $\mathbf{W}_{X,Z}$ are assumed to have weight 1, the scaling factor $\boldsymbol{\lambda}$ can be normalised at 2. Alternatively, a higher-order Euclidean or logistic sum combination function might be used. For a $\boldsymbol{n}$th-order Euclidean combination function it becomes:

$$\mathbf{eucl}_{n,\lambda}(W_1,\ W_2) = \sqrt[n]{\frac{W_1^n + W_2^n}{\lambda}} \tag{3.15}$$

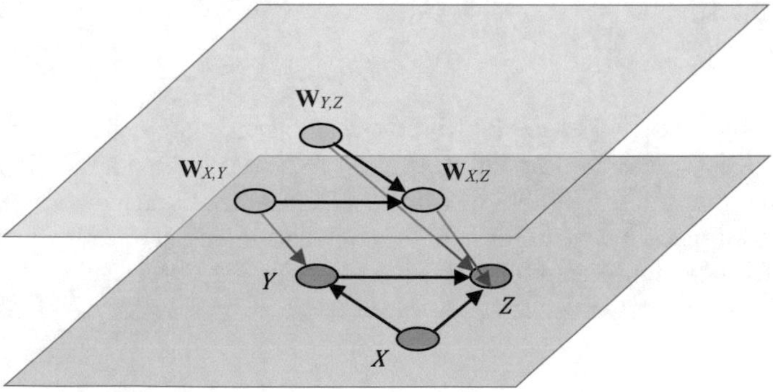

Fig. 3.5 Reified conceptual modeling of the triadic closure principle in Mental and Social Networks

There is also a counterpart of this principle in Mental Networks. It is a form of transitive closure which is implied indirectly by the Hebbian learning principle: strong connections from X to Y and from Y to Z will make more often X and Z active at the same time, and therefore their connection will become stronger by Hebbian learning. Box 3.3 shows mathematical analysis of emerging behaviour for this adaptation principle.

Box 3.3 Mathematical analysis of emerging behaviour for the triadic closure principle
For this case for a stationary point applying the criterion (3.5) of Lemma 1 provides the following linear equilibrium equation for $W = \mathbf{W}_{X,Z}(t)$:

$$W = \mathbf{ssum}_\lambda(W_1, W_2) = (W_1 + W_2)/\lambda \Leftrightarrow$$

$$W = (W_1 + W_2)/\lambda$$

For the Euclidean combination function, for a stationary point in the normalised case it holds

$$W = \sqrt{\frac{W_1^n + W_2^n}{\lambda}}$$

3.6.3 *Network Reification for a Preferential Attachment Adaptation Principle*

Another principle from Social Science is the principle of *preferential attachment* (Barabasi and Albert 1999). This principle states that connections to states that already have more or stronger connections will become more strong.

Connectivity of the reification states for the preferential attachment adaptation principle
This can be modeled by horizontal connections between the reification states, which can be applied to multiple other connection weight reification states $\mathbf{W}_{X_i,Y}$; see Fig. 3.6. The weights of these horizontal connections can be 1 for the sake of simplicity, or any other values.

Aggregation by a combination function for the preferential attachment principle
The combination function $\mathbf{c}_{\mathbf{W}_{X_i,Y}}(..)$ for the considered reification state $\mathbf{W}_{X_i,Y}$ can, for example, be a scaled sum function.

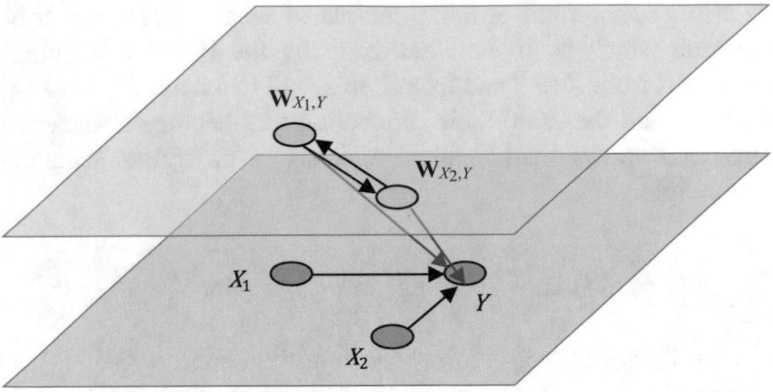

Fig. 3.6 Reified conceptual modeling of preferential attachment in Mental Networks and Social Networks

$$\mathbf{ssum}_{\lambda}(W_1, \ldots, W_k) = \frac{W_1 + \ldots + W_k}{\lambda} \tag{3.16}$$

where W_j is used for $\mathbf{W}_{X_j,Y}(t)$ and λ can be normalised at k. These $\mathbf{W}_{X_j,Y}$ represent the weights of all connections of base states X_j to Y, from which the considered X_i is one. Alternatively, a higher order Euclidean or logistic sum combination function might be used here. For a nth order Euclidean combination function it becomes

$$\mathbf{eucl}_{n,\lambda}(W_1, \ldots, W_k) = \sqrt[n]{\frac{W_1^n + \ldots + W_k^n}{\lambda}} \tag{3.17}$$

Also a logistic sum combination function can be used, in which case a higher number k of connections to Y more clearly leads to a higher weight $\mathbf{W}_{X_i,Y}$.

This principle has a counterpart in Mental Networks: for cases that X_1 and X_2 are conceptually related so that they often are activated in the same situations, a stronger connection from X_1 to Y leads to more activation of Y and by Hebbian learning also to a stronger connection from X_2 to Y.

Box 3.4 shows a mathematical analysis of emerging behaviour for this adaptation principle.

Box 3.4 Mathematical analysis of emerging behaviour for the preferential attachment principle

For the scaled sum case, for a stationary point applying the criterion (3.5) of Lemma 1 provides the following linear equation for $W = \mathbf{W}_{X_i,Y}(t)$:

$$W = \mathbf{ssum}_{\lambda}(W_1, \ldots, W_k) = \frac{W_1 + \ldots + W_k}{\lambda} \quad \Leftrightarrow$$
$$W = \frac{W_1 + \ldots + W_k}{k}$$

So the connection weight $\mathbf{W}_{X_i,Y}$ gets the average value of the weights $\mathbf{W}_{X_j,Y}$ for all j.
For the Euclidean case, a stationary point in the normalised case it holds

$$W = \sqrt[n]{\frac{W_1^n + \ldots + W_k^n}{k}}$$

3.6.4 Network Reification for the State-Connection Modulation Adaptation Principle

Yet another adaptation principle that applies both to Mental and Social Networks is the principle of *state- connection modulation.*

Connectivity of the reification states for the state-connection modulation adaptation principle
This can be modeled conceptually by upward arrows from control states Z_i in the base network to the reification states of connection weights; see Fig. 3.7. The weights of these upward connections may be 1 for the sake of simplicity, or any other value.

For a Mental Network Z_i can be a state of extreme stress (Sousa et al. 2012) or a chemical or medicine (e.g., a neurotransmitter). For a counterpart in a Social Network, Z_i can be a measure for the intensity of the actual interaction (e.g., taking into account frequency and emotional charge); this can be called the *interaction connects* principle (e.g., Treur 2016), Chap. 11.

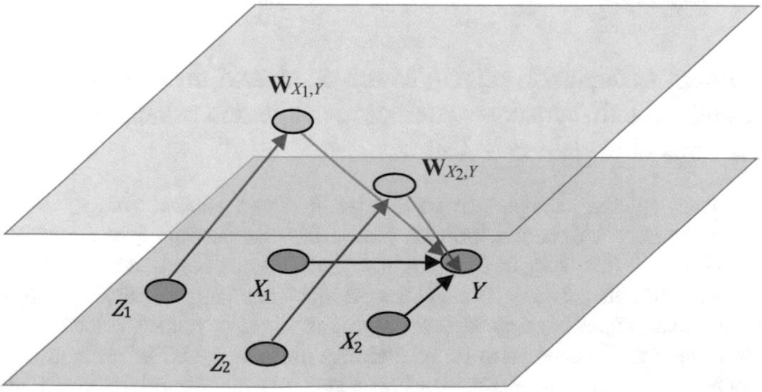

Fig. 3.7 Reified conceptual modeling of state-connection modulation by control state Z_i in a Mental Network or of the interaction connects principle in Social Network with state Z_i the intensity of the interaction

Aggregation by a combination function for the state-connection modulation principle
In this case the combination function $\mathbf{c}_{\mathbf{W}_{X_i,Y}}(..)$ for $\mathbf{W}_{X_i,Y}$ can, for example, be the state-connection modulation function $\mathbf{scm}_{\alpha}(..)$:

$$\mathbf{scm}_{\alpha}(W, V) = W + \alpha VW(1-W) \tag{3.18}$$

with α a modulation factor, which can be positive (amplifying effect) or negative (suppressing effect), where V is used for for $Z_i(t)$ and W for $\mathbf{W}_{X_i,Y}(t)$; for an application of this, see also Chap. 5 or (Treur and Mohammadi Ziabari 2018). Note that when this state-connection modulation function $\mathbf{scm}_{\alpha}(W, V)$ is used in combination with Hebbian learning, auxiliary variables V_1, V_2 (which are not actually used by the function) are included in this function, making it $\mathbf{scm}_{\alpha}(V_1, V_2, W, V)$ to get one shared sequence of values used by both functions. Box 3.5 shows a mathematical analysis of emerging behaviour for this adaptation principle.

Box 3.5 Mathematical analysis of emerging behavior for the state-connection modulation principle
For this case for a stationary point applying the criterion (3.5) of Lemma 1 provides the following quadratic equation for $W = \mathbf{W}_{X_i,Y}(t)$:

$$\begin{aligned} & W = \mathbf{scm}_{\alpha}(W, V) = W + \alpha VW(1-W) && \Leftrightarrow \\ & \alpha VW(1-W) = 0 && \Leftrightarrow \\ & V = 0 \quad \text{or} \quad W = 0 \quad \text{or} \quad W = 1 \end{aligned}$$

3.6.5 *Network Reification for Excitability Adaptation Principles*

Next, a number of adaptation principles are addressed that are not related to connection weights. The first concerns excitation adaptation principles. In Chandra and Barkai 2018) this is explained as follows:

> Learning-related cellular changes can be divided into two general groups: modifications that occur at synapses and modifications in the intrinsic properties of the neurons. While it is commonly agreed that changes in strength of connections between neurons in the relevant networks underlie memory storage, ample evidence suggests that modifications in intrinsic neuronal properties may also account for learning related behavioral changes. Long-lasting modifications in intrinsic excitability are manifested in changes in the neuron's response to a given extrinsic current (generated by synaptic activity or applied via the recording electrode). (Chandra and Barkai 2018, p. 30)

To address dynamic levels of excitability of base states, for a base state Y a logistic sum combination function is assumed, which has a threshold parameter τ. Decreasing the value of τ is increasing excitability of Y, as a lower threshold value will make that Y becomes more activated. Then a reification state $\mathbf{T}_Y$ can be included that represents the intrinsic excitability of Y, by the value of the threshold parameter τ_Y of its logistic sum combination function.

Connectivity of the reification state for the excitability adaptation principle
In Fig. 3.8 the basic connectivity pattern is shown that can be used to model excitability adaptation principles by network reification. Here c is a context factor that affects the excitability. Such a pattern is used in Chap. 4 as part of a more complex (multilevel) reified network model. The weights of the upward connections may be 1 for the sake of simplicity, or any other value.

Aggregation by a combination function for adaptive excitability
For this reification state $\mathbf{T}_Y$, a logistic sum combination function can be used, or an Euclidean function, for example, or any other specific form. In Chap. 4 the advanced logistic sum combination function is used for $\mathbf{T}_Y$; see that chapter for more details.

3.6.6 Network Reification for Response Speed Adaptation Principles

A person will not always respond to inputs with the same speed. Examples of factors affecting the response speed are workload (negative influence) or the availability of support staff (positive influence). A slightly different application is the presence of certain chemicals in the brain to stimulate or slow down transfer between neurons (for example, the effect of alcohol or a stress-suppressing medicine on reaction time).

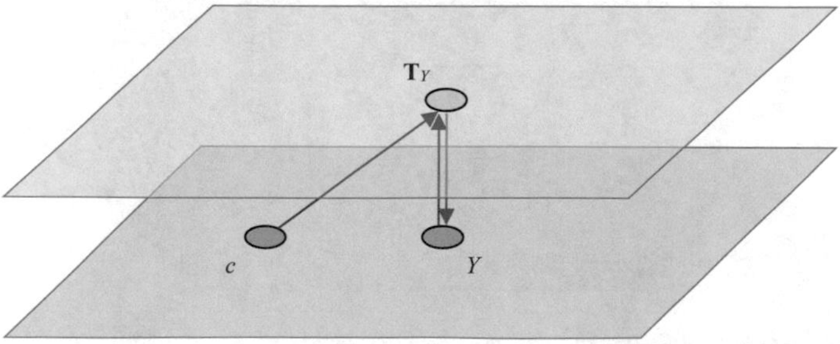

Fig. 3.8 Reified network model for excitability adaptation principles

Connectivity of the reification state for speed adaptation
The connectivity is modeled in Fig. 3.9 by two conditions Z_1 and Z_2 and their positive and negative connections to the speed factor reification $\mathbf{H}_Y$. The weights of the upward connections may be 1 for the sake of simplicity, or any other value.
Aggregation by a combination function for speed adaptation
For this, the combination function $\mathbf{c}_{\mathbf{H}_Y}(..)$ can be modeled by the scaled sum combination function

$$\mathbf{ssum}_\lambda(V_1, V_2) = \frac{V_1 + V_2}{\lambda} \tag{3.19}$$

where V_1 stands for $Z_1(t)$, and V_2 for $Z_2(t)$, and assuming the upward connections have weights 1, λ can be normalised at 2. Also here alternative options can be used such as ***n***th-order Euclidean combination functions or logistic combination functions. Box 3.6 shows mathematical analysis of emerging behaviour for this adaptation principle.

Box 3.6 Mathematical analysis of emerging behavior for adaptive speed factors
For the normalised case for a stationary point applying the criterion (3.5) of Lemma 1 provides the following linear equation for $H = \mathbf{H}_Y(t)$ in relation to the state values V_1 for Z_1 and V_2 for Z_2

$$H = 1/2(V_1 + V_2)$$

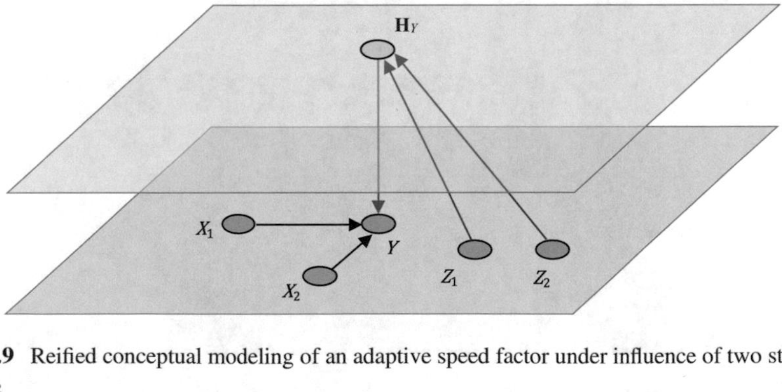

Fig. 3.9 Reified conceptual modeling of an adaptive speed factor under influence of two states Z_1 and Z_2

3.6.7 Network Reification for Aggregation Adaptation Principles

The following is an example of adaptive aggregation by adaptive combination functions. Suppose a manager wants to represent the opinions of her employees well within the organisation. She initially supports any proposal of any single individual employee about a certain issue. This can be modeled by the $\mathbf{smax}_\lambda$**(..)** combination function: if one of the input opinions is high, also the manager's opinion will become high. After bad experiences within the organisation, she may gradually move to a different way, based on averaging over the opinions of her group of employees, which can be modeled by a normalised scaled sum $\mathbf{ssum}_\lambda$**(..)** with $\boldsymbol{\lambda}$ as the sum of the weights of the incoming connections. Or eventually, she can decide only to support an idea when all employees share that opinion. This can be modeled by the $\mathbf{smin}_\lambda$**(..)** combination function. So, the following transitions can take place over time (gradually):

$$\mathbf{smax}_\lambda(..) \rightarrow \mathbf{ssum}_\lambda(..) \rightarrow \mathbf{smin}_\lambda(..)$$

This may look a bit extreme, but at least makes the idea clear. A more realistic version may be the following. First the manager aggregates the incoming opinions by averaging over the group, using the normalised scaled sum $\mathbf{ssum}_\lambda$**(..)**, but later on she gradually moves to using a logistic sum combination function $\mathbf{alogistic}_{\sigma,\tau}$**(..)** where she applies a certain threshold τ before she supports the opinion:

$$\mathbf{ssum}_\lambda(..) \rightarrow \mathbf{alogistic}_{\sigma,\tau}(..)$$

Connectivity of the reification state for the aggregation adaptation principle
A picture of a graphical representation of a network model for this is shown in Fig. 3.10. An example simulation for this scenario can be found in Sect. 3.7; see

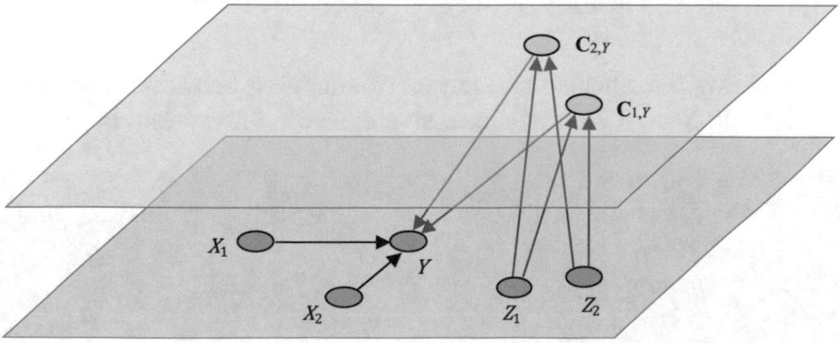

Fig. 3.10 Reified conceptual modeling of adaptive combination functions under influence of states Z_i

Figs. 3.12 and 3.13. Here the basic combination function $\mathrm{bcf}_1(..)$ relating to $\mathbf{C}_{1,Y}$ is $\mathbf{ssum}_\lambda(..)$ and $\mathrm{bcf}_2(..)$ relating to $\mathbf{C}_{2,Y}$ is $\mathbf{alogistic}_{\sigma,\tau}(..)$. Some condition Z_1 in the organisation starts to affect $\mathbf{C}_{1,Y}$ and $\mathbf{C}_{2,Y}$ by suppressing the former (negative connection weight) and increasing the latter (positive connection weight). This will make the above transition from $\mathbf{ssum}_\lambda(..)$ to $\mathbf{alogistic}_{\sigma,\tau}(..)$ happen. This scenario will be addressed in more detail in Sect. 3.7.

An example of adaptive combination functions for a Mental Network can be found by considering multicriteria decision-making. The valuations for the different criteria are aggregated according to some function. A person may adapt this function over time due to learning. For example, first a $\mathbf{smin}_\lambda(..)$ function is used to model that all criteria should be fulfilled to get a high overall score, but later, after it was experienced that based on that function almost no decisions were made, a $\mathbf{ssum}_\lambda(..)$ function is used to model that the decision should be based on the average of the valuations for the different criteria.

Aggregation by a combination function for aggregation adaptation
Suppose $Z_1, \ldots, Z_k$ are the states affecting the combination function reification states $\mathbf{C}_{i,Y}$. Note that some of their impacts can be positive and some can be negative according to the sign of their connection weight $\boldsymbol{\omega}_{Z_j,\mathbf{C}_{i,Y}}$ to $\mathbf{C}_{i,Y}$, and from the aggregation of their impacts it depends whether $\mathbf{C}_{i,Y}(t)$ will increase or decrease with t. A first option for the combination function $\mathbf{c}_{\mathbf{C}_{i,Y}}(..)$ is a scaled sum combination function to aggregate the impacts of $Z_1, \ldots, Z_k$ on $\mathbf{C}_{i,Y}$:

$$\mathbf{ssum}_\lambda(V_1, \ldots, V_k) = \frac{V_1 + \ldots + V_k}{\lambda} \tag{3.20}$$

where each V_j indicates $\boldsymbol{\omega}_{Z_j,\mathbf{C}_{i,Y}}\, Z_j(t)$. Alternatively, a higher order Euclidean or logistic sum combination function might be used here. For an nth-order Euclidean combination function it becomes

$$\mathbf{eucl}_{n,\lambda}(V_1, \ldots, V_k) = \sqrt[n]{\frac{V_1^n + \ldots + V_k^n}{\lambda}} \tag{3.21}$$

Box 3.7 shows a mathematical analysis of emerging behaviour for this adaptation principle. In Sect. 3.7 a more extensive example illustrates this.

Box 3.7 Mathematical analysis of emerging behavior for combination function adaptation
For a scaled sum combination function, by the criterion (3.5) of Lemma 1 the following equation for a stationary point is obtained (where $C = \mathbf{C}_{i,Y}(t)$):

$$C = \frac{V_1 + \ldots + V_k}{\lambda}$$

where each V_j indicates the value $\boldsymbol{\omega}_{Z_j,\mathbf{C}_{i,Y}}\ Z_j(t)$. Similarly for an $\boldsymbol{n}$th order Euclidean combination function:

$$C = \sqrt[n]{\frac{V_1^n + \ldots + V_k^n}{\lambda}}$$

3.7 A Reified Network Model for Response Speed Adaptation and Aggregation Adaptation

In this section, following Sects. 3.6.6 and 3.6.7 above, by an example scenario, the use of an adaptive speed factor and an adaptive combination function in a reified network is illustrated. Also, the network's emerging behaviour based on equilibrium values will be analysed. Consider, as also discussed in Sect. 3.6.6, within an organisation a manager of a group of 7 members with their opinions X_1, …, X_7. The adaptation focuses on the manager opinion; the manager adapts to the organization over time. She wants to represent the opinions of the group members well within the organization and therefore she initially uses a (normalized) scaled sum function $\mathbf{ssum}_\lambda$**(..)** to aggregate the opinions to some average. However, later on based on disappointing experiences within the organization, she decides to use a threshold τ through the logistic sum combination function $\mathbf{alogistic}_{\sigma,\tau}$**(..)**. Moreover, initially she is busy with other things and only later she gets more time to respond faster on the input she gets from her group members, so that her speed factor increases from that time point on.

3.7.1 Conceptual Graphical Representation of the Example Reified Network Model

In Fig. 3.11 an overview of this reified network is shown; see also Table 3.5 for the states and their explanation. Note that also group members X_1, …, X_7 have mutual connections, but this is not shown in the graph in Fig. 3.11 to keep the picture simple; see Box 3.8 for these connections. For the same reason the upward connections from all group members to combination function reification state $\mathbf{C}_{1,Y}$ have

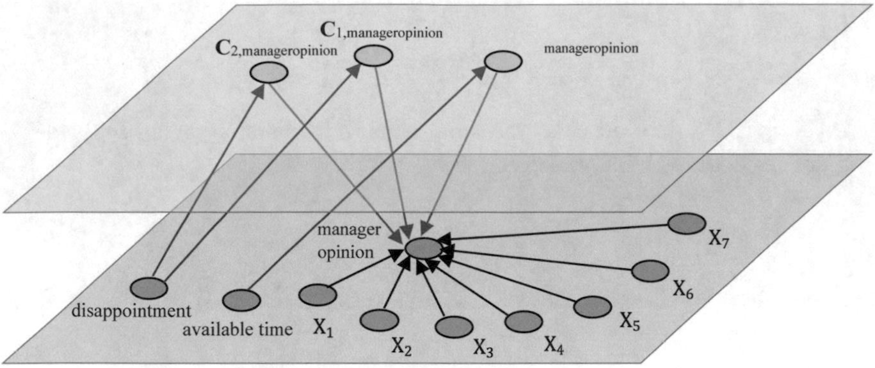

Fig. 3.11 The considered example adaptive reified network model

Table 3.5 States and their explanation

State		Explanation	Level
X_1			
X_2			
X_3			
X_4		These 7 states X_1 to X_7 represent the opinions of the 7 group members	
X_5			Base level
X_6			
X_7			
X_8	manageropinion	The opinion of the manager	
X_9	availabletime	The available time of the manager	
X_{10}	disappointment	The level of disappointment of the manager	
X_{11}	**H**manageropinion	The reified representation state for the speed of the manager's opinion	
X_{12}	**C**1,manageropinion	The reified representation state for the weight for the scaled sum combination function to aggregate the group members opinions for the manager's opinion	First reification level
X_{13}	**C**2,manageropinion	The reified representation state for the weight for the advanced logistic sum combination function to aggregate the group members opinions for the manager's opinion	

been left out; see Box 3.8 for them. Also, the connection from the independent state 'available time' to itself has been left out, and the same for independent state 'disappointment' representing the disappointing experiences; also see Box 3.8 for that. The reification states with their speed factors and combination functions, and connection weights for the incoming connections used to model this scenario are specified as shown in Box 3.8. Note that it is assumed:

$$\mathrm{bcf}_1(..) = \mathbf{ssum}_\lambda(..)$$
$$\mathrm{bcf}_2(..) = \mathbf{alogistic}_{\sigma,\tau}(..)$$

So, in the scenario the combination function used for the manager opinion state will change over time from basic combination function $\mathrm{bcf}_1(..)$ to $\mathrm{bcf}_2(..)$, using the reification states $\mathbf{C}_{1,\mathrm{manageropinion}}$ and $\mathbf{C}_{2,\mathrm{manageropinion}}$.

The concept of a *role matrix* as introduced in Chap. 2, Sect. 2.4 is used in a generalised form here to describe a reified network's structure. For the role matrices for this example network model, see Box 3.8. As can be seen, the main difference with the role matrices in Chap. 2 is that now in the cells in all role matrices there can be either a value (shaded in green) or the name of a state (shaded in red), whereas in Chap. 2 only values were used in the role matrices **mcw**, **ms**, **mcfw**, and **mcfp**. This is a relatively small step in terms of writing the role matrices, but it drastically changes the possibilities to model adaptive networks, as it makes that a non-reified network becomes a reified network. Indicating not a value but a state name in one of the cells, makes that the network characteristic described by this cell becomes adaptive and the indicated state becomes a reification state for this characteristic. For example by writing X_{11} in the cell for manager opinion in the speed factor role matrix **ms** in Box 3.8, the speed of change of the manager's opinion becomes adaptive according to the (dynamic) value of state X_{11}, and this makes X_{11} a reification state for this adaptive speed factor (and therefore X_{11} also is given the more informative name $\mathbf{H}_{\text{manageropinion}}$).

For the example reified network in the simulation the group members have mutual connections as specified by the role matrices **mb** and **mcw** in Box 3.8. This models a form of social contagion in a strongly connected network from which it is known that it eventually leads to a joint opinion; e.g., see Chap. 11. Note that in role matrices **mb** and **mcw** the cells correspond to all connections that in the picture in Fig. 3.11 are depicted by *upward or horizontal arrows*. There are no cells for the (pink) downward arrows in these two matrices, as these arrows concern special effects, and automatically get weight value 1 to keep the relation between a base network characteristic (as used in the dynamics) and its reification one-to-one. Therefore, the downward arrows from reification states are described in a different way in one of the other role matrices depending on which role these reification states describe, and in the cell of the concept they are reifying; see Table 3.6.

3.7.2 Conceptual Role Matrices Representation of the Example Reified Network Model

The role matrices for the example reified network model are shown in Box 3.8 together with the initial values in **iv**. Note that the independent dynamics of each of the states available time and disappointment, which serve as an external input to the model, were modeled by a logistic sum combination function applied to the connection of the state to itself with specific settings (steepness 18 and threshold 0.2) shown in role matrices **mcfw** and **mcfp**.

Table 3.6 Downward causal connections and role matrices: how and where specify what

In model picture	State name	State number	Role	In role matrix
Downward arrow from a reification state for an adaptive connection weight from state X to state Y	$\mathbf{W}_{X,Y}$	X_i	Connection weight reification state for $\boldsymbol{\omega}_{X,Y}$	**mcw** as notation X_i in the cell for the weight $\boldsymbol{\omega}_{X,Y}$ of the connection from X to Y
Downward arrow from a reification state for an adaptive speed factor for state Y	$\mathbf{H}_Y$	X_j	Speed factor reification state for $\boldsymbol{\eta}_Y$	**ms** as notation X_j in the cell for the value $\boldsymbol{\eta}_Y$ of the speed factor of Y
Downward arrow from a reification state for an adaptive combination function weight for state Y	$\mathbf{C}_{i,Y}$	X_k	Combination function weight reification state for $\gamma_{i,Y}$	**mcfw** as notation X_k in the cell for the weight $\gamma_{i,Y}$ of combination function i for state Y
Downward arrow from a reification state for an adaptive combination function parameter for state Y	$\mathbf{P}_{i,j,Y}$	X_l	Combination function parameter reification state for $\boldsymbol{\pi}_{i,j,Y}$	**mcfp** as notation X_l in the cell for the value $\boldsymbol{\pi}_{i,,j,Y}$ of parameter j of combination function i for state Y

3.7.3 Simulation Outcomes for the Example Reified Network Model

In Fig. 3.12 it is shown how in the simulation the manager's speed factor and combination function weights adapt over time, and in Fig. 3.13 the base states are shown: the group member opinions, the manager's opinion, and the change in

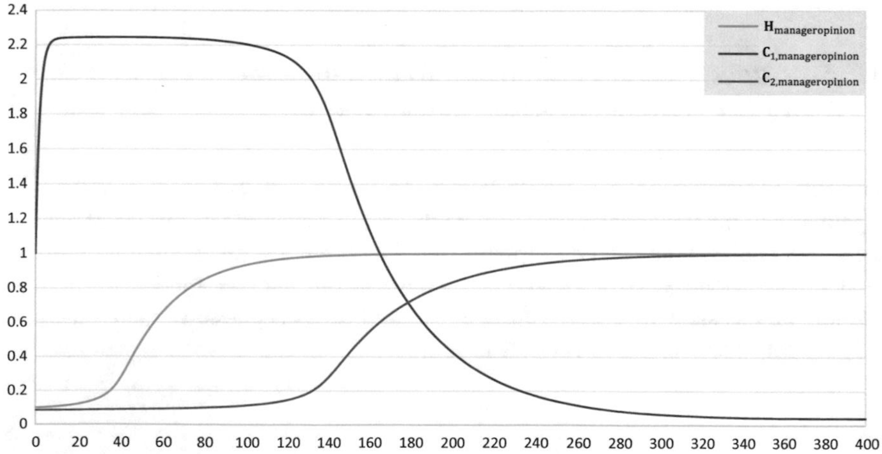

Fig. 3.12 Reified adaptive speed factor and combination function weights for the manager opinion state

available time and in disappointment. Note that this is one of those less usual cases in which state values can be outside the [0, 1] interval. That can also be modelled; in particular, in case of reification states for some role where values are used only for the special effects for that role. In this case, within the combination function a division by the sum of the weights takes place so that finally everything still comes in the [0, 1] interval. As an alternative, for reification state $\mathbf{C}_{1,\text{manageropinion}}$ a logistic sum combination function could have been used instead of a Euclidean function. Then the values would have stayed in the [0, 1] interval all the time.

Box 3.8 Role matrices for the example reified network model

mb	base connectivity	1	2	3	4	5	6	7	8
X_1		X_2	X_4	X_7					
X_2		X_1	X_3	X_5					
X_3		X_2	X_6						
X_4		X_2	X_7						
X_5		X_4	X_7						
X_6		X_1	X_3						
X_7		X_1	X_4						
X_8	manageropionion	X_1	X_2	X_3	X_4	X_5	X_6	X_7	
X_9	availabletime	X_9							
X_{10}	disappointment	X_{10}							
X_{11}	**H**manageropinion	X_9							
X_{12}	**C**1,manageropinion	X_1	X_2	X_3	X_4	X_5	X_6	X_7	X_{10}
X_{13}	**C**2,manageropinion	X_{10}							

mcw	connection weights	1	2	3	4	5	6	7	8
X_1		0.5	0.6	0.2					
X_2		0.5	0.6	0.5					
X_3		0.6	0.7						
X_4		0.3	0.7						
X_5		0.4	0.6						
X_6		0.6	0.4						
X_7		0.3	0.7						
X_8	manageropionion	1	1	1	1	1	1	1	
X_9	availabletime	1							
X_{10}	disappointment	1							
X_{11}	**H**manageropinion	1							
X_{12}	**C**1,manageropinion	0.01	0.01	0.01	0.01	0.01	0.01	0.01	-0.05
X_{13}	**C**2,manageropinion	1							

mcfw	combination function weights	1 eucl	2 alogistic
X_1		1	
X_2		1	
X_3		1	
X_4		1	
X_5		1	
X_6		1	
X_7		1	
X_8	manageropionion	X_{12}	X_{13}
X_9	availabletime		1
X_{10}	disappointment		1
X_{11}	**H**manageropinion	1	
X_{12}	**C**1,manageropinion	1	
X_{13}	**C**2,manageropinion	1	

mcfp	combination function	1 eucl		2 alogistic	
		1	2	1	2
	parameter	n	λ	σ	τ
X_1		1	1.3		
X_2		1	1.6		
X_3		1	1.3		
X_4		1	1		
X_5		1	1		
X_6		1	1		
X_7		1	1		
X_8	manageropionion	1	7	5	5.5
X_9	availabletime			18	0.2
X_{10}	disappointment			18	0.2
X_{11}	**H**manageropinion	1	1		
X_{12}	**C**1,manageropinion	1	0.02		
X_{13}	**C**2,manageropinion	1	1		

ms	speed factors	1
X_1		0.05
X_2		0.05
X_3		0.05
X_4		0.05
X_5		0.05
X_6		0.05
X_7		0.05
X_8	manageropionion	X_{11}
X_9	availabletime	0.04
X_{10}	disappointment	0.025
X_{11}	**H**manageropinion	0.5
X_{12}	**C**1,manageropinion	0.5
X_{13}	**C**2,manageropinion	0.5

iv	initial values	1
X_1		1
X_2		0.9
X_3		0.5
X_4		0.6
X_5		0.6
X_6		0.6
X_7		0.7
X_8	manageropionion	0.5
X_9	availabletime	0.1
X_{10}	disappointment	0.085
X_{11}	**H**manageropinion	0.1
X_{12}	**C**1,manageropinion	1
X_{13}	**C**2,manageropinion	0.085

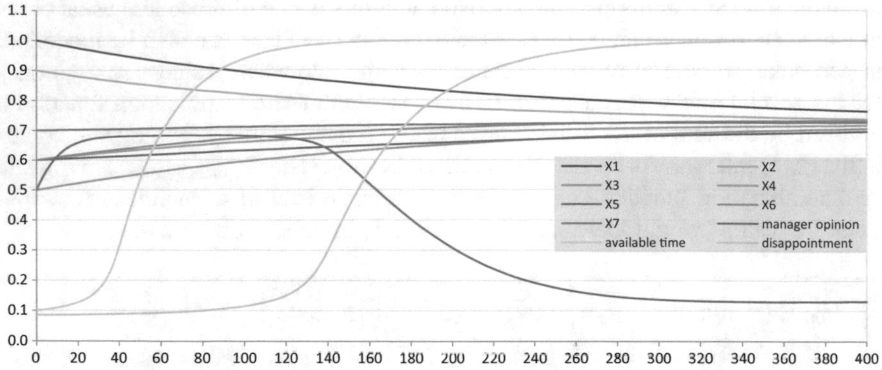

Fig. 3.13 The effect of the adaptive combination function of the manager on her opinion

It can be seen in Fig. 3.12 that after time point 40 the manager's speed factor increases (blue line; with as effect a shorter response time), due to more availability (the purple line in Fig. 3.12). After time point 140 in Fig. 3.12 a switch is shown from a dominant weight for the scaled sum function $\mathbf{ssum}_7$**(..)** (purple line) to a dominant weight for the logistic combination function $\mathbf{alogistic}_{5,5.5}(\mathrm{x})$ (red line), due to increasing disappointment (green line in Fig. 3.12). In Fig. 3.13 it is also shown how the manager's opinion is affected by the opinions of the group members. Here it can be seen that after time point 140 the manager's opinion becomes much lower due to the switch of combination function, which is resulting from the increase in disappointment (green line).

3.7.4 Analysis of the Equilibria for the Example Reified Network Model

Here the equilibrium equations for the different states are considered. First, the independent base states, next the reification states which depend on the independent states.

Equilibrium equations for the independent base states
Both independent base states have a circular causal relation and use the same combination function $\mathbf{alogistic}_{18,0.2}(..)$. From the criterion (3.5) in Lemma 1 it is derived that their equilibrium equations are:

$$\begin{aligned} \text{availabletime} &= \mathbf{alogistic}_{18,0.2}(\text{availabletime}) \\ \text{disappointment} &= \mathbf{alogistic}_{18,0.2}(\text{disapppointment}) \end{aligned} \tag{3.22}$$

So both have an equation of form $x = \mathbf{alogistic}_{18,0.2}(x)$. In Fig. 3.14 the graphs of both the functions x and $\mathbf{alogistic}_{18,0.2}(x)$ are shown. Given that the equilibrium values are not close to 0, it can be seen that they cross indeed very close to 1; the value can be approximately calculated in 12 digits as: x = 0.999999427374. This differs from 1 less than 10^{-6}. This applies both to available time and to disappointment. Indeed in Fig. 3.14 it is shown that in the simulation they end up very close to 1.

Equilibrium equations for the reification states

Applying the criterion (3.5) in Lemma 1 to the above specifications in Box 3.8, directly provides the equilibrium equations for the reification states by the following relations where H stands for the value of $\mathbf{H}_{\text{manageropinion}}$, C_1 for the value of $\mathbf{C}_{1,\text{manageropinion}}$ and C_2 for the value of $\mathbf{C}_{2,\text{manageropinion}}$:

$$\begin{aligned} H &= \text{availabletime} \\ C_1 &= 0.5\,X_1 + 0.5\,X_2 + 0.5\,X_3 + 0.5X_4 + 0.5\,X_5 \\ &\quad + 0.5X_6 + 0.5X_7 - 2.5\text{disapppointment} \\ C_2 &= \text{disapppointment} \end{aligned} \tag{3.23}$$

Assuming that in the equilibrium it holds $X_i(t) = 0.714$ for all $i = 1, \ldots, 7$ (this depends on the specification of the group members) the second equation can be rewritten as:

$$C_1 = 2.5 - 2.5\ \text{disapppointment} = 2.5\,(1 - \text{disapppointment}) \tag{3.24}$$

These equilibrium equations show how the equilibrium values of the reification states depend on the equilibrium values of the independent base states available time and disappointment. As discussed above, the equilibrium value for available time and disappointment are very close to 1 (difference from 1 less than 10^{-6}),

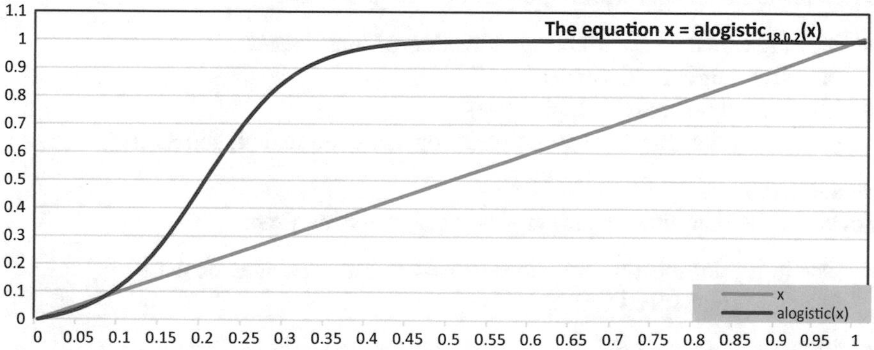

Fig. 3.14 Solving the equilibrium equations for the two independent states availble time and disappointment

therefore H and C_2 will also be very close to 1, which is confirmed in the simulation as shown in Fig. 3.12. Moreover, C_1 will be very close to 0 (by (3.24); difference from 0 less than 10^{-5}). This shows that the combination function for manager opinion indeed switched from the initial scaled sum **ssum**$_7$**(..)** to the logistic **alogistic**$_{5,5.5}$**(..)**; see how this is also confirmed in Figs. 3.12 and 3.13. Note that in Chap. 15, Sect. 15.3 more details are given of the difference and differential equations for this reified network model.

3.8 On Using a Joint Reification State for Multiple Base States or Multiple Roles

Usually a reification state has one role and is associated to one base state. This also supports transparency. However there may be specific cases in which one state can play the role of reification state for multiple states. There could even be situations in which one state plays multiple roles with respect to a given base state or multiple base states, but better be very careful to model like this. Be aware that the value of a reification state is where the characteristic related to its role is maintained, and this can only be one value. Therefore having a joint reification state for multiple base states only makes sense as all these base states should keep a joint value for that characteristic, for example, all have the same speed factor. As soon as there are differences in these values, the reification state has to split according to these differences. This will in general also stand in the way to use one reification state for multiple roles. For example, then the value of this reification state represents both a speed factor and a connection weight. As these are totally different concepts, it will, in general, not make sense to have the same value for them. Keeping this in mind, some examples in which joint reification states might make sense are as follows:

Mental Networks
Chemicals in the brain such as neurotransmitters or drugs or hormones or alcohol or stress related elements with a global effect on the brain:

- Slowing down or speeding up activation of brain states
  (adaptive joint speed factor)
- Lowering thresholds of brain states
  (adaptive joint threshold value for a logistic combination function)

Social networks
Social events that affect multiple persons at the same time:

- Due to a joint meeting no-one responds fast to messages or mail
  (adaptive joint speed factor)
- Within the meeting the threshold to respond publicly is higher
  (adaptive joint threshold value for a logistic combination function)

- Due to positive atmosphere in a meeting the scaling factor is lower, due to negative atmosphere scaling factor is higher
(adaptive joint threshold value for a logistic combination function)
- Due to positive atmosphere the threshold to accept/assimilate opinions is lower, due to negative atmosphere the threshold to accept is higher`
(adaptive joint threshold value for a logistic combination function).

But, as indicated, normally there are differences between individuals and between states, and to represent these differences, multiple reification states are needed.

3.9 On the Added Complexity for Reification

Note that, as for any dynamical system, by adding adaptivity to a network model, always complexity is added. In this section, it is discussed how the complexity of a network increases when reification is applied. It will at most be quadratic in the number of nodes N and linear in the number of connections M of the original network, as shown here. More specifically, if m the number of basic combination functions used in the given network model, then the number of *nodes* in the reified network is at most:

$$\begin{array}{ll} N & \text{(original nodes)} \\ +N & \text{(nodes for speed factors)} \\ +N^2 & \text{(nodes for connection weights)} \\ +mN & \text{(nodes for combination functions)} \end{array}$$

which adds to

$$(2+m+N)N \tag{3.25}$$

This is quadratic in the number of nodes.

If more general, not all N^2 connections are used for reification, but only a number M of them (which maximally can be N^2), the outcome is

$$(2+m)N+M \tag{3.26}$$

additional nodes. This is linear in the numbers of nodes and connections. Then the number of *connections* in the reified network is

$$\begin{array}{ll} M & \text{(original connection weights)} \\ +N & \text{(speed factors to their base states)} \\ +\sum_Y \text{indegree}(Y) = M & \text{(connection weights to their base states)} \\ +mN & \text{(combination function weights to their base states)} \end{array}$$

which adds to

$$(m+1)N+2M \tag{3.27}$$

Also, this is linear in number of nodes and connections. Note, however, that also connections to the reification states will be needed to get them adaptive. But these depend on the specific application. If at least one inward connection per reification state is assumed, this adds at least the number of additional nodes $(2 + m)N + M$ to the number of added connections. So, then the number of additional connections becomes

$$(2m+3)N+3M \tag{3.28}$$

which again is linear in numbers of nodes and connections.

3.10 Discussion

In this chapter, it was shown how network structure can be reified in a network by adding explicit network states representing the characteristics of the network structure, such as connection weights, combination functions and speed factors. Parts of this chapter were adopted from (Treur 2018a). This construction of network reification can provide advantages similar to those found for reification in modeling and programming languages in other areas of AI and Computer Science, in particular, substantially enhanced expressiveness (e.g., Weyhrauch 1980; Bowen and Kowalski 1982; Bowen 1985; Sterling and Shapiro 1986; Sterling and Beer 1989; Demers and Malenfant 1995; Galton 2006).

A reified network including an explicit representation of the network structure enables to model dynamics *of* the original network by dynamics *within* the reified network. In this way, an adaptive network can be represented by a non-adaptive network. The approach is applicable to the whole variety of adaptive processes as described in Chap. 1, Sect. 1.2.1. It was shown how the approach provides a unified manner of modeling network adaptation principles, and allows comparison of such principles across different domains. This was illustrated for known adaptation principles for Mental Networks and for Social Networks. Note that this approach to model network adaptation principles can be applied successfully to any adaptation principle that is described by (first-order) difference or differential equations (as usually is their format), as in (Treur 2017, Sect. 3.3), it is shown how any difference or differential equation can be modeled in the temporal-causal network format.

Note that in the description in this chapter the structure of the base network is reified but not the structure of the reified network as a whole. In a reification process always new structures are added which are themselves not reified. As the next step in Chap. 4 it will be shown how the structure of the reified network also can be reified by socalled second-order reification; see also (Treur 2018b). Then the network reification approach becomes applicable to all examples of second- or higher-order adaptation described in Chap. 1, Sects. 1.2.2 and 1.3. Structures in the first-order reified network used to model adaptation principles are not reified themselves in a first-order reification process. In a second-order reified network, they are reified as well: their structure is explicitly represented by second-order reification states and their connections, which allows modeling adaptive adaptation principles. It is possible for any n to repeat the construction n times and obtain nth-order *reification*. But still, there will be structures introduced in the step from $n - 1$ to n that have no reification. From a theoretical perspective it can also be considered to repeat the construction infinitely many times, for all natural numbers: ω-*order reification*, where ω is the ordinal for the natural numbers. Then an infinite network is obtained, which is theoretically well-defined; all structures in this network are reified within the network itself, but it may not be clear whether it can be applied in practice, or for theoretical questions. This also might be a subject for future research.

References

Banks, D.L., Carley, K.M.: Models for network evolution. J. Math. Sociol. **21**, 173–196 (1996)

Barabasi, A.-L., Albert, R.: Emergence of scaling in random networks. Science **286**, 509–512 (1999)

Bi, G., Poo, M.: Synaptic modification by correlated activity: Hebb's postulate revisited. Annu. Rev. Neurosci. **24**, 139–166 (2001)

Blankendaal, R., Parinussa, S., Treur, J.: A temporal-causal modelling approach to integrated contagion and network change in social networks. In: Proceedings of the 22nd European Conference on Artificial Intelligence, ECAI'16, pp. 1388–1396. IOS Press (2016)

Bowen, K.A., Kowalski, R.: Amalgamating language and meta-language in logic programming. In: Clark, K., Tarnlund, S. (eds.) Logic Programming, pp. 153–172. Academic Press, New York (1982)

Bowen, K.A.: Meta-level programming and knowledge representation. New Gener. Comput. **3**, 359–383 (1985)

Chandra, N., Barkai, E.: A non-synaptic mechanism of complex learning: Modulation of intrinsic neuronal excitability. Neurobiol. Learn. Mem. **154**(2018), 30–36 (2018)

Demers, F.N., Malenfant, J.: Reflection in logic, functional and objectoriented programming: a short comparative study. In: IJCAI'95Workshop on Reflection and Meta-Level Architecture and their Application in AI, pp. 29–38 (1995)

Galton, A.: Operators vs. arguments: the ins and outs of reification. Synthese **150**, 415–441 (2006)

Gerstner, W., Kistler, W.M.: Mathematical formulations of Hebbian learning. Biol. Cybern. **87**, 404–415 (2002)

Granovetter, M.S.: The strength of weak ties. Amer. J. Sociol. **78**(6), 1360–1380 (1973)

Hebb, D.: The organisation of behavior. Wiley (1949)

Hofstadter, D.R.: Gödel, Escher, Bach. Basic Books, New York (1979)

McPherson, M., Smith-Lovin, L., Cook, J.M.: Birds of a feather: homophily in social networks. Annu. Rev. Sociol. **27**, 415–444 (2001)

Pearl, J.: Causality. Cambridge University Press (2000)

Rapoport, A.: Spread of Information through a population with socio-structural bias: i. Assumption of transitivity. Bull. Math. Biophys. **15**, 523–533 (1953)

Smorynski, C.: The incompleteness theorems. In: Barwise, J. (ed.) Handbook of Mathematical Logic, vol. 4, pp. 821–865. North-Holland, Amsterdam (1977)

Sousa, N., Almeida, O.F.X.: Disconnection and reconnection: the morphological basis of (mal) adaptation to stress. Trends Neurosci. **35**(12), 742–751 (2012)

Sterling, L., Shapiro, E.: The Art of Prolog. MIT Press, (1986) (Ch 17, pp. 319–356)

Sterling, L., Beer, R.: Metainterpreters for expert system construction. J. Logic Program. **6**, 163–178 (1989)

Treur, J.: Network-Oriented Modeling: Addressing Complexity of Cognitive, Affective and Social Interactions. Springer, Berlin (2016)

Treur, J.: On the applicability of network-oriented modeling based on temporal-causal networks: why network models do not just model networks. J. Inf. Telecommun. **1**(1), 23–40 (2017)

Treur, J.: Network reification as a unified approach to represent network adaptation principles within a network. In: Proceedings of the 7th International Conference on Theory and Practice of Natural Computing, TPNC'18. Lecture Notes in Computer Science, vol 11324, pp. 344–358. Springer, Berlin (2018a)

Treur, J.: Multilevel network reification: representing higher order adaptivity in a network. In: Proceedings of the 7th International Conference on Complex Networks and their Applications, ComplexNetworks'18, vol. 1. Studies in Computational Intelligence, vol. 812, 635–651, Springer, Berlin (2018b)

Treur, J.: The ins and outs of network-oriented modeling: from biological networks and mental networks to social networks and beyond. In: LNCS Transactions on Computational Collective Intelligence. Paper on Keynote lecture at the 10th International Conference on Computational Collective Intelligence, ICCCI'18 vol. 32, pp. 120–139 (2019)

Treur, J., Mohammadi Ziabari, S.S.: An adaptive temporal-causal network model for decision making under acute stress. In: Proceedings of the 10th International Conference on Computational Collective Intelligence, ICCCI'18. Lecture Notes in Computer Science, Springer, Berlin (2018)

Weyhrauch, R.W.: Prolegomena to a theory of mechanized formal reasoning. Artif. Intell. **13**, 133–170 (1980)

Chapter 4
Modeling Higher-Order Network Adaptation by Multilevel Network Reification

Abstract In network models for real-world domains, often some form of network adaptation has to be incorporated, based on certain network adaptation principles. In some cases, also higher-order adaptation occurs: the adaptation principles themselves also change over time. To model such multilevel adaptation processes, it is useful to have some generic architecture. Such an architecture should describe and distinguish the dynamics within the network (base level), but also the dynamics of the network itself by certain adaptation principles (first-order adaptation), and also the adaptation of these adaptation principles (second-order adaptation), and maybe still more levels of higher-order adaptation. This chapter introduces a multilevel network architecture for this, based on the notion of network reification. Reification of a network occurs when a base network is extended by adding explicit reification states representing the characteristics of the structure of the base network (Connectivity, Aggregation, and Timing). In Chap. 3, it was shown how this construction can be used to explicitly represent network adaptation principles within a network. In the current chapter, it is discussed how, when the reified network is itself also reified, also second-order adaptation principles can be explicitly represented. For the multilevel network reification construction introduced here, it is shown how it can be used to model plasticity and metaplasticity as known from Cognitive Neuroscience. Here, plasticity describes how connections between neurons change over time, for example, based on a first-order adaptation principle for Hebbian learning, and metaplasticity describes second-order adaptation principles determining how the extent of plasticity is affected by certain circumstances; for example, under which circumstances plasticity will be accelerated or decelerated.

4.1 Introduction

Within the complex dynamical systems area, adaptive behaviour is an interesting and quite relevant challenge, addressed in various ways; see, for example, Helbing et al. (2015), Perc and Szolnoki (2010). In particular for network-oriented dynamic modeling approaches, network models for real-world domains often show some

J. Treur, *Network-Oriented Modeling for Adaptive Networks: Designing Higher-Order Adaptive Biological, Mental and Social Network Models*, Studies in Systems, Decision and Control 251, https://doi.org/10.1007/978-3-030-31445-3_4

form of network adaptation based on certain network adaptation principles. Such principles describe how certain characteristics of the network structure change over time, for example, the connection weights in Mental Networks with Hebbian learning (Hebb 1949) or in Social Networks with bonding based on homophily; e.g., Byrne (1986), McPherson et al. (2001), Pearson et al. (2006), Sharpanskykh and Treur 2014). Sometimes also higher-order adaptation occurs in the sense that the adaptation principles for a network themselves also change over time. For example, *plasticity* in Mental Networks as described, for example, by Hebbian learning is not a constant feature, but usually varies over time, according to what in Cognitive Neuroscience has been called *metaplasticity*; e.g., Abraham and Bear (1996), Magerl et al. (2018), Parsons (2018), Schmidt et al. (2013), Sehgal et al. (2013), Zelcer et al. (2006). For more examples of processes which are adaptive of different orders, see Chap. 1, Sects. 1.2 and 1.3.

To model such multilevel network adaptation processes in a principled manner it is useful to have some generic architecture. Such architecture should be able to distinguish and describe:

(1) the dynamics within the base network
(2) the dynamics of the base network structure by network adaptation principles (first-order adaptation)
(3) the adaptation of these adaptation principles (second-order adaptation)
(4) interactions between the levels
(5) and maybe still more levels of higher-order adaptation.

In the current chapter, it is shown how such distinctions indeed can be made within a Network-Oriented Modeling framework using the notion of *reified network architecture*.

As also described in Chap. 3, reification is known from different scientific areas. According to the Merriam-Webster and Oxford dictionaries, it literally means representing something abstract as a material or concrete thing, or making something abstract more concrete or real. Reification offers substantial advantages in modeling and programming languages, as shown for other areas of AI and Computer Science; e.g., Bowen and Kowalski (1982), Demers and Malenfant (1995), Galton (2006), Hofstadter (1979), Sterling and Shapiro (1996), Sterling and Beer (1989), Weyhrauch (1980). Modeling adaptivity and enhanced expressive power are some of these advantages. Network reification has similar advantages. In Chap. 3, it has been shown how network reification can be used to explicitly represent adaptation principles for networks in a transparent and unified manner. Examples of such adaptation principles are, among others, principles for Hebbian learning (to model plasticity in the brain) and for bonding based on homophily (to model adaptive social networks). Using network reification, adaptive Mental Networks and adaptive Social Networks can be addressed well, as shown by many examples in Chap. 3.

Including reification states for the characteristics of the base network structure (connection weights, speed factors, and combination functions and their parameters) in the extended network is one step. A next step is defining proper temporal-causal relations for them and relating them to the other states. Then a reified network is obtained that explicitly represents the characteristics of the base network, and, moreover, how this base network evolves over time, based on adaptation principles that change the causal network relations. In Chap. 3, it was shown how this can be used for a variety of adaptation principles known from Cognitive Neuroscience and Social Science.

Such reified adaptive networks form again a basic network structure defined by certain characteristics, such as learning rate or adaptation speed of connections. Adaptation principles may be adaptive themselves too, according to certain second-order adaptation principles. From recent literature, it has become clear that in real-world domains often these characteristics can still change over time, for example, in the case of metaplasticity. The notion of metaplasticity as already mentioned above, has become a focus of study in empirical literature such as Arnold et al. (2015), Chandra and Barkai (2018), Daimon et al. (2017), Magerl et al. (2018), Parsons (2018), Robinson et al. (2016), Sehgal et al. (2013), Schmidt et al. (2013), Zelcer et al. (2006). This area of higher-order adaptivity is a next challenge to be addressed. To this end, in the current chapter a construction of multilevel reification is illustrated for the Network-Oriented Modeling approach based on temporal-causal networks (Treur 2016, 2019). By an appropriate number of iterations, this multilevel reification construction introduced here can be used to model higher-order adaptivity of any level. The multilevel reification architecture has been implemented by the author in Matlab, as will be discussed in Chap. 9. The homophily context is used in this chapter as the first application of a second-order adaptive Social Network for the Social Science area, and the context of plasticity and metaplasticity as a second application for a second-order adaptive Mental Network in the Cognitive Neuroscience area.

In this chapter in Sect. 4.2, the Network-Oriented Modeling approach based on temporal-causal networks is briefly summarized. Next, in Sect. 4.3 the network reification concept is summarized, and in Sect. 4.4 the more general multilevel network reification construction is introduced. Moreover, it is shown how it can model examples of second-order network adaptivity. This is illustrated by a second-order adaptive network for plasticity and metaplasticity from Cognitive Neuroscience. In Sect. 4.5 example simulations for this multilevel network reification example are presented. Section 4.6 discusses the added complexity in a multilevel reification architecture. Section 4.7 is a discussion.

4.2 Structure and Dynamics of Temporal-Causal Networks

The network structure of a temporal-causal network model can be described conceptually by a graph with nodes and directed connections and a number of labels for such a graph for connectivity, aggregation, and timing:

(a) **Connectivity**

- In terms of connection weights $\boldsymbol{\omega}_{X,Y}$; see Table 4.1, upper part, and Fig. 4.1 for an example of a basic fragment of a network with states X_1, X_2 and Y, and labels $\boldsymbol{\omega}_{X_1,Y}$, $\boldsymbol{\omega}_{X_2,Y}$ for connection weights.

(b) **Aggregation**

- In terms of combination functions $\mathbf{c}_Y(..)$; a library with a number of standard combination functions is available, but also new functions can be added. Such functions are just declarative mathematical objects that relate real numbers to real numbers without any procedural elements involved, i.e., **c**: $R^k \rightarrow R$.

(c) **Timing**

- In terms of speed factors $\boldsymbol{\eta}_Y$.

In the lower part of Table 4.1, it is shown how the numerical representation of the network's dynamics is defined in terms of the above labels; see also

Table 4.1 Conceptual and numerical representation of a temporal-causal network structure

Concepts	Notation	Explanation
States and connections	X, Y, $X \rightarrow Y$	Describes the nodes and links of a network structure (e.g., in graphical or matrix format)
Connection weight	$\boldsymbol{\omega}_{X,Y}$	The *connection weight* $\boldsymbol{\omega}_{X,Y} \in [-1, 1]$ represents the strength of the causal impact of state X on state Y with $X \rightarrow Y$
Aggregating multiple impacts	$\mathbf{c}_Y(..)$	For each state Y a *combination function* $\mathbf{c}_Y(..)$ is chosen to combine the causal impacts of other states on state Y
Timing of the causal effect	$\boldsymbol{\eta}_Y$	For each state Y a *speed factor* $\boldsymbol{\eta}_Y \geq 0$ is used to represent how fast a state is changing upon causal impact

Concepts	Numerical representation	Explanation
State values over time t	$Y(t)$	At each time point t each state Y in the model has a real number value in [0, 1]
Single causal impact	$\mathbf{impact}_{X,Y}(t) = \boldsymbol{\omega}_{X,Y}X(t)$	At t state X with connection to state Y has an impact on Y, using weight $\boldsymbol{\omega}_{X,Y}$
Aggregating multiple impacts	$\mathbf{aggimpact}_Y(t) = \mathbf{c}_Y(\mathbf{impact}_{X_1,Y}(t), \ldots, \mathbf{impact}_{X_k,Y}(t)) = \mathbf{c}_Y(\boldsymbol{\omega}_{X_1,Y}X_1(t), \ldots, \boldsymbol{\omega}_{X_k,Y}X_k(t))$	The aggregated impact of multiple states X_i on Y at t, is determined using combination function $\mathbf{c}_Y(..)$
Timing of the causal effect	$Y(t+\Delta t) = Y(t) + \boldsymbol{\eta}_Y[\mathbf{aggimpact}_Y(t) - Y(t)]\Delta t = Y(t) + \boldsymbol{\eta}_Y[\mathbf{c}_Y(\boldsymbol{\omega}_{X_1,Y}X_1(t), \ldots, \boldsymbol{\omega}_{X_k,Y}X_k(t)) - Y(t)]\Delta t$	The causal impact on Y is exerted over time gradually, using speed factor $\boldsymbol{\eta}_Y$

Adopted from Treur (2019)

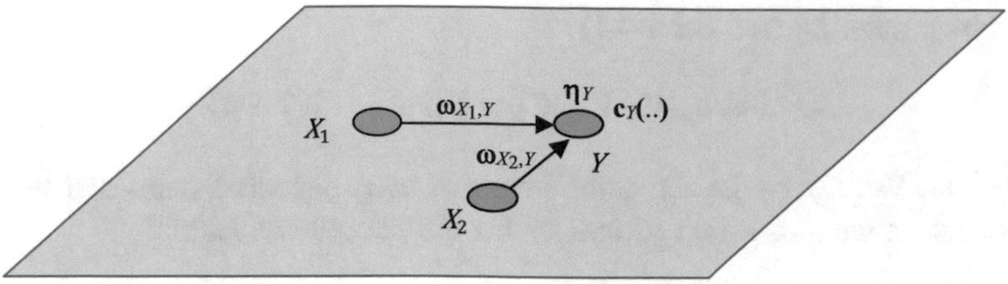

Fig. 4.1 Fragment of a temporal-causal network structure in a labeled graph representation. The basic elements are nodes and their connections, with for each node Y a speed factor η_Y and a combination function $\mathbf{c}_Y(..)$, and for each connection from X to Y a connection weight $\omega_{X,Y}$

Treur (2016), Chap. 2. Here $X_1, \ldots, X_k$ are the states from which state Y receives incoming connections. These formulas in the last row in Table 4.1 define the detailed dynamic semantics of a temporal-causal network. They can be used for mathematical analysis and for simulation, and can be written in differential equation format as follows:

$$\mathbf{d}Y(t)/\mathbf{d}t = \eta_Y\left[\mathbf{c}_Y\left(\omega_{X_1,Y}X_1(t), \ldots, \omega_{X_k,Y}X_k(t)\right) - Y(t)\right] \quad (4.1)$$

Examples of combination functions are the *identity* **id**(.) for states with impact from only one other state, the *scaled sum* $\mathbf{ssum}_\lambda$(.) with scaling factor λ, the scaled minimum function $\mathbf{smin}_\lambda$(..) and maximum function $\mathbf{smax}_\lambda$(..), and the *advanced logistic sum* combination function $\mathbf{alogistic}_{\sigma,\tau}$(..) with steepness σ and threshold τ; see also Treur (2016, Chap. 2, Table 2.10):

$$\begin{aligned}
&\mathbf{id}(V) = V\\
&\mathbf{ssum}_\lambda(V_1, \ldots, V_k) = \frac{V_1 + \cdots + V_k}{\lambda}\\
&\mathbf{smin}_\lambda(V_1, \ldots, V_k) = \frac{\min(V_1, \ldots, V_k)}{\lambda}\\
&\mathbf{smax}_\lambda(V_1, \ldots, V_k) = \frac{\max(V_1, \ldots, V_k)}{\lambda}\\
&\mathbf{alogistic}_{\sigma,\tau}(V_1, \ldots, V_k) = \left[\frac{1}{1+\mathrm{e}^{-\sigma(V_1 + \cdots + V_k - \tau)}} - \frac{1}{1+\mathrm{e}^{\sigma\tau}}\right](1+\mathrm{e}^{-\sigma\tau})
\end{aligned} \quad (4.2)$$

Note that for basic combination functions, specific parameters are considered. Examples are the scaling factor λ, the steepness σ, and the threshold τ above. These parameters can also be written as arguments in the function, for example, $\mathbf{alogistic}(\sigma, \tau, V_1, \ldots, V_k)$, or in lists as $\mathbf{alogistic}([\sigma, \tau], [V_1, \ldots, V_k])$; this actually is how they are represented in the software environment.

Examples of combination functions applied in particular for reification states in reified adaptive networks (introduced in Chap. 3, Sect. 3.6.1) are the following

- Hebbian learning (see Sect. 4.4)

$$\mathbf{hebb}_{\boldsymbol{\mu}}(V_1, V_2, W) = V_1V_2(1 - W) + \boldsymbol{\mu}W \tag{4.3}$$

Here V_1, V_2 refer to the activation levels of two connected states and W to their connection weight; $\boldsymbol{\mu}$ is a parameter for the persistence factor.

- Simple linear homophily (see Chap. 6)

$$\mathbf{slhomo}_{\boldsymbol{\alpha},\boldsymbol{\tau}}(V_1, V_2, W) = W + \boldsymbol{\alpha}W(1 - W)(\boldsymbol{\tau} - |V_1 - V_2|) \tag{4.4}$$

Here V_1, V_2 refer to the activation levels of the states of two connected persons and W to their connection weight; $\boldsymbol{\tau}$ is a tipping point parameter and $\boldsymbol{\alpha}$ is a homophily modulation parameter. This is applied to model bonding based on homophily; see Chap. 6.

The set of already available combination functions forms a *combination function library* (with at the time of writing 35 functions), which can be chosen as basic combination functions during the design of a network model. These functions are declarative mathematical functions relating real numbers to real numbers without any procedural or process elements.

4.3 Addressing Network Adaptation by Network Reification

Recall from Chap. 3 that network reification is a construction principle by which a base network is extended by extra states that represent the base network's structure. This construction principle is briefly summarized here.

4.3.1 Extending the Network by Reification States

The added states represent specific characteristics of the network structure. They are what are called *reification states* for these characteristics, in other words, the characteristics are reified by these states. More specifically, these reification states represent the labels for *connection weights*, *combination functions*, and *speed factors* shown in Table 4.1. For connection weights $\boldsymbol{\omega}_{X_i,Y}$ for the incoming connections from states X_i to state Y and speed factors $\boldsymbol{\eta}_Y$ for state Y, their reification states $\mathbf{W}_{X_i,Y}$ and $\mathbf{H}_Y$ represent the value of them, and the vector of reification states $\mathbf{C}_Y = (\mathbf{C}_{1,Y}, \mathbf{C}_{2,Y}, \ldots.)$ represents the weights for the chosen basic combination functions for state Y; moreover, reification states $\mathbf{P}_{i,j,Y}$ represent the adaptive parameters of combination functions. In Fig. 4.2, the reification states are depicted in the upper (blue) plane, whereas the states of the base network are in the lower (pink) plane.

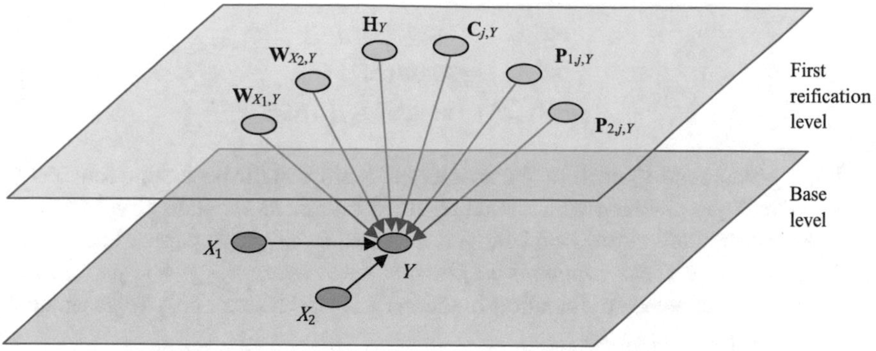

Fig. 4.2 Network reification for a temporal-causal network with in the upper, blue plane reification states $\mathbf{H}_Y$ for the speed factor of base state Y, $\mathbf{W}_{X_i,Y}$ for the weights of the connections from X_i to Y, $\mathbf{C}_{j,Y}$ for the basic combination function weights of Y, and $\mathbf{P}_{i,j,Y}$ for the parameter values of these basic combination functions. The downward connections from these reification states to state Y in the base network (in the lower, pink plane) indicate their special causal effect on Y

Within the reified network causal relations for the reification states for characteristics of a network can be defined: incoming connections affecting them, and outgoing connections from them to the related base network states. Such connections are the way in which adaptation principles are explicitly represented within the (reified) network; see also the many examples in Chap. 3. The downward pink arrows in Fig. 4.2 define how the reification states contribute their special effect to an aggregated impact on the related base network state.

These downward connections in Fig. 4.2 and the combination functions for the base states are defined in a generic manner. The general pattern is that the reification state roles $\mathbf{W}_{X_i,Y}$, $\mathbf{H}_Y$ and $\mathbf{C}_{j,Y}$ and $\mathbf{P}_{i,j,Y}$ for connection weights, speed factors, combination function weights, and combination function parameter values have a role-specific causal connection to state Y in the base network, as they all affect Y in their own role-dependent way. All depicted downward connections automatically get weight 1, so that there is a one-to-one correspondence between the base characteristic and its reification, and in the reified network the speed factors of the base states are set at 1 too. For the base states, new combination functions are needed that will be defined below (see also Chap. 3, Sect. 3.5). The different components

$$\mathbf{C}_{1,Y}, \mathbf{C}_{2,Y}, \ldots$$

for $\mathbf{C}_Y$ are explained as follows. During modeling a sequence of basic combination functions

$$\text{bcf}_1(..), \ldots, \text{bcf}_m(..)$$

is chosen from the function library discussed in Sect. 4.2 (in which also new functions can be added), to be used in the specific application addressed; for more details, see Chap. 9. For example,

$$\begin{aligned}\mathrm{bcf}_1(..) &= \mathbf{sum}(..),\\ \mathrm{bcf}_2(..) &= \mathbf{ssum}_\lambda(..)\\ \mathrm{bcf}_3(..) &= \mathbf{alogistic}_{\sigma,\tau}(..)\end{aligned}$$

For a given state *Y*, each of these selected basic combination functions $\mathrm{bcf}_j(..)$ gets a weight $\gamma_{j,Y}$ assigned which is represented by reification state $\mathbf{C}_{j,Y}$. Moreover, each basic combination function $\mathrm{bcf}_j(..)$ is assumed to have two parameters for each state: $\pi_{1,j,Y}$, $\pi_{2,j,Y}$. These *combination function parameters* $\pi_{1,1,Y}$, $\pi_{2,1,Y}$, …, $\pi_{1,m,Y}$, $\pi_{2,m,Y}$ in the *m* selected combination functions can also be explicitly represented by *parameter reification states*

$$\mathbf{P}_{1,1,Y}, \mathbf{P}_{2,1,Y}, \ldots, \mathbf{P}_{1,m,Y}, \mathbf{P}_{2,m,Y}$$

so that they also can become adaptive. Their values are considered as the first arguments in $\mathrm{bcf}_j(..)$, and also included as arguments in $\mathbf{c}_Y(...)$. Note that for applications, often more informative names are used for these parameters $\pi_{i,j,Y}$ and their reification states $\mathbf{P}_{i,j,Y}$; for example, reification state $\mathbf{H}_{\mathbf{W}_{\mathrm{srs}_s,\mathrm{ps}_a}}$ in Fig. 4.3 for the reified speed factor of the connection adaptation, and reification state $\mathbf{M}_{\mathbf{W}_{\mathrm{srs}_s,\mathrm{ps}_a}}$ for the persistence parameter μ for the Hebbian learning; this will be explained in more detail in Sect. 4.4.

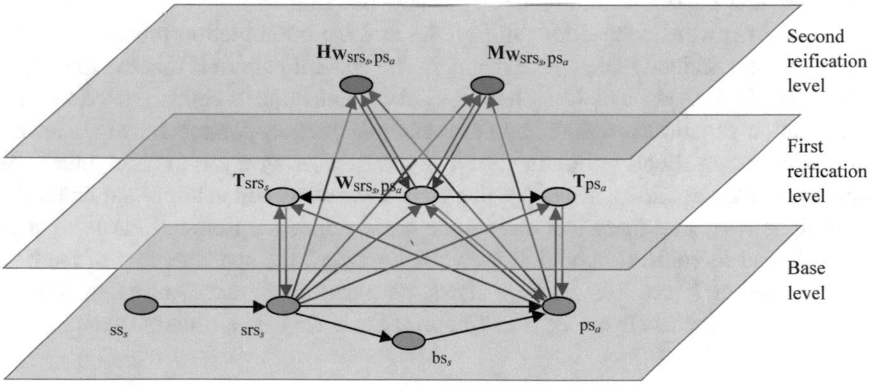

Fig. 4.3 Overview of the reified network architecture for plasticity and metaplasticity with base level (lower plane, pink), first reification level (middle plane, blue) and second reification level (upper plane, purple), and upward causal connections (blue) and downward causal connections (red) defining interlevel relations. The downward causal connections from the two **T**-states affect the excitability of the (presynaptic and postsynaptic) states srs_s and ps_a. The downward causal connections from the **H**-state and **M**-state affect the adaptation speed and the persistence factor of the connection weight reification state $\mathbf{W}_{\mathrm{srs}_s,\mathrm{ps}_a}$

So, in the base network for each state Y *combination function weights* γ are assumed: numbers $\gamma_{1,Y}$, $\gamma_{2,Y}$, ... ≥ 0 that may change over time such that the combination function $\mathbf{c}_Y(.)$ for Y is expressed by:

$$\mathbf{c}_Y(t, \pi_{1,1,Y}, \pi_{2,1,Y, \ldots}, \pi_{1,m,Y}, \pi_{2,m,Y}, V_1, \ldots, V_k) = \frac{\gamma_{1,Y}(t)\mathrm{bcf}_1(\pi_{1,1,Y}, \pi_{2,1,Y}, V_1, \ldots, V_k) + \cdots + \gamma_{m,Y}(t)\,\mathrm{bcf}_m(\pi_{1,m,Y}, \pi_{2,m,Y}, V_1, \ldots, V_k)}{\gamma_{1,Y}(t) + \cdots + \gamma_{m,Y}(t)} \tag{4.5}$$

The basic combination function weights $\gamma_{i,Y}(..)$ are represented by the reification states $\mathbf{C}_{i,Y}$ for Y. This describes that for Y a weighted average of basic combination functions is used. Note that, if exactly one of the $\mathbf{C}_{i,Y}(t)$ is nonzero, just one basic combination function is selected for $\mathbf{c}_Y(.)$. This approach makes it possible, for example, to gradually switch from one combination function $\mathrm{bcf}_i(..)$ to another one $\mathrm{bcf}_j(..)$ over time by decreasing the value of $\mathbf{C}_{i,Y}(t)$ and increasing the value of $\mathbf{C}_{j,Y}(t)$.

4.3.2 The Universal Combination Function and Universal Difference Equation for Reified Networks

In Chap. 3, Sect. 3.5 a *universal combination function* $\mathbf{c}^*_Y(..)$ has been found for any base state Y in the reified network. In cases of full reification, it has no parameters for network characteristics, only variables. Therefore it can be used in the same form for every base state as shown below; for a more detailed derivation, also see Chap. 10:

$$\begin{aligned} &\mathbf{c}^*_Y(H, C_1, \ldots, C_m, P_{1,1}, P_{2,1}, \ldots, P_{1,m}, P_{2,m}, W_1, \ldots, W_k, V_1, \ldots, V_k, V) \\ &= H\frac{C_1\,\mathrm{bcf}_1(P_{1,1,Y}, P_{2,1,Y}, W_1V_1, \ldots, W_kV_k) + \cdots + C_m\,\mathrm{bcf}_m(P_{1,m,Y}, P_{2,m,Y}, W_1V_1, \ldots, W_kV_k)}{C_1 + \cdots + C_m} + (1-H)V \\ &= H\left[\frac{C_1\,\mathrm{bcf}_1(P_{1,1,Y}, P_{2,1,Y}, W_1V_1, \ldots, W_kV_k) + \cdots + C_m\,\mathrm{bcf}_m(P_{1,m,Y}, P_{2,m,Y}, W_1V_1, \ldots, W_kV_k)}{C_1 + \cdots + C_m} - V\right] + V \end{aligned} \tag{4.6}$$

Here

- H refers to the speed factor reification $\mathbf{H}_Y(t)$
- C_i to the combination function weight reification $\mathbf{C}_{i,Y}(t)$
- $P_{i,j}$ to the parameter reification value $\mathbf{P}_{i,j,Y}(t)$ of parameter i = 1, 2 of basic combination function j = 1, ..., m
- W_i to the connection weight reification $\mathbf{W}_{X_i,Y}(t)$
- V_i to the state value $X_i(t)$ of base state X_i
- V to the state value $Y(t)$ of base state Y.

This combination function $\mathbf{c}^*_Y(..)$ in (4.4) makes that the dynamics of any base state Y within the reified network is described by the following *universal difference equation* in temporal-causal network format:

$$\begin{aligned} Y(t+\Delta t) = Y(t) + [\mathbf{c}^*_Y(&\mathbf{H}_Y(t), \mathbf{C}_{1,Y}(t), \ldots, \mathbf{C}_{m,Y}(t), \\ &\mathbf{P}_{1,1,Y}(t), \mathbf{P}_{2,1,Y}(t), \ldots, \mathbf{P}_{1,m,Y}(t), \\ &\mathbf{P}_{2,m,Y}(t), \mathbf{W}_{X_1,Y}(t), \ldots, \mathbf{W}_{X_k,Y}(t), \\ &X_1(t), \ldots, X_k(t), Y(t)) - Y(t)]\Delta t \end{aligned} \quad (4.7)$$

For more details, see Chap. 3, Sect. 3.5 and Chap. 10.

Structures added by the reification process are not reified themselves. However, the structure of the reified network can also be reified as another step: providing what is then called *second-order reification*. In the next section it is explored how such second-order reification can be done and how it can be used to model second-order adaptation for adaptive first-order adaptation principles.

4.4 Using Multilevel Network Reification for Higher-Order Adaptive Network Models

In this section, the multilevel reification architecture is introduced that allows modeling of networks with arbitrary orders of adaptation. In this architecture, the base network has its own internal dynamics, but it also evolves through one or more adaptation principles (called *first-order adaptation principles*). Moreover, these first-order adaptation principles themselves can change based on other adaptation principles (called *second-order adaptation principles*). So the architecture offers n reification levels for an arbitrary n where on reification level i adaptation principles are defined for ith-order adaptation. In this chapter, it is shown how the reified temporal-causal network modeling approach can be used to model important developments in empirical science, in particular concerning plasticity and metaplasticity. These are important notions in state of the art research on Cognitive Neuroscience, introduced not from a computational modeling perspective but by purely empirical researchers to clarify what was found empirically. This section shows how these notions can be connected to the reified temporal-network modeling approach described in the current chapter. This particular example shows the essential elements but is kept relatively simple; it can easily be extended by adding more states and connections.

4.4.1 Using Multilevel Network Reification for Plasticity and Metaplasticity from Cognitive Neuroscience

Mental networks equipped with a Hebbian learning mechanism (Hebb 1949) are able to adapt connection weights over time and learn or form memories in this way. Within Neuroscience this is usually called *plasticity*. In some circumstances it is better to learn (and change) fast, but in other circumstances, it is better to stay stable and persist what has been learnt in the past. To control this, by humans a type of (higher-order) adaptation called *metaplasticity* is used. It has become an important focus of study in Cognitive Neuroscience. In literature such as Abraham and Bear (1996), Chandra and Barkai (2018), Magerl et al. (2018), Parsons (2018), Robinson et al. (2016), Sehgal et al. (2013), Schmidt et al. (2013), Sjöström et al. (2008) various studies are reported which show how the adaptation of synapses (as described, for example, by Hebbian learning), is modulated by suppressing the adaptation process or amplifying it. Among the reported factors affecting synaptic plasticity are stimulus exposure, activation, previous experiences, and stress, which can accelerate or decelerate learning, or induce temporarily enhanced excitability of neurons which in turn positively affects learning; e.g., Chandra and Barkai (2018), Oh et al. (2003).

The reified network modeling approach was applied to a case involving both plasticity and metaplasticity, acquired from the literature mentioned above. A network picture of the designed reified network model for plasticity and metaplasticity is shown in Fig. 4.3. Table 4.2 displays the explanations of the states. Section 4.4.2 shows the complete specification. Here the plasticity of the response connection from srs_s to ps_a is considered, modeled by Hebbian learning. Note that the two **T**-states and the **M**-state are combination function parameter states here, respectively for excitability threshold τ of srs_s and ps_a and for persistence parameter $\boldsymbol{\mu}$ for the Hebbian learning of the connection from srs_s to ps_a. The alternative path via the belief state bs_s supports this learning by contributing to the activation of ps_a, thus relating to the original formulation in Hebb (1949):

Table 4.2 State names for the plasticity and metaplasticity model with their explanations

State nr	State name	Explanation	Level
X_1	ss_s	Sensor state for stimulus s	Base level
X_2	srs_s	Sensory representation state for stimulus s	
X_3	bs_s	Belief state for stimulus s	
X_4	ps_a	Preparation state for response a	
X_5	$\mathbf{W}_{\text{srs}_s,\text{ps}_a}$	Reified representation state for connection weight $\boldsymbol{\omega}_{\text{srs}_s,\text{ps}_a}$	First reification level
X_6	$\mathbf{T}_{\text{srs}_s}$	Reified representation state for threshold parameter $\boldsymbol{\tau}_{\text{srs}_s}$ of base state srs_s	
X_7	$\mathbf{T}_{\text{ps}_a}$	Reified representation state for threshold parameter $\boldsymbol{\tau}_{\text{ps}_a}$ of base state ps_a	
X_8	$\mathbf{HW}_{\text{srs}_s,\text{ps}_a}$	Reified representation state for speed factor $\boldsymbol{\eta}_{\mathbf{W}_{\text{srs}_s,\text{ps}_a}}$ for reified representation state $\mathbf{W}_{\text{srs}_s,\text{ps}_a}$	Second reification level
X_9	$\mathbf{MW}_{\text{srs}_s,\text{ps}_a}$	Reified representation state for persistence factor parameter $\boldsymbol{\mu}_{\mathbf{W}_{\text{srs}_s,\text{ps}_a}}$ for reified representation state $\mathbf{W}_{\text{srs}_s,\text{ps}_a}$	

> When an axon of cell A is near enough to excite B and repeatedly or persistently takes part in firing it, some growth process or metabolic change takes place in one or both cells such that A's efficiency, as one of the cells firing B, is increased. (Hebb 1949, p. 62)

In principle, this will start to work when the external stimulus s is sensed through sensor state ss_s. However, as discussed above, whether or not and to which extent learning actually takes place is controlled by a form of metaplasticity; this also relates to factors such as excitability characteristics of the involved states. To model metaplasticity, the model includes a second reification level with states $\mathbf{H}_{\mathbf{W}_{\mathrm{srs}_s,\mathrm{ps}_a}}$ representing the speed of the learning (learning rate) of $\boldsymbol{\omega}_{\mathrm{srs}_s,\mathrm{ps}_a}$, and $\mathbf{M}_{\mathbf{W}_{\mathrm{srs}_s,\mathrm{ps}_a}}$ representing the persistence $\boldsymbol{\mu}_{\mathbf{W}_{\mathrm{srs}_s,\mathrm{ps}_a}}$ of the connection weight $\boldsymbol{\omega}_{\mathrm{srs}_s,\mathrm{ps}_a}$. They have dynamic values depending on the other states. For example, if at some point in time the value of $\mathbf{H}_{\mathbf{W}_{\mathrm{srs}_s,\mathrm{ps}_a}}$ is 0, no learning will take place, and if $\mathbf{M}_{\mathbf{W}_{\mathrm{srs}_s,\mathrm{ps}_a}}$ has value 0, no learnt effects will persist; the value of these second-order reification states depend on activation of the presynaptic and postsynaptic states srs_s and ps_a, also see Robinson et al. (2016):

> Adaptation accelerates with increasing stimulus exposure. (Robinson et al. 2016, p. 2)

Note that a double level subscript notation for second-order reification states such as $\mathbf{H}_{\mathbf{W}_{\mathrm{srs}_s,\mathrm{ps}_a}}$ should be read as $\mathbf{H}_Y$ for a state Y at the first reification level, in this case, $Y = \mathbf{W}_{\mathrm{srs}_s,\mathrm{ps}_a}$. By substituting $\mathbf{W}_{\mathrm{srs}_s,\mathrm{ps}_a}$ for Y in $\mathbf{H}_Y$, this results in the double level subscript notation $\mathbf{H}_{\mathbf{W}_{\mathrm{srs}_s,\mathrm{ps}_a}}$; note that here for the sake of simplicity the subscripts in srs_s and ps_a are considered to be at the same subscript level as srs and ps. So, the subscript of $\mathbf{H}$ is $\mathbf{W}_{\mathrm{srs}_s,\mathrm{ps}_a}$ and this subscript itself has subscripts srs_s and ps_a; the notation should be interpreted as $\mathbf{H}_{(\mathbf{W}_{\mathrm{srs}_s,\mathrm{ps}_a})}$. In this way, the number of reification levels is reflected in the number of subscript levels. This applies to all states at the second reification level, so, for example, also to $\mathbf{M}_{\mathbf{W}_{\mathrm{srs}_s,\mathrm{ps}_a}}$. Up till now no cases of network adaptation of order higher than 2 have been addressed; however, see Chaps. 7 and 8 where more than 2 reification levels show up, and more subscript levels accordingly. From a modeling perspective there is nothing against adding a third reification level for the characteristics that define the second-order adaptation principles by the dynamics of the second-order reification states, for example, adding third-order reification states for their speed factors or their combination functions or the parameters of these functions.

To address dynamic levels of excitability of base states, first-order reification states $\mathbf{T}_{\mathrm{srs}_s}$ and $\mathbf{T}_{\mathrm{ps}_a}$ have been included that model the intrinsic excitability of the presynaptic and postsynaptic state srs_s and ps_a, respectively, by the value of the thresholds $\boldsymbol{\tau}_{\mathrm{srs}_s}$ and $\boldsymbol{\tau}_{\mathrm{ps}_a}$ of their logistic sum combination functions; also see Chandra and Barkai (2018):

> Learning-related cellular changes can be divided into two general groups: modifications that occur at synapses and modifications in the intrinsic properties of the neurons. While it is commonly agreed that changes in strength of connections between neurons in the relevant networks underlie memory storage, ample evidence suggests that modifications in intrinsic neuronal properties may also account for learning related behavioral changes. Long-lasting modifications in intrinsic excitability are manifested in changes in the neuron's response to a given extrinsic current (generated by synaptic activity or applied via the recording electrode). (Chandra and Barkai 2018, p. 30)

For most of the states, the combination function used below is the $\mathbf{alogistic}_{\sigma,\tau}$**(..)** function. The only exceptions are the sensor state ss_s which uses the Euclidean combination function $\mathbf{eucl}_{1,\lambda}$**(..)** and $\mathbf{W}_{\text{srs}_s,\text{ps}_a}$ which uses the Hebbian combination function $\mathbf{hebb}_{\mu_{\mathbf{W}_{\text{srs}_s,\text{ps}_a}}}$**(..)**.

4.4.2 *Role Matrices Covering Plasticity and Metaplasticity*

The multilevel reified network model described in Sect. 4.4.1 by a conceptual graphical representation, is described in the current section by a conceptual role matrices representation. The role matrix **mb** specifies for this network model on each row for a given state which states at the same or a lower level have outgoing connections to that state. This plays the role of *base connectivity*. This matrix contains the information depicted in Fig. 4.3 by upward (blue) or leveled (black) arrows, and includes for each state a numbering of the incoming base connections (the 1–4 in the top row), and for some of the states a connection from the state itself. The latter applies to all (first- and second-order) reification states, as can be seen in **mb**. For example, in the third row, it is indicated that state X_3 (=bs_s) only has one incoming base connection, from state X_2 (=srs_s). As another example, the fifth row indicates that state X_5 (=$\mathbf{W}_{\text{srs}_s,\text{ps}_a}$) has incoming base connections from X_2 (=srs_s), X_4 (=ps_a), X_5 (=$\mathbf{W}_{\text{srs}_s,\text{ps}_a}$) itself, and in that order. This order is important as the Hebbian combination function $\mathbf{hebb}_\mu$**(.)** used is not symmetric in its arguments. Note that the second column with more informative state names in each of the role matrices depicted in Box 4.1 is not part of the specification but has just been added for human understanding.

Box 4.1 Role matrices for the second-order reified network for plasticity and metaplasticity

mb	**base connectivity**	1	2	3	4
X_1	ss_s	X_1			
X_2	srs_s	X_1			
X_3	bs_s	X_2			
X_4	ps_a	X_2	X_3		
X_5	$\mathbf{W}_{srs_s,ps_a}$	X_2	X_4	X_5	
X_6	$\mathbf{T}_{srs_s}$	X_2	X_4	X_5	X_6
X_7	$\mathbf{T}_{ps_a}$	X_2	X_4	X_5	X_7
X_8	$\mathbf{HW}_{srs_s,ps_a}$	X_2	X_4	X_5	X_8
X_9	$\mathbf{MW}_{srs_s,ps_a}$	X_2	X_4	X_5	X_9

mcw	**connection weights**	1	2	3	4
X_1	ss_s	1			
X_2	srs_s	1			
X_3	bs_s	1			
X_4	ps_a	X_5	1		
X_5	$\mathbf{W}_{srs_s,ps_a}$	1	1	1	
X_6	$\mathbf{T}_{srs_s}$	-0.4	-0.4	1	1
X_7	$\mathbf{T}_{ps_a}$	-0.4	-0.4	1	1
X_8	$\mathbf{HW}_{srs_s,ps_a}$	1	1	-0.1	1
X_9	$\mathbf{MW}_{srs_s,ps_a}$	1	1	1	1

mcfw	**combination function weights**	1	2	3
		eucl	**alogistic**	**hebb**
X_1	ss_s	1		
X_2	srs_s		1	
X_3	bs_s		1	
X_4	ps_a		1	
X_5	$\mathbf{W}_{srs_s,ps_a}$			1
X_6	$\mathbf{T}_{srs_s}$		1	
X_7	$\mathbf{T}_{ps_a}$		1	
X_8	$\mathbf{HW}_{srs_s,ps_a}$		1	
X_9	$\mathbf{MW}_{srs_s,ps_a}$		1	

mcfp	**function**	1		2		3	
		eucl		**alogistic**		**hebb**	
		1	2	1	2	1	2
	parameter	n	λ	σ	τ	μ	
X_1	ss_s	1	1				
X_2	srs_s			5	X_6		
X_3	bs_s			5	0.2		
X_4	ps_a			5	X_7		
X_5	$\mathbf{W}_{srs_s,ps_a}$					X_9	
X_6	$\mathbf{T}_{srs_s}$			5	0.7		
X_7	$\mathbf{T}_{ps_a}$			5	0.7		
X_8	$\mathbf{HW}_{srs_s,ps_a}$			5	1		
X_9	$\mathbf{MW}_{srs_s,ps_a}$			5	1		

ms	**speed factors**	1
X_1	ss_s	0.5
X_2	srs_s	0.5
X_3	bs_s	0.2
X_4	ps_a	0.5
X_5	$\mathbf{W}_{srs_s,ps_a}$	X_8
X_6	$\mathbf{T}_{srs_s}$	0.3
X_7	$\mathbf{T}_{ps_a}$	0.3
X_8	$\mathbf{HW}_{srs_s,ps_a}$	0.5
X_9	$\mathbf{MW}_{srs_s,ps_a}$	0.1

In a similar way the four types of role matrices for *non-base roles* (showing either values or reification states to play that role; in the later case the downward arrows in Fig. 4.3 are defined here), were defined; see Box 4.1: role matrices **mcw** for connection weights, **ms** for speed factors, **mcfw** for combination function weights, and **mcfp** for combination function parameters. As before, within each role matrix, cell entries in red indicate a reference to the name of another state that as a form of reification represents in a dynamic manner an adaptive network characteristic, while entries indicating in green indicate fixed values for nonadaptive characteristics. The red cells represent the downward causal connections from the reification states in pictures as shown in Fig. 4.3, with their specific roles **W**, **H**, **C**, **P** indicated by the type of role matrix. The type of role matrix in which they are represented actually defines the roles of the reification states so that there is no need to computationally use information from the names **W**, **H**, **C**, **P** of them; they may have any own given names.

For example, in Box 4.1 the name X_5 in the red cell row-column (4, 1) in role matrix **mcw** indicates that the value of the connection weight from srs_s to ps_a is the value of state X_5. In contrast, the 1 in green cell (5, 1) of **mcw** indicates the static value of the connection weight from X_2 (=srs_s) to X_5 (=$\mathbf{W}_{srs_s,ps_a}$). Similarly, role matrix **ms** indicates (in red) that X_8 represents the adaptive speed factor of X_5, and (in green) that the speed factors of all other states have fixed values.

For a given application a limited fixed sequence of combination functions is specified by **mcf** = [1 2 3], where the numbers 1, 2, 3 refer to the numbering in the function library which currently contains 35 combination functions, the first three being $\mathbf{eucl}_{n,\lambda}(..)$, $\mathbf{alogistic}_{\sigma,\tau}(..)$, $\mathbf{hebb}_{\mu}(..)$. In Box 4.1 the role matrices **mcfw** and **mcfp** are shown for combination function weights and parameters, respectively. Here the matrix **mcfp** is a 3D matrix with first dimension for the states, second dimension for the two combination function parameters and third dimension for the combination functions.

4.5 Simulation for a Second-Order Reified Network Model for Plasticity and Metaplasticity

Following what is reported in the literature on metaplasticity, a number of simulation experiments have been performed. In particular, a scenario is shown here in which the focus was on the effect of activation of the postsynaptic state ps_a on plasticity; the effect of the presynaptic state srs_s on reification states was blocked (weights of upward links from srs_s were set 0). In Fig. 4.4 the simulation results are shown. For settings, see the specification in Sect. 4.4.2, Box 4.1. The upper graph shows the activation levels of the base states and how the weight of the connection from srs_s to ps_a is learnt. Here the activation levels and the exact shape of the learning curve also depend on controlling factors shown in the lower graph in Fig. 4.4. As can be seen there, following exposure to stimulus s, the threshold values $\mathbf{T}_{srs_s}$ and $\mathbf{T}_{ps_a}$ for the activation of srs_s and ps_a are decreasing to low levels. This substantially increases the excitability of srs_s and ps_a conform (Chandra and Barkai 2018) and therefore gives a boost to the activation levels of these base states, which in turn strengthens the Hebbian learning. Also, it is shown that following exposure to stimulus s the learning speed $\mathbf{H}_{\mathbf{W}_{srs_s,ps_a}}$ strongly increases, conform (Robinson et al. 2016). These controlling measures together result in a quite steep increase of the connection weight reification state. However, after the learnt level of the weight has become high, the thresholds increase again, and the learning speed decreases again. This makes the excitability of srs_s and ps_a lower and stops the boosts on learning; this has a positive effect on stabilising the situation, in accordance with what, e.g., in Sjöström et al. (2008) is called 'The Plasticity Versus Stability Conundrum' (p. 773).

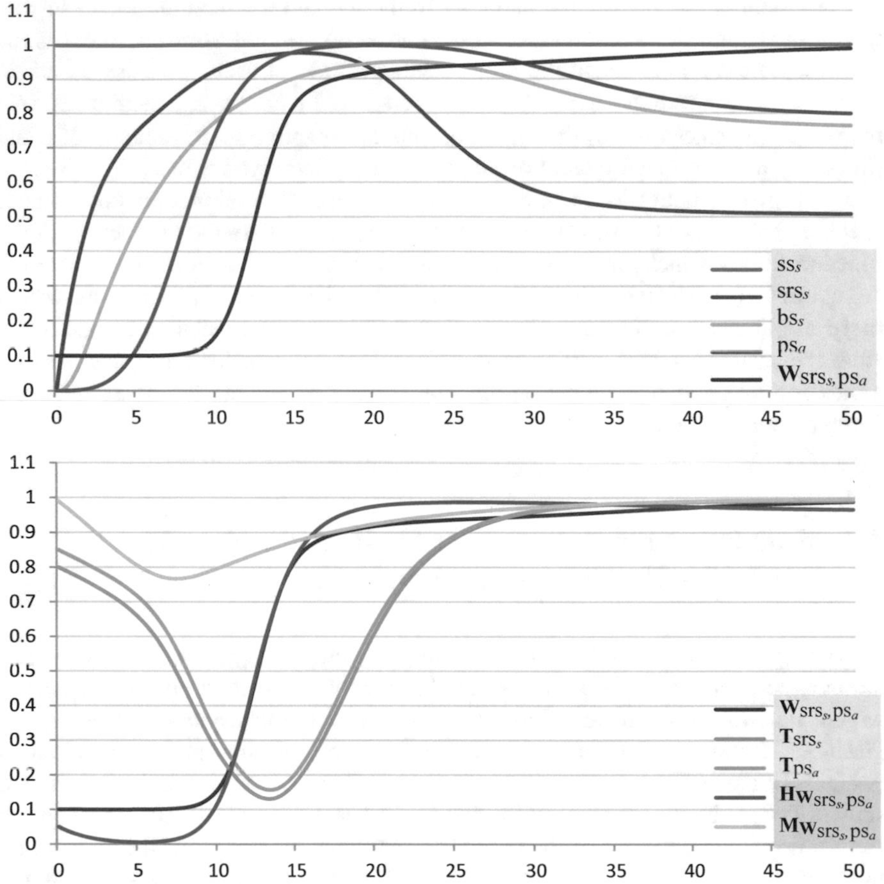

Fig. 4.4 Upper graph: dynamics of base states and the adaptive connection weight represented by $\mathbf{W}_{\text{srs}_s,\text{ps}_a}$. Lower graph: dynamics of the reification states including the first-order reification state $\mathbf{W}_{\text{srs}_s,\text{ps}_a}$ for the adaptive connection weight, and $\mathbf{T}_{\text{srs}_s}$ and $\mathbf{T}_{\text{ps}_a}$ for the activation threshold for the presynaptic and postsynaptic states srs$_s$ and ps$_a$, and the second-order reification states $\mathbf{H}_{\mathbf{W}_{\text{srs}_s,\text{ps}_a}}$ and $\mathbf{M}_{\mathbf{W}_{\text{srs}_s,\text{ps}_a}}$ for the adaptation speed and persistence factor of the connection weight reification state $\mathbf{W}_{\text{srs}_s,\text{ps}_a}$

4.6 On the Added Complexity for Higher-Order Network Reification

Note that, as for any dynamical system, by adding adaptivity to a network always complexity is added. In this section, it is discussed how complexity of a network increases when reification is applied. The added complexity for first-order network reification was addressed in Chap. 3, Sect. 3.9. The outcome will be briefly summarized and next, the step to higher-order reification is made. To start with the outcome, network reification will increase complexity, but this will at most be quadratic in the number of nodes N and linear in the number of connections M of the original network. More specifically, if m the number of basic combination functions considered, then the number of nodes in the reified network is at most $(2 + m + N)N$. If not all connections are used but only a number M of them, the outcome is $(2 + m)N + M$. This is linear in the number of nodes and connections. The number of connections in the reified network is $(m + 1)N + 2M$. Again this is linear in the number of nodes and connections.

If this analysis is applied in an iterative manner for second-order network reification, then the increase in complexity is still polynomial: at most in the fourth power of the number of nodes:

$$\left(N^2\right)^2 = N^4 \tag{4.8}$$

Can this iteration still be continued further, thus obtaining nth-order reification for any n? Yes, theoretically there is no end in this. But also practically, for example, in the case used as illustration in the current chapter, the parameter $\mathbf{v}_{\mathrm{T}_{\Omega X_i,X_j}}$ for the norm of the average connection weight for the tipping point adaptation used as characteristic at the second reification level still could be made adaptive (e.g., related to how busy someone is) and reified at a third reification level. For third-order reification, the increase in complexity is still polynomial: at most in the order of

$$\left(\left(N^2\right)^2\right)^2 = N^8 \tag{4.9}$$

If n reification levels are added, then it is in the order of

$$N^{(2^n)} \tag{4.10}$$

which is still polynomial in N, but double exponential in n. The latter may suggest limiting the number of reification levels in practical applications to just a few, or, alternatively, in each reification step add only a few new reification states: for each step reification can be done in a partial manner as well. For example, if only speed factors are reified, the number of states will only increase in a linear way: one extra state for each existing state. Recall the double negative exponential pattern of hits in the order of

$$\mathrm{e}^{35.19\,\mathrm{e}^{-0.8684n}} \tag{4.11}$$

discussed in Chap. 1, Sect. 1.3. In the current literature, an adaptation of order higher than 2 is extremely rare and of order higher than 3 practically absent. This supports the idea that for now adaptation of order >3 is not considered interesting enough to be addressed. As shown above, for adaptation of order 3 the added complexity is in the order of an 8th degree polynomial in n, and for order 2 a 4th degree polynomial. In this context, note Chap. 10, Sect. 10.7 pointing out how efficient simulation of large-scale reified networks of thousands or even millions of states can be achieved by applying a form of compilation.

4.7 Discussion

The multilevel network reification architecture described here has advantages similar to those found for reification in modeling and programming languages in other areas of AI and Computer Science; e.g., Bowen and Kowalski (1982), Demers and Malenfant (1995), Galton (2006), Hofstadter (1979), Sterling and Shapiro (1996), Sterling and Beer (1989), Weyhrauch (1980). Some parts of this chapter were adopted from Treur (2018). A reified network enables to model dynamics of the original network by dynamics within the reified network, thus representing an adaptive network by a non-adaptive network. Network reification provides a unified manner of modelling adaptation principles, and allows comparison of such principles across different domains, as has been illustrated in Chap. 3. In the current chapter it was shown how a multilevel reified network architecture enables a structured and transparent manner to model network adaptation of any order, illustrated for second-order adaptive networks.

In this chapter, the introduced modeling environment for reified temporal-causal networks was applied to model a second-order adaptive Mental Network showing plasticity and metaplasticity as known from the empirical neuroscientific literature. Although some specific computational models for metaplasticity have been put forward with interesting perspectives for artificial neural networks, for example in Marcano-Cedeno et al. (2011), Andina et al. (2007, 2009), Fombellida et al. (2017), the modeling environment proposed here provides a more general architecture. Applications may extend well beyond the neuro-inspired area (as will be shown in Chap. 6 for a second-order adaptive Social Network).

The causal modeling area has a long history in AI; e.g., Kuipers and Kassirer (1983), Kuipers (1984). The current chapter can be considered a new branch in this causal modeling area. It adds dynamics to causal models, making them temporal, but the main contribution in the current chapter is that it adds a way to specify (multi-order) adaptivity in causal models, thereby conceptually using ideas on meta-level architectures that also have a long history in AI; e.g., Weyhrauch (1980), Bowen and Kowalski (1982), Sterling and Beer (1989). So the modeling approach connects two different areas with a long tradition in AI, thereby strongly extending the applicability of causal modeling to dynamic and adaptive notions such as plasticity and metaplasticity of any order, which otherwise are out of reach of causal modeling.

In the modeling approach, combination functions play a crucial role. They are declarative mathematical functions relating real numbers to real numbers. The functionality of an overall reified network is determined mainly by the choice of these functions and their use within a reified network architecture. In this sense, they are the powerful building blocks that enable to model in an easy manner dynamic processes which are adaptive of any order.

This construction can be continued to obtain a network architecture which is adaptive up to any order *n*. In Chap. 1, Sect. 1.3 it was discussed in how far adaptation principles of order 3 or higher are considered to be useful in the current literature, and a double negative exponential pattern was found for the number of hits in Google Scholar against the order of adaptation. However, in Chap. 7 an example network for evolutionary processes will be described of order higher than 2, and in Chap. 8 one or two inspired by ideas from Hofstadter (1979, 2007).

In an *n*th-order reified network there still will be network structures introduced in the last step from $n - 1$ to n that have no reification within the *n*th-order reified network. From a theoretical perspective, the construction can be repeated (countable) infinitely many times, for all natural numbers *n*; then ω-order reification is obtained, where ω is the ordinal for the natural numbers. This is theoretically well-defined as a mathematical structure. All network structures in this ω-order reified network are reified within the network itself, so it is closed under reification. Whether or not such an ω-order construction has a useful application in practice, or can be used to explore theoretical research questions is still an open question, another subject for future research.

References

Abraham, W.C., Bear, M.F.: Metaplasticity: the plasticity of synaptic plasticity. Trends Neurosci. **19**(4), 126–130 (1996)

Andina, D., Jevtic, A., Marcano, A., Adame, J.M.B.: Error weighting in artificial neural networks learning interpreted as a metaplasticity model. In: Proceedings of IWINAC'07, Part I. Lecture Notes in Computer Springer Science, pp. 244–252 (2007)

Andina, D., Alvarez-Vellisco, A., Jevtic, A., Fombellida, J.: Artificial metaplasticity can improve artificial neural network learning. Intell. Autom. Soft Comput. **15**(4), 681–694 (2009)

Arnold, S., Suzuki, R., Arita, T.: Selection for representation in higher-order adaptation. Mind. Mach. **25**(1), 73–95 (2015)

Bowen, K.A., Kowalski, R.: Amalgamating language and meta-language in logic programming. In: Clark, K., Tarnlund, S. (eds.) Logic Programming. Academic Press, New York, pp. 153–172 (1982)

Byrne, D.: The attraction hypothesis: do similar attitudes affect anything? J. Pers. Soc. Psychol. **51** (6), 1167–1170 (1986)

Chandra, N., Barkai, E.: A non-synaptic mechanism of complex learning: modulation of intrinsic neuronal excitability. Neurobiol. Learn. Mem. **154**, 30–36 (2018)

Daimon, K., Arnold, S., Suzuki, R., Arita, T.: The emergence of executive functions by the evolution of second-order learning. Artif. Life Rob. **22**, 483–489 (2017)

Demers, F.N., Malenfant, J.: Reflection in logic, functional and object-oriented programming: a short comparative study. In IJCAI'95 Workshop on Reflection and Meta-Level Architecture and their Application in AI, pp. 29–38 (1995)

Fombellida, J., Ropero-Pelaez, F.J., Andina, D.: Koniocortex-like network unsupervised learning surpasses supervised results on WBCD breast cancer database. In: Proceedings of IWINAC'17, Part II, LNCS, vol. 10338, pp. 32–41. Springer Publishers (2017)

Galton, A.: Operators vs. Arguments: The Ins and Outs of Reification. Synthese **150**, 415–441 (2006)
Hebb, D.O.: The Organization of Behavior: A Neuropsychological Theory (1949)
Helbing, D., Brockmann, D., Chadefaux, T., Donnay, K., Blanke, U., Woolley-Meza, O., Moussaid, M., Johansson, A., Krause, J., Schutte, S., Perc, M.: Saving human lives: what complexity science and information systems can contribute. J. Stat. Phys. **158**, 735–781 (2015)
Hofstadter, D.R.: Gödel, Escher, Bach. Basic Books, New York (1979)
Hofstadter, D.R.: I Am a Strange Loop. Basic Books, New York (2007)
Kuipers, B.J.: Commonsense reasoning about causality: deriving behavior from structure. Artif. Intell. **24**, 169–203 (1984)
Kuipers, B.J., Kassirer, J.P.: How to discover a knowledge representation for causal reasoning by studying an expert physician. In: Proceedings of the Eighth International Joint Conference on Artificial Intelligence, IJCAI'83. William Kaufman, Los Altos, CA (1983)
Magerl, W., Hansen, N., Treede, R.D., Klein, T.: The human pain system exhibits higher-order plasticity (metaplasticity). Neurobiol. Learn. Mem. **154**, 112–120 (2018)
Marcano-Cedeno, A., Marin-De-La-Barcena, A., Jimenez-Trillo, J., Pinuela, J.A., Andina, D.: Artificial metaplasticity neural network applied to credit scoring. Int. J. Neural Syst. **21**(4), 311–317 (2011)
McPherson, M., Smith-Lovin, L., Cook, J.M.: Birds of a feather: homophily in social networks. Annu. Rev. Sociol. **27**, 415–444 (2001)
Oh, M.M., Kuo, A.G., Wu, W.W., Sametsky, E.A., Disterhoft, J.F.: Watermaze learning enhances excitability of CA1 pyramidal neurons. J. Neurophysiol. **90**(4), 2171–2179 (2003)
Parsons, R.G.: Behavioral and neural mechanisms by which prior experience impacts subsequent learning. Neurobiol. Learn. Mem. **154**, 22–29 (2018)
Pearson, M., Steglich, C., Snijders, T.: Homophily and assimilation among sport-active adolescent substance users. Connections **27**(1), 47–63 (2006)
Perc, M., Szolnoki, A.: Coevolutionary games—a mini review. BioSystems **99**, 109–125 (2010)
Robinson, B.L., Harper, N.S., McAlpine, D.: Meta-adaptation in the auditory midbrain under cortical influence. Nat. Commun. **7**, 13442 (2016)
Sehgal, M., Song, C., Ehlers, V.L., Moyer Jr., J.R.: Learning to learn—intrinsic plasticity as a metaplasticity mechanism for memory formation. Neurobiol. Learn. Mem. **105**, 186–199 (2013)
Schmidt, M.V., Abraham, W.C., Maroun, M., Stork, O., Richter-Levin, G.: Stress-induced metaplasticity: from synapses to behavior. Neuroscience **250**, 112–120 (2013)
Sharpanskykh, A., Treur, J.: Modelling and analysis of social contagion in dynamic networks. Neurocomputing **146**, 140–150 (2014)
Sjöström, P.J., Rancz, E.A., Roth, A., Hausser, M.: Dendritic excitability and synaptic plasticity. Physiol. Rev. **88**(769–840), 2008 (2008)
Sterling, L., Beer, R.: Metainterpreters for expert system construction. J. Logic Program. **6**, 163–178 (1989)
Sterling, L., Shapiro, E.: The Art of Prolog, Chap. 17, pp. 319–356. MIT Press (1996)
Treur, J.: Network-Oriented Modeling: Addressing Complexity of Cognitive, Affective and Social Interactions. Springer Publishers (2016)
Treur, J.: Multilevel network reification: representing higher-order adaptivity in a network. In: Proceedings of the 7th International Conference on Complex Networks and their Applications, Complex Networks'18, vol. 1. Studies in Computational Intelligence, vol. 812, 635–651. Springer (2018)
Treur, J.: The ins and outs of network-oriented modeling: from biological networks and mental networks to social networks and beyond. In: Transactions on Computational Collective Intelligence. Contents of Keynote Lecture at ICCCI'18, vol. 32, pp. 120–139. Springer Publishers (2019a)
Treur, J.: Design of a software architecture for multilevel reified temporal-causal networks (2019b). https://doi.org/10.13140/rg.2.2.23492.07045. https://www.researchgate.net/publication/333662169

Weyhrauch, R.W.: Prolegomena to a theory of mechanized formal reasoning. Artif. Intell. **13**, 133–170 (1980)

Zelcer, I., Cohen, H., Richter-Levin, G., Lebiosn, T., Grossberger, T., Barkai, E.: A cellular correlate of learning-induced metaplasticity in the hippocampus. Cereb. Cortex **16**, 460–468 (2006)

Part III
Applications of Higher-Order Adaptive Network Models

Chapter 5
A Reified Network Model for Adaptive Decision Making Based on the Disconnect-Reconnect Adaptation Principle

Abstract In recent literature from Neuroscience, the adaptive role of the effects of stress on decision making is highlighted. In this chapter, it is addressed how that role can be modelled computationally using a reified adaptive temporal-causal network architecture. The presented network model addresses the so-called disconnect-reconnect adaptation principle. In the first phase of the acute stress suppression of the existing network connections takes place (disconnect), and then in a second phase after some time there is a relaxation of the suppression. This gives room to quickly get rid of old habits that are not applicable anymore in the new stressful situation and start new learning (reconnect) of better decision making, more adapted to this new stress-triggering context.

Keywords Network reification · Adaptive temporal-causal network model · Hebbian learning · Stress · Decision making

5.1 Introduction

Stress has a strong impact on both cognitive and affective processes. This impact can be experienced as disturbing, but recent findings suggest that its main goal is to improve coping with challenging situations. Stress has bad health effects, as it may cause disorders like depression, anxiety or schizophrenia. But from the positive side, it supports individuals to respond to specific types of threats more adequately, keeping an individual's homeostasis up to date and ready for future threats of similar types. In the very first moment of facing the acute stress, an emotional response is triggered which elevates surveillance, perception, and attention on threat-related stimuli (Quaedflieg et al. 2015). Humans initially experience that they do not have the power to control the occurring stress. They are not able to change the situation and make it better (Glass et al. 1971). But it turns out that stress also has a positive effect on learning new decision making behaviour that is better adapted to the new situations encountered. To get rid of old decision making, as a first step existing connections are suppressed as a kind of reset by which more room

J. Treur, *Network-Oriented Modeling for Adaptive Networks: Designing Higher-Order Adaptive Biological, Mental and Social Network Models*, Studies in Systems, Decision and Control 251, https://doi.org/10.1007/978-3-030-31445-3_5

is created for learning new connections (Sousa and Almeida 2012); this adaptation principle is called the *disconnect-reconnect principle*. The current chapter addresses the question of how this can be modelled by a reified adaptive network model. Moreover, it is also explored how this principle works in conjunction with second-order adaptation as described by metaplasticity.

The chapter is organized as follows. In Sect. 5.2 the neurological principles of the suppressing and adaptive effects of stress and the parts of the brain which deal with stress in that way are addressed. In Sect. 5.3 the reified adaptive temporal-causal network model is introduced, and illustrated by simulation of two example scenarios in Sect. 5.4: one for the disconnect-reconnect principle without using metaplasticity and one in conjunction with metaplasticity. Section 5.5 addresses the verification of the reified network model by mathematical analysis.

5.2 Neurological Principles

Acute stress is considered to involve interaction with the amygdala (Quaedflieg et al. 2015). The more activity in the amygdala, the more a human becomes sensitive and respondent to the threat (Radley and Morrison 2005). Stress reactions on stressors often deteriorate homeostasis in organisms (de Kloet et al. 2005). But stressors also stimulate a constructive reaction which makes physiological and psychological alterations in the body that are advantageous for the organism. Recovery from a stressor is accompanied by a decreasing negative coupling between the amygdala and the frontal Anterior Cingulate Cortex (ACC) and pre-Supplementary Motor Area (preSMA) (Hermans et al. 2014). This has been found, for example, in stress-related psychiatric disorders (Etkin et al. 2010; Johnstone et al. 2007). It has been found that the left Prefrontal Cortex (PFC) is relevant to stress adaptation, and individuals with stronger Hypothalamic Pituitary Adrenal (HPA) axis reactivity show reduced amygdala-left dlPFC functional connectivity (Quaedflieg et al. 2015). In a safe environment, it is advantageous that the cortex can suppress the stress response but in a harmful environment, this may cause a false idea of security (Reser 2016). The reaction of the amygdala to new stressors makes animals behave the same as before the new stress arrived.

In Barsegyan et al. (2010), it is claimed that the executive control network is suppressed in the very starting period of the inducing of stress (by dlPFC). Stress handling is viewed as adaptive (e.g., Sousa and Almeida 2012), due to the fact that when the decision making is improved, humans can learn how to handle the situations where the stress comes from. In the case of acute stress, at the very first period of stress-induction, the salience network starts working and executive control is suppressed. After a while, executive control starts performing functionality and suppress the salience network. Due to this, plasticity, for example, based on Hebbian learning will work more efficiently; this is called the *disconnect-reconnect adaptation principle* (Sousa and Almeida 2012).

5.3 The Reified Adaptive Network Model

To simulate how stress adaptively affects decision making, the following scenario is addressed here. Person A is working (performing action a_1) in a convenient condition with her colleague B until B's context causes extreme stress for person A. This condition disturbs her normal functioning. A's brain has a neurocognitive mechanism to overcome this by learning new decision making to cope with that situation. By this mechanism, first as a form of resetting her existing connections are suppressed to create more room for new learning of connections, better adapted to the new conditions. Next, after some time, A's suppression is ending, and new Hebbian learning to cope with the new situation takes place. Finally, after this learning how to cope with the situation has led to improved decision making, person A decides for a different option (action a_2) in which B and his context do not play a dominant role anymore.

The Network-Oriented Modeling approach based on multilevel reified temporal-causal network models presented in Chap. 4 is used to model this process; see also Treur (2018a, b). Recall that a temporal-causal network model in the first place involves representing in a declarative manner states and connections between them that represent (causal) impacts of states on each other, as assumed to hold for the application domain addressed. The states are assumed to have (activation) levels in the interval [0, 1] that vary over time. These three notions form the defining part of a conceptual representation of a temporal-causal network model structure:

- **Connectivity**
  - Each incoming connection of a state *Y*, from a state *X* has a *connection weight value* $\boldsymbol{\omega}_{X,Y}$ representing the strength of the connection.
- **Aggregation**
  - For each state a *combination function* $\mathbf{c}_Y(..)$ is chosen to combine the causal impacts state *Y* receives from other states.
- **Timing**
  - For each state *Y* a *speed factor* $\boldsymbol{\eta}_Y$ is used to represent how fast state *Y* is changing upon causal impact.

In Figs. 5.1 and 5.2. the conceptual representation of the reified temporal-causal network model is depicted. A brief explanation of the states used is shown in Table 5.1.

5.3.1 The Base Network

The base states are as follows (see also Fig. 5.1). The state srs_s stands for the sensory representation of stimulus *s* from the world. The state srs_c is the sensory

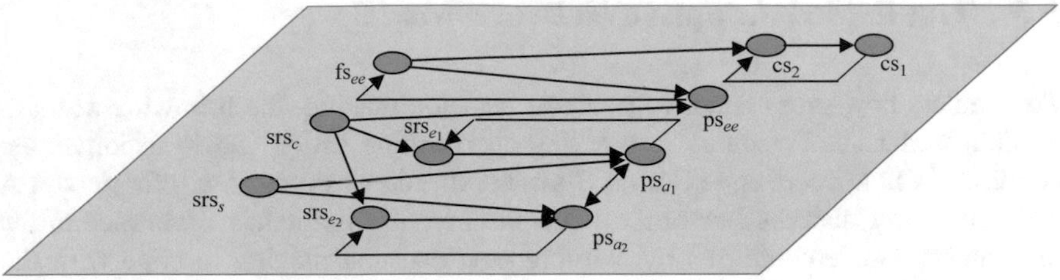

Fig. 5.1 The base network model

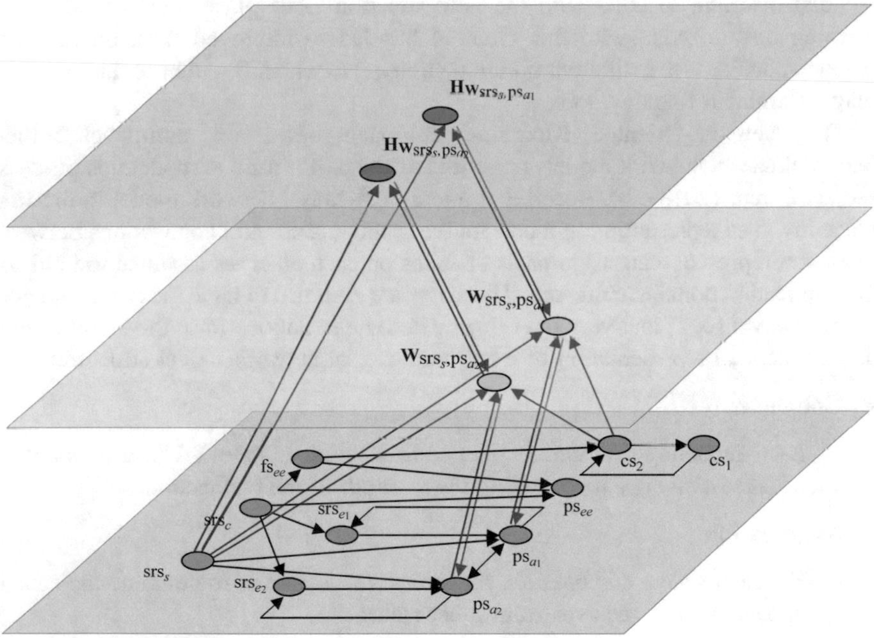

Fig. 5.2 The reified network model with plasticity and metaplasticity

representation of the stressful context c. The sensory representation srs_s of the stimulus is a trigger for the preparation state of one or both of two actions a_1 or a_2. The sensory representation of the predicted feeling effects of the preparation states ps_{a_1} and ps_{a_2} of the actions are represented by srs_{e_1} and srs_{e_2}, respectively.

Furthermore, ps_{ee} is the preparation state of a stressful emotional response on the sensory representation srs_c of the disturbing context c, and fs_{ee} denotes the feeling state associated to this extreme emotion. Finally, cs_2 stands for a control state for suppression of connections (here from srs_s to ps_{a_1} and to ps_{a_2}) and cs_1 for a control state to limit this suppression in time.

Table 5.1 Explanation of the states in the model

State nr	State name	Explanation	Level
X_1	srs_s	Sensory representation of stimulus s	Base level
X_2	srs_c	Sensory representation of context c	
X_3	srs_{e_1}	Sensory representation of action effect e_1	
X_4	srs_{e_2}	Sensory representation of action effect e_2	
X_5	fs_{ee}	Feeling state for extreme emotion ee	
X_6	ps_{a_1}	Preparation state for action a_1	
X_7	ps_{a_2}	Preparation state for action a_2	
X_8	ps_{ee}	Preparation state for response of extreme emotion ee	
X_9	cs_1	Control state for timing of suppression of connections	
X_{10}	cs_2	Control state for suppression of connections	
X_{11}	$\mathbf{W}_{srs_s,ps_{a_1}}$	Reification state for the weight of the connection from srs_s to ps_{a_1}	First reification level
X_{12}	$\mathbf{W}_{srs_s,ps_{a_2}}$	Reification state for the weight of the connection from srs_s to ps_{a_2}	
X_{13}	$\mathbf{H}_{\mathbf{W}_{srs_s,ps_{a_1}}}$	Reification state for speed factor $\boldsymbol{\eta}$ of reification state $\mathbf{W}_{srs_s,ps_{a_1}}$	Second reification level
X_{14}	$\mathbf{H}_{\mathbf{W}_{srs_s,ps_{a_2}}}$	Reification state for speed factor $\boldsymbol{\eta}$ of reification state $\mathbf{W}_{srs_s,ps_{a_2}}$	

The connections for the base network depicted in Fig. 5.1 are as follows. Preparation state ps_{a_1} has three incoming connections from srs_s, srs_{e_1}, ps_{a_2}. The first connection is for triggering preparation for action a_1, based on stimulus s. The second connection from srs_{e_1} amplifies the preparation due to a positive feeling for the predicted effect of the action. The third connection from ps_{a_2} is a negative connection to achieve that in general only one of the actions is chosen. Whether or not this action a_1 is actually chosen depends on these three connections and the activation of the three connected states. From these three connections, the connection from srs_s to ps_{a_1} is considered to be adaptive. Similarly, the other action option a_2 is handled (but with different values for the three connections). Note that by the connections from srs_c to srs_{e_1} and srs_{e_2} the stressful context c affects the predicted effects of the two actions; also these connections will have different weights as they represent how the actions differ in their suitability for this context.

The extreme stressful emotion is modeled by the as-if body loop between ps_{ee} and fs_{ee}, whereby ps_{ee} is triggered by the context representation srs_c. An effect of these stress states goes via the connection from fs_{ee} to control state cs_2. This control state is limited in its value over time by another control state cs_1 via the two mutual links; to achieve that, the connection from cs_1 to cs_2 is negative, and activation of cs_1 is slower than of cs_2.

5.3.2 *Modeling First- and Second-Order Adaptation of the Connection Weights by Reification States*

Control state cs_2 has an adaptive effect on the two network connections from srs_s in the base network. This adaptive effect combines with the Hebbian learning effect for the same adaptive connections. This is where the first reification level (see Fig. 5.2) comes in with reification states $\mathbf{W}_{srs_s,ps_{a_1}}$ and $\mathbf{W}_{srs_s,ps_{a_2}}$ for the weights of these two adaptive connections.

The blue upward links from cs_2 model the effect of cs_2 and the blue upward links from srs_s, ps_{a_1} and ps_{a_2} model the Hebbian learning on the same connections. The pink downward links from the reification states model the effect of the adaptive connection weights on ps_{a_1} and ps_{a_2}.

Yet another adaptive effect is modeled by the second reification level in Fig. 5.2. This is a form of metaplasticity that increases the adaptation speed of the two adaptive connection weights, which is triggered by experiencing srs_s; also see: 'Adaptation accelerates with increasing stimulus exposure' (Robinson et al. 2016, p. 2). Second-order reification states $\mathbf{H}_{\mathbf{W}_{srs_s,ps_{a_1}}}$ and $\mathbf{H}_{\mathbf{W}_{srs_s,ps_{a_2}}}$ achieve this second-order effect, using the upward blue links from srs_s and the downward pink links to $\mathbf{W}_{srs_s,ps_{a_1}}$ and $\mathbf{W}_{srs_s,ps_{a_2}}$.

5.4 Combination Functions and Role Matrices for the Reified Network Model

In this section, first, the combination functions used are discussed. Next, the role matrices are presented.

5.4.1 *The Combination Functions Used*

For base states the following combination functions $\mathbf{c}_Y(\ldots)$ were used:

$$\mathbf{ssum}_{\lambda}(V_1, \ldots, V_k) = \frac{V_1 + \cdots + V_k}{\lambda} \tag{5.1}$$

$$\mathbf{eucl}_{n,\lambda}(V_1, \ldots, V_k) = \sqrt[n]{\frac{V_1^n + \cdots + V_k^n}{\lambda}} \tag{5.2}$$

$$\mathbf{alogistic}_{\sigma,\tau}(V_1, \ldots V_k) = \left[\frac{1}{1 + \mathrm{e}^{-\sigma(V_1 + \cdots + V_k - \tau)}} - \frac{1}{1 + \mathrm{e}^{\sigma\tau})}\right](1 + \mathrm{e}^{-\sigma\tau}) \tag{5.3}$$

Similarly, for the reification states of the connection weights, the following two combination functions were used. *Hebbian learning* of a connection from state X_i to state X_j with connection weight reification state **W** makes use of:

$$\mathbf{hebb}_{\boldsymbol{\mu}}(V_1, V_2, W) = V_1 V_2 (1-W) + \boldsymbol{\mu}\, W \tag{5.4}$$

where $\boldsymbol{\mu}$ is the persistence factor with value 1 as full persistence and 0 as none. For *state-connection modulation* with control state cs_2 for connection weight reification state **W**:

$$\mathbf{scm}_{\boldsymbol{\alpha}}(V_1, V_2, W, V) = W + \boldsymbol{\alpha}\, V\, W(1-W) \tag{5.5}$$

where $\boldsymbol{\alpha}$ (or $\boldsymbol{\alpha}_{cs_2,\mathbf{W}}$) is the modulation parameter for **W** from cs_2, and V is the single impact from cs_2. Note that the first two variables of $\mathbf{scm}_{\boldsymbol{\alpha}}(V_1, V_2, W, V)$ are auxiliary variables that are not used in Formula (5) for $\mathbf{scm}_{\boldsymbol{\alpha}}(V_1, V_2, W, V)$. These auxiliary variables are included to be able to combine this function with the Hebbian learning function while using the same sequence of variables (see 6 below). More specifically, this combination is done as follows. These two adaptive combination functions are used as a weighted average with $\boldsymbol{\gamma_1}$ and $\boldsymbol{\gamma_2}$ as combination function weights (with sum 1) for $\mathbf{hebb}_{\boldsymbol{\mu}}(V_1, V_2, W)$ and $\mathbf{scm}_{\boldsymbol{\alpha}}(V_1, V_2, W, V)$, respectively, based on (4) and (5) as follows:

$$\mathbf{c}_{\mathbf{W}}(V_1, V_2, W, V) = \boldsymbol{\gamma_1}\mathbf{hebb}_{\boldsymbol{\mu}}(V_1, V_2, W) + \boldsymbol{\gamma_2}\mathbf{scm}_{\boldsymbol{\alpha}}(V_1, V_2, W, V) \tag{5.6}$$

5.4.2 The Role Matrices

Based on the graphical representations from Fig. 5.2 and the specific values for the intended scenario the role matrices shown in Boxes 5.1 and 5.2 have been specified. Note that the reification states $\mathbf{W}_{\text{srs}_s,\text{ps}_{a_1}}$ and $\mathbf{W}_{\text{srs}_s,\text{ps}_{a_2}}$ for the connection weights also have a connection to themselves, as can be seen in role matrix **mb**. This is because Hebbian learning needs that; these connections are not shown in Fig. 5.2.

Box 5.1 Role matrices **mb** for base connections, **mcw** for connection weights, **mcfw** for combination function weights, and **mcfp** for combination function parameters.

mb	base connectivity	1	2	3	4
X_1	srs_s	X_1			
X_2	srs_c	X_2			
X_3	srs_{e1}	X_2	X_6		
X_4	srs_{e2}	X_2	X_7		
X_5	fs_{ee}	X_8			
X_6	ps_{a1}	X_1	X_3	X_7	
X_7	ps_{a2}	X_1	X_4	X_6	
X_8	ps_{ee}	X_5	X_2		
X_9	cs_1	X_{10}			
X_{10}	cs_2	X_5	X_9		
X_{11}	$\mathbf{W}_{srs_s,ps_{u_1}}$	X_1	X_6	X_{11}	X_{10}
X_{12}	$\mathbf{W}_{srs_s,ps_{a_2}}$	X_1	X_7	X_{12}	X_{10}
X_{13}	$\mathbf{Hw}_{srs_s,ps_{a_1}}$	X_1	X_{11}		
X_{14}	$\mathbf{Hw}_{srs_s,ps_{a_2}}$	X_1	X_{12}		

mcw	connection weights	1	2	3	4
X_1	srs_s	1			
X_2	srs_c	1			
X_3	srs_{e1}	-0.1	0.7		
X_4	srs_{e2}	0.3	0.7		
X_5	fs_{ee}	1			
X_6	ps_{a1}	X_{11}	0.7	-0.2	
X_7	ps_{a2}	X_{12}	0.7	-0.2	
X_8	ps_{ee}	1	1		
X_9	cs_1	1			
X_{10}	cs_2	1	-0.9		
X_{11}	$\mathbf{W}_{srs_s,ps_{a_1}}$	1	1	1	-0.7
X_{12}	$\mathbf{W}_{srs_s,ps_{a_2}}$	1	1	1	-0.7
X_{13}	$\mathbf{Hw}_{srs_s,ps_{a_1}}$	1	-0.4		
X_{14}	$\mathbf{Hw}_{srs_s,ps_{a_2}}$	1	-0.4		

mcfw	combination function weights	1 eucl	2 alo-gistic	3 hebb	4 scm
X_1	srs_s	1			
X_2	srs_c		1		
X_3	srs_{e1}	1			
X_4	srs_{e2}	1			
X_5	fs_{ee}	1			
X_6	ps_{a1}	1			
X_7	ps_{a2}	1			
X_8	ps_{ee}	1			
X_9	cs_1	1			
X_{10}	cs_2	1			
X_{11}	$\mathbf{W}_{srs_s,ps_{a_1}}$			0.85	0.15
X_{12}	$\mathbf{W}_{srs_s,ps_{a_2}}$			0.85	0.15
X_{13}	$\mathbf{Hw}_{srs_s,ps_{a_1}}$		1		
X_{14}	$\mathbf{Hw}_{srs_s,ps_{a_2}}$		1		

mcfp combination function parameters	function	1 eucl		2 alo-gistic		3 hebb		4 scm	
	parameter	1 n	2 λ	1 σ	2 τ	1 μ	2	1 α	2
X_1	srs_s	1	1						
X_2	srs_c			18	0.2				
X_3	srs_{e1}	1	0.7						
X_4	srs_{e2}	1	1						
X_5	fs_{ee}	1	1						
X_6	ps_{a1}	1	2						
X_7	ps_{a2}	1	2						
X_8	ps_{ee}	1	2						
X_9	cs_1	1	1						
X_{10}	cs_2	1	1						
X_{11}	$\mathbf{W}_{srs_s,ps_{a_1}}$					0.8		0.5	
X_{12}	$\mathbf{W}_{srs_s,ps_{a_2}}$					0.8		0.5	
X_{13}	$\mathbf{Hw}_{srs_s,ps_{a_1}}$			5	0.8				
X_{14}	$\mathbf{Hw}_{srs_s,ps_{a_2}}$			5	0.8				

In matrix **mcfw** in Box 5.1, it is indicated which states get which combination functions. As shown, almost all base states get the Euclidean combination function. The only exception is state srs_c which only has a link to itself. The chosen logistic sum combination function allows to get some pattern over time for the environment in which the stressor occurs after some time. Also the second-order reification states $\mathbf{H}_{\mathbf{W}_{srs_s,ps_{a_1}}}$ and $\mathbf{H}_{\mathbf{W}_{srs_s,ps_{a_2}}}$ have a logistic combination function. Note that the first-order reification states $\mathbf{W}_{srs_s,ps_{a_1}}$ and $\mathbf{W}_{srs_s,ps_{a_2}}$ get a weighted average (with weights $\gamma_1 = 0.85$ and $\gamma_2 = 0.15$) of two combination functions: $\mathbf{hebb}_{\mu}(..)$ for

Hebbian learning, and $\mathbf{scm}_\alpha(..)$ for state-connection modulation as described in (6) above. All other states have single combination functions. In role matrix **mcfp** the parameter values for the chosen combination functions are shown. As can be seen, all Euclidean combination functions have order 1, which actually makes them scaled sum functions.

Box 5.2 Role matrix **ms** for speed factors and vector **iv** of initial values.

ms	**speed factors**	1
X_1	srs_s	0
X_2	srs_c	0.05
X_3	srs_{e1}	0.5
X_4	srs_{e2}	0.5
X_5	fs_{ee}	0.5
X_6	ps_{a1}	0.5
X_7	ps_{a2}	0.5
X_8	ps_{ee}	0.5
X_9	cs_1	0.02
X_{10}	cs_2	0.6
X_{11}	$\mathbf{W}_{srs_s,ps_{a_1}}$	X_{13}
X_{12}	$\mathbf{W}_{srs_s,ps_{a_2}}$	X_{14}
X_{13}	$\mathbf{H}_{\mathbf{W}srs_s,ps_{a_1}}$	0.5
X_{14}	$\mathbf{H}_{\mathbf{W}srs_s,ps_{a_2}}$	0.5

iv	**initial values**	1
X_1	srs_s	1
X_2	srs_c	0.1
X_3	srs_{e1}	0
X_4	srs_{e2}	0
X_5	fs_{ee}	0
X_6	ps_{a1}	0
X_7	ps_{a2}	0
X_8	ps_{ee}	0
X_9	cs_1	0
X_{10}	cs_2	0
X_{11}	$\mathbf{W}_{srs_s,ps_{a_1}}$	0.9
X_{12}	$\mathbf{W}_{srs_s,ps_{a_2}}$	0.3
X_{13}	$\mathbf{H}_{\mathbf{W}srs_s,ps_{a_1}}$	0.05
X_{14}	$\mathbf{H}_{\mathbf{W}srs_s,ps_{a_2}}$	0.05

In Box 5.2 the speed factors and the initial values for the chosen scenario are shown. Note that initially the weight of the incoming connection for action a_1 is high and for a_2 low. This models that initially the preferred action is a_1.

5.5 Example Simulation Scenarios

Two example simulation scenarios of this process are shown in Figs. 5.3, 5.4, 5.5 and 5.6. Boxes 5.1 and 5.2 show the reified network characteristics used. The step size was $\Delta t = 0.4$. In the two scenarios, coping with an extremely stressful condition c (disturbing context due to person B) takes place. The trigger for doing one of the actions is the sensory representation state srs_s (also denoted by X_1), which has value 1 all the time. In the first scenario, only first-order adaptation takes place according to the disconnect-reconnect principle (Sousa and Almeida 2012). In the second scenario, in addition, also metaplasticity is assumed, which leads to second-order adaptation.

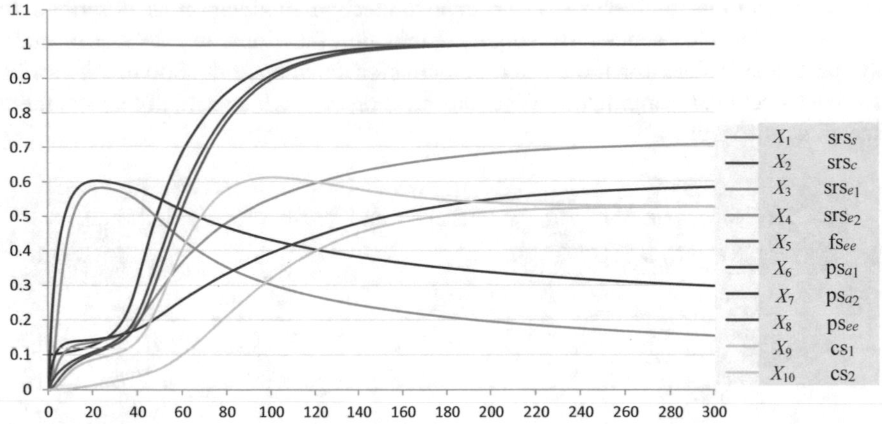

Fig. 5.3 Simulation results of working under an extremely stressful condition without metaplasticity: base states

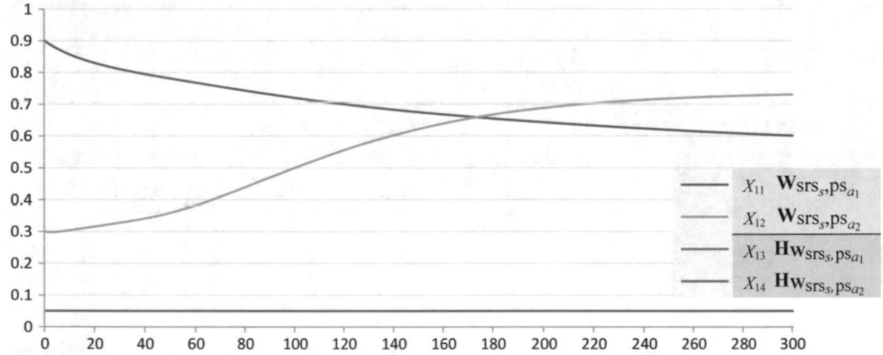

Fig. 5.4 Simulation results for state-connection modulation and Hebbian learning (without metaplasticity) for X_{11} (connection weight reification $\mathbf{W}_{srs_s,ps_{a_1}}$) and X_{12} (connection weight reification $\mathbf{W}_{srs_s,ps_{a_2}}$)

5.5.1 Scenario 1: First-Order Adaptation; No Adaptive Speed of Connection Weight Change

In Scenario 1, the speed factors for the second-order reification states for metaplasticity have been set at 0, so that the speed of adaptation of the connection weights $\mathbf{W}_{srs_s,ps_{a_1}}$ (X_{11}) and $\mathbf{W}_{srs_s,ps_{a_2}}$ (X_{12}) was constant, equal to the initial values 0.05. Note that the initial preference for action a_1 over a_2 is shown by a high initial value 0.9 for $\mathbf{W}_{srs_s,ps_{a_1}}$ (X_{11}) and a low initial value 0.3 for $\mathbf{W}_{srs_s,ps_{a_2}}$ (X_{12}); see Fig. 5.4. At the early time of working, there is a convenient condition in the working place. Therefore, as can be seen in Fig. 5.3, srs_c (also denoted by X_2) has a

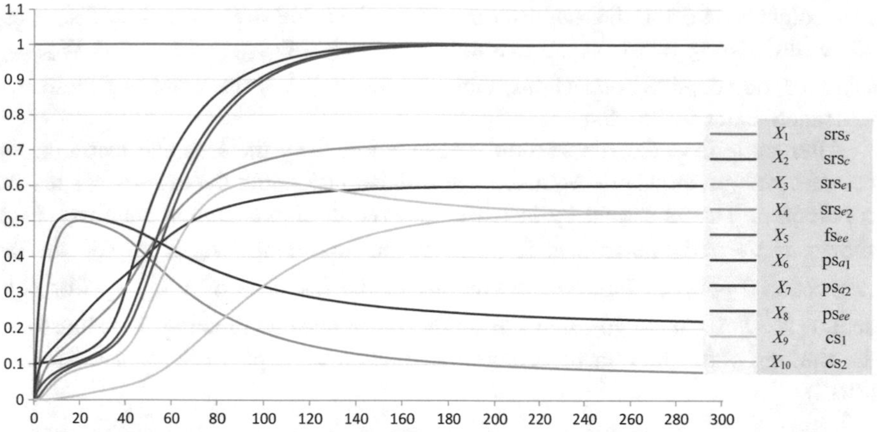

Fig. 5.5 Simulation results of working under an extremely stressful condition with metaplasticity: base states

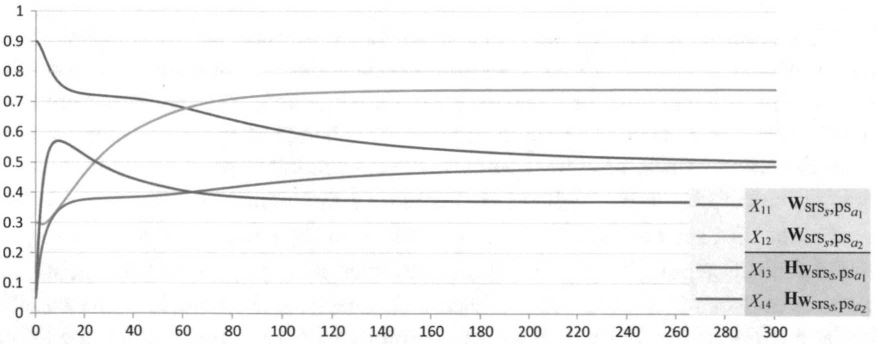

Fig. 5.6 Simulation results for state-connection modulation and Hebbian learning (with metaplasticity) for X_{11} (connection weight reification $\mathbf{W}_{srs_s,ps_{a_1}}$) and X_{12} (connection weight reification $\mathbf{W}_{srs_s,ps_{a_2}}$)

low value, which, however, strongly increases after some time (around time point 30 in Fig. 5.3) when the disturbances start.

Sensory representation state srs_{e_1} (X_3) shows that person A initially has a good feeling for the effect of action a_1, which strengthens the preparation ps_{a_1} (X_6) for this action in the first phase; see the purple line in Fig. 5.3 that steeply goes up to 0.6. However, after time point 30 when the acute stress occurs, this changes. Both the feeling srs_{e_1} for the action effect and the preparation ps_{a_1} for the action a_1 drop. Moreover, after this time point control state cs_2 (X_{10}), which stands for the control state for suppression of the connections, starts to go up but after some time the other control state cs_1 (X_9) in turn begins to play a role in suppressing cs_2. Therefore, the actual suppression of the connections mainly takes place between time points 50 and 150. During that time due to the acute stress, the control state cs_2 has a suppressing effect on the two reification states $\mathbf{W}_{srs_s,ps_{a_1}}$ (X_{11}) and $\mathbf{W}_{srs_s,ps_{a_2}}$ (X_{12}) for the adaptive connections, which is illustrated in the graphs of these two reification states in Fig. 5.4.

After some time the suppression is released, and by the Hebbian learning, the person develops another decision to cope with her task under the extremely stressful condition c. This is shown by increased activation of preparation state ps_{a_2} (X_7), shown as the red line in Fig. 5.3, and by an increasing $\mathbf{W}_{srs_s,ps_{a_2}}$ (X_{12}, for the connection X_1–X_7) in Fig. 5.4, in contrast to the decrease of $\mathbf{W}_{srs_s,ps_{a_1}}$ (for connection X_1–X_6). Due to this, now the action a_2 becomes dominating. This illustrates the working of the disconnect-reconnect adaptation principle of Sousa and Almeida (2012).

5.5.2 *Scenario 2: Second-Order Adaptation; Adaptive Speed of Connection Weight Change*

In this second scenario, the speed factors of the connection weight adaptation are themselves adaptive, so there is second-order adaptation. As in Scenario 1, it can be seen in Fig. 5.5 that also for Scenario 2 after a while action a_2 becomes dominant over a_1. However, as can be seen in Fig. 5.6, the change of the connection weights is much earlier, so now after time point 60 the connection to ps_{a_2} is already stronger than the connection to ps_{a_1}, whereas in Scenario 1 that point was only reached after time point 180; see Fig. 5.4. This has also effect on the pattern for ps_{a_1} and ps_{a_2} in Fig. 5.5. In Scenario 2, ps_{a_2} is already stronger than ps_{a_1} after time point 60, whereas in Scenario 1 this is only the case after time point 110. So, in both cases the initial preference for action a_1 changes to a preference for action a_2, but due to the second-order adaptation, this adaptation happens much faster in Scenario 2, which is an advantage in urgent, stressful situations. This illustrates once more the relevance of second-order adaptation or metaplasticity.

5.6 Verification of the Network Model by Mathematical Analysis

For temporal-causal network models, dedicated methods have been developed enabling to verify whether the implemented model shows behavior as expected; see Treur (2016a) or Treur (2016b), Chap. 12. In this section, in particular, the focus is on equilibria: they are determined by Mathematical Analysis and then used for verification by comparison to simulation results. First a definition

Definition (stationary point and equilibrium)
A state Y in an adaptive temporal-causal network model has a *stationary point* at t if $\mathbf{d}Y(t)/\mathbf{d}t = 0$.
A temporal-causal network model is in an *equilibrium state* at t if all states have a stationary point at t.
In that case, the above equations $\mathbf{d}Y(t)/\mathbf{d}t = 0$ for all states Y provide the *equilibrium equations*.

The above definition is quite general. However, for adaptive temporal-causal network models the following simple criterion was obtained in terms of the basic characteristics defining the network structure, in particular (besides the states X_i and Y), speed factor $\boldsymbol{\eta}_Y$, connection weights $\boldsymbol{\omega}$ and the combination function $\mathbf{c}(..)$; see Treur, (2016a) or Treur, (2016b), Chap. 12.

Criterion for stationary points and equilibria in temporal-causal network models A state Y in an adaptive temporal-causal network model has a stationary point at t if and only if

$$\boldsymbol{\eta}_Y = 0 \quad \text{or} \quad \mathbf{c}_Y(\boldsymbol{\omega}_{X_1,Y}(t)X_1(t), \ldots, \boldsymbol{\omega}_{X_k,Y}(t)X_k(t)) = Y(t)$$

where X_1, …, X_k are the states with outgoing connections to Y.

An adaptive temporal-causal network model is in an equilibrium state at t if and only if for all states with nonzero speed factor, the above criteria hold at t.

Note that in the case of reification of characteristics $\boldsymbol{\eta}_Y$, $\boldsymbol{\omega}_{X_k,Y}$, $\mathbf{c}_Y(..)$ that occur in this criterion, in principle, the universal combination function $\mathbf{c}^*_Y(..)$ and the related difference equation has to be considered. However, the universal differential equation can be written in the following form, as has been mentioned in Chap. 3, Sect. 3.5 (for more details, see also Chap. 10, Sect. 10.4.2 and 10.5):

$$\mathbf{d}Y/\mathbf{d}t = \mathbf{H}_Y\left[\frac{\mathbf{C}_{1,Y}\text{bcf}_1(\mathbf{P}_{1,1,Y}, \mathbf{P}_{2,1,Y}, \mathbf{W}_{X_1,Y}X_1, \ldots, \mathbf{W}_{X_k,Y}X_k) + \cdots + \mathbf{C}_{m,Y}\text{bcf}_m(\mathbf{P}_{1,m,Y}, \mathbf{P}_{2,m,Y}, \mathbf{W}_{X_1,Y}X_1, \ldots, \mathbf{W}_{X_k,Y}X_k)}{\mathbf{C}_{1,Y} + \cdots + \mathbf{C}_{m,Y}} - Y\right]$$

The right hand side of this is 0 if and only if

$$\mathbf{H}_Y = 0 \quad \text{or}$$
$$\frac{\mathbf{C}_{1,Y}\text{bcf}_1(\mathbf{P}_{1,1,Y}, \mathbf{P}_{2,1,Y}, \mathbf{W}_{X_1,Y}X_1, \ldots, \mathbf{W}_{X_k,Y}X_k) + \cdots + \mathbf{C}_{m,Y}\text{bcf}_m(\mathbf{P}_{1,m,Y}, \mathbf{P}_{2,m,Y}, \mathbf{W}_{X_1,Y}X_1, \ldots, \mathbf{W}_{X_k,Y}X_k)}{\mathbf{C}_{1,Y} + \cdots + \mathbf{C}_{m,Y}} = Y$$

where the left-hand side is the combination function. This shows that the above criterion also can be used when some or all characteristics are adaptive.

5.6.1 Solving the Linear Equilibrium Equations for the Base Network

The criterion for an equilibrium for a scaled sum function

$$\mathbf{c}_Y(\boldsymbol{\omega}_{X_1,Y}(t)X_1(t),\ldots,\boldsymbol{\omega}_{X_k,Y}(t)X_k(t)) = (\boldsymbol{\omega}_{X_1,Y}(t)X_1(t) + \cdots + \boldsymbol{\omega}_{X_k,Y}(t)X_k(t))/\boldsymbol{\lambda}_Y = Y(t) \tag{5.7}$$

provides a linear equilibrium equation (leaving out t):

$$\boldsymbol{\omega}_{X_1,Y}X_1 + \cdots + \boldsymbol{\omega}_{X_k,Y}X_k = \boldsymbol{\lambda}_Y Y \tag{5.8}$$

in the state values $X_i(t)$ and $Y(t)$ involved. In this way for the network model introduced here the equilibrium equations for the states were obtained as shown in Box 5.3, where the values for srs_s and srs_c are indicated by A_1 and A_2 (here to simplify notation the reference to t has been left out, and underlining is used to indicate that this concerns state values, not state names). Moreover, the scaling factor for state X_i is denoted by $\boldsymbol{\lambda}_i$, and in Table 5.2 numbered connection weight names are indicated.

Then the linear equilibrium equations are obtained as shown in Box 5.3, in the left half in terms of the informative state names, and in the right half in terms of the numbered X_i as state names; see Table 5.1.

Box 5.3 General equilibrium equations for the base network

$\underline{\text{srs}_s} = A_1$	$X_1 = A_1$
$\underline{\text{srs}_c} = A_2$	$X_2 = A_2$
$\boldsymbol{\lambda}_3\underline{\text{srs}_{e_1}} = \boldsymbol{\omega}_7\underline{\text{ps}_{a_1}} + \boldsymbol{\omega}_9\underline{\text{srs}_c}$	$\boldsymbol{\lambda}_3X_3 = \boldsymbol{\omega}_7X_6 + \boldsymbol{\omega}_9X_2$
$\boldsymbol{\lambda}_4\underline{\text{srs}_{e_2}} = \boldsymbol{\omega}_8\underline{\text{ps}_{a_2}} + \boldsymbol{\omega}_{10}\underline{\text{srs}_c}$	$\boldsymbol{\lambda}_4X_4 = \boldsymbol{\omega}_8X_7 + \boldsymbol{\omega}_{10}X_2$
$\underline{\text{fs}_{ee}} = \boldsymbol{\omega}_{13}\underline{\text{ps}_{ee}}$	$X_5 = \boldsymbol{\omega}_{13}X_8$
$\boldsymbol{\lambda}_6\underline{\text{ps}_{a_1}} = \boldsymbol{\omega}_1\underline{\text{srs}_s} + \boldsymbol{\omega}_3\underline{\text{srs}_{e_1}} + \boldsymbol{\omega}_5\underline{\text{ps}_{a_2}}$	$\boldsymbol{\lambda}_6X_6 = \boldsymbol{\omega}_1X_1 + \boldsymbol{\omega}_3X_3 + \boldsymbol{\omega}_5X_7$
$\boldsymbol{\lambda}_7\underline{\text{ps}_{a_2}} = \boldsymbol{\omega}_2\underline{\text{srs}_s} + \boldsymbol{\omega}_4\underline{\text{srs}_{e_2}} + \boldsymbol{\omega}_6\underline{\text{ps}_{a_1}}$	$\boldsymbol{\lambda}_7X_7 = \boldsymbol{\omega}_2X_1 + \boldsymbol{\omega}_4X_4 + \boldsymbol{\omega}_6X_6$
$\boldsymbol{\lambda}_8\underline{\text{ps}_{ee}} = \boldsymbol{\omega}_{11}\underline{\text{srs}_c} + \boldsymbol{\omega}_{12}\underline{\text{fs}_{ee}}$	$\boldsymbol{\lambda}_8X_8 = \boldsymbol{\omega}_{11}X_2 + \boldsymbol{\omega}_{12}X_5$
$\underline{\text{cs}_1} = \boldsymbol{\omega}_{14}\underline{\text{cs}_2}$	$X_9 = \boldsymbol{\omega}_{14}X_{10}$
$\boldsymbol{\lambda}_{10}\underline{\text{cs}_2} = \boldsymbol{\omega}_{15}\underline{\text{fs}_{ee}} + \boldsymbol{\omega}_{16}\underline{\text{cs}_1}$	$\boldsymbol{\lambda}_{10}X_{10} = \boldsymbol{\omega}_{15}X_5 + \boldsymbol{\omega}_{16}X_9$

Table 5.2 Numbering of connection weights

From	To	Connection weight	From	To	Connection weight
srs_s	ps_{a_1}	ω_1	srs_c	srs_{e_1}	ω_9
srs_s	ps_{a_2}	ω_2	srs_c	srs_{e_2}	ω_{10}
srs_{e_1}	ps_{a_1}	ω_3	srs_c	ps_{ee}	ω_{11}
srs_{e_2}	ps_{a_2}	ω_4	fs_{ee}	ps_{ee}	ω_{12}
ps_{a_2}	ps_{a_1}	ω_5	ps_{ee}	fs_{ee}	ω_{13}
ps_{a_1}	ps_{a_2}	ω_6	fs_{ee}	cs_2	ω_{14}
ps_{a_1}	srs_{e_1}	ω_7	cs_2	cs_1	ω_{15}
ps_{a_2}	srs_{e_2}	ω_8	cs_1	cs_2	ω_{16}

Using the WIMS Linear Solver,[1] the (unique) algebraic solution was obtained for the general case of these equations as shown in Box 5.4.

Box 5.4 General solution of the equilibrium equations for the base network as generated by the WIMS Linear Solver (see Footnote 1)

$$X_1 = A_1$$
$$X_2 = A_2$$
$$\begin{aligned} X_3 = &-(A_2(\lambda_6(\lambda_4\lambda_7\omega_9 - \omega_4\omega_8\omega_9) + \omega_5(\omega_{10}\omega_4\omega_7 - \lambda_4\omega_6\omega_9)) \\ &+A_1(\omega_1(\lambda_4\lambda_7\omega_7 - \omega_4\omega_7\omega_8) + \lambda_4\omega_2\omega_5\omega_7)) \\ &/(\omega_3(\lambda_4\lambda_7\omega_7 - \omega_4\omega_7\omega_8) + \lambda_6(\lambda_3\omega_4\omega_8 - \lambda_3\lambda_4\lambda_7) + \lambda_3\lambda_4\omega_5\omega_6) \end{aligned}$$
$$\begin{aligned} X_4 = &(A_2(\omega_3(\lambda_7\omega_{10}\omega_7 - \omega_6\omega_8\omega_9) + \lambda_3\omega_{10}\omega_5\omega_6 - \lambda_3\lambda_6\lambda_7\omega_{10}) \\ &+A_1(\omega_2\omega_3\omega_7\omega_8 - \lambda_3\omega_1\omega_6\omega_8 - \lambda_3\lambda_6\omega_2\omega_8)) \\ &/(\omega_3(\lambda_4\lambda_7\omega_7 - \omega_4\omega_7\omega_8) + \lambda_6(\lambda_3\omega_4\omega_8 - \lambda_3\lambda_4\lambda_7) + \lambda_3\lambda_4\omega_5\omega_6) \end{aligned}$$
$$X_5 = -A_2\omega_{11}\omega_{13}/(\omega_{12}\omega_{13} - \lambda_8)$$
$$\begin{aligned} X_6 = &-(A_2(\omega_3(\lambda_4\lambda_7\omega_9 - \omega_4\omega_8\omega_9) + \lambda_3\omega_{10}\omega_4\omega_5) \\ &+A_1(\omega_1(\lambda_3\lambda_4\lambda_7 - \lambda_3\omega_4\omega_8) + \lambda_3\lambda_4\omega_2\omega_5)) \\ &/(\omega_3(\lambda_4\lambda_7\omega_7 - \omega_4\omega_7\omega_8) + \lambda_6(\lambda_3\omega_4\omega_8 - \lambda_3\lambda_4\lambda_7) + \lambda_3\lambda_4\omega_5\omega_6) \end{aligned}$$
$$\begin{aligned} X_7 = &(A_2(\omega_3(\omega_{10}\omega_4\omega_7 - \lambda_4\omega_6\omega_9) - \lambda_3\lambda_6\omega_{10}\omega_4) \\ &+A_1(\lambda_4\omega_2\omega_3\omega_7 - \lambda_3\lambda_4\omega_1\omega_6 - \lambda_3\lambda_4\lambda_6\omega_2)) \\ &/(\omega_3(\lambda_4\lambda_7\omega_7 - \omega_4\omega_7\omega_8) + \lambda_6(\lambda_3\omega_4\omega_8 - \lambda_3\lambda_4\lambda_7) + \lambda_3\lambda_4\omega_5\omega_6) \end{aligned}$$

[1]https://wims.unice.fr/wims/wims.cgi?session=K06C12840B.2&+lang=nl&+module=tool%2Flinear%2Flinsolver.en.

$$X_8 = -A_2\omega_{11}/(\omega_{12}\omega_{13} - \lambda_8)$$
$$X_9 = A_2\omega_{11}\omega_{13}\omega_{14}\omega_{15}/(\omega_{12}\omega_{13}(\omega_{14}\omega_{16} - \lambda_{10}) + \lambda_8(\lambda_{10} - \omega_{14}\omega_{16}))$$
$$X_{10} = A_2\omega_{11}\omega_{13}\omega_{15}/(\omega_{12}\omega_{13}(\omega_{14}\omega_{16} - \lambda_{10}) + \lambda_8(\lambda_{10} - \omega_{14}\omega_{16}))$$

To compare these outcomes with simulation outcomes, in particular, the ones depicted in Figs. 5.3 and 5.4, that specific scenario has been addressed, with parameter values as indicated in Sect. 4, and $A_1 = A_2 = 1$.

$$\begin{aligned} X_1 &= 1 & X_6 &= -0.1040\underline{\omega}_2 + 0.7852\underline{\omega}_1 - 0.1004 \\ X_2 &= 1 & X_7 &= 0.6760\underline{\omega}_2 - 0.1040\underline{\omega}_1 + 0.1524 \\ X_3 &= -0.1040\underline{\omega}_2 + 0.7852\underline{\omega}_1 - 0.2432 & X_8 &= 1 \\ X_4 &= 0.4732\underline{\omega}_2 - 0.07280\underline{\omega}_1 + 0.4067 & X_9 &= 0.5263 \\ X_5 &= 1 & X_{10} &= 0.5263 \end{aligned}$$

Now, from the simulation it turns out that in the equilibrium state $\omega_1 = 0.5025559926$ and $\omega_2 = 0.7428984649$. Substituting this in the above expressions provides:

$$\begin{array}{lllll} X_1 = 1 & X_3 = 0.0741 & X_5 = 1 & X_7 = 0.6023 & X_9 = 0.5263 \\ X_2 = 1 & X_4 = 0.7216 & X_6 = 0.2170 & X_8 = 1 & X_{10} = 0.5263 \end{array}$$

For verification, these state values found by analysis have been compared (in more precision) with the equilibrium state values found in the simulation for $\Delta t = 0.25$. The results are shown in Table 5.3. As can be seen, all deviations (differences between value from the simulation and value from the analysis) are in absolute value less than 0.001, which provides evidence that the model does what is expected.

Table 5.3 Comparing analysis and simulation

State	srs_s X_1	srs_c X_2	srs_{e_1} X_3	srs_{e_2} X_4	fs_{ee} X_5
Simulation	1.0000000000	0.9999866365	0.0750260707	0.7214675588	0.9999771825
Analysis	1.0000000000	1.0000000000	0.0741370620	0.7216223422	1.0000000000
Deviation	0	−1.33635E−05	0.000889009	0.000154783	-2.28175E-05
State	ps_{a_1} X_6	ps_{a_2} X_7	ps_{ee} X_8	cs_1 X_9	cs_2 X_{10}
Simulation	0.2175996003	0.6021532734	0.9999794070	0.5257977548	0.5267889246
Analysis	0.2169942048	0.6023176317	1.0000000000	0.5263157895	0.5263157895
Deviation	0.000605395	−0.000164358	−2.0593E−05	−0.000518035	0.000473135

5.6.2 *Addressing the Nonlinear Equilibrium Equations for the Reification States*

Also for the two reification states $\mathbf{W}_{\mathrm{srs}_s,\mathrm{ps}_{a_1}}$ (now indicated by $\mathbf{W}_1$ with value W_1) and $\mathbf{W}_{\mathrm{srs}_s,\mathrm{ps}_{a_2}}$ (now indicated by $\mathbf{W}_2$ with value W_2) for the two adaptive connection weights equilibrium equations can be found. They are (as long as the speed factor of them is not 0):

$$
\begin{aligned}
W_1 &= c_{\mathbf{W}_1}(X_1, X_6, W_1, -0.7\,\mathrm{cs}_2) \\
W_2 &= c_{\mathbf{W}_2}(X_1, X_7, W_2, -0.7\,\mathrm{cs}_2)
\end{aligned}
\tag{5.9}
$$

Or in terms of X_i:

$$
\begin{aligned}
X_{11} &= c_{\mathbf{W}_1}(X_1, X_6, X_{11}, -0.7\,X_{10}) \\
X_{12} &= c_{\mathbf{W}_2}(X_1, X_7, X_{12}, -0.7\,X_{10})
\end{aligned}
$$

Filling in $c_{\mathbf{W}_1}(..)$ and $c_{\mathbf{W}_2}(..)$ the two combination functions $\mathbf{hebb}_\mu$**(..)** and $\mathbf{scm}_\alpha$**(..)** and combination function weights $\boldsymbol{\gamma}_1$ and $\boldsymbol{\gamma}_2$ provides:

$$
\begin{aligned}
X_{11} &= \boldsymbol{\gamma}_1 \mathbf{hebb}_\mu(X_1, X_6, X_{11}) + \boldsymbol{\gamma}_2 \mathbf{scm}_\alpha(X_1, X_6, X_{11}, -0.7\,X_{10}) \\
X_{12} &= \boldsymbol{\gamma}_1 \mathbf{hebb}_\mu(X_1, X_7, X_{12}) + \boldsymbol{\gamma}_2 \mathbf{scm}_\alpha(X_1, X_7, X_{12}, -0.7\,X_{10})
\end{aligned}
$$

$$
\begin{aligned}
X_{11} &= \boldsymbol{\gamma}_1(X_1X_6(1 - X_{11}) + \boldsymbol{\mu}\,X_{11}) + \boldsymbol{\gamma}_2(X_{11} - 0.7\,\boldsymbol{\alpha}\,X_{11}(1 - X_{11})X_{10}) \\
X_{12} &= \boldsymbol{\gamma}_1(X_1X_7(1 - X_{12}) + \boldsymbol{\mu}\,X_{12}) + \boldsymbol{\gamma}_2(X_{12} - 0.7\,\boldsymbol{\alpha}\,X_{12}(1 - X_{12})X_{10})
\end{aligned}
$$

However, these equations are not linear and more difficult to be solved algebraically. Nevertheless, they still can be used for verification by substitution of values found in simulations. The values used are

$$
\boldsymbol{\gamma}_1 = 0.85 \quad \boldsymbol{\gamma}_2 = 0.15 \quad \boldsymbol{\mu} = 0.8 \quad \boldsymbol{\alpha} = 0.5 \quad X_1 = A_1 = 1
$$

Then the instantiated equations are

$$
\begin{aligned}
X_{11} &= 0.85(X_6(1 - X_{11}) + 0.8X_{11}) + 0.15(X_{11} - 0.35\,X_{11}(1 - X_{11})X_{10}) \\
X_{12} &= 0.85(X_7(1 - X_{12}) + 0.8X_{12}) + 0.15(X_{12} - 0.35\,X_{12}(1 - X_{12})X_{10})
\end{aligned}
$$

The relevant equilibrium values obtained in the simulation are

$$
\begin{aligned}
&X_6 = 0.217599600251934 \quad && X_{11} = 0.502555992556694 \\
&X_7 = 0.602153273449135 \quad && X_{12} = 0.742898464938841 \\
&X_{10} = 0.526788924575543 &&
\end{aligned}
$$

After substitution of these in the instantiated equations, the following is obtained:

$$0.502214624461369 = 0.502555992556694$$
$$0.742915691976951 = 0.742898464938841$$

The deviations of these equations are −0.0003413680953248120 and 0.0000172270381099127 respectively, which both are less than 0.001. This provides evidence that also for the reification states used for adaptation of the connections the model does what is expected.

5.7 Discussion

In this chapter, a second-order adaptive reified temporal-causal network model was presented for the adaptive role of stress in decision making based on the disconnect-reconnect principle (Sousa and Almeida 2012). Parts of this chapter are based on Treur and Mohammadi Ziabari (2018). In this computational network model, connections developed in the past are suppressed due to acute stress as a form of reset (disconnect) and Hebbian learning takes place to adapt the decision making to the stressful conditions (reconnect). A number of simulations were performed two of which were presented in the chapter. Findings from Neuroscience were taken into account in the design of the adaptive model (Quaedflieg et al. 2015; Hermans et al. 2014; Reser 2016; Sousa and Almeida 2012). This literature reports experiments and measurements for stress-induced conditions as addressed from a computational perspective in the current chapter. In addition to this, also metaplasticity has been incorporated in the adaptive network model. This makes it a second-order adaptive network model. In the simulations, it has been shown how this second-order adaptation accelerates the learning effect, on top of the disconnect-reconnect principle, as also claimed in another context by Robinson et al. (2016). Also, a precise mathematical analysis has been done to verify that behaviour of the reified network model is as expected.

In other, more applied literature, such as Gok and Atsan (2016), not the Neuroscience perspective is followed, but a more general psychological perspective on decision making applied to a manager's context. This may seem to contrast with the Neuroscience perspective followed in the current chapter which is mainly based on Sousa and Almeida (2012). However, the more refined approach on decision making and its subprocesses in Gok and Atsan (2016), such as the generation of decision options and selection of an option, may provide interesting inspiration for future research in making a more refined version of the current model, and place it in a more applied context. Another future extension may address an explicit role for cortisol in the development of stress. The current states used to model the extreme emotion (ps_{ee} and fs_{ee}) can be seen as aggregate states for a number of brain states, including the cortisol level. In a more refined approach different substates may be distinguished, under which a separate state for the cortisol level.

This network model can be used as the basis of a virtual agent model to get insight in such processes and to consider certain support or treatment of individuals to handle extreme emotions when they have to work in a stressful context condition and prevent some stress-related disorders that otherwise might develop. In future research, other scenarios will be addressed and simulated for individuals with different characteristics.

References

Barsegyan, A., Mackenzie, S.M., Kurose, B.D., McGaugh, J.L., Roozendaal, B.: Glucocorticoids in the prefrontal cortex enhance memory consolidation and impair working memory by a common neural mechanism. Proc. Natl. Acad. Sci. USA **107**, 16655–16660 (2010)

de Kloet, E.R., Joëls, M., Holsboer, F.: Stress and the brain: from adaptation to disease. Nat. Rev. Neurosci. **6**, 463–475 (2005)

Etkin, A., Prater, K.E., Hoeft, F., Menon, V., Schatzberg, A.F.: Failure of anterior cingulate activation and connectivity with amygdala during implicit regulation of emotional processing in generalized anxiety disorder. Am. J. Psychiatry **167**, 545–554 (2010). https://doi.org/10.1176/ajp.2009.09070931 PMID: 201123913

Glass, D.C., Reim, B., Singer, J.E.: Behavioral consequences of adaptation to controllable and uncontrollable noise. J. Exp.Soc. Psychol. **7**, 244–257 (1971)

Gok, K., Atsan, N.: Decision-making under stress and its implications for managerial decision-making: a review of literature. Int. J. Bus. Soc. Res. **6**(3), 38–47 (2016)

Hermans, E.J., Hencknes, M.J.A.G., Joels, M.: Guillen Fernandes: dynamic adaption of large-scale brain networks in response to acute stressors, Trends Neurosci. **37**(6), 304–14 (2014). https://doi.org/10.1016/j.tins.2014.03.006. Epub

Johnstone, T., van Reekum, C.M., Ury, H.L., Klain, N.H., Davidson, R.J.: Failure to regulate: counterproductive recruitment of top-down prefrontal-subcortical circuitry in major depression. J. Neurosci. **27**, 8877–8884 (2007). PMID: 17699669

Quaedflieg, C.W.E.M., van de Ven, V., Meyer, T., Siep, N., Merckelbach, H., Smeets, T.: Temporal dynamics of stress-induced alternations of intrinsic amygdala connectivity and neuroendocrine levels. PLoS ONE **10**(5), e0124141 (2015). https://doi.org/10.1371/journal.pone.0124141

Radley, J., Morrison, J.: Repeated stress and structural plasticity in the brain. Ageing Res. Rev. **4**, 271–287 (2005)

Reser, J.E.: Chronic stress, cortical plasticity and neuroecology. Behave Process. (2016). https://doi.org/10.1016/j.beproc.2016.06.010. Epub

Robinson, B.L., Harper, N.S., McAlpine, D.: Meta-adaptation in the auditory midbrain under cortical influence. Nat. Commun. **7**, 13442 (2016)

Sousa, N., Almeida, O.F.X.: Disconnection and reconnection: the morphological basis of (mal) adaptation to stress. Trends in Neurosci. **35**(12), 742–51 (2012). https://doi.org/10.1016/j.tins.2012.08.006. Epub 2012 Sep 21

Treur, J.: Verification of temporal-causal network models by mathematical analysis. Vietnam J. Comput. Sci. **3**, 207–221 (2016a)

Treur, J.: Network-Oriented Modeling: Addressing Complexity of Cognitive, Affective and Social Interactions. Springer Publishers, Berlin (2016b)

Treur, J.: Network reification as a unified approach to represent network adaptation principles within a network. In: Proceedings of the 7th International Conference on Natural Computing. Lecture Notes in Computer Science, vol. 11324, pp. 344–358. Springer Publishers, Berlin (2018a)

Treur, J.: Multilevel network reification: representing higher order adaptivity in a network. In: Proceedings of the 7th International Conference on Complex Networks and their Applications, ComplexNetworks'18, vol. 1. Studies in Computational Intelligence, vol. 812, pp. 635–651, Springer Publishers, Berlin (2018b)

Treur, J., Mohammadi Ziabari, S.S.: An adaptive temporal-causal network model for decision making under acute stress. In: Nguyen, N.T., Trawinski, B., Pimenidis, E., Khan, Z. (eds.) Computational Collective Intelligence: Proceedings of the 10th International Conference, ICCCI'18, vol. 2. Lecture Notes in Computer Science, vol. 11056, pp. 13–25. Springer Publishers, Berlin (2018)

Chapter 6
Using Multilevel Network Reification to Model Second-Order Adaptive Bonding by Homophily

Abstract The concept of multilevel network reification introduced in the previous chapters enables representation within a network not only of first-order adaptation principles, but also of second-order adaptation principles expressing change of characteristics of first-order adaptation principles. In the current chapter, this approach is illustrated for an adaptive Social Network. This involves a first-order adaptation principle for bonding by homophily represented at the first reification level, and a second-order adaptation principle describing change of characteristics of this first-order adaptation principle, and represented at the second reification level. The second-order adaptation addresses adaptive change of two of the characteristics of the first-order adaptation, specifically similarity tipping point and connection adaptation speed factor.

6.1 Introduction

In Chaps. 4 and 5, it was illustrated how second-order adaptive Mental Network models can be designed using second-order network reification. In the current chapter, it will be shown how to do this for Social Network models. This example addresses the way in which connections between two persons change over time based on the similarities or differences between the persons. This concerns the bonding-by-homophily adaptation principle as a first-order adaptation principle, that can be represented at the first reification level of a multilevel reification architecture; it is also explained as 'birds of a feather flock together'; see also (Byrne 1986; McPherson et al. 2001; Pearson et al. 2006). In this first-order adaptation principle, there is an important role for the *homophily similarity tipping point* τ. This indicates the value such that

- when the dissimilarity between two persons is less than this value, their connection will become stronger
- when the dissimilarity is more, their connection will become weaker.

J. Treur, *Network-Oriented Modeling for Adaptive Networks: Designing Higher-Order Adaptive Biological, Mental and Social Network Models*, Studies in Systems, Decision and Control 251, https://doi.org/10.1007/978-3-030-31445-3_6

Earlier work (Sharpanskykh and Treur 2014; Holme and Newman 2006; Vazquez 2013; Vazquez et al. 2007) and (Treur 2016), Chap. 11, addressed the interaction (co-evolution) of social contagion (Levy and Nail 1993) and bonding. Such tipping points were usually considered constant, but this may not be realistic. For example, someone who already has many strong connections perhaps will be much more critical in strengthening connections than someone who has only a very few and only very weak connections. Such differences can be modeled when the tipping point value is modeled adaptively, depending on a person's circumstances, for example, on how many (and how strong) connections are already there. This makes the first-order adaptation principle of bonding based on homophily itself adaptive, where the tipping point changes over time by a second-order adaptation principle. This second-order adaptation principle can be represented at the second reification level in the multilevel reification architecture. Yet, another factor that may better be modeled adaptively is the speed of change of connections. Also, this may depend on how many (and how strong) connections someone has at some point in time. In the multilevel reified network model described here, also a second-order adaptation principle is included for the speed factor of the connection weight adaptation based on the first-order adaptation principle for bonding by homophily.

6.2 Conceptual Representation of the Second-Order Adaptive Social Network Model

In the second-order reified network model introduced here, the two main adaptation principles addressed are a first-order adaptation principle on bonding by homophily based on a tipping point τ, and one second-order adaptation principle for adaptation of this tipping point τ based on the extent of available connections and a norm $\mathbf{v}$ for this, and another second-order adaptation principle for the first-order adaptation speed. For an overview of the states and their levels, see Table 6.1.

First-order adaptation principle

- when the dissimilarity between two persons is less than tipping point τ, their connection will become stronger
- when the dissimilarity is more than tipping point τ, their connection will become weaker.

Second-order adaptation principle 1

- when for a person the number and strength of the connections is less than norm value $\mathbf{v}$, the tipping point will become higher
- when the number and strength of the total number of connections is more than norm value $\mathbf{v}$, the tipping point will become lower.

This second-order adaptation principle will have as effect that in the end, a person will have a number and strength of connections according to this norm value $\mathbf{v}$. Yet

Table 6.1 State names with their explanations

State		Explanation	Level
X_1			
X_2			
X_3			
X_4			
X_5		The 10 members of the example Social Network	Base level
X_6			
X_7			
X_8			
X_9			
X_{10}			
X_{11}	$\mathbf{W}_{X_2,X_1}$		
..	..	Reified representation states for the adaptive weights of the connections from member X_i to member X_j where $i \neq j$	First reification level
..	..		
X_{100}	$\mathbf{W}_{X_9,X_{10}}$		
X_{101}	$\mathbf{TPW}_{X_2,X_1}$		
..	..	Reified representation states for the adaptive tipping points for the reified representation states $\mathbf{W}_{X_i,X_j}$ of the adaptive connections from member X_i to member X_j	
..	..		
X_{190}	$\mathbf{TPW}_{X_9,X_{10}}$		Second reification level
X_{191}	$\mathbf{HW}_{X_2,X_1}$		
..	..	Reified representation states for the adaptive speed factors for the reified representation states $\mathbf{W}_{X_i,X_j}$ of the adaptive weights of the connections from member X_i to member X_j	
..	..		
X_{280}	$\mathbf{HW}_{X_9,X_{10}}$		

another second-order adaptation principle is considered which makes the speed of change of the connections dependent on how many and how strong contacts there already are.

Second-order adaptation principle 2

- when a person has more and stronger connections than norm value **v**, the speed of change for the connections will become lower
- when a person has less and weaker connections than norm value **v**, the speed of change for the connections will become higher.

Note that for a model phrases such as 'more and stronger connections' still have to be made more precise. The 'more' can be related in some way to a number of connections, and the 'stronger' can be related to (average) connection weights.

Recall that a temporal-causal network model represents in a declarative manner states and connections between them that represent (causal) impacts of states on each other, as assumed to hold for the application domain addressed. The states are assumed to have (activation) levels in the interval [0, 1] that vary over time. The following three notions form the defining part of a conceptual representation of a temporal-causal network model structure; they apply both to the base network and the added reification states:

- **Connectivity**
  - Each incoming connection of a state Y, from a state X has a *connection weight value* $\boldsymbol{\omega}_{X,Y}$ representing the strength of the connection.

- **Aggregation**
  - For each state a *combination function* $\mathbf{c}_Y(..)$ is chosen to combine the causal impacts state *Y* receives from other states.
- **Timing**
  - For each state *Y* a *speed factor* $\boldsymbol{\eta}_Y$ is used to represent how fast state *Y* is changing upon causal impact.

The conceptual graphical representation of the multilevel reified network model is shown in Fig. 6.1. The following reification states are included.

6.2.1 Reification States at the First Reification Level

The middle (blue) plane shows how the reification states $\mathbf{W}_{X_i,Y}$ are used for the first-order reification of the connection weights $\boldsymbol{\omega}_{X_i,Y}$. The downward arrows show the network relations of these reification states $\mathbf{W}_{X_i,Y}$ to the states X_i and *Y* in the base network. Such network relations (including their labels, such as combination functions; see below) for reification states define the first-order adaptation principle for bonding by homophily. For this example, the speed factors and the combination functions for all base level states *Y* are considered constant; therefore speed factor

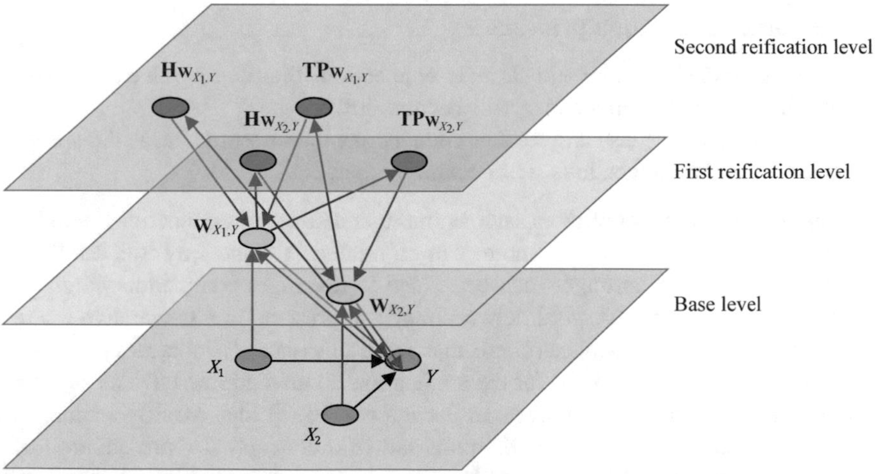

Fig. 6.1 Multilevel Reified Network model picture of a fragment of a second-order adaptive Social Network model based on Homophily. At the first reification level (middle, blue plane) the reification states $\mathbf{W}_{X_1,Y}$ and $\mathbf{W}_{X_2,Y}$ represent the adaptive connection weights for *Y*; they are changing based on a homophily principle. At the second reification level (upper, purple plane) the reified tipping point states $\mathbf{TPw}_{X_1,Y}$ and $\mathbf{TPw}_{X_2,Y}$ represent the adaptive tipping point values for the second-order connection adaptation based on homophily; similarly, the second-order **H**-states address the first-order adaptation speed

reification states $\mathbf{H}_Y$ and combination function reification states $\mathbf{C}_{i,Y}$ for base level states Y are not shown in the role matrices and the simulation graphs.

6.2.2 Reification States at the Second Reification Level

On top of the first-order reified network, a second reification level has been added (the upper, purple plane), in order to define second-order adaptation principles. The following reification states are added:

- reification states $\mathbf{TP}_{\mathbf{W}_{X_i,Y}}$ for the similarity tipping point parameter of the homophily adaptation principle for the connection weight reified by state $\mathbf{W}_{X_i,Y}$
- reification states $\mathbf{H}_{\mathbf{W}_{X_i,Y}}$ for the speed factor characteristic of the homophily adaptation principle.

Note that a double level subscript notation for second-order reification states such as $\mathbf{TP}_{\mathbf{W}_{X_i,Y}}$ should be read as tipping point for state Y, i.e., $\mathbf{TP}_Y$, for a state Y at the first reification level, in this case, $Y = \mathbf{W}_{X_i,Y}$. By substituting $\mathbf{W}_{X_i,Y}$ for Y in $\mathbf{TP}_Y$, this results in the double level subscript notation $\mathbf{TP}_{\mathbf{W}_{X_i,Y}}$; note that here for the sake of simplicity the index i of X_i is considered to be at the same subscript level as X. So, the subscript of $\mathbf{TP}$ is $\mathbf{W}_{X_i,Y}$ and this subscript itself has subscripts X_i and Y; the notation should be interpreted as $\mathbf{TP}_{(\mathbf{W}_{X_i,Y})}$. In this way, the number of reification levels is reflected in the number of subscript levels. This applies to all states at the second reification level, so, for example, also to $\mathbf{H}_{\mathbf{W}_{X_i,Y}}$. Up till now no cases of network adaptation of order higher than 2 have been explored in this book, so that more than 2 subscript levels did not show up. However, from a modeling perspective, there is nothing against adding a third reification level for the characteristics that define the second-order adaptation principles based on $\mathbf{TP}_{\mathbf{W}_{X_i,Y}}$ and $\mathbf{H}_{\mathbf{W}_{X_i,Y}}$. An example in the current context can be the addition of third-order reification states for their speed factors or their combination functions, or the parameters of these functions. In Chaps. 7 and 8 other examples of order higher than two will be shown.

Also for the above second-order reification states (upward and downward), connections have been added. These connections (together with the relevant combination functions discussed in Sect. 6.3 below) define the second-order adaptation principles based on them.

After having defined the overall conceptual representation of the reified network model, the combination functions for the new network are defined. Note that for the example simulations, speed factors and combination functions for all base states Y are considered constant; therefore speed factor reification states $\mathbf{H}_Y$ and combination function reification states $\mathbf{C}_{i,Y}$ for base states Y are not shown in the role matrices in Sect. 6.4 and in the simulation graphs in Sect. 6.5. However, for the purpose of illustration, in the general model described in Sect. 6.3, speed factor reification states $\mathbf{H}_{\mathbf{W}_{X_i,Y}}$ for the connection weight reification states $\mathbf{W}_{X_i,Y}$ are still also discussed at a general level.

6.3 Combination Functions Used at the Three Levels

In this section, the combination functions for the three levels are described. Note that all these illustrate the important role of these declarative mathematical functions:

(a) These declarative mathematical functions are powerful building blocks to define a network's structure, and this network structure determines the network's dynamics in a well-defined and unique manner
(b) By applying network reification this also covers networks that are adaptive of any order; all adaptation principles of any order are also specified just using declarative mathematical functions as building blocks.

Note that here the term declarative means that no algorithmic or procedural elements are involved in their specification, they just relate real numbers to real numbers in a time or process independent manner; they may be considered relational specifications that are functional.

A few mathematical details for the explanation below are shown in Box 6.1 (for the base level and the first reification level) and in Boxes 6.2 and 6.3 (for the second reification level). Note that the specific difference or differential equations as used in the software environment are not discussed in Chaps. 1–9. More details about them can be found in Chap. 15, Sect. 15.5, and in Chap. 10 for the universal difference and differential equation.

6.3.1 Base Level and First Reification Level Combination Functions

Base level combination functions (the lower, pink plane)

For this level, the combination functions for base states *Y* are chosen as the advanced logistic sum function $\mathbf{alogistic}_{\sigma,\tau}(\ldots)$; together with the adaptive connection weights this obtains the combination functions shown in Box 6.1.

First reification level combination functions (the middle, blue plane)

At this first reification level, the combination function for the homophily adaptation principle (at the middle plane) is needed, as shown in Box 6.1. As a measure of dissimilarity $|V_1 - V_2|$ is used. This makes that an increase or decrease in connection weight will depend on whether $\tau - |V_1 - V_2|$ is positive (less dissimilarity than τ) or negative (more dissimilarity than τ). Here τ is reified by $\mathbf{TP}_{\mathbf{W}_{X_i,Y}}$ at the second reification level. So, when the difference in states $|X_i(t) - Y(t)|$ of two persons is less than the tipping point $\mathbf{TP}_{\mathbf{W}_{X_i,Y}}(t)$, increase of $\mathbf{W}_{X_i,Y}$ will take place, and when this difference $|X_i(t) - Y(t)|$ is more than $\mathbf{TP}_{\mathbf{W}_{X_i,Y}}(t)$, decrease of $\mathbf{W}_{X_i,Y}$ will take place.

Box 6.1 Combination functions for the base level and first reification level
Base level (lower, pink plane)
Combination function for each of the base states X_i is the advanced logistic sum function $\mathbf{alogistic}_{\sigma,\tau}(..)$
First reification level (middle, blue plane)
See Sects. 4.3.2, and 4.2, or Chap. 3, Sect. 3.6.1, or (Treur 2016), Chap 11, Sect. 11.7, Combination function for connection weight reification state $\mathbf{W}_{X_i,Y}$ is $\mathbf{slhomo}_{\alpha}(..)$:

$$\mathbf{slhomo}_{\alpha}(V_1, V_2, W) = W + \alpha W(1 - W)(\tau - |V_1 - V_2|)$$

where for the variables used to define this function

- W refers to connection weight reification state $\mathbf{W}_{X_i,Y}(t)$
- V_1 refers to $X_i(t)$ and V_2 to $Y(t)$
- α is a homophily modulation factor for $\mathbf{W}_{X_i,Y}$
- τ is a homophily tipping point for $\mathbf{W}_{X_i,Y}$, which actually is reified by $\mathbf{TP}_{\mathbf{W}_{X_i,Y}}$ at the second reification level.

6.3.2 Second Reification Level Combination Functions

At this level, it is defined how the tipping points should be adapted according to circumstances, and similarly the first-order adaptation speed. The principle is used that the tipping point of a person will become higher if the person lacks strong connections (the person becomes less strict) and will become lower if the person already has strong connections (the person becomes more strict). This is handled using an *average norm weight* **v** for connections. This can be considered to relate to the amount of time or energy available for social contacts but also the desired extent of social contacts. So the effect is:

- if the connections of a person are (on average) stronger than **v**, downward regulation takes place: the tipping point will become lower
- when the connections of this person are (on average) weaker than **v**, upward regulation takes place: the tipping point will become higher.

This is expressed in the combination function for the homophily tipping point reification state $\mathbf{TP}_{\mathbf{W}_{X_i,Y}}$, a function called *simple linear tipping point function* $\mathbf{sltip}_{\mathbf{v},\alpha}(..)$, for the second reification level shown in Box 6.2. As an alternative, not the average but total cumulative weight of connections could be used. Similarly, a combination function $\mathbf{slspeed}_{\mathbf{v},\alpha}(..)$ for *simple linear speed function* is found for the speed factor reification state $\mathbf{H}_{\mathbf{W}_{X_i,Y}}$, as shown in Box 6.3. Note that the combination functions for the first- and second-order reification states all include an argument for their current value. Therefore, in role matrix **mb**, for each state at the first or second reification level, the state itself is included in the row, as can be seen in Box 6.4.

Box 6.2 Combination function for the homophily tipping point reification state at the second reification level

The following combination function called *simple linear tipping point function* $\mathbf{sltip}_{\mathbf{v},\alpha}(..)$ can be used for the second-order reification state $\mathbf{TP}_{\mathbf{W}_{X_i,Y}}$ at the second reification level (upper, purple plane):

$$\mathbf{sltip}_{\mathbf{v},\alpha}(W_1, \ldots, W_k, T) = T + \alpha T(1 - T)(\mathbf{v} - (W_1 + \ldots + W_k)/k)$$

where for the variables used to define this function

- T refers to the homophily tipping point reification value $\mathbf{TP}_{\mathbf{W}_{X_i,Y}}(t)$ for $\mathbf{W}_{X_i,Y}$
- W_j to connection weight reification value $\mathbf{W}_{X_i,,Y}(t)$
- α is a modulation factor for the tipping point $\mathbf{TP}_{\mathbf{W}_{X_i,Y}}$
- $\mathbf{v}$ is a norm for Y for average (incoming) connection weight from the $X_{i.}$

This function can be explained as follows. The norm parameter $\mathbf{v}$ indicates the preferred average level of the connection weights $\mathbf{W}_{X_i,Y}$ for person Y. The part ($\mathbf{v}$ − $(W_1+ \ldots + W_k)/k$) in the formula is positive when the current average connection weight $(W_1+ \ldots + W_k)/k$ is lower than this norm, and negative when it is higher than the norm. When T is not 0 or 1, in the first case, the combination function provides a value higher than T, which makes that the tipping point value T is increased, and therefore more connections are strengthened by the homophily adaptation. So, in this case, the average connection weight will become more close to the norm $\mathbf{v}$. In the second case, the opposite takes place: the combination function provides a value lower than T, which makes that the tipping point value T is decreased, and as a consequence, more connections are weakened by the homophily adaptation. So also now the average connection weight will become more close to the norm $\mathbf{v}$. Together this makes that in principle (unless in the meantime other factors change) the average connection weight will approximate the norm $\mathbf{v}$. The factor $T(1-T)$ in the formula takes care that the tipping point values T stay within the [0, 1] interval.

As an alternative, note that as a slightly different variant the division by k can be left out. Then the norm does not concern the average but the cumulative connection weights; also a logistic sum function could be used here.

Box 6.3 Combination function for the connection weight speed factor reification state at the second reification level
For the adaptive speed factor for the connection weight adaptation, the following combination function called *simple linear speed function* $\mathbf{slspeed}_{\mathbf{v},\alpha}(..)$ can be considered; it makes use of a similar mechanism using a norm for connection weights:

$$\mathbf{slspeed}_{\mathbf{v},\alpha}(W_1, \ldots, W_k, H) = H + \alpha H(1 - H)(\mathbf{v} - (W_1 + \ldots + W_k)/k)$$

where for the variables used to define this function

- H refers to $\mathbf{W}_{X_i,Y}$ speed factor reification value $\mathbf{H}_{\mathbf{W}_{X_i,Y}}(t)$
- W_j to connection weight reification value $\mathbf{W}_{X_i,Y}(t)$
- α is a modulation factor for $\mathbf{H}_{\mathbf{W}_{X_i,Y}}$
- $\mathbf{v}$ is a norm for an average of (outgoing) connection weights for Y.

This function can be explained as follows. Also here the norm parameter $\mathbf{v}$ indicates the preferred average level of the connection weights $\mathbf{W}_{X_i,Y}$ for person Y. The part $(\mathbf{v} - (W_1 + \ldots + W_k)/k)$ in the formula is positive when the current average connection weight $(W_1 + \ldots + W_k)/k$ is lower than this norm, and negative when it is higher than the norm. When H is not 0 or 1, in the first case, the combination function provides a value higher than H, thus the speed factor is increased, and the connection weights are changing faster by the homophily adaptation. In the second case, the combination function provides a value lower than H; this decrease in the speed factor value makes the homophily adaptation slower. The factor $H(1-H)$ in the formula takes care that the values for H stay within the [0, 1] interval.

6.4 Role Matrices for the Reified Social Network Model

The role matrices are shown in Boxes 6.4 and 6.5. The specific numbers shown here relate, in particular, to Scenario 3 described in Sect. 6.5. Note that in the simulated scenarios a relatively large number of states was taken into account:

- 10 base level states for the base Social Network
- 90 first-order reification states for the first-order adaptation principle for bonding based on homophily for all of the base level connections
- 90 second-order reification states for the second-order adaptive tipping point adaptation principle for the tipping point parameter used in the first-order adaptation principle for bonding based on homophily.

For reasons of space limitations, not all these states have been written down explicitly in the role matrices here. For the same reason the second-order **H**-states have been left out here. First, in Box 6.4 role matrices **mb** and **mcw** are shown. In Box 6.5 the other three role matrices are shown, and the initial values.

Box 6.4 Role matrices **mb** for connected base states (=states of same or lower level with outgoing connections to the given state) and **mcw** for connection weights of these connections

mb	**base connectivity**	1	2	3	4	5	6	7	8	9	10
X_1		X_2	X_3	X_4	X_5	X_6	X_7	X_8	X_9	X_{10}	
X_2		X_1	X_3	X_4	X_5	X_6	X_7	X_8	X_9	X_{10}	
X_3		X_1	X_2	X_4	X_5	X_6	X_7	X_8	X_9	X_{10}	
X_4		X_1	X_2	X_3	X_5	X_6	X_7	X_8	X_9	X_{10}	
X_5		X_1	X_2	X_3	X_4	X_6	X_7	X_8	X_9	X_{10}	
X_6		X_1	X_2	X_3	X_4	X_5	X_7	X_8	X_9	X_{10}	
X_7		X_1	X_2	X_3	X_4	X_5	X_6	X_8	X_9	X_{10}	
X_8		X_1	X_2	X_3	X_4	X_5	X_6	X_7	X_9	X_{10}	
X_9		X_1	X_2	X_3	X_4	X_5	X_6	X_7	X_8	X_{10}	
X_{10}		X_1	X_2	X_3	X_4	X_5	X_6	X_7	X_8	X_9	
X_{11}	$\mathbf{W}_{X_2,X_1}$	X_1	X_2	X_{11}							
..	..	..	..	..							
..	..	..	..	..							
X_{100}	$\mathbf{W}_{X_9,X_{10}}$	X_{10}	X_9	X_{100}							
X_{101}	$\mathbf{TPW}_{X_2,X_1}$	X_{11}	X_{12}	X_{13}	X_{14}	X_{15}	X_{16}	X_{17}	X_{18}	X_{19}	X_{101}
..	..	..	..	..	..	..	..	..	..	..	..
..	..	..	..	..	..	..	..	..	..	..	..
X_{190}	$\mathbf{TPW}_{X_9,X_{10}}$	X_{92}	X_{93}	X_{94}	X_{95}	X_{96}	X_{97}	X_{98}	X_{99}	X_{100}	X_{190}
X_{191}	$\mathbf{HW}_{X_2,X_1}$	X_{11}	X_{12}	X_{13}	X_{14}	X_{15}	X_{16}	X_{17}	X_{18}	X_{19}	X_{191}
..	..	..	..	..	..	..	..	..	..	..	..
..	..	..	..	..	..	..	..	..	..	..	..
X_{280}	$\mathbf{HW}_{X_9,X_{10}}$	X_{92}	X_{93}	X_{94}	X_{95}	X_{96}	X_{97}	X_{98}	X_{99}	X_{100}	X_{280}
mcw	**connection weights**	1	2	3	4	5	6	7	8	9	10
X_1		X_{11}	X_{12}	X_{13}	X_{14}	X_{15}	X_{16}	X_{17}	X_{18}	X_{19}	
X_2		X_{20}	X_{21}	X_{22}	X_{23}	X_{24}	X_{25}	X_{26}	X_{27}	X_{28}	
X_3		X_{29}	X_{30}	X_{31}	X_{32}	X_{33}	X_{34}	X_{35}	X_{36}	X_{37}	
X_4		X_{38}	X_{39}	X_{40}	X_{41}	X_{42}	X_{43}	X_{44}	X_{45}	X_{46}	
X_5		X_{47}	X_{48}	X_{49}	X_{50}	X_{51}	X_{52}	X_{53}	X_{54}	X_{55}	
X_6		X_{56}	X_{57}	X_{58}	X_{59}	X_{60}	X_{61}	X_{62}	X_{63}	X_{64}	
X_7		X_{65}	X_{66}	X_{67}	X_{68}	X_{69}	X_{70}	X_{71}	X_{72}	X_{73}	
X_8		X_{74}	X_{75}	X_{76}	X_{77}	X_{78}	X_{79}	X_{80}	X_{81}	X_{82}	
X_9		X_{83}	X_{84}	X_{85}	X_{86}	X_{87}	X_{88}	X_{89}	X_{90}	X_{91}	
X_{10}		X_{92}	X_{93}	X_{94}	X_{95}	X_{96}	X_{97}	X_{98}	X_{99}	X_{100}	
X_{11}	$\mathbf{W}_{X_2,X_1}$	1	1	1							
..	..	..	..	..							
..	..	..	..	..							
X_{100}	$\mathbf{W}_{X_9,X_{10}}$	1	1	1							
X_{101}	$\mathbf{TPW}_{X_2,X_1}$	1	1	1	1	1	1	1	1	1	1
..	..	..	..	..	..	..	..	..	..	..	..
..	..	..	..	..	..	..	..	..	..	..	..
X_{190}	$\mathbf{TPW}_{X_9,X_{10}}$	1	1	1	1	1	1	1	1	1	1
X_{191}	$\mathbf{HW}_{X_2,X_1}$	1	1	1	1	1	1	1	1	1	1
..	..	..	..	..	..	..	..	..	..	..	..
..	..	..	..	..	..	..	..	..	..	..	..
X_{280}	$\mathbf{HW}_{X_9,X_{10}}$	1	1	1	1	1	1	1	1	1	1

Box 6.5 Role matrices **mcfw** and **mcfp** for combination function weights and combination function parameters, and for speed factors, and the initial values

mcfw combination function weights		1 alogistic	2 slhomo	3 sltip	4 slspeed
X_1		1			
X_2		1			
X_3		1			
X_4		1			
X_5		1			
X_6		1			
X_7		1			
X_8		1			
X_9		1			
X_{10}		1			
X_{11}	$\mathbf{W}_{X_2,X_1}$		1		
..	..		..		
..	..		..		
X_{100}	$\mathbf{W}_{X_9,X_{10}}$		1		
X_{101}	$\mathbf{TPW}_{X_2,X_1}$			1	
..	..			..	
..	..			..	
X_{190}	$\mathbf{TPW}_{X_9,X_{10}}$			1	
X_{191}	$\mathbf{HW}_{X_2,X_1}$				1
..	..				..
..	..				..
X_{280}	$\mathbf{HW}_{X_9,X_{10}}$				1

ms speed factors		1
X_1		0.5
X_2		0.5
X_3		0.5
X_4		0.5
X_5		0.5
X_6		0.5
X_7		0.5
X_8		0.5
X_9		0.5
X_{10}		0.5
X_{11}	$\mathbf{W}_{X_2,X_1}$	X_{191}
..	..	..
..	..	..
X_{100}	$\mathbf{W}_{X_9,X_{10}}$	X_{280}
X_{101}	$\mathbf{TPW}_{X_2,X_1}$	1
..	..	..
..	..	..
X_{190}	$\mathbf{TPW}_{X_9,X_{10}}$	1
X_{191}	$\mathbf{HW}_{X_2,X_1}$	1
..	..	..
..	..	..
X_{280}	$\mathbf{HW}_{X_9,X_{10}}$	1

mcfp	**function**	1 alogistic		2 slhomo		3 sltip		3 slspeed	
		1	2	1	2	1	2	1	2
	parameter	σ	τ_{log}	α_{homo}	τ_{homo}	α_{tip}	τ_{tip}	α_{speed}	τ_{speed}
X_1		1	1.5						
X_2		1	1.5						
X_3		1	1.5						
X_4		1	1.5						
X_5		1	1.5						
X_6		1	1.5						
X_7		1	1.5						
X_8		1	1.5						
X_9		1	1.5						
X_{10}		1	1.5						
X_{11}	$\mathbf{W}_{X_2,X_1}$			1	X_{101}				
..	..			..	..				
..	..			..	..				
X_{100}	$\mathbf{W}_{X_9,X_{10}}$			1	X_{190}				
X_{101}	$\mathbf{TPW}_{X_2,X_1}$					0.1	0.4		
..	..					..	..		
..	..					..	..		
X_{190}	$\mathbf{TPW}_{X_9,X_{10}}$					0.1	0.4		
X_{191}	$\mathbf{HW}_{X_2,X_1}$							1	0.4
..	..							..	..
..	..							..	..
X_{280}	$\mathbf{HW}_{X_9,X_{10}}$							1	0.4

iv initial values		1
X_1		0.8
X_2		0.9
X_3		0.5
X_4		0.6
X_5		0.6
X_6		0.6
X_7		0.7
X_8		0.5
X_9		0.6
X_{10}		0.8
X_{11}	$\mathbf{W}_{X_2,X_1}$	0.5
..	..	..
..	..	..
X_{100}	$\mathbf{W}_{X_9,X_{10}}$	0.6
X_{101}	$\mathbf{TPW}_{X_2,X_1}$	0.4
..	..	..
..	..	..
X_{190}	$\mathbf{TPW}_{X_9,X_{10}}$	0.55
X_{191}	$\mathbf{HW}_{X_2,X_1}$	0.6
..	..	..
..	..	..
X_{280}	$\mathbf{HW}_{X_9,X_{10}}$	0.4

For role matrix **mcw** (see Box 6.4), note again, as in the previous chapters, that there are two types of cells here: those for constant values (in green) as also in the role matrices discussed in Chap. 2, but here it can be seen that at the base level in all cells adaptive connection weights are indicated (in peach colour), which means that all weights for all 90 connections at the base level are adaptive. In these cells, there are no values written but only names X_k of other states one reification level higher: the names X_{11} to X_{100} of the reification states $\mathbf{W}_{X_i X_j}$ for these connection weights. These refer to the first-order adaptation principle for bonding by homophily, which is further specified by the characteristics for the 90 reification states $\mathbf{W}_{X_i X_j}$ (also indicated by X_{11} to X_{100}): characteristics like their speed factors, their incoming connections, and their combination functions and the parameters thereof.

The role matrices **ms** for the speed factors and **mcfw** for the combination function weights (in Box 6.5) only contain values (in green cells) as all of them are non-adaptive. But in role matrix **mcfp** (also in Box 6.5) there are peach-coloured cells that indicate that the tipping point parameters for the first-order bonding by homophily adaptation principle for all base connections are adaptive. So, for them the names X_k of the corresponding reification states $\mathbf{TP}_{\mathbf{W}_{X_i X_j}}$ at one level higher are indicated (in red cells). This refers to the second-order adaptation principle, which is further specified by the characteristics of the reification states $\mathbf{TP}_{\mathbf{W}_{X_i X_j}}$ for these tipping point parameters (their speed factors, their incoming connections, and their combination functions and the parameters thereof). The initial values (also shown in Box 6.5) were varying from 0.1 to 0.9 for the reification states $\mathbf{W}_{X_i X_j}$ for the nonzero base connection weights and from 0.15 to 0.7 for the 90 reification states $\mathbf{TP}_{\mathbf{W}_{X_i X_j}}$ for the tipping points. For specific initial values for the scenarios, see also below, in Sect. 6.5, Tables 6.2 and 6.4.

6.5 Simulation of the Social Network Model for Adaptive Bonding by Homophily

The example simulation scenarios introduced in Sects. 6.5.1–6.5.3 concern adaptive social network scenarios with 10 persons X_1 to X_{10}. The first two scenarios address a case in which only the outgoing connections of X_1 are adaptive, the other connection weights are kept constant. For all simulations, $\Delta t = 1$ was used, and the focus in all three scenarios was on the homophily adaptation with constant connection weight speed factor $\mathbf{H}_{\mathbf{W}_{X_j,X_i}} = \boldsymbol{\eta}_{\mathbf{W}_{X_j,X_i}} = 1$. In Table 6.2 the main parameter values for Scenarios 1 and 2 can be found. In Table 6.3 the initial connection weights and tipping points for Scenario 1 and 2 are shown. Note that when connection weights are 0 or 1, they will not change due to the specific combination function for bonding by homophily chosen (see also the analysis in Sect. 6.6).

Table 6.2 Main parameter values for Scenario 1/Scenario 2

Base level		First reification level		Second reification level	
Steepness $\boldsymbol{\sigma}$ for **alogistic(..)** for X_i	1	Homophily modulation factor $\boldsymbol{\alpha}$ for $\mathbf{W}_{X_1,X_i}$	1	Tipping point speed factors $\boldsymbol{\eta}$ for $\mathbf{TP}_{\mathbf{W}_{X_1,X_i}}$	1
Threshold $\boldsymbol{\tau}$ for **alogistic(..)** for X_i	1.5	Connection weight speed factor $\boldsymbol{\eta}$ for $\mathbf{W}_{X_1,X_i}$	1	Tipping point modulation factors $\boldsymbol{\alpha}$ for $\mathbf{TP}_{\mathbf{W}_{X_1,X_i}}$	0.1/ 0.9
Speed factor $\boldsymbol{\eta}$ for base state X_i	0.5			Tipping point connection norms $\mathbf{v}$ for $\mathbf{TP}_{\mathbf{W}_{X_1,X_i}}$	0.6

Table 6.3 Scenarios 1 and 2: Initial values for connection weights and tipping points

connections	X_1	X_2	X_3	X_4	X_5	X_6	X_7	X_8	X_9	X_{10}
X_1		0.5	0.3	0.1	0.2	0.6	0.5	0.2	0.3	0.4
X_2	0.5		0.6	0.3	0.4	0.7	0.7	0.9	0.5	
X_3	0.3	0.6		0.7		0.4	0.4		0.6	0.8
X_4	0.6	0.4	0.6		0.4	0.6	0.7	0.8		
X_5	0.2	0.5		0.7	0.6	0.4			0.9	
X_6	0.6	0.6	0.7	0.5			0.7	0.7	0.5	0.7
X_7	0.2	0.8	0.6	0.7	0.6	0.7		0.7		
X_8	0.6	0.5				0.6	0.5		0.4	0.5
X_9	0.6		0.6	0.7	0.4		0.7			0.6
X_{10}	0.6	0.7		0.7	0.4	0.6		0.8		

	$\mathbf{TP}_{\mathbf{W}_{X_1,X_2}}$	$\mathbf{TP}_{\mathbf{W}_{X_1,X_3}}$	$\mathbf{TP}_{\mathbf{W}_{X_1,X_4}}$	$\mathbf{TP}_{\mathbf{W}_{X_1,X_5}}$	$\mathbf{TP}_{\mathbf{W}_{X_1,X_6}}$	$\mathbf{TP}_{\mathbf{W}_{X_1,X_7}}$	$\mathbf{TP}_{\mathbf{W}_{X_1,X_8}}$	$\mathbf{TP}_{\mathbf{W}_{X_1,X_9}}$	$\mathbf{TP}_{\mathbf{W}_{X_1,X_{10}}}$
$\mathbf{TP}_{\mathbf{W}_{X_1,X_i}}(0)$	0.4	0.35	0.5	0.65	0.2	0.3	0.25	0.55	0.6

So, the connections which have weight 0 initially, will keep weight 0 forever. In Chap. 13, various other combination functions for bonding by homophily are explored, among which a class of them which do allow connections with weight 0 or 1 to change.

6.5.1 Scenario 1: Adaptive Connections for One Person; Tipping Point Modulation Factor 0.1

For this scenario (with modulation factor $\boldsymbol{\alpha} = 0.1$), the initial values for connection weights and tipping points can be found in Table 6.2. The average of the initial values of $\mathbf{W}_{X_1,X_i}$ is 0.344, which is below the norm $\mathbf{v}$ which is 0.6. The example simulation for this scenario shown in Figs. 6.2, 6.3, 6.4, 6.5 and 6.6 may look a bit chaotic where some connections seem to meander between high and low. However, in this scenario, it can be seen that the average connection weight, indicated by the pink line converges to 0.60145 (at time point 1750), which is close to 0.6, which was chosen as the norm $\mathbf{v}$ for the average connection weight. So at least this convergence of the average connection weight to $\mathbf{v}$ makes sense. As can be seen in Figs. 6.2 and 6.3 there is some variation of the connection weights around the average connection weight 0.60145 at time 1750. Note that the connection weights at time 1750 do not correlate to the initial values for the connections weights; they are determined by the similarity in states via the homophily principle (see Fig. 6.3). With all of these 9 persons, X_1 initially developed very strong connections (above 0.97) around time 50, but that turned out too much. Therefore 6 of the 9 were reduced between time 100 and 500, while 3 stayed high all the time: $\mathbf{W}_{X_1,X_3}$, $\mathbf{W}_{X_1,X_5}$ and $\mathbf{W}_{X_1,X_9}$. Two of these 6 stayed very low (staying at or around 0): $\mathbf{W}_{X_1,X_8}$ and $\mathbf{W}_{X_1,X_{10}}$.

As with six very low connections, this made the average of connections too low, from these six, three were increased after time 750, and the fourth one after time 1000. Eventually, two of them, $\mathbf{W}_{X_1,X_2}$ and $\mathbf{W}_{X_1,X_7}$, are around 0.6, one, $\mathbf{W}_{X_1,X_4}$, is around 0.8, and one, $\mathbf{W}_{X_1,X_6}$, is around 0.35. So what has emerged is that the person eventually has developed and kept three very good contacts with X_3, X_5, and X_9, has lost two contacts with X_8, X_{10}, and has kept the other four contacts with an intermediate type of different strengths. Figure 6.4 shows the variation in tipping point reification states over time.

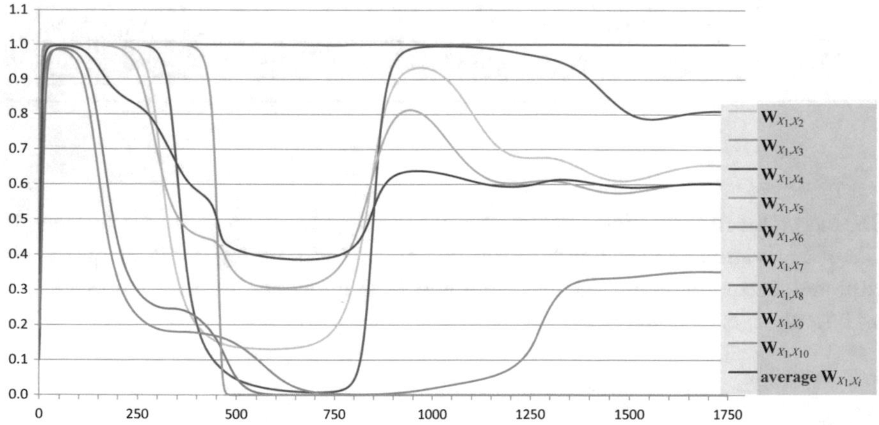

Fig. 6.2 Scenario 1: adaptive weights $\mathbf{W}_{X_1,X_i}$ of outgoing connections from X_1 over time, with the thick pink line showing the average weight of them. This average weight initially is 0.344, which is below the desired value 0.6 of the norm $\mathbf{v}$, but finally, it approximates the value 0.6 of this norm. Before that is achieved, over-controlling reactions cause strong fluctuations of having too intensive connections until time point 300, and again too low intensivity of connections between time point 400 and time point 750

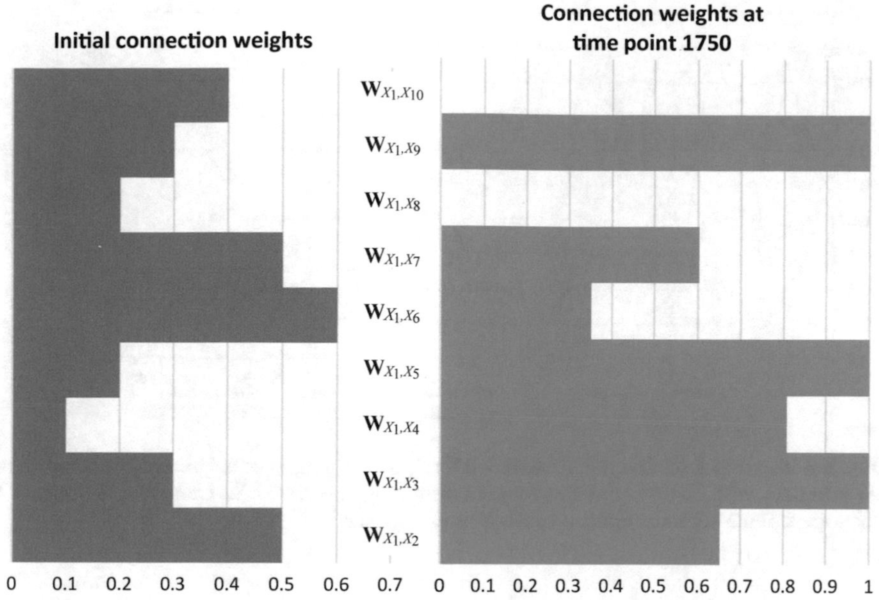

Fig. 6.3 Scenario 1 (modulation factor $\alpha = 0.1$): Resulting connection weights $\mathbf{W}_{X_1,X_i}$ at time point 1750 compared to their initial values. As by the homophily principle the similarity of state values have an important effect on the dynamics of these connection weights, it can be seen that there is not much correlation between initial and final values of the weights

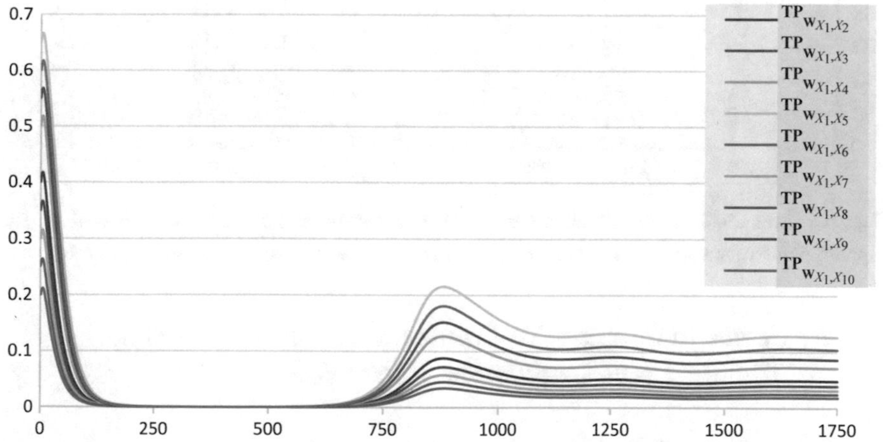

Fig. 6.4 Scenario 1 (modulation factor $\alpha = 0.1$): The adaptive tipping points $\mathbf{TP}_{\mathbf{W}_{X_1,X_j}}$ over time. Initially, the tipping point values were much too high, which explains why in the first phase (until time 400) too many connections were strengthened. After the short initial phase, the tipping point values have been adapted so that they became very low so that the connections were weakened, and finally, they reached some equilibrium values

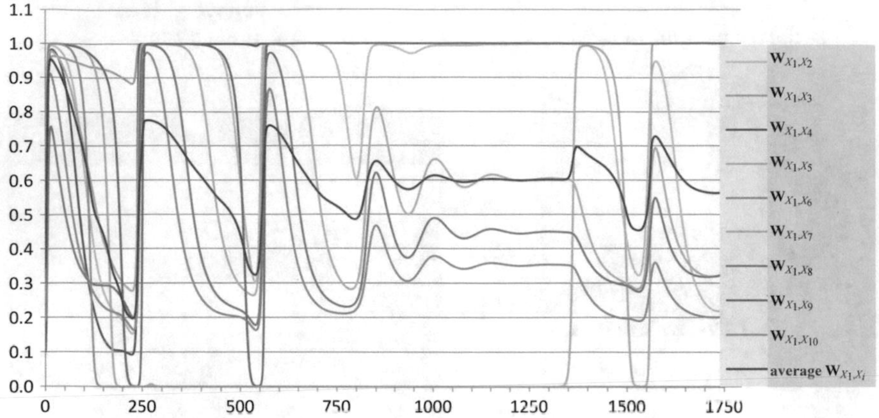

Fig. 6.5 Scenario 2 (modulation factor $\boldsymbol{\alpha}$ = 0.9): Adaptive weights of outgoing connections from X_1 over time, with the thick pink line showing the average weight for X_1. Compared to Scenario 1 this time it turns out more difficult to reach an equilibrium

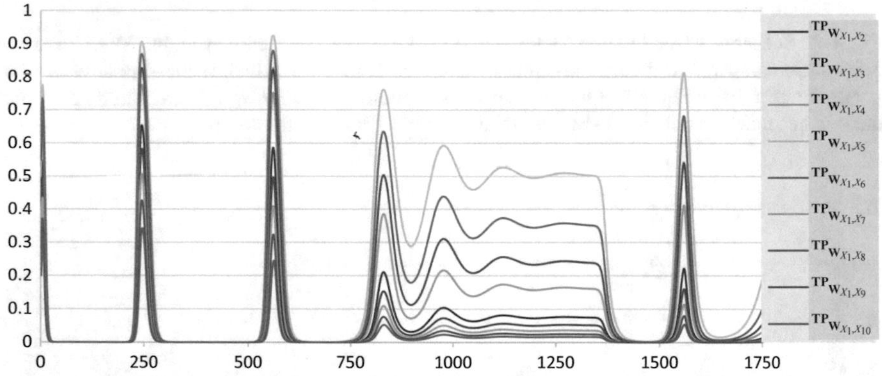

Fig. 6.6 Scenario 2 (modulation factor $\boldsymbol{\alpha}$ = 0.9): Adaptive tipping points $\mathbf{TP}_{\mathbf{W}_{X_1,X_j}}$ over time. The tipping point values react in a rather sensitive manner to the fluctuations shown in Fig. 6.5

6.5.2 Scenario 2: Adaptive Connections for One Person; Tipping Point Modulation Factor 0.9

Scenario 1 shown above is actually not one of the most chaotic scenarios; some other scenarios show a much more chaotic pattern. As an example, when for the tipping point adaptation a much higher modulation factor $\boldsymbol{\alpha}$ = 0.9 is chosen (instead of the 0.1 in Scenario 1; all other values stay the same) the pattern is still more chaotic, as shown in Figs. 6.5 and 6.6. Yet on the long term, the average connection weight moves around the set point 0.6; but notice that around time point 1250, it seemed that the process was close to an equilibrium, but that was violated by later fluctuations. Moreover, the fluctuating pattern of the tipping points in Fig. 6.6 also does not suggest that it will become stable.

6.5.3 Scenario 3: Adaptive Connections for All Persons

In Scenario 3 all connections are adaptive with main parameters shown in Table 6.4 and initial connection weight values shown in Table 6.5. The norm for average connection weight is 0.4 this time.

In Figs. 6.7, 6.8, 6.9 and 6.10, the simulation results are shown for Scenario 3. As can be seen in Fig. 6.10 eventually all connection weights converge to 0 or 1. Figure 6.7 shows, in particular, the values of the connection weights from X_1, and their average, and Fig. 6.8 shows the corresponding tipping points.

Figure 6.9 shows that in the process eventually the average connection weights per person converge in some unclear manner to a discrete set of equilibrium values: 0.111111 (connections of X_{10}), 0.222222 (connections of X_5), 0.333333 (connections of X_3, X_9), and 0.555555 (connections of X_1, X_2, X_4, X_6, X_7, X_8), all multiples of 0.111111; the overall average ends up in 0.433333 (recall that the norm $\mathbf{v}$ of $\mathbf{TP}_{\mathbf{W}_{X_i X_j}}$ for average connection weight for each person was 0.4). Also in other simulations, this discrete set of multiples of 0.111111 shows up. These specific values will be discussed as part of the analysis in Sect. 6.6.

Figure 6.10 shows that all connection weights $\mathbf{W}_{X_1,X_1}$ to $\mathbf{W}_{X_{10},X_{10}}$ (1–100) converge to 0 or 1. Note that connections from one state to itself were 0 by definition (in Fig. 6.10 the numbers 1, 12, 23, 34, 45, 56, 67, 78, 89, 100). Also this will be analysed in Sect. 6.6. The tipping points for all outgoing connections of X_1 converge to 0 (see also Fig. 6.8), and for all outgoing connections of the other persons, they converge to 1.

Table 6.4 Scenario 3: Main parameter values

Base level		First reification level		Second reification level	
Contagion alogistic steepness $\boldsymbol{\sigma}$ for X_i	0.8	Homophily modulation factor $\boldsymbol{\alpha}$ for $\mathbf{W}_{X_j,X_i}$	1	Tipping point speed factor $\boldsymbol{\eta}$ for $\mathbf{TP}_{\mathbf{W}_{X_j,X_i}}$	0.5
Contagion alogistic threshold $\boldsymbol{\tau}$ for X_i	0.15	Connection weight speed factor $\boldsymbol{\eta}$ for $\mathbf{W}_{X_j,X_i}$	1	Tipping point modulation factor $\boldsymbol{\alpha}$ for $\mathbf{TP}_{\mathbf{W}_{X_j,X_i}}$	0.4
Speed factor $\boldsymbol{\eta}$ for base state X_i	0.5			Tipping point connection norm $\mathbf{v}$ for $\mathbf{TP}_{\mathbf{W}_{X_j,X_i}}$	0.4

Table 6.5 Scenario 3: Initial connection weights

connections	X_1	X_2	X_3	X_4	X_5	X_6	X_7	X_8	X_9	X_{10}
X_1		0.5	0.3	0.1	0.2	0.6	0.5	0.2	0.3	0.4
X_2	0.5		0.6	0.3	0.4	0.7	0.7	0.9	0.5	
X_3	0.3	0.6		0.7	0.7	0.4	0.4		0.6	0.8
X_4	0.6	0.4	0.6		0.4	0.6	0.7	0.8		0.9
X_5	0.2	0.5		0.7		0.4		0.4	0.9	0.4
X_6	0.6	0.6	0.7	0.5			0.7	0.7	0.5	0.7
X_7	0.2	0.8	0.6	0.7	0.6	0.7		0.7		
X_8	0.6	0.5		0.4		0.6	0.5		0.4	0.5
X_9	0.6		0.6	0.7	0.4		0.7			0.6
X_{10}	0.6	0.7		0.7	0.4	0.6		0.8		

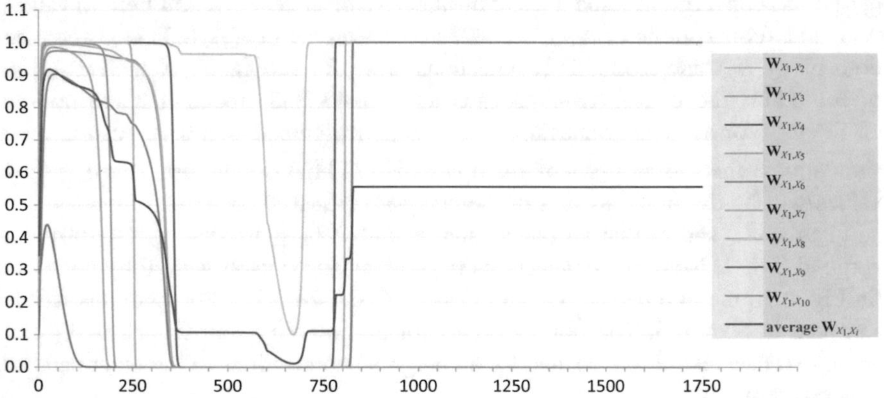

Fig. 6.7 Scenario 3: Adaptive weights of outgoing connections from X_1 over time, with the pink line showing the average weight for X_1. Here, after time 750 all connection weights become 0 or 1

6.6 Analysis of the Equilibria of the Reification States

In this section, the possible values to which certain states in the second-order reified network may converge are analysed. Recall the following definition and criterion for stationary points and equilibria.

Definition (stationary point and equilibrium)

A state Y has a *stationary point* at t if $\mathbf{d}Y(t)/\mathbf{d}t = 0$. The network is in *equilibrium* at t if every state Y of the model has a stationary point at t.

Note that this Y also applies to the reification states. Given the differential equation for a temporal-causal network model, a more specific criterion can be found:

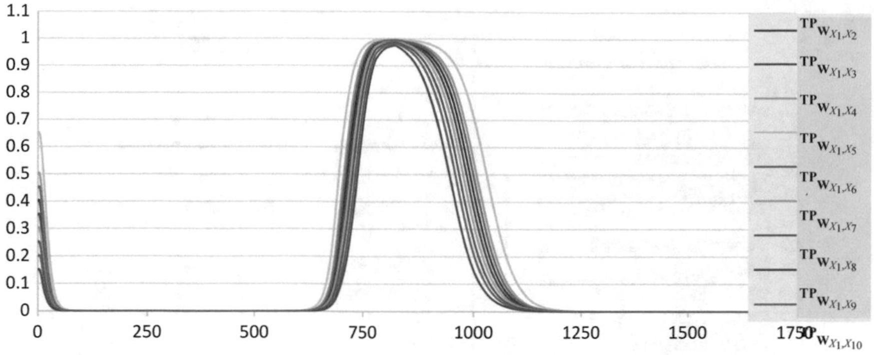

Fig. 6.8 Scenario 3: Adaptive tipping points $\mathbf{TP}_{\mathbf{W}_{X_1,X_j}}$ over time. Due to the very low connection weights between 500 and 750 (see Fig. 6.7), the tipping point values show a strong temporary increase to enable strengthening of connections

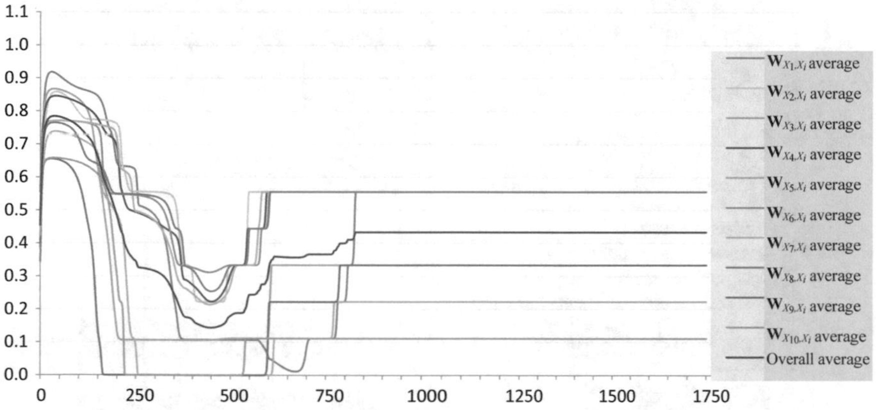

Fig. 6.9 Scenario 3: Average connection weights for each of X_1 to X_{10} and an overall average of all connections over time

Criterion for a stationary point in a temporal-causal network

Let Y be a state with speed factor $\boldsymbol{\eta}_Y$ and $X_1, \ldots, X_k$ the states from which state Y has incoming connections. Then Y has a stationary point at t iff

$$\boldsymbol{\eta}_Y = 0 \text{ or } \mathbf{c}_Y(\boldsymbol{\omega}_{X_1,Y}X_1(t), \ldots, \boldsymbol{\omega}_{X_k,Y}X_k(t)) = Y(t) \quad (6.1)$$

For an equilibrium these are called *equilibrium equations*.

This can be applied to the states at all levels in the second-order reified network, in particular to the first and second reification level. The solutions found (assuming parameters $\boldsymbol{\alpha}_1$, $\boldsymbol{\alpha}_2$ and $\boldsymbol{\alpha}_3$ nonzero) are

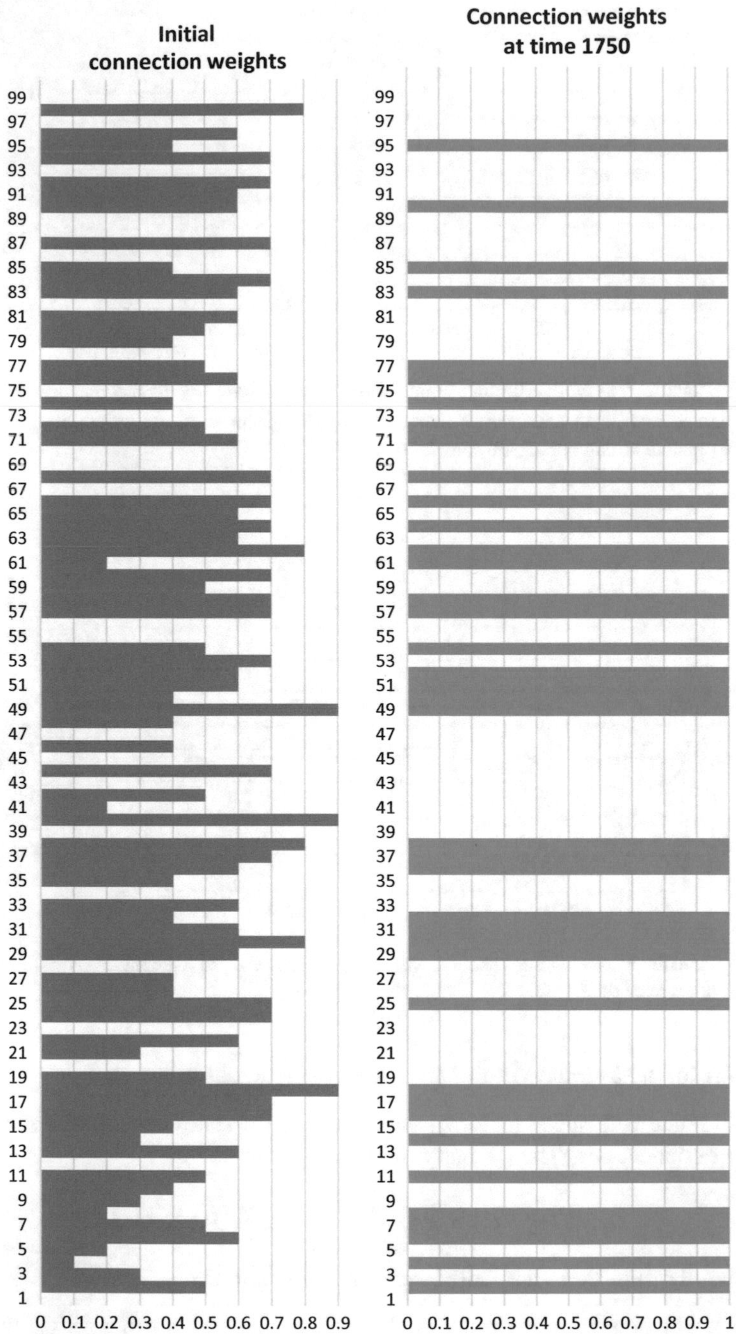

Fig. 6.10 Scenario 3: The connection weights from $\mathbf{W}_{X_1,X_1}$ to $\mathbf{W}_{X_{10},X_{10}}$ (1 to 100) initially, and finally they are 0 or 1

$$\begin{array}{lll} \underline{\mathbf{T}} = 0 \text{ or } \underline{\mathbf{T}} = 1 & \text{or} & (\underline{\mathbf{W}}_1 + \ldots \underline{\mathbf{W}}_k)/k = \mathbf{v} \\ \underline{\mathbf{H}} = 0 \text{ or } \underline{\mathbf{H}} = 1 & \text{or} & (\underline{\mathbf{W}}_1 + \ldots \underline{\mathbf{W}}_k)/k = \mathbf{v} \\ \underline{\mathbf{H}} = 0 \text{ or } \underline{\mathbf{W}} = 0 & \text{or} & \underline{\mathbf{W}} = 1 \text{ or } |X_i - X_j| = \underline{\mathbf{T}} \end{array}$$

More details about this can be found in Box 6.6. In principle, this gives $4*3^2 = 36$ combined solutions, but in practice only 8. For the first two, by combining the different cases, the following five options can be found:

$$\begin{array}{l} \underline{\mathbf{T}} = 0 \text{ and } \underline{\mathbf{H}} = 0 \\ \underline{\mathbf{T}} = 1 \text{ and } \underline{\mathbf{H}} = 0 \\ \underline{\mathbf{T}} = 0 \text{ and } \underline{\mathbf{H}} = 1 \\ \underline{\mathbf{T}} = 1 \text{ and } \underline{\mathbf{H}} = 1 \\ (\underline{\mathbf{W}}_1 + \ldots + \underline{\mathbf{W}}_k)/k = \mathbf{v} \end{array}$$

In cases that $\underline{\mathbf{H}}$ is nonzero and $|X_i(t) - X_j(t)| \neq \underline{\mathbf{T}}$ (which was the case for the simulations displayed in Sect. 6.5), all $\underline{\mathbf{W}}_j$ are 0 or 1, so with $k = 9$ the average $(\underline{\mathbf{W}}_1 + \ldots + \underline{\mathbf{W}}_k)/k$ is a multiple of $\frac{1}{9}$. In the simulation case the norm $\mathbf{v} = 0.4$ is not a multiple of $\frac{1}{9}$. Therefore the cases $(\underline{\mathbf{W}}_1 + \ldots + \underline{\mathbf{W}}_k)/k = \mathbf{v}$ cannot actually occur, so then $\underline{\mathbf{T}} = 0$ or $\underline{\mathbf{T}} = 1$ and $\underline{\mathbf{W}} = 0$ or $\underline{\mathbf{W}} = 1$ are the only solutions for the first two; then all solutions are:

$$\begin{array}{l} \underline{\mathbf{T}} = 0 \text{ and } \underline{\mathbf{H}} = 0 \\ \underline{\mathbf{T}} = 1 \text{ and } \underline{\mathbf{H}} = 0 \end{array}$$

$$\begin{array}{l} \underline{\mathbf{T}} = 0 \text{ and } \underline{\mathbf{H}} = 1 \text{ and } \underline{\mathbf{W}} = 0 \\ \underline{\mathbf{T}} = 0 \text{ and } \underline{\mathbf{H}} = 1 \text{ and } \underline{\mathbf{W}} = 1 \\ \underline{\mathbf{T}} = 0 \text{ and } \underline{\mathbf{H}} = 1 \text{ and } |X_i - X_j| = \underline{\mathbf{T}} \end{array}$$

$$\begin{array}{l} \underline{\mathbf{T}} = 1 \text{ and } \underline{\mathbf{H}} = 1 \text{ and } \underline{\mathbf{W}} = 0 \\ \underline{\mathbf{T}} = 1 \text{ and } \underline{\mathbf{H}} = 1 \text{ and } \underline{\mathbf{W}} = 1 \\ \underline{\mathbf{T}} = 1 \text{ and } \underline{\mathbf{H}} = 1 \text{ and } |X_i - X_j| = \underline{\mathbf{T}} \end{array}$$

This is also shown by the simulations (e.g., see Fig. 6.10); indeed all averages are multiples of $\frac{1}{9} = 0.111111$, as found above (see Fig. 6.9). This explains the discrete set of numbers 0.111111, 0.222222, 0.333333, … observed in the simulations; as shown here, this strongly depends on the number of states. Note that although in general the reified speed factor $\mathbf{H}_{\mathbf{W}_{X_i X_j}}(t)$ may be assumed nonzero, there may also be specific processes in which it converges to 0, for example, like the temperature in simulated annealing.

Box 6.6 Solving the equilibrium equations for the second-order reification states $\mathbf{TP}_{\mathbf{W}_{X_iX_j}}$ and $\mathbf{H}_{\mathbf{W}_{X_iX_j}}$ and the first-order reification state $\mathbf{W}_{X_iX_j}$

As a first step, the above criterion applied to tipping point reification states at the second reification level is as follows where $\underline{\mathbf{T}}$ is the considered equilibrium value for $\mathbf{TP}_{\mathbf{W}_{X_iX_j}}$, and $\underline{\mathbf{W}}_j$ for $\mathbf{W}_{X_iX_j}$.

$$\mathbf{sltip}_{\mathbf{v},\alpha_1}(\underline{\mathbf{W}}_1,\ldots,\underline{\mathbf{W}}_k,\underline{\mathbf{T}}) = \underline{\mathbf{T}}$$

This provides the following equation

$$\underline{\mathbf{T}} + \alpha_1\underline{\mathbf{T}}(1-\underline{\mathbf{T}})(\mathbf{v} - (\underline{\mathbf{W}}_1 + \ldots + \underline{\mathbf{W}}_k)/k) = \underline{\mathbf{T}}$$

which by subtracting $\underline{\mathbf{T}}$ can be rewritten as follows

$$\alpha_1\,\underline{\mathbf{T}}(1-\underline{\mathbf{T}})(\mathbf{v} - (\underline{\mathbf{W}}_1 + \ldots + \underline{\mathbf{W}}_k)/k) = 0$$

Assuming $\boldsymbol{\alpha_1}$ nonzero, this equation has three solutions:

$$\underline{\mathbf{T}} = 0 \text{ or } \underline{\mathbf{T}} = 1 \text{ or } (\underline{\mathbf{W}}_1 + \ldots + \underline{\mathbf{W}}_k)/k = \mathbf{v}$$

The same can be done for the combination function for the speed:

$$\mathbf{slspeedopp}_{\mathbf{v},\alpha_2}(\underline{\mathbf{W}}_1,\ldots,\underline{\mathbf{W}}_k,\underline{\mathbf{H}}) = \underline{\mathbf{H}}$$
$$\underline{\mathbf{H}} + \alpha_2\underline{\mathbf{H}}(1-\underline{\mathbf{H}})(\mathbf{v} - (\underline{\mathbf{W}}_1 + \ldots + \underline{\mathbf{W}}_k)/k) = \underline{\mathbf{H}}$$
$$\alpha_2\underline{\mathbf{H}}(1-\underline{\mathbf{H}})(\mathbf{v} - (\underline{\mathbf{W}}_1 + \ldots + \underline{\mathbf{W}}_k)/k) = 0$$

Assuming $\boldsymbol{\alpha_2}$ is nonzero this has the following solutions:

$$\underline{\mathbf{H}} = 0 \text{ or } \underline{\mathbf{H}} = 1 \text{ or } (\underline{\mathbf{W}}_1 + \ldots + \underline{\mathbf{W}}_k)/k = \mathbf{v}$$

Similarly, the criterion can be applied to connection weights at the first reification level:

$$\mathbf{adapslhomo}_{\alpha_3}(\underline{\mathbf{H}}, X_i, X_j, \underline{\mathbf{T}}, \underline{\mathbf{W}}) = \underline{\mathbf{W}}$$

This provides the following equation, where $\underline{\mathbf{H}}$ is the considered equilibrium value for $\mathbf{H}_{\mathbf{W}_{X_iX_j}}(t)$ and $\underline{\mathbf{W}}$ the considered equilibrium value for $\mathbf{W}_{Xi,\ Xj}$:

$$\underline{\mathbf{H}}\big(\underline{\mathbf{W}} + \alpha_3\underline{\mathbf{W}}(1-\underline{\mathbf{W}})\big(\underline{\mathbf{T}} - |X_i - X_j|\big)\big) + (1-\underline{\mathbf{H}})\underline{\mathbf{W}} = \underline{\mathbf{W}}$$

which by subtracting $\underline{\mathbf{W}}$ can be rewritten into

$$\underline{\mathbf{H}}\,\alpha\,\underline{\mathbf{W}}(1 - \underline{\mathbf{W}})\left(\underline{\mathbf{T}} - |X_i - X_j|\right) = 0$$

Assuming $\boldsymbol{\alpha}_3$ nonzero, this equation has four solutions:

$$\underline{\mathbf{H}} = 0 \text{ or } \underline{\mathbf{W}} = 0 \text{ or } \underline{\mathbf{W}} = 1 \text{ or} |X_i - X_j| = \underline{\mathbf{T}}$$

6.7 Discussion

In this chapter, a second-order reified adaptive Social Network model was presented. The reified network model is based on a first-order adaptation principle for bonding-by-homophily from Social Science (Byrne 1986; McPherson et al. 2001; Pearson et al. 2006; Sharpanskykh and Treur 2014) represented at the first reification level, and in addition for a second-order adaptation principle describing change of the characteristics 'similarity tipping point' and 'speed factor' of this first-order adaptation principle.

First-order adaptive network models for bonding by homophily, that can be considered precursors of the current second-order network model, have been compared to empirical data sets in (van Beukel et al. 2019; Blankendaal et al. 2016; Boomgaard et al. 2018). It is an interesting challenge to compare the new second-order network model described here to empirical data. This is left for a future enterprise. An extension to a first-order adaptive network for bonding by multicriteria homophily was addressed in (Kozyreva et al. 2018). This also could be developed further to a second-order adaptive network model.

Also in further Social Science literature, cases are reported where network adaptation is itself adaptive. For example, in (Carley et al. 2001; Carley 2002, 2006) the second-order adaptation concept called 'inhibiting adaptation' for network organisations is described. For further work, it would be interesting to explore the applicability of the introduced modeling environment for such domains further.

References

Blankendaal, R., Parinussa, S., Treur, J.: A temporal-causal modelling approach to integrated contagion and network change in social networks. In: Proceeding of the 22nd European Conference on Artificial Intelligence, ECAI'16. IOS Press, Frontiers in Artificial Intelligence and Applications, vol. 285, pp. 1388–1396 (2016)

Boomgaard, G., Lavitt, F., Treur, J.: Computational analysis of social contagion and homophily based on an adaptive social network model. In: Proceedings of the 10th International Conference on Social Informatics, SocInfo'18. Lecture Notes in Computer Science, vol. 11185, pp. 86–101, Springer Publishers (2018)

Byrne, D.: The attraction hypothesis: do similar attitudes affect anything? J. Pers. Soc. Psychol. **51** (6), 1167–1170 (1986)

Carley, K.M.: Inhibiting adaptation. In: Proceedings of the 2002 Command and Control Research and Technology Symposium, pp. 1–10. Naval Postgraduate School, Monterey, CA (2002)

Carley, K.M.: Destabilization of covert networks. Comput. Math. Organ. Theor. **12**, 51–66 (2006)

Carley, K.M., Lee, J.-S., Krackhardt, D.: Destabilizing networks. Connections **24**(3), 31–34 (2001)

Holme, P., Newman, M.E.J.: Nonequilibrium phase transition in the coevolution of networks and opinions. Phys. Rev. E **74**(5), 056108 (2006)

Kozyreva, O., Pechina, A. Treur, J.: Network-oriented modeling of multi-criteria homophily and opinion dynamics in social media. In: Koltsova, O., Ignatov, D.I., Staab, S. (eds.) Social Informatics: Proceedings of the 10th International Conference on Social Informatics, SocInfo'18, vol. 1. Lecture Notes in AI, vol. 11185, pp. 322–335 Springer (2018)

Levy, D.A., Nail, P.R.: Contagion: a theoretical and empirical review and reconceptualization. Genet. Soc. Gen. Psychol. Monogr. **119**(2), 233–284 (1993)

McPherson, M., Smith-Lovin, L., Cook, J.M.: Birds of a feather: homophily in social networks. Annu. Rev. Soc. **27**, 415–444 (2001)

Pearson, M., Steglich, C., Snijders, T.: Homophily and assimilation among sport-active adolescent substance users. Connections **27**(1), 47–63 (2006)

Sharpanskykh, A., Treur, J.: Modelling and analysis of social contagion in dynamic networks. Neurocomputing **146**, 140–150 (2014)

Treur, J.: Network-oriented modeling: addressing complexity of cognitive, affective and social interactions. Springer Publishers (2016)

van Beukel, S., Goos, S., Treur, J.: An adaptive temporal-causal network model for social networks based on the homophily and more-becomes-more principle. Neurocomputing **338**, 361–371 (2019)

Vazquez, F.: Opinion dynamics on coevolving networks. In: Mukherjee, A., Choudhury, M., Peruani, F., Ganguly, N., Mitra, B. (eds.) Dynamics On and Of Complex Networks, Volume 2, Modeling and Simulation in Science, Engineering and Technology, pp. 89–107. Springer, New York (2013)

Vazquez, F., Gonzalez-Avella, J.C., Eguíluz, V.M., San Miguel, M.: Time-scale competition leading to fragmentation and recombination transitions in the coevolution of network and states. Phys. Rev. E **76**, 046120 (2007)

Chapter 7
Modeling Higher-Order Adaptive Evolutionary Processes by Reified Adaptive Network Models

Abstract In this chapter, a fourth-order reified network model is introduced to describe different orders of adaptivity found in a case study on evolutionary processes. The network model describes how the causal pathways for newly developed features in this case study affect the causal pathways of already existing features, which makes the pathways of these new features one order of adaptivity higher than the existing ones, as they adapt a previous adaptation. The network reification approach is shown to be an adequate means to model this transparently.

7.1 Introduction

In the literature, many examples can be found relating to first-order adaptive networks, in different (e.g., cognitive, mental, social) domains. For second-order adaptive networks there is at least a substantial amount of Cognitive Neuroscience literature on metaplasticity that controls under which circumstances and to which extent plasticity should occur; e.g., Abraham and Bear (1996), Schmidt et al. (2013). This has been used in Chap. 4, Sect. 4.4 to design a second-order adaptive Mental Network model, which also has been applied in Chap. 5 to model adaptive decision making under extreme emotions. Also for Social Science, some literature refers to second-order adaptive networks, but only a modest amount; e.g., Carley (2002, 2006). In Chap. 6, an example second-order reified adaptive Social Network model was addressed. So far about second-order adaptive networks; but how about adaptive networks of order higher than two? Recall that in Chap. 1, Sect. 1.3 such a question was discussed concerning the occurrence or relevance for real world applications of adaptive networks of order higher than two in the literature. The outcome seemed a bit disappointing. In the current and next chapter answers are obtained on this question, positive answers: the two chapters are discussing three examples of reified networks of order higher than two.

The current chapter focuses on a case study of evolutionary processes, and the orders of adaptation that are recognized in them or attributed to them; e.g., Fessler et al. (2005, 2015), Jones et al. (2005), Fleischman and Fessler (2011). In Chap. 1,

J. Treur, *Network-Oriented Modeling for Adaptive Networks: Designing Higher-Order Adaptive Biological, Mental and Social Network Models*, Studies in Systems, Decision and Control 251, https://doi.org/10.1007/978-3-030-31445-3_7

Sects. 1.2.2 and 1.3.2 some input from that side was already pointed out. One of these case studies addresses how the existence of pathogens has led to the adaptation of developing a defense system with an internal component (internal immune system) and (maybe not at the same time) an external component based on disgust, sometimes called a behavioural immune system (Aarøe et al. 2017; Schaller 2016; Schaller and Park 2011). On top of that, (first trimester) pregnancy led to the adaptation of temporary suppression of this defense system to give the half-foreign conceptus a chance to get embedded. Moreover, above that, as another adaptation, for the first trimester of pregnancy, a strong feeling of disgust was developed to still strengthen the overall defense system by strengthening, in particular, the external component of it. For more information about this interesting research area, see Aarøe et al. (2017), Curtis et al. (2011), Fleischman and Fessler (2018), Jones et al. (2017), Lieberman et al. (2018), Oaten et al. (2009), Schaller and Park (2011), Schaller (2016), Tybur and Lieberman (2016).

The case study pointed out above is chosen here, analysed in some depth and modeled by a fourth-order reified network model. For this model, different scenarios were simulated, and mathematical analysis of its emerging behaviour in terms of equilibria was performed, part of which was also used to verify the model. In Sect. 7.2, the case study itself is briefly discussed. Section 7.3 introduces the fourth-order reified network model, and Sect. 7.4 the simulations with it. Section 7.5 addresses the mathematical analysis of the model's emerging behaviour, and verification of the model based on this analysis and simulations for the case study.

7.2 Higher-Order Adaptation in Evolutionary Processes

Viewed from a distance, evolutionary processes are adaptation processes that are changing the physical world by creating new causal pathways or blocking existing causal pathways. This can be described as changing the causal connections in such causal pathways from 0 or very low to high, or conversely. The adaptations are driven by changing environmental circumstances, making that organisms with more favourable causal pathways for these circumstances become more dominant. Then they determine the average causal pathways of the population to a higher extent: this leads to a shift in the average pathways by changes in the causal connections in these pathways. These circumstances can be considered as environmental properties. In particular, consider the quote shown in Chap. 1, Sect. 1.3.2:

> Also of relevance here, one form of disgust, pathogen disgust, functions in part as a third-order adaptation, as disease-avoidance responses are up-regulated in a manner that compensates for the increases in vulnerability to pathogens that accompany pregnancy and preparation for implantation – changes that are themselves a second-order adaptation addressing the conflict between maternal immune defenses and the parasitic behavior of the half-foreign conceptus (Fessler et al. 2005; Jones et al. 2005; Fleischman and Fessler 2011). (Fessler et al. 2015)

From this quote, it is suggested that three levels of adaptation might be considered applicable for the first trimester of pregnancy. However, also the occurrence of pathogens can be considered a form of adaptation for the wider ecological context. Therefore the following four adaptation orders can be distinguished:

- First-order adaptation:
  Pathogens occur, with causal pathways negatively affecting the causal pathways for good health.
- Second-order adaptation:
  An internal defense system occurs, with causal pathways which negatively affects the causal pathways used by pathogens.
- Third-order adaptation:
  For pregnancy, causal pathways are added to make the defense system's causal pathways less strong as the half-foreign conceptus might easily be identified as a kind of parasite and attacked.
- Fourth-order adaptation:
  Disgust during (first trimester) pregnancy adds causal pathways by which potential pathogens in the external world are avoided so that fewer risks are taken for entering pathogens while the internal defense system (the internal immune system) is low functioning. This strengthens the overall defense system by strengthening the external defense system (the behavioural immune system) by which the pathogens are addressed outside the body; this makes the causal pathway from (first trimester) pregnancy to suppress the causal pathways of the overall defense system less strong as the external component of the defense system strengthened by disgust is not addressed by it.

So, can this be used as a basis for a fourth-order reified adaptive network model? This will be addressed in Sect. 7.3.

7.3 A Reified Network Model for Fourth-Order Adaptive Evolutionary Processes

The Network-Oriented Modeling approach used to model these evolutionary processes is based on reified temporal-causal network models as presented in Chaps. 3 and 4; see also (Treur 2018a, b). Recall that a temporal-causal network model in the first place involves representing in a declarative manner states and connections between them that represent (causal) impacts of states on each other, as assumed to hold for the application domain addressed. The states are assumed to have (activation) levels, usually in the interval [0, 1], that vary over time. The following three notions form the defining part of a conceptual representation of a temporal-causal network model structure (Treur 2016, 2019a):

- **Connectivity**
  - Each incoming connection of a state *Y*, from a state *X* has a *connection weight value* $\boldsymbol{\omega}_{X,Y}$ representing the strength of the connection.
- **Aggregation**
  - For each state a *combination function* $\mathbf{c}_Y(..)$ is chosen to combine the causal impacts state *Y* receives from other states
- **Timing**
  - For each state *Y* a *speed factor* $\boldsymbol{\eta}_Y$ is used to represent how fast state *Y* is changing upon causal impact.

The notion of *network reification* as introduced in Chaps. 3 and 4 is a conceptual tool to model adaptive networks more transparently within a Network-Oriented Modelling perspective. Reification literally means representing something abstract as a material or more real concrete thing (Merriam-Webster and Oxford dictionaries). This concept is used in different scientific areas in which it has been shown to provide substantial advantages in expressivity and transparency of models, and, in particular, within AI; e.g., Davis and Buchanan (1977), Davis (1980), Galton (2006), Sterling and Beer (1989), Weyhrauch (1980). Specific cases of reification from a linguistic or logical perspective are representing relations between objects by objects themselves, or representing more complex statements by objects or numbers.

For network models, reification can be applied by reifying network structure characteristics (such as the $\boldsymbol{\omega}_{X,Y}, \mathbf{c}_Y(..)$, $\boldsymbol{\eta}_Y$ indicated above) in the form of additional network states (called *reification states*, indicated by $\mathbf{W}_{X,Y}, \mathbf{C}_Y, \mathbf{H}_Y$, respectively) within an extended network. According to the specific network structure characteristic represented, *roles* **W**, **C**, **H** are assigned to reification states (or to values): *connection weight reification*, *combination function reification*, *speed factor reification*, respectively. Also, a role **P** for *combination function parameters* is used. A format based on *role matrices* **mb** (for base role), **mcw** (for connection weight role **W**), **mcfw** (for combination function weight role **C**), **mcfp** (for combination function parameter role **P**), and **ms** (for speed factor role **H**), is used to specify a reified network model according to these roles. Multilevel reified networks can be used to model networks which are adaptive of different orders. For more details, see Chaps. 3 and 4, or (Treur 2018a, b).

Inspired by the information on evolutionary processes in Sect. 7.2 but abstracting from specific details, a fourth-order reified adaptive network for these evolutionary processes has been designed. First, the general blueprint is discussed (Sect. 7.3.1), and next, the more specific network model (Sect. 7.3.2).

7.3.1 Adaptive Causal Modeling of Changing Causal Pathways in Evolutionary Processes

As pointed out in Sect. 7.2, first paragraph, evolutionary adaptation usually concerns affecting existing causal pathways by adding new causal pathways that weaken or strengthen the existing causal pathways. This indicates that levels of adaptation can be created where the causal pathways at one adaptation level are adapted by the causal pathways at the next level. The adaptation of a causal pathway can be done by strengthening or weakening one or more causal connections within such a causal pathway. This fits well in a reified network architecture where for each level, for connection weights in causal pathways at that level, reification states are introduced at the next level. The general representation then becomes in a simple form:

- **Base level**:
  causal pathway by a causal connection from a to b
- **First adaptation level**:
  causal pathway by a causal connection from a_1 to $\mathbf{W}_{a,b}$; this $\mathbf{W}_{a,b}$ represents the causal connection from a to b from the base level
- **Second adaptation level**:
  causal pathway by a causal connection from a_2 to $\mathbf{W}_{a_1,\mathbf{W}_{a,b}}$; this $\mathbf{W}_{a_1,\mathbf{W}_{a,b}}$ represents the causal connection from a_1 to $\mathbf{W}_{a,b}$ from the first adaptation level
- **Third adaptation level**:
  causal pathway by a causal connection from a_3 to $\mathbf{W}_{a_2,\mathbf{W}_{a_1,\mathbf{W}_{a,b}}}$; this $\mathbf{W}_{a_2,\mathbf{W}_{a_1,\mathbf{W}_{a,b}}}$ represents the causal connection from a_2 to $\mathbf{W}_{a_1,\mathbf{W}_{a,b}}$ from the second adaptation level
- **Fourth adaptation level**:
  causal pathway by a causal connection from a_4 to $\mathbf{W}_{a_3,\mathbf{W}_{a_2,\mathbf{W}_{a_1,\mathbf{W}_{a,b}}}}$; this $\mathbf{W}_{a_3,\mathbf{W}_{a_2,\mathbf{W}_{a_1,\mathbf{W}_{a,b}}}}$ represents the causal connection from a_3 to $\mathbf{W}_{a_2,\mathbf{W}_{a_1,\mathbf{W}_{a,b}}}$ from the third adaptation level
- **Fifth adaptation level**:
  etc.

This general pattern for hierarchical adaptation processes for causal pathways will be used to obtain a more specific reified network model for the multilevel adaptation processes described in Sect. 7.2.

7.3.2 The Reified Adaptive Network Model for the Described Fourth-Order Adaptation Case

In the considered reified network model four reification levels are considered, where for each level its causal pathway can be changed by causal pathways at one

Table 7.1 The states and their explanations

State		Explanation	level
X_1	s_1	Occurrence of pathogens	
X_2	s_2	Occurrence of internal defense system	
X_3	s_3	Occurrence of pregnancy	Base level
X_4	s_4	Occurrence of disgust	
X_5	s_5	Contextual circumstances	
X_6	e_1	Health level; on a causal pathway with a connection from s_5 for context	
X_7	$\mathbf{W}_{s5,e_1}$	Reified representation state for the weight of the base level connection from s_5 for context to $e1$ for health level; on a causal pathway with a connection from s_1 for pathogens	First reification level
X_8	$\mathbf{W}_{s1,\mathbf{W}_{s5,e_1}}$	Reified representation state for the weight of the first reification level connection from s_1 for pathogens to $\mathbf{W}_{s5,e_1}$; on a causal pathway with a connection from s_2 for interrnal defense system	Second reification level
X_9	$\mathbf{W}_{s2,\mathbf{W}_{s1,\mathbf{W}_{s5,e_1}}}$	Reified representation state for the weight of the second reification level connection from s_2 for internal defense system to $\mathbf{W}_{s1,\mathbf{W}_{s5,e_1}}$; on a causal pathway with a connection from s_3 for pregnancy	Third reification level
X_{10}	$\mathbf{W}_{s3,\mathbf{W}_{s2,\mathbf{W}_{s1,\mathbf{W}_{s5,e_1}}}}$	Reified representation state for the weight of the third reification level connection from s_3 for pregnancy to $\mathbf{W}_{s2,\mathbf{W}_{s1,\mathbf{W}_{s5,e_1}}}$; on a causal pathway with a connection from s_4 for disgust	Fourth reification level

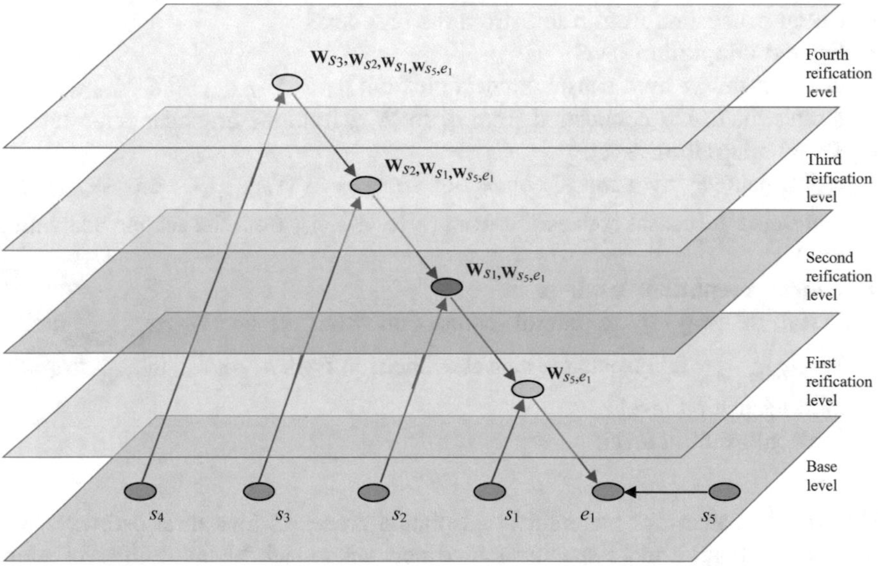

Fig. 7.1 Fourth-order reified network model for fourth-order adaptation in an evolutionary context

level higher. To limit the complexity of the overall model, the causal pathways at each level are kept simple; they will have just one causal connection. Table 7.1 explains the states of the network model.

Figure 7.1 shows a picture of the conceptual graphical representation of the reified network model, and Box 7.1 shows the role matrices **mb** (base connectivity),

mcw (connection weights), **ms** (speed factors), **mcfw** (combination function weights), **mcfp** (combination function parameters). Each role matrix has a format in which in each row for the indicated state it is specified which other states (red cells) or values (green cells) affect it and according to which role. In particular, in role matrix **mcw** the red cells indicate which states X_i play the role of the reification states for the weights of the connection indicated in that cell in **mb**. Note that this time there are no connections from a reification state to itself. Instead, the input states s_1, s_2, and s_3 have connections to themselves; these are used to give them appropriately timed entrance times, so that not everything happens at once.

The reified network model includes four reification states at four levels which each reify the connection weight of the causal pathway one level lower:

- $\mathbf{W}_{s_5,e_1}$ a first reification level state representing the causal connection from s_5 to e_1 from the base level
- $\mathbf{W}_{s_1,\mathbf{W}_{s_5,e_1}}$ a second reification level state representing the causal connection from s_1 to $\mathbf{W}_{s_5,e_1}$ from the first reification level
- $\mathbf{W}_{s_2,\mathbf{W}_{s_1,\mathbf{W}_{s_5,e_1}}}$ a third reification level state representing the causal connection from s_2 to $\mathbf{W}_{s_1,\mathbf{W}_{s_5,e_1}}$ from the second reification level
- $\mathbf{W}_{s_3,\mathbf{W}_{s_2,\mathbf{W}_{s_1,\mathbf{W}_{s_5,e_1}}}}$ a fourth reification level state representing the causal connection from s_3 to $\mathbf{W}_{s_2,\mathbf{W}_{s_1,\mathbf{W}_{s_5,e_1}}}$ from the third reification level.

For the specific context described in Sect. 7.2, these elements are associated with

- for environmental context s_5 a causal pathway leads to a good health e_1
- pathogen state s_1 leads to disturbing the above causal pathway to a good health effect e_1
- well functioning internal defense system s_2 blocks the causal pathway for the effect of pathogens s_1 on the health pathway to e_1
- pregnancy in the first trimester s_3 needs less blocking of the effect of pathogens
- disgust s_4 is needed to compensate for the less blocking of foreign material.

Box 7.1 Role matrices for the fourth-order adaptive network model

mb	base connectivity	1
X_1	s_1	X_1
X_2	s_2	X_2
X_3	s_3	X_3
X_4	s_4	X_4
X_5	s_5	X_5
X_6	e_1	X_5
X_7	$\mathbf{W}_{s_5,e_1}$	X_1
X_8	$\mathbf{W}_{s_1,\mathbf{W}_{s_5,e_1}}$	X_2
X_9	$\mathbf{W}_{s_2,\mathbf{W}_{s_1,\mathbf{W}_{s_5,e_1}}}$	X_3
X_{10}	$\mathbf{W}_{s_3,\mathbf{W}_{s_2,\mathbf{W}_{s_1,\mathbf{W}_{s_5,e_1}}}}$	X_4

mcw	connection weights	1
X_1	s_1	1
X_2	s_2	1
X_3	s_3	1
X_4	s_4	1
X_5	s_5	1
X_6	e_1	X_7
X_7	$\mathbf{W}_{s_5,e_1}$	X_8
X_8	$\mathbf{W}_{s_1,\mathbf{W}_{s_5,e_1}}$	X_9
X_9	$\mathbf{W}_{s_2,\mathbf{W}_{s_1,\mathbf{W}_{s_5,e_1}}}$	X_{10}
X_{10}	$\mathbf{W}_{s_3,\mathbf{W}_{s_2,\mathbf{W}_{s_1,\mathbf{W}_{s_5,e_1}}}}$	1

ms	speed factors	1
X_1	s_1	0.08
X_2	s_2	0.05
X_3	s_3	0.015
X_4	s_4	0.008
X_5	s_5	0.2
X_6	e_1	0.5
X_7	$\mathbf{W}_{s_5,e_1}$	0.05
X_8	$\mathbf{W}_{s_1,\mathbf{W}_{s_5,e_1}}$	0.05
X_9	$\mathbf{W}_{s_2,\mathbf{W}_{s_1,\mathbf{W}_{s_5,e_1}}}$	0.004
X_{10}	$\mathbf{W}_{s_3,\mathbf{W}_{s_2,\mathbf{W}_{s_1,\mathbf{W}_{s_5,e_1}}}}$	0.004

mcfw		1	2
	combination function weights	alogistic	compid
X_1	s_1	1	
X_2	s_2	1	
X_3	s_3	1	
X_4	s_4	1	
X_5	s_5	1	
X_6	e_1	1	
X_7	$\mathbf{W}_{s_5,e_1}$		1
X_8	$\mathbf{W}_{s_1,\mathbf{W}_{s_5,e_1}}$		1
X_9	$\mathbf{W}_{s_2,\mathbf{W}_{s_1,\mathbf{W}_{s_5,e_1}}}$		1
X_{10}	$\mathbf{W}_{s_3,\mathbf{W}_{s_2,\mathbf{W}_{s_1,\mathbf{W}_{s_5,e_1}}}}$		1

mcfp combination function parameters		1		2	
	function	alogistic		compid	
	parameter	1	2	1	2
		σ	τ		
X_1	s_1	18	0.2		
X_2	s_2	18	0.2		
X_3	s_3	18	0.2		
X_4	s_4	18	0.2		
X_5	s_5	18	0.2		
X_6	e_1	8	0.5		
X_7	$\mathbf{W}_{s_5,e_1}$				
X_8	$\mathbf{W}_{s_1,\mathbf{W}_{s_5,e_1}}$				
X_9	$\mathbf{W}_{s_2,\mathbf{W}_{s_1,\mathbf{W}_{s_5,e_1}}}$				
X_{10}	$\mathbf{W}_{s_3,\mathbf{W}_{s_2,\mathbf{W}_{s_1,\mathbf{W}_{s_5,e_1}}}}$				

iv	initial values	1
X_1	s_1	0.2
X_2	s_2	0.1
X_3	s_3	0.11
X_4	s_4	0.1
X_5	s_5	0.5
X_6	e_1	0
X_7	$\mathbf{W}_{s_5,e_1}$	0.8
X_8	$\mathbf{W}_{s_1,\mathbf{W}_{s_5,e_1}}$	0.8
X_9	$\mathbf{W}_{s_2,\mathbf{W}_{s_1,\mathbf{W}_{s_5,e_1}}}$	0.8
X_{10}	$\mathbf{W}_{s_3,\mathbf{W}_{s_2,\mathbf{W}_{s_1,\mathbf{W}_{s_5,e_1}}}}$	0.8

7.4 Simulation Experiments

Simulations have been performed using the dedicated software environment for reified network models described in Chap. 9 and Treur (2019b). To focus on different phases, three scenarios are considered successively, each with a longer time duration.

7.4.1 *Simulation for Scenario 1: Occurrence of Pathogens and Defense System*

This scenario focuses on a period in which pathogens occur, and subsequently a defense system against them is developed. Speed factors of X_3 (pregnancy) and X_4 (disgust) are set at 0 for this scenario. There are two orders of adaptation:

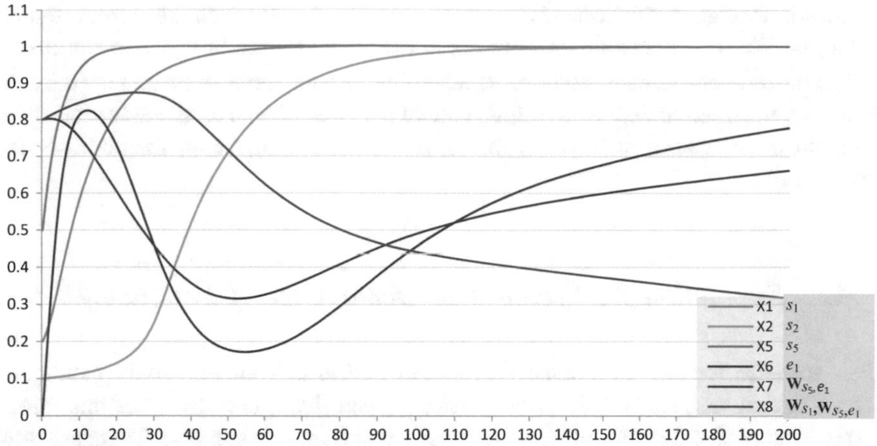

Fig. 7.2 Simulation for Scenario 1 with pathogens and internal defense system occurring, but no pregnancy nor disgust (yet)

• Adaptation 1 Pathogens are introduced	First-order adaptation
• Adaptation 2 Internal defense system is developed	Second-order adaptation

Before adaptation 1 health is good; after adaptation 1 health becomes bad, and after adaptation 2 health becomes good again. In Fig. 7.2 the simulation is shown.

The evolutionary story here is as follows. The red line starting at 0 displays the health level X_6 or e_1, which initially due to environmental context X_5 or s_5 (the pink line starting at 0.5) gets a high level above 0.8 as no pathogens have developed yet. After some time the first evolutionary adaptation (in the environment) is that pathogens X_1 (or s_1) develop, displayed by the blue curve starting at 0.2 and increasing up to 1. When these pathogens reach higher levels they negatively affect X_7 or $\mathbf{W}_{s_5,e_1}$ which represents the strength of the causal pathway from s_5 to e_1 for a good health level. Due to this negative effect on that causal pathway the health level e_1 goes down to a level as low as 0.2. However, after time 50 the internal defense system X_2 or s_2 develops, displayed by the green line starting at 0.1. When it reaches high levels after time 100, by attacking the causal pathway from s_1 to X_7 by which the pathogens have their bad effect (represented by X_8 or $\mathbf{W}_{s_1,\mathbf{W}_{s_5,e_1}}$), it more and more positively affects the health level e_1, which in this case is rising to levels above 0.9.

Note that X_7 or $\mathbf{W}_{s_5,e_1}$ represents the connection weight in the causal pathway from environmental context X_5 (the pink line starting at 0.5) to health level X_6 or e_1. The effect of the pathogen on that causal connection is that it is blocked or at least decreased. This connection weight is displayed in the graph by the purple line for X_7 starting at 0.8 which indeed goes down after the pathogens appear. However, when the internal defense system X_2 or s_2 develops, after time 100 X_7 or $\mathbf{W}_{s_5,e_1}$ goes up again. The reason for this is that the defense system X_2 or s_2 causes that the connection weight for the causal pathway of the effect of the pathogens (represented by X_8 or $\mathbf{W}_{s_1,\mathbf{W}_{s_5,e_1}}$) goes down. This X_8 is displayed by the brown line starting at 0.8; it indeed goes down after time 70 when the internal defense system appears. In this way, the internal defense system's causal pathway blocks or at least reduces the strength of the causal pathway of the effect of the pathogens on the pathway for good health.

7.4.2 *Simulation for Scenario 2: Occurrence of Pregnancy*

This scenario focuses on a somewhat longer period in which not only pathogens occur, and subsequently a defense system against them, but also after this, pregnancy occurs. Speed factor of X_4 (disgust) is set at 0 for this scenario. There are now three orders of adaptation:

• Adaptation 1 Pathogens are introduced	First-order adaptation
• Adaptation 2 Internal defense system is developed	Second-order adaptation
• Adaptation 3 Pregnancy introduced	Third-order adaptation

Before adaptation 1 health is good, after adaptation 1 health becomes bad, after adaptation 2 health becomes good again, and after adaptation 3 health becomes worse again. The simulation for this is shown in Fig. 7.3.

This scenario includes two states that were not displayed in Fig. 7.2, namely X_3 (which is s_3 for the pregnancy), and X_9 (which is $\mathbf{W}_{s_2,\mathbf{W}_{s_1,\mathbf{W}_{s_5,e_1}}}$). For the first phase until around time point 100 Scenario 2 is almost the same as Scenario 1. The main difference is that X_3 starts to increase after time point 70, and gets its causal effect

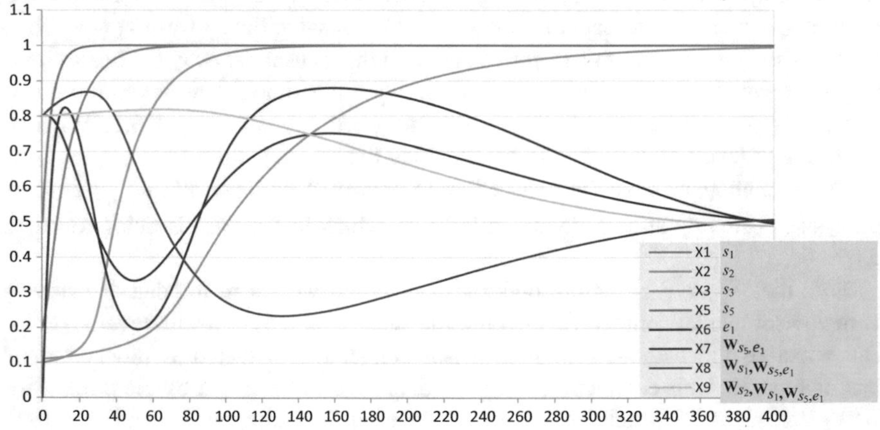

Fig. 7.3 Simulation for Scenario 2 with pathogens, internal defense system, and pregnancy occurring, but no disgust (yet)

on X_9, which starts to decrease. This makes the effect of the defense system s_2 on the causal pathway of the pathogens represented by $\mathbf{W}_{s_1,\mathbf{W}_{s_5,e_1}}$ weaker. As a result of this weakened defense system the health level e_1 starts a downward trend from time point 140 on, whereas in Scenario 1 it is keeping an upward trend until time point 200 (and even further). This makes sense, as in Scenario 1 the well functioning defense system is able to maintain a good health level till the end of days. This is not the case from the moment that (first trimester) pregnancy occurred, hence the health level e_1 keeps a downward trend in Fig. 7.3.

7.4.3 Simulation for Scenario 3: Occurrence of Disgust

This scenario focuses on a still longer time period in which not only pathogens occur, a defense system against them is developed, and pregnancy occurs, but also disgust (in the first trimester of pregnancy) occurs. Now there are four orders of adaptation:

• Adaptation 1 Pathogens are introduced	First-order adaptation
• Adaptation 2 Defense system is developed	Second-order adaptation
• Adaptation 3 Pregnancy	Third-order adaptation
• Adaptation 4 Disgust	Fourth-order adaptation

Before adaptation 1 health is good, after adaptation 1 health becomes bad, after adaptation 2 health becomes good again, after adaptation 3 health becomes worse again, and after adaptation 4 health becomes better again. The simulation results for this scenario are shown in Fig. 7.4.

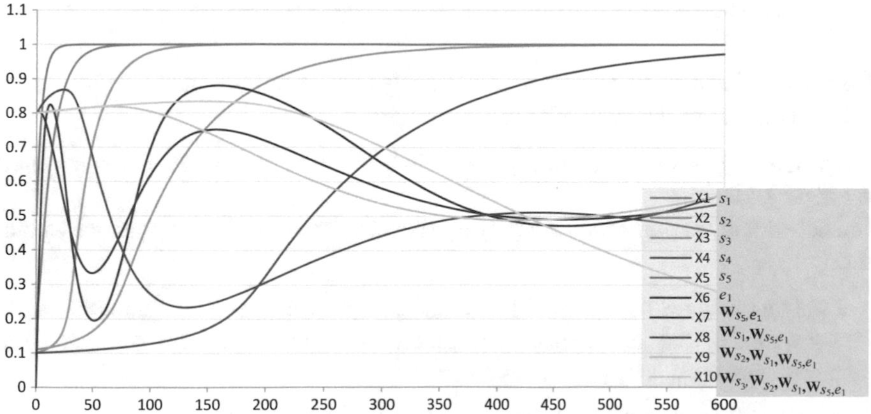

Fig. 7.4 Simulation for Scenario 3 with pathogens, internal defense system, pregnancy, and disgust occurring

This scenario includes two states that were not displayed in Fig. 7.3, namely X_4 (which is s_4 for the disgust), and X_{10} (which is $\mathbf{W}_{s_3,\mathbf{W}_{s_2,\mathbf{W}_{s_1,\mathbf{W}_{s_5,e_1}}}}$). For the first phase until around time point 300 Scenario 3 is almost the same as Scenario 2. A difference is that X_4 starts to increase after time point 200, and gets its causal effect on X_{10}, which then starts to decrease. This makes the effect of the pregnancy s_3 on the causal pathway of the defense system (represented by $\mathbf{W}_{s_2,\mathbf{W}_{s_1,\mathbf{W}_{s_5,e_1}}}$) weaker. As a result of this weakened effect of pregnancy, the health level e_1 starts an upward trend again from time point 450 on, whereas in Scenario 2 it kept a downward trend until time point 400 (and even further). This makes sense, as in Scenario 2 the reduced functioning of the defense system is not able to maintain a good health level. This is different from the moment that disgust occurred, hence the health level e_1 finally gets an upward trend in Fig. 7.4.

For each of the discussed scenarios, one may wonder where the displayed trends, for example for the health level e_1, end when $t \to \infty$. This emerging behaviour will be analysed in Sect. 7.5.

7.5 Mathematical Analysis of the Emerging Behaviour and Verification of the Model

In this section the emerging behaviour of the fourth-order network model is analysed, first in Sect. 7.5.1 from a general perspective, and next in Sect. 7.5.2 for each of the scenarios. In Sect. 7.5.3, this analysis is used to perform verification of the designed network model.

7.5.1 General Approach to the Mathematical Analysis of Equilibria

First, a definition and criterion used for analysis of equilibria and verification of reified temporal-causal network models are given.

Definition 1 (stationary point and equilibrium) A state Y has a *stationary point* at t if $\mathbf{d}Y(t)/\mathbf{d}t = 0$.
The network is in *equilibrium* at t if every state Y of the model has a stationary point at t.

Applying this for the specific differential equation format for a temporal-causal network model, the more specific criterion in terms of the network structure characteristics $\boldsymbol{\omega}_{X,Y}, \mathbf{c}_Y(..), \boldsymbol{\eta}_Y$ is as follows:

Lemma 1 (Criterion for a stationary point in a temporal-causal network) Let Y be a state and $X_1, \ldots, X_k$ the states from which state Y gets its incoming connections. Then Y has a stationary point at t if and only if

$$\boldsymbol{\eta}_Y = 0 \quad \text{or} \quad \mathbf{c}_Y\left(\boldsymbol{\omega}_{X_1,Y}X_1(t), \ldots, \boldsymbol{\omega}_{X_k,Y}X_k(t)\right) = Y(t)$$

The latter equation is called a *stationary point equation*, or in case of an equilibrium an *equilibrium equation*.

This criterion will be used here to analyse the state values in an equilibrium. The states X_1 to X_5 are independent states as they have no incoming connections from other states. They can be handled separately. States X_7 to X_{10} depend on the independent states X_1 to X_4, respectively; they have a single incoming arrow with the complemental identity function **compid(..)** as combination function $\mathbf{c}_Y(V)$ for Y, defined by

$$\mathbf{compid}(V) = 1 - V \tag{7.1}$$

where V is a variable used for the single impact $\boldsymbol{\omega}_{X_i,X_j}X_i$. Then for state X_j depending on state X_i based on the above criterion (where now $k = 1$), the form of the equilibrium equation for X_j becomes

$$\mathbf{compid}(\boldsymbol{\omega}_{X_i,X_j}X_i) = X_j \tag{7.2}$$

which is

$$1 - \boldsymbol{\omega}_{X_i,X_j}X_i = X_j$$

where $\boldsymbol{\omega}_{X_i,X_j}$ is the value of the reification state $\mathbf{W}_{X_i,X_j}$ (which is also one of the X_k, actually for $k = j+1$) for this connection weight. So the equation becomes

$$1 - X_{j+1}X_i = X_j$$

From this, the following equilibrium equations can be found for X_7 to X_{10}.

$$\begin{aligned} X_7 &= 1 - X_8X_1 \\ X_8 &= 1 - X_9X_2 \\ X_9 &= 1 - X_{10}X_3 \\ X_{10} &= 1 - X_4 \end{aligned} \tag{7.3}$$

State X_6 depends on independent state X_5, but also on the connection weight represented by X_7. Based on the above criterion the equilibrium equation for X_6 is

$$X_6 = \mathbf{alogistic}_{8,0.5}(X_7X_5) \tag{7.4}$$

Box 7.2 Solving the general equilibrium equations by expressing the values for all states X_6 to X_{10} in the values of the independent states X_1 to X_5.
Based on the general equilibrium equations, the equilibrium values can be expressed in terms of the equilibrium values for the independent states X_1 to X_5:

$$\begin{aligned}
X_{10} &= 1 - X_4 \\
X_9 &= 1 - (1 - X_4)X_3 \\
X_8 &= 1 - (1 - (1 - X_4)X_3)X_2 \\
X_7 &= 1 - (1 - (1 - (1 - X_4)X_3)X_2)X_1
\end{aligned}$$

These expressions can also be rewritten as

$$\begin{aligned}
X_{10} &= 1 - X_4 \\
X_9 &= 1 - X_3 + X_4X_3 \\
X_8 &= 1 - X_2 + X_3X_2 - X_4X_3X_2 \\
X_7 &= 1 - X_1 + X_2X_1 - X_3X_2X_1 + X_4X_3X_2X_1
\end{aligned}$$

Substituting the expression found for X_7 obtains:

$$X_6 = \mathbf{alogistic}_{8,0.5}((1 - X_1 + X_2X_1 - X_3X_2X_1 + X_4X_3X_2X_1)X_5)$$

So, also this is a function of the values of the independent states X_1 to X_5. For the nonzero independent states among X_1 to X_5 the equilibrium equation is

$$X_i = \mathbf{alogistic}_{18,0.2}(X_i)$$

and the equilibrium value is numerically approximated by 0.999999427374 (see Chap. 3, Sect. 3.7.4) which has a distance to 1 less than 10^{-6}.

7.5.2 *Mathematical Analysis of the Different Scenarios*

Each of the scenarios has different values for the independent states. Therefore, based on the general relations in Sect. 7.5.1, Box 7.2 above, for each of the scenarios, it is analysed which are the equilibrium values for all states. Also, a Scenario 0 is added in which only pathogens X_1 occur and no internal defense system X_2.

Scenario 0 (internal defense system $X_2 = 0$, pregnancy $X_3 = 0$, disgust $X_4 = 0$) In this scenario after substitution of the above values the outcomes are:

$$\begin{aligned} X_{10} &= 1 \\ X_9 &= 1 \\ X_8 &= 1 \\ X_7 &= 1 - X_1 \\ X_6 &= \textbf{alogistic}_{8,0.5}((1 - X_1)X_5) \end{aligned} \tag{7.5}$$

As X_1 and X_5 are nonzero their equilibrium value is 0.999999427374 (see Chap. 3, Sect. 3.7.4) which has a distance to 1 less than 10^{-6}. So, approximately (deviation in the order of 10^{-6}) the values are obtained as shown in Table 7.2. The 0 for X_6 indicates bad health, which is to be expected as in this scenario pathogens have occurred but no defense system has developed yet.

Scenario 1 (pregnancy X_3 = 0, disgust X_4 = 0) In this scenario, after substitution of the above values the outcomes are:

$$\begin{aligned} X_{10} &= 1 \\ X_9 &= 1 \\ X_8 &= 1 - X_2 \\ X_7 &= 1 - X_1 + X_2X_1 \\ X_6 &= \textbf{alogistic}_{8,0.5}((1 - X_1 + X_2X_1)X_5) \end{aligned} \tag{7.6}$$

Again, as X_1, X_2 and X_5 are nonzero, their equilibrium value is 0.999999427374 (see Chap. 3, Sect. 3.7.4) which has distance to 1 less than 10^{-6}. So, approximately the values are as shown in Table 7.2 (deviation in the order of 10^{-6}). The value 0.981684 for X_6 (=e_1) indicates good health, which is to be expected as in this scenario not only pathogens have occurred but also an internal defense system that

Table 7.2 The equilibrium values for the 4 scenarios based on the mathematical analysis

State		Scenario 0	Scenario 1	Scenario 2	Scenario 3
X_1	s_1	1	1	1	1
X_2	s_2	0	1	1	1
X_3	s_3	0	0	1	1
X_4	s_4	0	0	0	1
X_5	s_5	1	1	1	1
X_6	e_1	0	0.981684	0	0.981684
X_7	$\mathbf{W}_{s_5,e_1}$	0	1	0	1
X_8	$\mathbf{W}_{s_1},\mathbf{W}_{s_5,e_1}$	1	0	1	0
X_9	$\mathbf{W}_{s_2},\mathbf{W}_{s_1},\mathbf{W}_{s_5,e_1}$	1	1	0	1
X_{10}	$\mathbf{W}_{s_3},\mathbf{W}_{s_2},\mathbf{W}_{s_1},\mathbf{W}_{s_5,e_1}$	1	1	1	0

attacks them. This shows that the upward trend for the health level e_1 displayed in Fig. 7.2 after time point 50 will eventually end up in a very high health level.

Scenario 2 (disgust $X_4 = 0$) In this scenario after substitution of the above values the outcomes are:

$$\begin{aligned} X_{10} &= 1 \\ X_9 &= 1 - X_3 \\ X_8 &= 1 - X_2 + X_3X_2 \\ X_7 &= 1 - X_1 + X_2X_1 - X_3X_2X_1 \end{aligned} \tag{7.7}$$

Again, as X_1, X_2, X_3 and X_5 are nonzero, their equilibrium value is 0.999999427374 (see Chap. 3, Sect. 3.7.4) which has a distance to 1 less than 10^{-6}. So, approximately (deviation in the order of 10^{-6}) the values are as shown in Table 7.2.

The value 0 for X_6 indicates bad health, which is to be expected as in this scenario not only pathogens and an internal defense system have occurred but also a facility that suppresses that due to pregnancy. This shows that the downward trend for the health level e_1 displayed in Fig. 7.3 after time point 150 will eventually end up in a very low health level.

Scenario 3 (all independent states X_1 to X_5 are nonzero) In this scenario the general equations apply

$$\begin{aligned} X_{10} &= 1 - X_4 \\ X_9 &= 1 - X_3 + X_4X_3 \\ X_8 &= 1 - X_2 + X_3X_2 - X_4X_3X_2 \\ X_7 &= 1 - X_1 + X_2X_1 - X_3X_2X_1 + X_4X_3X_2X_1 \\ X_6 &= \mathbf{alogistic}_{8,0.5}((1 - X_1 + X_2X_1 - X_3X_2X_1 + X_4X_3X_2X_1)X_5) \end{aligned} \tag{7.8}$$

Again, as X_1, X_2, X_3, X_4 and X_5 are all nonzero, their equilibrium value is 0.999999427374 (see Chap. 3, Sect. 3.7.4) which has a distance to 1 less than 10^{-6}. So, approximately (deviation in the order of 10^{-6}) for the independent states the values 1 can be substituted in the above equations and the values shown in the last column in Table 7.2 are obtained. So, indeed also for this case a good health e_1, as expected. This shows that the upward trend for the health level e_1 displayed in Fig. 7.4 after time point 450 will eventually end up in a very high health level.

7.5.3 Verification of the Reified Network Model

For verification for Scenario 3, approximate equilibrium values have been determined by running the model (in about 10 s on an ordinary laptop) for 30,000 steps

Table 7.3 Outcomes at t = 3000 for a simulation of Scenario 3 with Δt = 0.1, compared to the equilibrium values from the analysis

state		simulation	analysis	deviation
X_1	s_1	0.999999	1	$-1\ 10^{-6}$
X_2	s_2	0.999999	1	$-1\ 10^{-6}$
X_3	s_3	0.999999	1	$-1\ 10^{-6}$
X_4	s_4	0.999999	1	$-1\ 10^{-6}$
X_5	s_5	0.999999	1	$-1\ 10^{-6}$
X_6	e_1	0.981646	0.981684	$-3.8\ 10^{-5}$
X_7	$\mathbf{W}_{s_5,e_1}$	0.999738	1	−0.00026
X_8	$\mathbf{W}_{s_1,\mathbf{W}_{s_5,e_1}}$	0.000243	0	0.000243
X_9	$\mathbf{W}_{s_2,\mathbf{W}_{s_1,\mathbf{W}_{s_5,e_1}}}$	0.999775	1	−0.00023
X_{10}	$\mathbf{W}_{s_3,\mathbf{W}_{s_2,\mathbf{W}_{s_1,\mathbf{W}_{s_5,e_1}}}}$	$2.11\ 10^{-5}$	0	$2.11\ 10^{-5}$

with Δt = 0.1 in the dedicated software environment described in Chap. 9. The outcomes at t = 3000 are shown in Table 7.3, and compared there to the equilibrium values from the analysis in Sect. 7.5.2. The deviation is in the order of 10^{-4} to 10^{-6}, which is fair and gives evidence that the implemented model is correct.

7.6 Discussion

In this chapter, a fourth-order reified network model was introduced to describe different orders of adaptivity found in a case study on evolutionary processes; e.g., Fessler et al. (2005, 2015), Jones et al. (2005), Fleischman and Fessler (2011). The network model describes how the causal pathways for newly developed features in this case study affect the causal pathways of already existing features, which makes the pathways of these new features one order of adaptivity higher than the existing ones, as they adapt the previous adaptation. The network reification approach has shown to be an adequate means to model this in a transparent manner. In future research, it can be explored how the network model introduced here can be extended and whether this also works for other evolutionary case studies. For more literature, see, for example, Aarøe et al. (2017), Curtis et al. (2011), Fleischman and Fessler (2018), Jones et al. (2017), Lieberman et al. (2018), Oaten et al. (2009), Schaller and Park (2011), Schaller (2016), Tybur and Lieberman (2016).

References

Aarøe, L., Petersen, M.B., Arceneaux, K.: The behavioral immune system shapes political intuitions: why and how individual differences in disgust sensitivity underlie opposition to immigration. Am. Polit. Sci. Rev. **111**(2), 277–294 (2017)

Abraham, W.C., Bear, M.F.: Metaplasticity: the plasticity of synaptic plasticity. Trends Neurosci. **19**(4), 126–130 (1996)

Carley, K.M.: Inhibiting adaptation. In: Proceedings of the 2002 Command and Control Research and Technology Symposium, pp. 1–10. Naval Postgraduate School, Monterey, CA (2002)

Carley, K.M.: Destabilization of covert networks. Comput. Math. Organiz. Theor. **12**, 51–66 (2006)

Curtis, V., de Barra, M., Aunger, R.: Disgust as an adaptive system for disease avoidance behaviour. Phil. Trans. R. Soc. B **2011**(366), 389–401 (2011)

Davis, R.: Meta-rules: reasoning about control. Artif. Intell. **15**, 179–222 (1980)

Davis, R., Buchanan, B.G.: Meta-level knowledge: overview and applications. In: Proceedings of 5th IJCAI, pp. 920–927 (1977)

Galton, A.: The ins and outs of reification. Synthese **150**, 415–441 (2006)

Fessler, D.M.T., Clark, J.A., Clint, E.K.: Evolutionary psychology and evolutionary anthropology. In: Buss, D.M. (ed.) The Handbook of Evolutionary Psychology, pp. 1029–1046. Wiley, Hoboken (2015)

Fessler, D.M.T., Eng, S.J., Navarrete, C.D.: Elevated disgust sensitivity in the first trimester of pregnancy: evidence supporting the compensatory prophylaxis hypothesis. Evol. Hum. Behav. **26**(4), 344–351 (2005)

Fleischman, D.S., Fessler, D.M.T.: Progesterone's effects on the psychology of disease avoidance: support for the compensatory behavioral prophylaxis hypothesis. Horm. Behav. **59**(2), 271–275 (2011)

Fleischman, D.S., Fessler, D.M.T.: Response to "hormonal correlates of pathogen disgust: testing the compensatory prophylaxis hypothesis". Evol. Hum. Behav. **39**, 468–469 (2018)

Jones, B.C., Perrett, D.I., Little, A.C., Boothroyd, L., Cornwell, R.E., Feinberg, D.R., Tiddeman, B.P., Whiten, S., Pitman, R.M., Hillier, S.G., Burt, D.M., Stirrat, M.R., Law Smith, M.J., Moore, F.R.: Menstrual cycle, pregnancy and oral contraceptive use alter attraction to apparent health in faces. Proc. R. Soc. B **5**(272), 347–354 (2005)

Jones, B.C., Hahn, A.C., Fisher, C.I., Wang, H., Kandrik, M., Lee, A.J., Tybur, J.M., DeBruine, L. M.: Hormonal correlates of pathogen disgust: testing the compensatory prophylaxis hypothesis. Evol. Hum. Behav. **39**, 166–169 (2018)

Lieberman, D., Billingsley, J., Patrick, C.: Consumption, contact and copulation: how pathogens have shaped human psychological adaptations. Phil. Trans. R. Soc. B **373**, 20170203 (2018)

Oaten, M., Stevenson, R.J., Case, T.I.: Disgust as a disease-avoidance mechanism. Psychol. Bull. **135**(2), 303–321 (2009)

Schaller, M., Park, J.H.: The behavioral immune system (and why it matters). Curr. Dir. Psychol. Sci. **20**(2), 99–103 (2011)

Schaller, M.: The behavioral immune system. In: Buss, D.M. (ed.) The Handbook of Evolutionary Psychology, vol. 1, 2nd edn, pp. 206–224. Wiley, New York (2016)

Schmidt, M.V., Abraham, W.C., Maroun, M., Stork, O., Richter-Levin, G.: Stress-induced metaplasticity: from synapses to behavior. Neuroscience **250**, 112–120 (2013)

Sterling, L., Beer, R.: Metainterpreters for expert system construction. J. Logic Program. **6**, 163–178 (1989)

Treur, J.: Network-Oriented Modeling: Addressing Complexity of Cognitive, Affective and Social Interactions. Springer Publishers, Berlin (2016)

Treur, J.: Network reification as a unified approach to represent network adaptation principles within a network. In: Proceedings of the 7th International Conference on Theory and Practice of Natural Computing, TPNC'18. Lecture Notes in Computer Science, vol. 11324, pp. 344–358. Springer Publishers, Berlin (2018)

Treur, J.: Multilevel network reification: representing higher-order adaptivity in a network. In: Proceedings of the 7th International Conference on Complex Networks and their Applications, ComplexNetworks'18, vol. 1. Studies in Computational Intelligence, vol. 812, 635–651. Springer, Berlin (2018b)

Treur, J.: The ins and outs of network-oriented modeling: from biological networks and mental networks to social networks and beyond. Trans. Comput. Collective Intell. **32**, 120–139 (2019a) (Springer Publishers. Contents of Keynote Lecture at ICCCI'18)

Treur, J.: Design of a software architecture for multilevel reified temporal-causal networks (2019b). https://doi.org/10.13140/rg.2.2.23492.07045, https://www.researchgate.net/publication/333662169

Tybur, J.M., Lieberman, D.: Human pathogen avoidance adaptations. Curr. Opin. Psychol. 7, 6–11 (2016)

Weyhrauch, R.W.: Prolegomena to a theory of mechanized formal reasoning. Artif. Intell. **13**, 133–170 (1980)

Chapter 8
Higher-Order Reified Adaptive Network Models with a Strange Loop

Abstract In this chapter, as in Chap. 7, the challenge of exploring plausible reified network models of order higher than two is addressed. This time another less usual option for application was addressed: the notion of Strange Loop which from a philosophical perspective sometimes is claimed to be at the basis of human intelligence and consciousness. This notion will be illustrated by examples from music, graphic art and paradoxes, and by Hofstadter's claims about how Strange Loops apply to the brain. A reified adaptive network model of order higher than 2 was found, that even can be considered as being of infinite order. An example simulation shows the upward and downward interactions between the different levels, together with the processes within the levels. Another example addresses adaptive decision making according to two levels that are mutually reifying each other, as in Escher's Drawing Hands lithograph.

8.1 Introduction

Like in Chap. 7, the challenge addressed in the current chapter relates to the open question left from Sect. 1.3: are there good examples of adaptive networks of third-order? Or even higher? The aim of this chapter is to get answers "yes" on both questions for a domain as described by the idea of Strange Loop in (Hofstadter 1979, 2006, 2007).

Hofstadter (1979) describes a Strange Loop as the phenomenon that going upward through a hierarchy of levels after a while you find yourself back at the level where you started; a hierarchy of levels that turns out to form of cycle. He illustrates this notion for different domains, such as music (Bach), graphic art (Escher), and paradoxes and logic (Gödel). For example, in logic the idea was exploited by Gödel by defining a coding by natural numbers of all logical statements on arithmetic, as formalised by logical formulae. Such coding is a form of reification as statements become numbers, which has some similarity with the reification used for combination functions in reified temporal-causal networks. Using this coding, he was able to prove his famous incompleteness theorems in

J. Treur, *Network-Oriented Modeling for Adaptive Networks: Designing Higher-Order Adaptive Biological, Mental and Social Network Models*, Studies in Systems, Decision and Control 251, https://doi.org/10.1007/978-3-030-31445-3_8

logic; e.g., (Nagel and Newman 1965; Smorynski 1977). Other literature relates Strange Loops, for example, to architectural education (Gannon 2017), advertising (Hendlin 2019), self-representation in consciousness (Kriegel and Williford 2006), and psychotherapeutic understanding (Strijbos and Glas 2018). Hofstadter's claim is that the brain also makes use of Strange Loops as an essential ingredient for human intelligence and consciousness. This has been used as a source of inspiration in the current chapter.

In the current chapter, it is shown how a Strange Loop can be modeled by a reified network. Here the lowest (base) level also functions as a reification level for the highest level, so that the reification levels form a cycle. Since a cycle has no end, this can be considered an example of a reified adaptive network of order ∞.

Hofstadter (1979)'s illustration of the notion of Strange Loop by its application in music (Bach), graphic art (Escher), and paradoxes and logic (Gödel) will be very briefly summarized in Sect. 8.2, and Hofstadter's ideas about Strange Loops in the brain are discussed. Section 8.3 presents an example 12-level reified network model for a Strange Loop, and a simplified 4-level version of it. Section 8.4 shows an example simulation of it; this adaptive network model follows Escher (1960)'s design of his Ascending and Descending lithograph (Fig. 8.2). Another example is addressed in Sect. 8.5, of a two-level reified network where each level reifies the other level, following Escher (1948)'s design of his Drawing Hands lithograph (Fig. 8.3). It shows how a Strange Loop can be used to model adaptive decision making. Section 8.6 shows an example simulation of this mutually reified network model.

8.2 The Notion of Strange Loop

In this chapter, like in Chap. 7, the inspiration for examples of reified network models may feel a bit out of the ordinary, as the aim is to use the concept of reified network to explore the idea of Strange Loop described by Douglas Hofstadter in (Hofstadter 1979, 2007).

8.2.1 Strange Loops in Music, Graphic Art and Paradoxes

Hofstadter illustrates the idea first metaphorically from a music context as follows:

> There is one canon in the Musical Offering which is particularly unusual. Labeled simply "Canon per Tonos", it has three voices. The uppermost voice sings a variant of the Royal Theme, while underneath it, two voices provide a canonic harmonization based on a second theme. The lower of this pair sings its theme in C minor (which is the key of the canon as a whole), and the upper of the pair sings the same theme displaced upwards in pitch by an interval of a fifth. What makes this canon different from any other, however, is that when it concludes-or, rather, seems to conclude-it is no longer in the key of C minor, but now is in

> D minor. Somehow Bach has contrived to modulate (change keys) right under the listener's nose. And it is so constructed that this "ending" ties smoothly onto the beginning again; thus one can repeat the process and return in the key of E, only to join again to the beginning. These successive modulations lead the ear to increasingly remote provinces of tonality, so that after several of them, one would expect to be hopelessly far away from the starting key. And yet magically, after exactly six such modulations, the original key of C minor has been restored! All the voices are exactly one octave higher than they were at the beginning, and here the piece may be broken off in a musically agreeable way. (Hofstadter 1979), p. 18

Then a short description of the idea is:

> In this canon, Bach has given us our first example of the notion of Strange Loops. The "Strange Loop" phenomenon occurs whenever, by moving upwards (or downwards) through the levels of some hierarchical system, we unexpectedly find ourselves right back where we started. (Hofstadter 1979), p. 18

Another metaphoric illustration from (Hofstadter 2007) considers 10 persons sitting on your lap, on top of each other on each other's laps, where it turns out you yourself sit on the lap of the 10th person sitting on top of you; magically, no collapse occurs, see Fig. 8.1, adopted from (Hofstadter 2007).

In (Hofstadter 1979, 2007) also from another context metaphorical illustrations for the Strange Loop idea were obtained: from graphic art. Maurits Cornelis Escher

Fig. 8.1 Lap loop, sitting on each other's laps. Adopted from (Hofstadter 2007), Chap 8: embarking on a strange loop safari (Douglas Hofstadter is the guy smiling at you)

Fig. 8.2 Ascending and descending (Lithograph, Escher, 1960)

(1898–1972) was a famous Dutch graphic artist who made mathematically-inspired woodcuts, lithographs, and mezzotints. In Figs 8.2, 8.3 and 8.4 three examples of his lithographs are shown that were used as an illustration in (Hofstadter 1979, 2007). For example, in Ascending and Descending (Fig. 8.2) two lines of persons are seen, one of which walks in circles upstairs all the time and one downstairs, both all the time after one cycle returning to the same level. In Drawing Hands (Fig. 8.3) the right hand is drawing the left hand which in turn is drawing the right hand. Print Gallery (Fig. 8.4) shows a more complex situation in which the art and the reality that it depicts are mixed up, not unlike having the idea that Douglas Hofstadter smiles at you in Fig. 8.1.

Hofstadter (1979, 2007) also illustrates Strange Loops by examples of paradoxes. A simple example is:

This sentence is not true

Or a similar one:

I lie

The paradox here is that when such a sentence is true, it is false, and conversely. In these cases what is referred to from within the sentence is the sentence itself, which has many similarities with some of Escher's work.

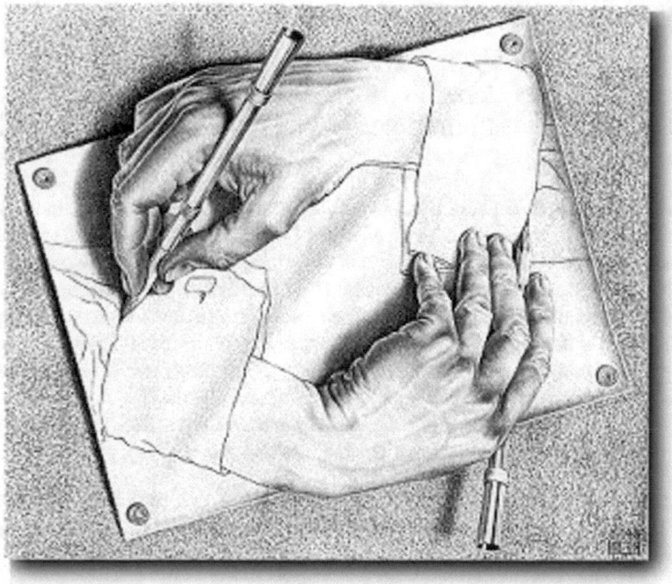

Fig. 8.3 Drawing hands (Lithograph, Escher, 1948)

Fig. 8.4 Print gallery (Lithograph, Escher, 1956)

8.2.2 Strange Loops in the Brain

A description that comes closest to the scope of the current book is the following quote, where it is analysed how different levels of rules can generate intelligent behaviour:

> What sorts of "rules" could possibly capture all of what we think of as intelligent behavior, however? Certainly there must be rules on all sorts of different levels. There must be many "just plain" rules. There must be "metarules" to modify the "just plain" rules; then "metametarules" to modify the metarules, and so on. The flexibility of intelligence comes from the enormous number of different rules, and levels of rules. The reason that so many rules on so many different levels must exist is that in life, a creature is faced with millions of situations of completely different types. In some situations, there are stereotyped responses which require "just plain" rules. Some situations are mixtures of stereotyped situations - thus they require rules for deciding which of the "just plain" rules to apply. Some situations cannot be classified - thus there must exist rules for inventing new rules … and on and on. (Hofstadter 1979), pp. 34–35

This view may be related to the approach described in (Davis and Buchanan 1977; Davis 1980); note that in that time 'rules' was an often used representation format to specify functionality in AI and knowledge-based system models. If in the above quote the word 'rule' is replaced by 'causal connection' as a different format to specify functionality, this comes even more close to the topic of the current book:

> What sorts of "causal connections" could possibly capture all of what we think of as intelligent behavior, however? Certainly there must be causal connections on all sorts of different levels. There must be many "just plain" causal connections. There must be "metacausal connections" to modify the "just plain" causal connections; then "metameta-causal connections" to modify the metacausal connections, and so on. The flexibility of intelligence comes from the enormous number of different causal connections, and levels of causal connections. The reason that so many causal connections on so many different levels must exist is that in life, a creature is faced with millions of situations of completely different types. In some situations, there are stereotyped responses which require "just plain" causal connections. Some situations are mixtures of stereotyped situations - thus they require causal connections for deciding which of the "just plain" causal connections to apply. Some situations cannot be classified - thus there must exist causal connections for inventing new causal connections … and on and on. Adaptation of (Hofstadter 1979), pp. 34–35, replacing 'rule' by 'causal connection'

In this way, it can be seen as a rough sketch of a multilevel reified network architecture where from each level, the level beneath is adapted. That is indeed exactly what happens in a reified network model, so this can be used as inspiration for a multilevel reified temporal-causal network model. However, the text continues as follows:

> Without doubt, Strange Loops involving rules that change themselves, directly or indirectly, are at the core of intelligence. (Hofstadter 1979), pp. 34–35

This suggests that such a multilevel reified temporal-causal network should get some kind of cyclic level structure as meant for Strange Loops, comparable to the metaphorical illustrations discussed above. The quote below also suggests this in some more detail:

> My belief is that the explanations of "emergent" phenomena in our brains - for instance, ideas, hopes, images, analogies, and finally consciousness and free will - are based on a kind of Strange Loop, an interaction between levels in which the top level reaches back down towards the bottom level and influences it, while at the same time being itself determined by the bottom level. In other words, a self-reinforcing "resonance" between different levels (…) The self comes into being at the moment it has the power to reflect itself. (Hofstadter 1979), p. 704

Moreover, he also explicitly points at the role of causality in such an architecture, in particular for the upward and downward interactions between the levels, which also brings the Strange Loop idea closer to the domain of reified temporal-causal networks:

> … we will have to admit various types of "causality": ways in which an event at one level of description can "cause" events at other levels to happen. Sometimes event A will be said to "cause" event B simply for the reason that the one is a translation, on another level of description, of the other. Sometimes "cause" will have its usual meaning: physical causality. Both types of causality - and perhaps some more - will have to be admitted in any explanation of mind, for we will have to admit causes that propagate both upwards *and* downwards in the Tangled Hierarchy of mentality … (Hofstadter 1979), p. 704

8.3 A Twelve- and Four-Level Reified Adaptive Network Model Based on a Strange Loop

So, all in all, based on these ideas, examples of reified network architectures have been designed. The Network-Oriented Modeling approach used to model this process is based on reified temporal-causal network models described in Chaps. 3 and 4; see also (Treur 2018a, b) or for a software architecture (Treur 2019b). Recall that a temporal-causal network model in the first place involves representing in a declarative manner states and connections between them that represent (causal) impacts of states on each other. The states are assumed to have (activation) levels in the interval [0, 1] that vary over time. The following three notions form the defining part of a conceptual representation of a temporal-causal network model (Treur 2016, 2019a):

- **Connectivity**
  - Each incoming connection of a state *Y*, from a state *X* has a *connection weight value* $\boldsymbol{\omega}_{X,Y}$ representing the strength of the connection.
- **Aggregation**
  - For each state a *combination function* $\mathbf{c}_Y(..)$ is chosen to combine the causal impacts state *Y* receives from other states.
- **Timing**
  - For each state *Y* a *speed factor* $\boldsymbol{\eta}_Y$ is used to represent how fast state *Y* is changing upon causal impact.

The notion of *network reification* introduced in Chaps. 3 and 4 is a means to model adaptive networks more transparently from a Network-Oriented Modelling perspective. The concept of reification has been shown to provide substantial advantages in expressivity and transparency of models within AI; e.g., (Davis and Buchanan 1977; Davis 1980; Galton 2006; Hofstadter 1979; Sterling and Beer 1989; Weyhrauch 1980). For network models, reification can be applied by reifying network structure characteristics (such as $\boldsymbol{\omega}_{X,Y}$, $\mathbf{c}_Y(..)$, $\boldsymbol{\eta}_Y$) in the form of additional network states (called *reification states*, indicated by $\mathbf{W}_{X,Y}$, $\mathbf{C}_Y$, $\mathbf{H}_Y$, respectively) within an extended network. In addition, reification states $\mathbf{P}_{i,j,Y}$ for parameters of combination functions are used. *Roles* **W**, **C**, **H**, and **P** are assigned to reification states according to the specific network structure characteristic they represent: *connection weight reification*, *combination function reification*, *speed factor reification*, or values, respectively. Also, a role **P** for *combination function parameters* is used. A specification format based on *role matrices* is used to specify for each state which role it is playing and in relation to which other states (see Box 8.1). Role matrices are **mb** (the base connectivity role), **mcw** (connection weights, role **W**), **ms** (speed factors, role **H**), **mcfw** (combination function weights, role **C**), **mcfp** (combination function parameters, role **P**). Multilevel reified networks can be used to model networks which are adaptive of different orders. For more details, see Chaps. 3 and 4, or (Treur 2018a, b).

8.3.1 A Twelve-Level Reified Adaptive Network Model Forming a Cycle of Levels

A first reified network model depicted in Fig. 8.5 illustrates how the notion of Strange Loop can be modeled in a reified network of 12 levels, by shaping the levels not straight upward, but in a cyclic form. This looks like a crazy network structure, even a bit Escher-ish, in which the 12 levels of the reified network seem to be juggled around, thereby forming a circle (depicted here in the shape of a dodecagon). As a cycle never ends, the levels actually go on forever; then it can be considered to be adaptive of order ∞. It looks similar to Escher's Ascending and Descending where after four upward (or four downward) stairs the persons find themselves at the same level again. Is that what intelligent behaviour or consciousness is about, as Hofstadter suggests?

To model this, in Fig. 8.5 starting from the horizontal plane at the left hand side in the picture, the pink arrows are downward causal connections (going anti-clockwise) and the blue arrows are upward causal connections (going clockwise). So far, so good. But on the right hand side of the picture the pink downward causal connections are displayed as upward and the blue upward causal connections are displayed as downward. So you have to keep your head upside down to see it

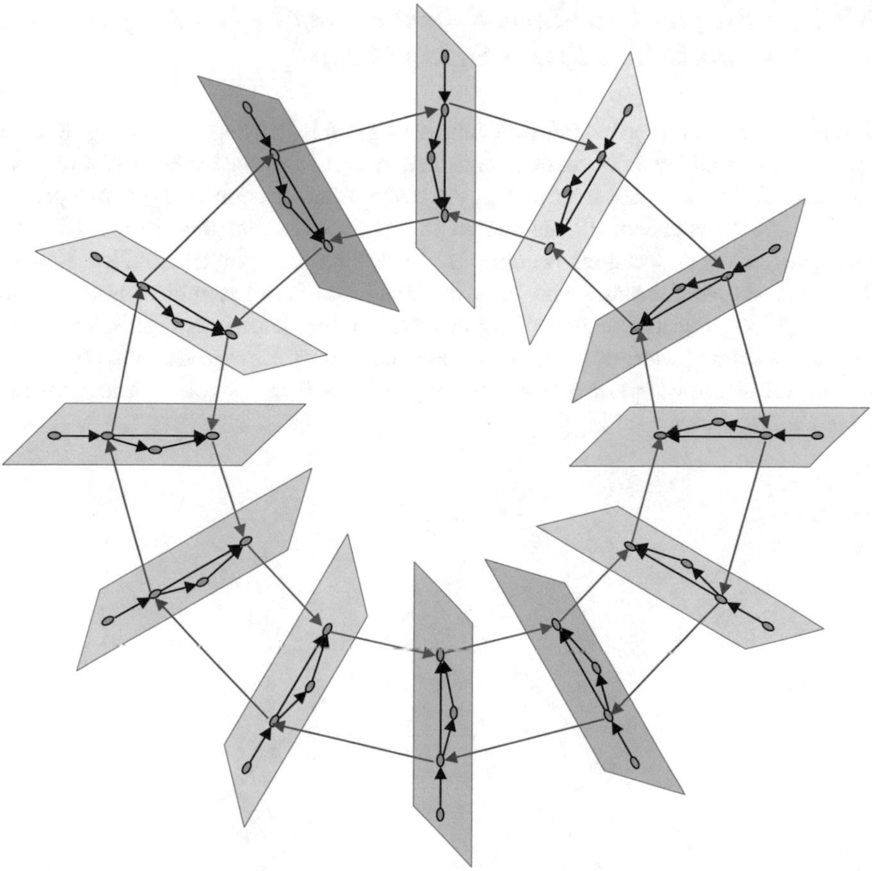

Fig. 8.5 Multilevel reified network of 11th-order forming a strange loop of 12 levels, which actually makes the order ∞

correctly. In fact, locally, keeping your head (or the page) in the right position, for every two adjacent levels the model is quite normal, just a first-order adaptive network. The only strange thing is that globally it is connected as a cycle; but you cannot point at a specific part where there would be something wrong locally, exactly like usually is the case in Escher's work.

This does not look like a conventional reified network as the usual examples in this book. However, is it impossible? If this is modeled and simulated at a computer, will there not be an inconsistency, like in many paradoxes where levels are mixed up? The answer to this is 'no, this is not impossible'. One main reason why no inconsistency occurs in a dynamic network modeling setting is that time keeps opposite views separate, they will not occur at the same instant.

8.3.2 A Simpler Four-Level Reified Example of an Adaptive Network Model for a Strange Loop

It seems that such a reified network model as graphically depicted in Fig. 8.5 can still be described by role matrices and then it will just run in the software environment. To test this, consider in Fig. 8.6 a simplified version of the above picture with 4 levels instead of 12, as shown in Fig. 8.5, so that the regular 12 sided polygon (called dodecagon) becomes a 4 sided regular polygon (called square). Table 8.1 shows the states and their explanations. This resembles more closely Escher (1960)'s design of Ascending and Descending shown in Fig. 8.2, with the blue arrows the upward walking line of persons, and the downward pink arrows the downward walking line of persons; see Fig. 8.9. In Box 8.1 role matrices for this example are shown.

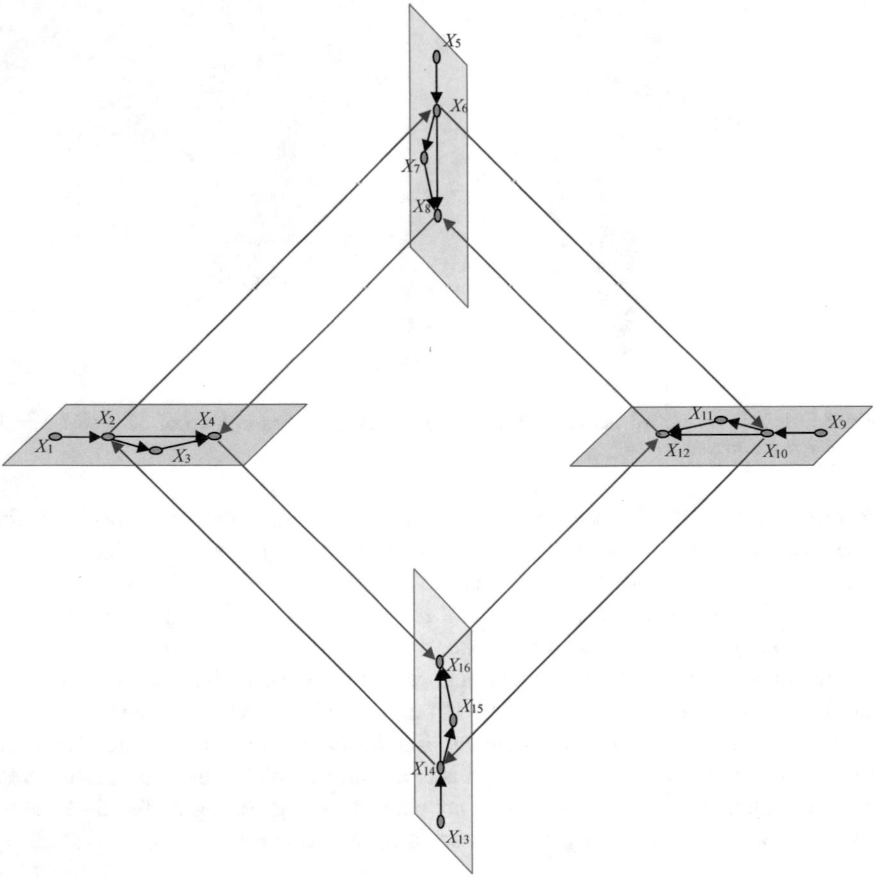

Fig. 8.6 A third-order multilevel reified network forming a strange loop, which actually makes the order ∞

Table 8.1 The states and their explanations

State		Explanation	Level
X_1			
X_2			Base level
X_3			
X_4	$\mathbf{W}_{X_{15},X_{16}}$	Reified representation state for the weight of the connection from X_{15} to X_{16}	
X_5			
X_6			First reification level
X_7			
X_8	$\mathbf{W}_{X_3,X_4}$	Reified representation state for the weight of the connection from X_3 to X_4	
X_9			
X_{10}			Second reification level
X_{11}			
X_{12}	$\mathbf{W}_{X_7,X_8}$	Reified representation state for the weight of the connection from X_7 to X_8	
X_{13}			
X_{14}			Third reification level
X_{15}			
X_{16}	$\mathbf{W}_{X_{11},X_{12}}$	Reified representation state for the weight of the connection from X_{11} to X_{12}	

The summary of this example reified network model is as follows. As an example, within the level of X_5 to X_8, the first state is an independent state X_5 for external contextual input, which can be considered a kind of sensor state for that level. It affects state X_6 which can be considered as a kind of representation for the context. That state is also affected by state X_2 at one level lower, so that this context representation X_6 covers more contextual aspects than the one via X_5. Next, state X_7 can be considered a form of interpretation or belief about this context. Finally, state X_8 can be considered some form of (preparation for) a concluding action. It has incoming connections from both X_6 and X_7. The connection from X_7 to X_8 is adaptive; its weight is determined by state X_{12} at the next level. This description applies to each of the four levels, where at each level a different aspect of the context may be used as an external input. Through the upward interlevel connections (the upward blue arrows) this contextual information is integrated in some way.

The states X_4, X_8, X_{12}, or X_{16} function as reification states for the connection weights indicated below. More specifically, the downward interlevel causal connections (pink arrows) determine the weights of one of the connections at the target level in the following way:

the pink arrow from X_8	determines	the weight of the connection from X_3 to X_4
the pink arrow from X_4	determines	the weight of the connection from X_{15} to X_{16}
the pink arrow from X_{16}	determines	the weight of the connection from X_{11} to X_{12}
the pink arrow from X_{12}	determines	the weight of the connection from X_7 to X_8

Note that within each level the determined connection weight itself affects the state to which the connection is pointing to: X_4, X_8, X_{12}, or X_{16}. These are exactly the states that in turn determine the weights at the next level.

All this is specified in the role matrices shown in Box 8.1. Note that each level has a similar structure; this can easily be varied at any level separately as well.

Each of the role matrices **mb** (base connectivity), **mcw** (connection weights), **ms** (speed factors), **mcfw** (combination function weights), **mcfp** (combination function parameters) has a format in which in each row for the indicated state it is specified which other states (red cells) or values (green cells) causally affect it and according to which role. In role matrix **mcw**, in particular, the red cells indicate which states X_i play the role of the reification states for the weights of the connection indicated in that cell in **mb**.

Box 8.1 Role matrices for the first Strange Loop reified network example

mb	**base connectivity**	1	2
X_1		X_1	
X_2		X_1	X_{14}
X_3		X_2	
X_4	$\mathbf{W}_{X_{15},X_{16}}$	X_2	X_3
X_5		X_5	
X_6		X_5	X_2
X_7		X_6	
X_8	$\mathbf{W}_{X_3,X_4}$	X_6	X_7
X_9		X_9	
X_{10}		X_9	X_6
X_{11}		X_{10}	
X_{12}	$\mathbf{W}_{X_7,X_8}$	X_{10}	X_{11}
X_{13}		X_{13}	
X_{14}		X_{13}	X_{10}
X_{15}		X_{14}	
X_{16}	$\mathbf{W}_{X_{11},X_{12}}$	X_{14}	X_{15}

mcw	**connection weights**	1	2
X_1		1	
X_2		1	1
X_3		1	
X_4	$\mathbf{W}_{X_{15},X_{16}}$	1	X_8
X_5		1	
X_6		1	1
X_7		1	
X_8	$\mathbf{W}_{X_3,X_4}$	1	X_{12}
X_9		1	
X_{10}		1	1
X_{11}		1	
X_{12}	$\mathbf{W}_{X_7,X_8}$	0.5	X_{16}
X_{13}		1	
X_{14}		1	1
X_{15}		1	
X_{16}	$\mathbf{W}_{X_{11},X_{12}}$	1	X_4

ms	**speed factors**	1
X_1		0.05
X_2		0.5
X_3		0.5
X_4	$\mathbf{W}_{X_{15},X_{16}}$	0.5
X_5		0.03
X_6		0.5
X_7		0.5
X_8	$\mathbf{W}_{X_3,X_4}$	0.5
X_9		0.02
X_{10}		0.5
X_{11}		0.5
X_{12}	$\mathbf{W}_{X_7,X_8}$	0.5
X_{13}		0.015
X_{14}		0.5
X_{15}		0.5
X_{16}	$\mathbf{W}_{X_{11},X_{12}}$	0.5

mcfw	**combination function weights**	1 **alogistic**
X_1		1
X_2		1
X_3		1
X_4	$\mathbf{W}_{X_{15},X_{16}}$	1
X_5		1
X_6		1
X_7		1
X_8	$\mathbf{W}_{X_3,X_4}$	1
X_9		1
X_{10}		1
X_{11}		1
X_{12}	$\mathbf{W}_{X_7,X_8}$	1
X_{13}		1
X_{14}		1
X_{15}		1
X_{16}	$\mathbf{W}_{X_{11},X_{12}}$	1

mcfp combination function parameters	function	1 **alogistic**	
	parameter	1 σ	2 τ
X_1		18	0.2
X_2		3	1.6
X_3		3	0.3
X_4	$\mathbf{W}_{X_{15},X_{16}}$	3	0.6
X_5		18	0.2
X_6		3	1.3
X_7		3	0.3
X_8	$\mathbf{W}_{X_3,X_4}$	3	1
X_9		18	0.2
X_{10}		3	1.3
X_{11}		3	0.3
X_{12}	$\mathbf{W}_{X_7,X_8}$	3	1
X_{13}		18	0.2
X_{14}		3	1.3
X_{15}		3	0.3
X_{16}	$\mathbf{W}_{X_{11},X_{12}}$	3	0.6

iv	**initial values**	1
X_1		0.13
X_2		0
X_3		0
X_4	$\mathbf{W}_{X_{15},X_{16}}$	0
X_5		0.12
X_6		0
X_7		0
X_8	$\mathbf{W}_{X_3,X_4}$	0
X_9		0.11
X_{10}		0
X_{11}		0
X_{12}	$\mathbf{W}_{X_7,X_8}$	0
X_{13}		0.1
X_{14}		0
X_{15}		0
X_{16}	$\mathbf{W}_{X_{11},X_{12}}$	0

8.4 Simulation Example of the Four Level Strange Loop Reified Network Model

Using the dedicated software environment described in Chap. 9, simulation has been performed based on the data shown in Box 8.1. The overall graph is shown in Fig. 8.7.

To get a bit more insight four groups of states are depicted separately in Fig. 8.8. Each group consists of four states from each of the four levels. The first, upper graph in Fig. 8.8 shows the independent external input each level gets at different points in time (the lower levels earlier than the higher levels). The second graph shows the states that are mutually connected by the upward connections (the blue arrows in Fig. 8.6). The third graph shows the states for which the outgoing connection is adapted from one reification level higher.

The fourth graph at the bottom in Fig. 8.8 shows the states with incoming connection the one that is adapted from one reification level higher, and, besides, as reification states they represent the corresponding connection weight for one level lower; that's because these are the states at the different levels that have the mutual downward connections (the pink arrows in Fig. 8.6).

Note that the upward connections (the blue arrows) make that X_2, X_6, X_{10}, X_{14} in principle cyclically affect each other (clockwise interlevel interaction): upward cyclic causality. However, it can be seen that at the lowest level state X_2 does not respond much on state X_1 (due to a relatively high threshold of X_2), so without the other levels that lowest level would not have become very active. It can also be seen that at the second and the third level states X_6 and X_{10} and later X_{14} at the fourth

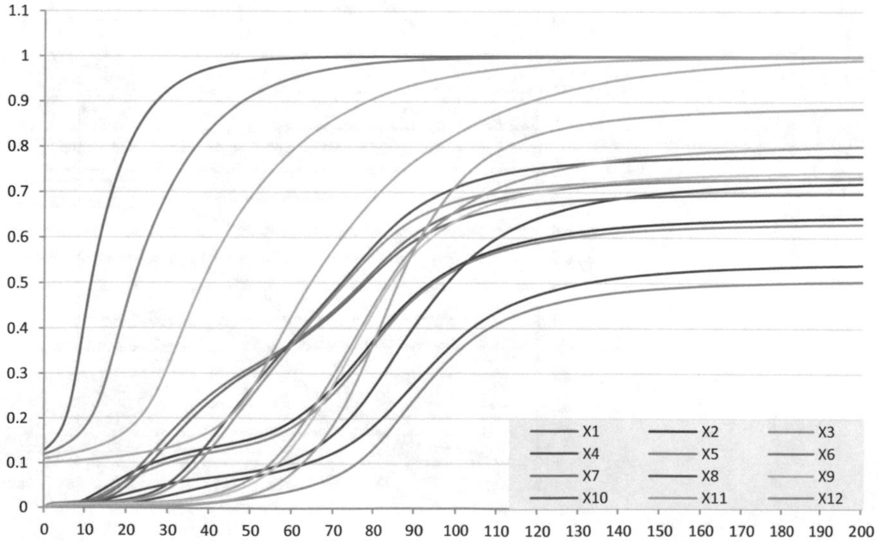

Fig. 8.7 Overall view on the strange loop example simulation

Fig. 8.8 Simulations for groups of states from each of the four levels in separate graphs

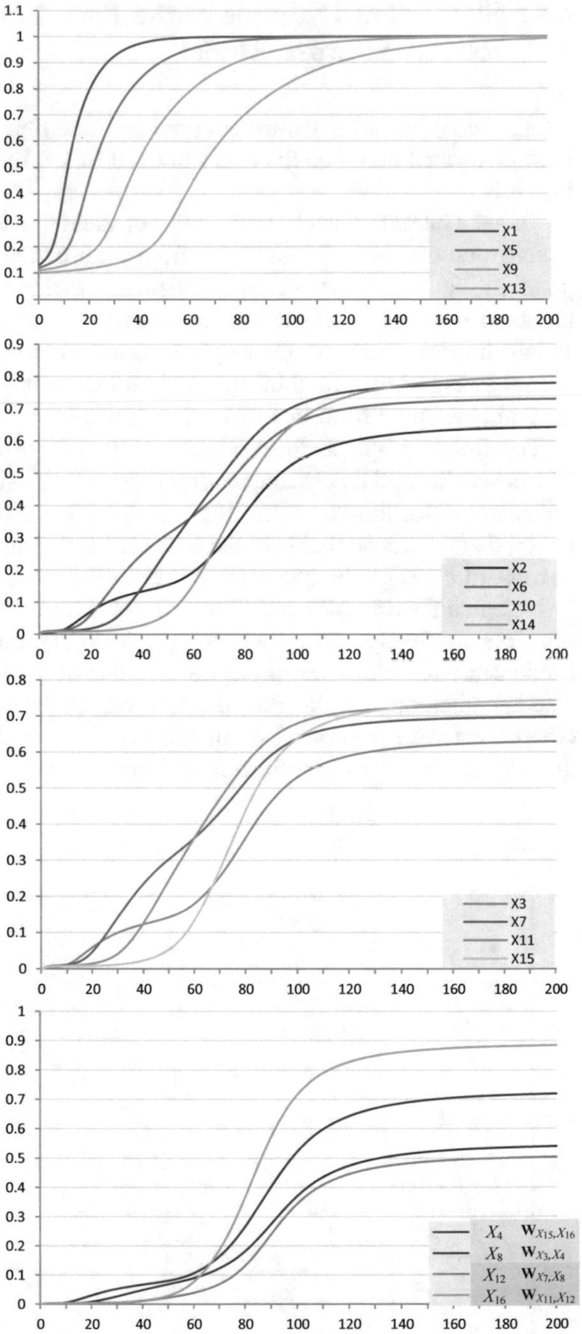

level respond much more actively on their input (their threshold is a bit lower). Finally, also X_2 becomes more active because of the other levels, in particular via X_{14} at the fourth level. Within that first level X_2 will affect X_4 more now; this is a first causal path by which X_4 is affected (via clockwise interlevel interaction via X_2).

The downward connections (the pink arrows) make that X_4, X_{16}, X_{12}, X_8 in principle cyclically affect each other in that order (anti-clockwise interlevel interaction), via the connection weights of their adaptive incoming connection that is affected from one level higher: downward causality. In the second graph also the effect of the less responsive X_2 (because of its higher threshold value $\tau = 1.6$) is seen. It only goes up after time point 70, later that the other three states in the second graph. What happens is that X_{16} at the fourth level (having a relatively low threshold value $\tau = 0.6$) saves the situation, and in turn, in a domino effect pushes the other states in this group X_4, X_{16}, X_{12}, X_8 upward by first increasing the weight of the incoming connection to X_{12} one level lower which then in turns affect X_8 which then affects X_4; this is a second causal path by which X_4 is affected (via anti-clockwise interlevel interaction).

In Fig. 8.9 the match of this four level cyclic reified network and Escher (1960)'s design of his Ascending and Descending lithograph is shown.

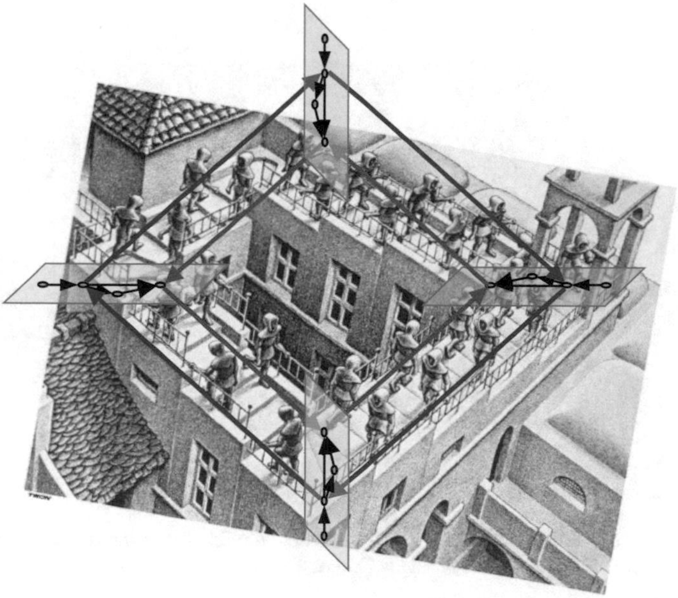

Fig. 8.9 The match between Escher (1960)'s ascending and descending design and the design of the four level cyclic reified network model

8.5 A Drawing Hands Reified Network Model for Adaptive Decision Making

As another example of a reified network model for a Strange Loop, an example adaptive decision making model is considered, designed according to Escher's Drawing Hands lithograph shown in Fig. 8.3. This decision making takes place in the lower (base) plane in Fig. 8.10, where two options a_1 and a_2 are modeled. Recall the quote of Hofstadter in Sect. 8.2.2 about rules at different levels, and the adaptation made by replacing the word 'rule' by 'causal connection'. The idea of rules at different levels was also explored in (Davis 1980), although in that case no Strange Loop was modeled. For the network model in the current section, a Strange Loop is incorporated, and following Escher's Drawing Hands design, just two levels are used, where one level is a reification level for (and controlling) the other and vice versa. And, of course, no rules are used but causal connections. So, instead of priorities for rules, here the connection weights used in the decision making process (at the lower level) are adaptive and controlled by the other (upper) level. In turn, the connections in the upper level are controlled by the lower decision making level, in particular, based on the extent of successfulness of a decision in the given context. The network model's conceptual representation is depicted in Fig. 8.10. In Table 8.2 an overview of all states is shown.

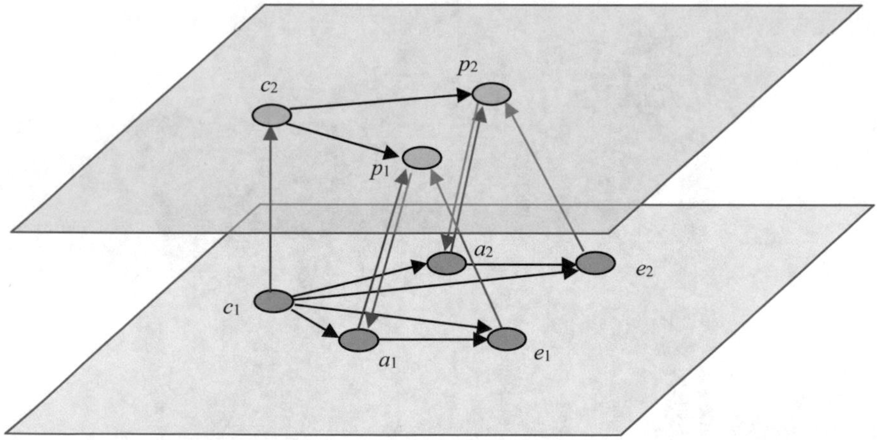

Fig. 8.10 Mutually reified temporal-causal network model for adaptive decision making according to Escher (1948)'s drawing hands design

Table 8.2 Overview of the states in the drawing hands adaptive decision model

State		Explanation	Level
X_1	c_1	First level context factor	Base level
X_2	a_1	Action option 1	
X_3	a_2	Action option 2	
X_4	$e_1 = \mathbf{W}_{X_6,X_7}$	Positive evaluation for a_1 / Reified representation state for weight from c_2 to p_1	
X_5	$e_2 = \mathbf{W}_{X_6,X_8}$	Positive evaluation for a_2 / Reified representation state for weight from c_2 to p_2	
X_6	c_2	Second level context factor	First reification level
X_7	$p_1 = \mathbf{W}_{X_1,X_2}$	Priority for a_1 / Reified representation state for weight from c_1 to a_1	
X_8	$p_2 = \mathbf{W}_{X_1,X_3}$	Priority for a_2 / Reified representation state for weight from c_1 to a_2	

In the base plane, a context state c_1 is shown with two decision options a_1 and a_2, triggered in this context according to the two connection weights. Whether or not an option a_i is successful, is modeled by state e_i, which as a kind of reality check gets an incoming connection from the context state c_1. For the adaptation process, the upper plane uses three states c_2, p_1 and p_2. Here c_2 is another state representing (aspects of) the current context, and p_1 and p_2 can be considered as indicating priorities for the choice for a_1 or a_2, respectively. These states p_1 and p_2 play the role of connection weight reification states for the incoming connections to a_1 and a_2, as indicated by X_7 and X_8 in the red cells in the second and third row in role matrix **mcw** in Box 8.2. In the picture the pink or red arrows indicate what is usually called a downward causal connection. This is a connection from a reification state to the related base state.

The pattern described this far can be considered an adaptive hierarchical decision model. However, a Strange Loop comes in by assigning states e_1 and e_2 the role of reification state for the connections from c_2 to p_1 and p_2, as indicated in role matrix **mcw** in Box 8.2 by X_4 and X_5 in the red cells in the last two rows for p_1 and p_2. In this way evaluation of decisions made at the 'lower' level provides feedback to the 'higher' level, so that at that level adaptation can take place. However, due to the cyclic structure of the levels, there is no lower and higher level anymore.

The role matrices and initial values for this mutually reified network model are shown in Box 8.2. The simulations shown in Sect. 8.6 are based on these role matrices and initial values.

Box 8.2 Role matrices for the mutually reified network for adaptive decision making

mb	**base connectivity**	1	2
X_1	c_1	X_1	
X_2	a_1	X_1	
X_3	a_2	X_1	
X_4	$e_1 = \mathbf{W}_{X_6,X_7}$	X_2	X_1
X_5	$e_2 = \mathbf{W}_{X_6,X_8}$	X_3	X_1
X_6	c_2	X_1	
X_7	$p_1 = \mathbf{W}_{X_1,X_2}$	X_6	X_2
X_8	$p_2 = \mathbf{W}_{X_1,X_3}$	X_6	X_3

mcw	**connection weights**	1	2
X_1	c_1	1	
X_2	a_1	X_7	
X_3	a_2	X_8	
X_4	$e_1 = \mathbf{W}_{X_6,X_7}$	1	0.6
X_5	$e_2 = \mathbf{W}_{X_6,X_8}$	0.2	0.5
X_6	c_2	1	
X_7	$p_1 = \mathbf{W}_{X_1,X_2}$	X_4	1
X_8	$p_2 = \mathbf{W}_{X_1,X_3}$	X_5	1

ms	**speed factors**	1
X_1	c_1	0.05
X_2	a_1	0.5
X_3	a_2	0.5
X_4	$e_1 = \mathbf{W}_{X_6,X_7}$	0.005
X_5	$e_2 = \mathbf{W}_{X_6,X_8}$	0.007
X_6	c_2	0.5
X_7	$p_1 = \mathbf{W}_{X_1,X_2}$	0.004
X_8	$p_2 = \mathbf{W}_{X_1,X_3}$	0.005

mcfw	**combination function weights**	1 **alogistic**
X_1	c_1	1
X_2	a_1	1
X_3	a_2	1
X_4	$e_1 = \mathbf{W}_{X_6,X_7}$	1
X_5	$e_2 = \mathbf{W}_{X_6,X_8}$	1
X_6	c_2	1
X_7	$p_1 = \mathbf{W}_{X_1,X_2}$	1
X_8	$p_2 = \mathbf{W}_{X_1,X_3}$	1

mcfp	function	1 **alogistic**	
combination function parameters	parameter	1 σ	2 τ
X_1	c_1	18	0.2
X_2	a_1	4	0.5
X_3	a_2	4	0.5
X_4	$e_1 = \mathbf{W}_{X_6,X_7}$	3	1.2
X_5	$e_2 = \mathbf{W}_{X_6,X_8}$	3	1.2
X_6	c_2	18	0.4
X_7	$p_1 = \mathbf{W}_{X_1,X_2}$	6	0.5
X_8	$p_2 = \mathbf{W}_{X_1,X_3}$	6	0.5

iv	**initial values**	1
X_1	c_1	0.1
X_2	a_1	0
X_3	a_2	0
X_4	$e_1 = \mathbf{W}_{X_6,X_7}$	0.3
X_5	$e_2 = \mathbf{W}_{X_6,X_8}$	0.5
X_6	c_2	0
X_7	$p_1 = \mathbf{W}_{X_1,X_2}$	0.3
X_8	$p_2 = \mathbf{W}_{X_1,X_3}$	0.6

Figure 8.11 shows how the mutually reified network model matches to Escher (1948)'s fully symmetric Drawing Hands design.

8.6 An Example Simulation of the Drawing Hands Reified Network Model

In this section, an example is shown of a simulation of the adaptive decision making process modeled by the mutually reified network model. For a relatively short term the outcome looks as shown in Fig. 8.12.

Here it can be seen that action option a_2 (the green line) is preferred over option a_1 (the red line), as the latter seems not to reach a level that is much higher than 0.2. However, in a longer term, for the same simulation, the adaptation process does its work. This is shown in Fig. 8.13 (note the 10 times longer time scale). Now it is

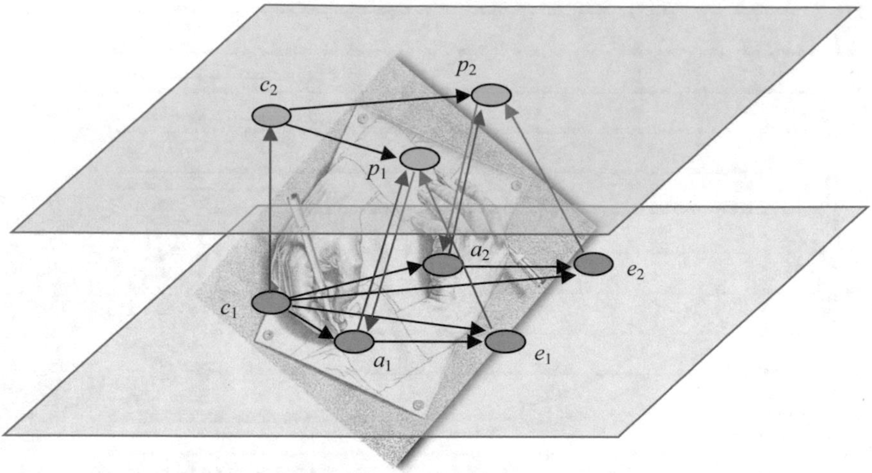

Fig. 8.11 The match between Escher's drawing hands design and the design of the mutually reified network for adaptive decision making

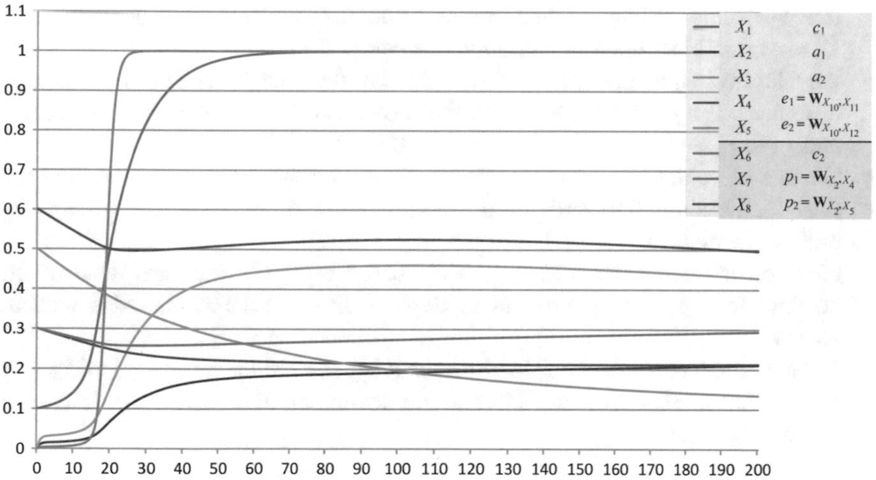

Fig. 8.12 Simulation for a relatively short term

clear that ultimately action option a_1 is the preferred one, as its value reaches a level above 0.8, whereas action option a_2 drops to 0.1.

So, although initially, a responsive process triggers the options with a preference for a_2, the evaluation e_2 based on the reality of the context shows that for the given context this choice is not adequate (whereas a choice for a_1 would be more adequate). This leads after time point 150 or 200 subsequently to the following adaptations:

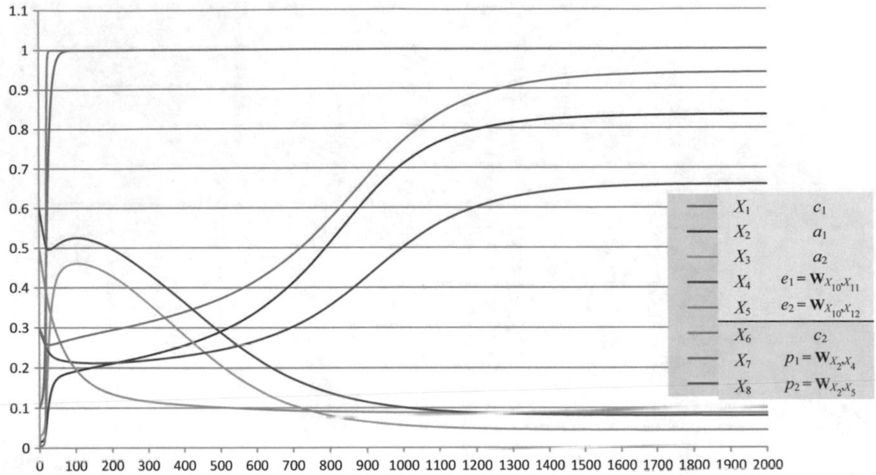

Fig. 8.13 Simulation for a longer term

- Change of values of evaluation states e_1 and e_2; see the upward trend of $X_4 = e_1$ (the grey line ending up between 0.8 and 0.7) and the downward trend of $X_5 = e_2$ (the blue-green line ending just below 0.1)
- An adaptive adjustment of the connections in the blue plane from context state c_2 to the states p_1 and p_2, based on the 'downward' causal connections from e_1 and e_2 to p_1 and p_2, depicted in Fig. 8.10 by the pink upward arrows
- Change of values of p_1 and p_2; see the upward trend of $X_7 = p_1$ (the blue line finally exceeding 0.9) and the downward trend of $X_8 = p_2$ (the brown line ending below 0.1)
- effect on the connections to a_1 and a_2, based on the 'downward' causal connections from p_1 and p_2 to a_1 and a_2, depicted in Fig. 8.10 by the red downward arrows
- a changed effect on the choices for a_1 and a_2; see the upward trend of $X_2 = a_1$ (the red line ending up above 0.8) and the downward trend of $X_3 = a_2$ (the green line ending below 0.1).

8.7 Discussion

In this chapter, like in Chap. 7, the challenge to find and explore a plausible reified network model of order higher than 2 was addressed. To this end, the notion of Strange Loop was considered, which from a philosophical perspective is claimed by Hofstadter (1979, 2007) to be at the basis of human intelligence and consciousness. Reified adaptive network models using a Strange Loop of order higher than 3 were shown. Due to the Strange Loop they can even be interpreted as being of infinite order, as by a Strange Loop the reification levels form a (closed) cycle with no beginning or end.

The first network model as presented is mainly meant to show how the interaction between the levels works. The processes within each level were kept a bit toylike. Its design follows Escher (1960)'s design of his lithograph Ascending and Descending. The second example addresses adaptive decision making in which two levels have mutual reification relations. This second design follows Escher (1948)'s design of his lithograph Drawing Hands.

A future step may be to explore more complex real-world reified network models using a Strange Loop, for example, by using further inspiration from the literature relating to architectural design such as (Gannon 2017), to advertising such as (Hendlin 2019), or to psychotherapeutic understanding such as (Strijbos and Glas 2018).

Another challenge to explore adaptive networks of order higher than 2 was to focus the application scope for reified networks on evolutionary processes, which was addressed in Chap. 7 for the adaptive evolutionary processes leading to disgust in the first trimester of pregnancy (Fessler et al. 2005, 2015; Fleischman and Fessler 2011; Jones et al. 2005); also see Chap. 1, Sect. 1.3.2. Together Chaps. 7 and 8 provide different positive answers on the question of whether or not adaptive networks of order higher than two may have interesting applications as posed in Chap. 1, Sect. 1.3.

References

Davis, R.: Meta-rules: reasoning about control. Artif. Intell. **15**, 179–222 (1980)

Davis, R., Buchanan, B.G.: Meta-level knowledge: overview and applications. In: Proceedings of 5th IJCAI, pp. 920–927 (1977)

Escher, M.C.: Drawing hands (1948)

Escher, M.C.: Print gallery (1956)

Escher, M.C.: Ascending and Descending (1960)

Fessler, D.M.T., Clark, J.A., Clint, E.K.: Evolutionary psychology and evolutionary anthropology. In: Buss, D.M. (ed.) The Handbook of Evolutionary Psychology, pp. 1029–1046, Wiley and Sons (2015)

Fessler, D.M.T., Eng, S.J., Navarrete, C.D.: Elevated disgust sensitivity in the first trimester of pregnancy: evidence supporting the compensatory prophylaxis hypothesis. Evol. Hum. Behav. **26**(4), 344–351 (2005)

Fleischman, D.S., Fessler, D.M.T.: Progesterone's effects on the psychology of disease avoidance: support for the compensatory behavioral prophylaxis hypothesis. Horm. Behav. **59**(2), 271–275 (2011)

Galton, A.: Operators vs. arguments: the ins and outs of reification. Synthese **150**, 415–441 (2006)

Gannon, T.: Strange loops: toward an aesthetics for the anthropocene. J. Arch. Educ. **71**(2), 142–145 (2017)

Hendlin, Y.H.: I am a fake loop: the effects of advertising-based artificial selection. Biosemiotics **12**(1), 131–156 (2019)

Hofstadter, D.R.: Gödel, Escher, Bach. Basic Books, New York (1979)

Hofstadter, D.R.: What is it like to be a strange loop? In: Kriegel, U., Williford, K. (eds.) Self-Representational Approaches to Consciousness. MIT Press, Cambridge (2006)

Hofstadter, D.R.: I Am a Strange Loop. Basic Books, New York (2007)

Jones, B.C., Perrett, D.I., Little, A.C., Boothroyd, L., Cornwell, R.E., Feinberg, D.R., Tiddeman, B.P., Whiten, S., Pitman, R.M., Hillier, S.G., Burt, D.M., Stirrat, M.R., Law Smith, M.J., Moore, F.R.: Menstrual cycle, pregnancy and oral contraceptive use alter attraction to apparent health in faces. Proc. R. Soc. B **5**(272), 347–354 (2005)

Kriegel, U., Williford, K. (eds.).: Self-Representational Approaches to Consciousness. MIT Press, Cambridge (2006)

Nagel, E., Newman, J.: Gödel's Proof. New York University Press, New York (1965)

Smorynski, C.: The incompleteness theorems. In: Barwise, J. (ed.) Handbook of Mathematical Logic, vol. 4, pp. 821–865. North-Holland, Amsterdam (1977)

Sterling, L., Beer, R.: Metainterpreters for expert system construction. J. Logic Program. **6**, 163–178 (1989)

Strijbos, D., Glas, G.: Self-knowledge in personality disorder: self-referentiality as a stepping stone for psychotherapeutic understanding. J. Pers. Dis. **32**(3), 295–310 (2018)

Treur, J.: Network-Oriented Modeling: Addressing Complexity of Cognitive, Affective and Social Interactions. Springer Publishers (2016)

Treur, J.: Network reification as a unified approach to represent network adaptation principles within a network. In: Proceedings of the 7th International Conference on Theory and Practice of Natural Computing, TPNC'18. Lecture Notes in Computer Science, vol. 11324, pp. 344–358. Springer Publishers (2018a)

Treur, J.: Multilevel network reification: representing higher-order adaptivity in a network. In: Proceedings of the 7th International Conference on Complex Networks and their Applications, Complex Networks' 18, vol. 1. Studies in Computational Intelligence, vol. 812, pp. 635–651. Springer (2018b)

Treur, J.: The ins and outs of network-oriented modeling: from biological networks and mental networks to social networks and beyond. Trans. Comput. Collective Intell. **32**, 120–139 (Springer Publishers. Contents of Keynote Lecture at ICCCI'18) (2019a)

Treur, J.: Design of a Software Architecture for Multilevel Reified Temporal-Causal Networks. Doi: https://doi.org/10.13140/rg.2.2.23492.07045. Url: https://www.researchgate.net/publication/333662169 (2019b)

Weyhrauch, R.W.: Prolegomena to a theory of mechanized formal reasoning. Artif. Intell. **13**, 133–170 (1980)

Part IV
A Modeling Environment for Reified Networks

Chapter 9
A Modeling Environment for Reified Temporal-Causal Network Models

Abstract The introduced multilevel reified (temporal-causal) network architecture is the basis of the implementation of a dedicated software environment developed by the author in Matlab. The environment includes a combination function library and a generic computational reified network engine. It uses role matrices specifying the characteristics for the designed network model as input. Based on this input, the computational reified network engine can be used to generate simulations for the network model, thereby using combination functions from the library. In this chapter, this software environment is described in more detail.

9.1 Introduction

The notion of temporal-causal network model is a point of departure (Treur 2016, 2019a). Based on this notion, in Chaps. 3 and 4 the architecture of a reified network model has been introduced and illustrated by, respectively

- a first-order adaptive example Social Network with adaptive responding speed and adaptive aggregation for contagion
- a second-order adaptive example Mental Network showing plasticity and metaplasticity.

See also (Treur 2018a, b). In the subsequent Chaps. 5–8, five more applications of this architecture have been shown, respectively,
for

- a second-order adaptive Mental Network for adaptive decision making based on a state-connection adaptation principle
- a second-order adaptive Social Network for Bonding by Homophily
- a fourth-order adaptive Biological Network for evolutionary processes in handling pathogens
- a third-order adaptive Mental Network based on a Strange Loop, and a first-order mutually reified Mental Network for adaptive decision making based on a Strange Loop.

J. Treur, *Network-Oriented Modeling for Adaptive Networks: Designing Higher-Order Adaptive Biological, Mental and Social Network Models*, Studies in Systems, Decision and Control 251, https://doi.org/10.1007/978-3-030-31445-3_9

To fully exploit the potential of this approach for such applications for adaptive networks, a dedicated modeling environment has been developed. This modeling environment uses a specific implementation-independent modeling format for a reified network architecture based on role matrices (see Chaps. 2–4, and many later examples in the book). This format has a basis in the implementation-independent mathematical notions of matrix and function. The notion of role matrix provides an alternative and more compact specification format for networks, in contrast to connection matrices that are often used as main means to specify networks at an implementation-independent level. Section 9.2 briefly summarizes role matrices as specification format for reified adaptive temporal-causal network models. This role matrix format can be used for implementation in different ways; one example implementation is the computational reified temporal-causal network engine discussed here.

The basic elements of the specification format are network characteristics that are declarative by nature: connection weights, combination functions, speed factors, grouped in role matrices. Together these elements define in a standard manner a set of first-order difference or differential equations (as shown in Chap. 2, Tables 2.1 and 2.2), which are used computationally in the software environment. The model's behavior is fully determined and explainable by these declarative temporal specifications (given some initial values). The modeling process is strongly supported by using these declarative building blocks. By using a reified network architecture, the scope is substantially extended to adaptive networks of any order. As shown in earlier chapters, very complex adaptive patterns can be modeled easily in temporal declarative form. The universal combination function and universal difference equation as described in Chap. 3, Sect. 3.5 (see also Chap. 10) form a basis to address multiple orders of adaptation.

Combination functions play a central role in the modeling approach. The software environment includes a combination function library (currently with 35 functions), which is discussed in Sect. 9.3. In Sect. 9.4 the computational reified temporal-causal network engine is described that has been developed. This is the software that takes the declarative specification of Sect. 9.2 in terms of role matrices and makes it run. The design of this software is structure-preserving in relation to the mathematical description and can be easily translated into any other software environment that supports the mathematical notions of function and matrix.

9.2 Role Matrices as a Specification Format for Reified Network Models

For networks, matrices are often considered a way that is easily accessible computationally and vastly used in network modeling and analysis. The choice was made to use specific types of matrices (called *role matrices*) to define a specification format to design reified network models, and as a well-defined basis for implementation. The implementation described here is based on Matlab ('Matrix

Laboratory') which handles matrices well; but any other software environment supporting matrices can be used as well. Role matrices were introduced in Chap. 2 as matrices grouping the relevant values for the characteristics of the network structure, for example as used in a given simulation scenario. For a nonadaptive network, role matrices are indeed just neatly structuring these data in a table format (e.g., in Word or in Excel) that provides a complete description of the design of the network model. When these role matrices are copied to Matlab, the network model can be executed (see Sect. 9.4). In Chap. 3 the use of role matrices was extended to reified networks. In that case, some of the cells of these matrices indicate adaptive network structure characteristics, which means that they do not contain a constant value, but a variable value instead; these cells were shaded in red, whereas the value cells were in green. This 'variable value' is indicated in that cell by putting the name of the reification state that plays the role for that characteristic. This indication specifies the standard downward causal connection for that reification state. The specific role matrix in which this reification state name is indicated defines for which role it is a reification state; see also Table 9.1.

More specifically, the role matrices have rows for all states of the network. For an example, borrowed from Chap. 4, see Box 9.1; for the picture of the model, see Chap. 4, Sect. 4.4.1, Fig. 4.3. In the row for state *Y*, in base role matrix **mb** for *base connectivity* it is specified which other states have an impact on *Y*. This corresponds to the upward and horizontal incoming arrows for state *Y* in a conceptual graphical representation of the model. In the other (non-base) role matrices, other roles are

Table 9.1 Downward causal connections and role matrices: how and where specify what

In conceptual graphical representation of the model	State name	State number	Role	In role matrix
Downward arrow from a reification state for an adaptive connection weight from state *X* to state *Y*	$\mathbf{W}_{X,Y}$	X_i	Connection weight reification state for $\boldsymbol{\omega}_{X,Y}$	**mcw** as notation X_i in the cell for the weight $\boldsymbol{\omega}_{X,Y}$ of the connection from *X* to *Y*
Downward arrow from a reification state for an adaptive speed factor for state *Y*	$\mathbf{H}_Y$	X_j	Speed factor reification state for $\boldsymbol{\eta}_Y$	**ms** as notation X_j in the cell for the value $\boldsymbol{\eta}_Y$ of the speed factor of *Y*
Downward arrow from a reification state for an adaptive combination function weight for state *Y*	$\mathbf{C}_{i,Y}$	X_k	Combination function weight reification state for weight $\gamma_{i,Y}$	**mcfw** as notation X_k in the cell for the weight $\gamma_{i,Y}$ of combination function *i* for state *Y*
Downward arrow from a reification state for an adaptive combination function parameter for state *Y*	$\mathbf{P}_{i,j,Y}$	X_l	Combination function parameter reification state for parameter $\pi_{i,j,Y}$	**mcfp** as notation X_l in the cell for the value $\pi_{i,j,Y}$ of parameter *j* of combination function *i* for state *Y*

indicated (either values or reification states playing that role): in **mcw** *connection weights*, in **ms** *speed factors,* in **mcfw** *combination function weights*, and in **mcfp** *combination function parameters*. In the case of reification states, these (non-base) role indications correspond to downward arrows in a conceptual graphical representation of the model. Each of these downward arrows in such a picture gets a special effect assigned due to the role matrix in which it is indicated: the connected reification state plays the role of connection weight, the role of speed factor, the role of combination function weight, or the role of a combination function parameter. The special causal effect of a reification state on the related base state *Y* has to be according to this role. For example, the value of a reification state with role speed factor has to be used for the speed factor characteristic in the processing for *Y*, and not for anything else. See Table 9.1 for an overview of these non-base roles.

Box 9.1 Example specification of a second order reified network model for plasticity and metaplasticity by role matrices

mb	**base connectivity**	1	2	3	4
X_1	ss_s	X_1			
X_2	srs_s	X_1			
X_3	bs_s	X_2			
X_4	ps_a	X_2	X_3		
X_5	$\mathbf{W}_{srs_s,ps_a}$	X_2	X_4	X_5	
X_6	$\mathbf{T}_{srs_s}$	X_2	X_4	X_5	X_6
X_7	$\mathbf{T}_{ps_a}$	X_2	X_4	X_5	X_7
X_8	$\mathbf{HW}_{srs_s,ps_a}$	X_2	X_4	X_5	X_8
X_9	$\mathbf{MW}_{srs_s,ps_a}$	X_2	X_4	X_5	X_9

mcw	**connection weights**	1	2	3	4
X_1	ss_s	1			
X_2	srs_s	1			
X_3	bs_s	1			
X_4	ps_a	X_5	1		
X_5	$\mathbf{W}_{srs_s,ps_a}$	1	1	1	
X_6	$\mathbf{T}_{srs_s}$	-0.4	-0.4	1	1
X_7	$\mathbf{T}_{ps_a}$	-0.4	-0.4	1	1
X_8	$\mathbf{Hw}_{srs_s,ps_a}$	1	1	-0.1	1
X_9	$\mathbf{Mw}_{srs_s,ps_a}$	1	1	1	1

mcfw	**combination function weights**	1 **eucl**	2 **alogistic**	3 **hebb**
X_1	ss_s	1		
X_2	srs_s		1	
X_3	bs_s		1	
X_4	ps_a		1	
X_5	$\mathbf{W}_{srs_s,ps_a}$			1
X_6	$\mathbf{T}_{srs_s}$		1	
X_7	$\mathbf{T}_{ps_a}$		1	
X_8	$\mathbf{HW}_{srs_s,ps_a}$		1	
X_9	$\mathbf{MW}_{srs_s,ps_a}$		1	

mcfp	**function**	1 **eucl**		2 **alogistic**		3 **hebb**	
	parameter	1 n	2 λ	1 σ	2 τ	1 μ	2
X_1	ss_s	1	1				
X_2	srs_s			5	X_6		
X_3	bs_s			5	0.2		
X_4	ps_a			5	X_7		
X_5	$\mathbf{W}_{srs_s,ps_a}$					X_9	
X_6	$\mathbf{T}_{srs_s}$			5	0.7		
X_7	$\mathbf{T}_{ps_a}$			5	0.7		
X_8	$\mathbf{Hw}_{srs_s,ps_a}$			5	1		
X_9	$\mathbf{Mw}_{srs_s,ps_a}$			5	1		

ms	**speed factors**	1
X_1	ss_s	0.5
X_2	srs_s	0.5
X_3	bs_s	0.2
X_4	ps_a	0.5
X_5	$\mathbf{W}_{srs_s,ps_a}$	X_8
X_6	$\mathbf{T}_{srs_s}$	0.3
X_7	$\mathbf{T}_{ps_a}$	0.3
X_8	$\mathbf{HW}_{srs_s,ps_a}$	0.5
X_9	$\mathbf{MW}_{srs_s,ps_a}$	0.1

Table 9.2 Role matrices **mcwa** and **mcwv** for the connection weights for the specification in Box 9.1

	mcwa	1	2	3	4
X_1	ss_s				
X_2	srs_s				
X_3	bs_s				
X_4	ps_a	5			
X_5	$\mathbf{W}_{srs_s, ps_a}$				
X_6	$\mathbf{T}_{srs_s}$				
X_7	$\mathbf{T}_{ps_a}$				
X_8	$\mathbf{H}\mathbf{W}_{srs_s, ps_a}$				
X_9	$\mathbf{M}\mathbf{W}_{srs_s, ps_a}$				

	mcwv	1	2	3	4
X_1	ss_s	1			
X_2	srs_s	1			
X_3	bs_s	1			
X_4	ps_a		1		
X_5	$\mathbf{W}_{srs_s, ps_a}$	1	1	1	
X_6	$\mathbf{T}_{srs_s}$	-0.4	-0.4	1	
X_7	$\mathbf{T}_{ps_a}$	-0.4	-0.4	1	
X_8	$\mathbf{H}\mathbf{W}_{srs_s, ps_a}$	1	1	-0.1	1
X_9	$\mathbf{M}\mathbf{W}_{srs_s, ps_a}$	1	1	1	1

Moreover, notice that the connected reification states are given standard names to reflect their role **H**, **W**, **C** or **P**, meant for human understanding, and these are not used in the execution of the model. State names used in the implementation are just the X_1, X_2, … based on the numbering of the states; see also Table 9.1. In the execution of the model, the role matrices fully define which are the downward causal connections and which of the network structure characteristics for target state Y they affect as their special effect.

Note that the role matrices used in the reified network model specification format have a compact format, and also specify an ordering, which is important as combination functions used to integrate the impact from multiple connections which are not always symmetric in their arguments.

Here, the red cells represent the downward causal connections from the reification states in pictures as shown in Chap. 4, Fig. 4.3, with their specific roles **W**, **H**, **C**, **P** indicated by the type of role matrix in which such a red cell occurs. The static values are in the green cells, and the remaining cells are empty. According to this red/green/empty partition, each role matrix can be split into two matrices: one for the green part (and the other cells empty) and one for the red part (and the other cells empty). The former matrix is called the *adaptation matrix* (with an **a** added to the name) and the latter the *values matrix* (with a **v** added to the name); for example, see Table 9.2. This will be discussed in more detail in Sect. 9.4.

9.3 The Combination Function Library

The combination function library currently consists of 35 functions. Three main groups are briefly discussed first, after which the standard format for these functions is discussed. More details can be found at Treur (2019b) and at https://www.researchgate.net/publication/336681331.

9.3.1 Different Classes of Combination Functions

As a first class, the library currently includes the following functions that in principle are suitable to model social contagion in social networks, but can also be used in other network models (for further analysis of these functions and the emerging behaviour entailed by them, see also Chap. 11, Sects. 11.4–11.6):

```
eucl(p,v), alogistic(p,v), slogistic(p,v), invtan(p,v),
smin(p,v),smax(p,v), aminmax(p,v), sgeomean(p,v)
```

Here, and below, `p` is the vector of parameters and `v` the vector of values for the function. As a second class, to model reification states for Hebbian learning, currently available options are:

```
hebb(p,v), sconnhebb(p,v), srconnhebb(p,v),
srstateshebb(p,v), sstateshebb(p,v)
```

These functions are analysed further in Chap. 14, Sect. 14.6, and satisfy the relevant properties of Hebbian learning combination functions discussed in Chap. 14, Sect. 14.4.1. Similarly, another class of functions is available to model reification states for bonding by homophily (for analysis of these functions, see also Chap. 13):

```
slhomo(p,v), sqhomo(p,v), alhomo(p,v), aqhomo(p,v),
cubehomo(p,v), exphomo(p,v), log1homo(p,v), log2homo(p,v),
sinhomo(p,v), tanhomo(p,v), multicriteriahomo(p,v)
```

These functions are shown in particular in Chap. 13, Sect. 13.3.2 and satisfy the relevant properties of bonding by homophily combination functions discussed in Chap. 13, Sect. 13.5.

Note that when more than one combination function is selected for one given state with nonzero combination function weights in matrix **mcfw**, then these functions should use the same shared sequence of values. This is automatically the case with functions of the first group, which are symmetric in their arguments and have a variable number of *k* arguments. For functions from the other groups this not automatically the case, as they are not symmetric in their arguments; then auxiliary variables may have to be added to be able to use the same sequence of values for multiple functions, from which each function only actually uses the subsequence that is relevant for it. For an illustration of how this is done, see Chap. 5, Sect. 5.4.1
.

9.3.2 The Standard Format of Combination Functions

A complete specification of these combination functions can be found at (Treur 2019b). To obtain a general format easily usable within the software environment, these functions were numbered and rewritten in the standard *basic combination function* form

bcf(i, **p**, **v**)

where i is the number of the basic function, **p** is its vector of parameters and **v** is a vector of values. This was implemented in Matlab (and stored as bcf.m) by the definition for bcf shown in Box 9.2.

Box 9.2 Specification of bcf getting the basic combination functions in a standard format.

```
function x = bcf(i,p,v)
% bcf = basic combination functions; this function combi-
nes all basic
combination functions in one format
if    i ==1          x = eucl(p,v);
elseif i ==2         x = alogistic(p,v);
elseif i ==3         x = hebb(p,v);
elseif i ==4         x = scm(p,v);
elseif i ==5         x = slhomo(p,v);
elseif i ==6         x = sqhomo(p,v);
elseif i ==7         x = alhomo(p,v);
elseif i ==8         x = aqhomo(p,v);
elseif i ==9         x = sconnhebb(p,v);
elseif i ==10        x = srconnhebb(p,v);
elseif i ==11        x = srstateshebb(p,v);
elseif i ==12        x = sstateshebb(p,v);
elseif i ==13        x = slogistic(p,v);
elseif i ==14        x = cubehomo(p,v);
elseif i ==15        x = exphomo(p,v);
elseif i ==16        x = log1homo(p,v);
elseif i ==17        x = log2homo(p,v);
elseif i ==18        x = sinhomo(p,v);
elseif i ==19        x = tanhomo(p,v);
elseif i ==20        x = invtan(p,v);
elseif i ==21        x = id(p,v);
elseif i ==22        x = complementid(p,v);
elseif i ==23        x = product(p,v);
```

```
elseif i ==24     x = coproduct(p,v);
elseif i ==25     x = sminimum(p,v);
elseif i ==26     x = smaximum(p,v);
elseif i ==27     x = aproduct(p,v);
elseif i ==28     x = aminmax(p,v);
elseif i ==29     x = multicriteriahomo(p,v);
elseif i ==30     x = ssum(p,v);
elseif i ==31     x = adnormsum(p,v);
elseif i ==32     x = adnormeucl(p,v);
elseif i ==33     x = sgeomean(p,v);
elseif i ==34     x = stepmod(p,v);
elseif i ==35     x = stepmodopp(p,v);
end
end
```

The selection of a sequence of combination functions from the library for a given network model is indicated by

mcf = [1 2 3]

where the sequence 1, 2, 3 (referring to the numbering of the library specified by bcf in Box 9.2) can be replaced by any sequence of any length of numbers from 1 to 35. After this specification, for the specific network model the numbering from this subsequence is used. For example, if in a network model specification **mcf** = [5 9 12 18] is indicated, then for that network model the 5th function from the library (**slhomo**) is indicated as basic combination function number 1, the 9th (**sconnhebb**) as number 2, and so on. It should be kept in mind that in principle this (local) numbering is different for each network model.

9.4 The Computational Reified Temporal-Causal Network Engine

The designed computational reified network engine takes a specification in the role matrices format as described in Sect. 9.2 and generates simulations for the model. It uses two main steps: first retrieving the characteristics of the network model from the role matrices (Sect. 9.4.2), and next performing the actual execution (Sect. 9.4.3). But before that, the role matrices have to be entered (Sect. 9.4.1).

9.4.1 Splitting the Role Matrices and Copying Them to Matlab

First, each role matrix (which can be specified easily by a table in Word or in Excel, for example) is copied or read in Matlab in two variants:

- a *value matrix* for the static values (adding the letter **v** to the name) from the green cells, and
- an *adaptivity matrix* for the adaptive values represented by reification states (adding the letter **a** to the name) from the red cells, thereby replacing X_j by the index j.

Note that states X_j are represented in Matlab by their index number j. For example, from **mcw** two matrices **mcwa** (adaptive connection weights matrix) and **mcwv** (connection weight values matrix) are derived in this way (see Table 9.2). The (index) numbers in **mcwa** indicate the state numbers of the reification states where the values can be found (so, the 5 in **mcwa** should be interpreted as state X_5, not as a value 5), and in **mcwv** the numbers indicate the static values directly.

Before copying, the empty cells are filled for Matlab with NaN (Not a Number) indications, to get a neat rectangular matrix structure. After copying to Matlab this results in the matrices in Matlab representation as shown in Table 9.3.

Note that for a given network model the values in the value matrices can differ per scenario addressed, they represent the specific settings for the simulation scenario. As another example, for the 3D role matrices **mcfpa** and **mcfpv** their representations in Matlab are as depicted in Table 9.4.

9.4.2 Retrieving Information from the Role Matrices

During execution of a simulation, for each step from k to $k + 1$ (with step size Δt, in Matlab denoted by `dt`), above role matrices are used. As a first step, for each state X_j the right values (either the fixed value, or the adaptive value found in the indicated reification state) are assigned to:

Table 9.3 Role matrices **mcwa** and **mcwv** for connection weights as represented within Matlab

mcwa = [NaN NaN NaN NaN	**mcwv** = [1 NaN NaN NaN
NaN NaN NaN NaN	1 NaN NaN NaN
NaN NaN NaN NaN	1 NaN NaN NaN
5 NaN NaN NaN	NaN 1 NaN NaN
NaN NaN NaN NaN	1 1 1 NaN
NaN NaN NaN NaN	0 -0.4 1 1
NaN NaN NaN NaN	0 -0.4 1 1
NaN NaN NaN NaN	0 1 -0.4 1
NaN NaN NaN NaN	0 1 1 1
]	]

Table 9.4 Role matrices **mcfpa** and **mcfpv** for combination function parameters as represented in Matlab

mcfpa = cat(3,[NaN NaN	**mcfpv** = cat(3,[1 1
NaN NaN	NaN NaN
NaN NaN	NaN NaN
NaN NaN	NaN NaN
NaN NaN	NaN NaN
NaN NaN	NaN NaN
NaN NaN	NaN NaN
NaN NaN	NaN NaN
NaN NaN	NaN NaN
],[NaN NaN	],[NaN NaN
NaN 6	5 NaN
NaN NaN	5 0.2
NaN 7	5 NaN
NaN NaN	NaN NaN
NaN NaN	5 0.7
NaN NaN	5 0.7
NaN NaN	5 1
NaN NaN	5 1
],[NaN NaN	],[NaN NaN
NaN NaN	NaN NaN
NaN NaN	NaN NaN
NaN NaN	NaN NaN
9 NaN	NaN NaN
NaN NaN	NaN NaN
NaN NaN	NaN NaN
NaN NaN	NaN NaN
NaN NaN	NaN NaN
])	])

These are basically three 2D matrices each with a first (vertical) dimension for the states and a second (horizontal) dimension for the parameters (2 in this case), indicated in Box 9.1 in **mcfpv** as three subboxes for eucl, alogistic, and hebb, respectively. These 2D matrices are grouped in the vertical direction according to the third dimension for the 3 combination functions selected for this specific model (**eucl(..)**, **alogistic(..)**, and **hebb(..)**, respectively), which results in the following 3D matrix representation in Matlab

`s(j, k)`	speed factor for state X_j
`b(j, p, k)`	value for the *p*th state with outgoing base connection to state X_j
`cw(j, p, k)`	connection weight for the *p*th state with outgoing base connection to state X_j
`cfw(j, m, k)`	weight for the *m*th combination function for X_j
`cfp(j, p, m, k)`	the *p*th parameter value of the *m*th combination function for X_j.

This is done by the code shown in Box 9.3.

Box 9.3 First part of the Reified Temporal-Causal Network Engine developed in Matlab; see also Treur (2019b)

```
if not(isnan(msa(j, 1)))
      s(j, k)  = X(msa(j, 1), k);
    elseif not(isnan(msv(j, 1)))
      s(j, k)  = msv(j, 1);
    elseif isnan(msv(j, 1))
         s(j, k)  = 0;
end
    %  This extracts the speed factor value from the
appropriate role matrix msa or msv;
    %  if none of them gives a value, then default value 0 is as-
signed
for p = 1:1:size(mb,2)
    if not(isnan(mb(j, p)))
         b(j, p, k)  = X(mb(j,p), k);
    elseif isnan(mb(j, p))
         b(j, p, k)  = 0;
    end
end
    %  This extracts the relevant base state values
from the appropriate role matrix mb;
    %  if none of them gives a value, then default value 0
is assigned
for p = 1:1:size(mcwa,2)
    if not(isnan(mcwa(j, p)))
         cw(j, p, k)  = X(mcwa(j,p), k);
    elseif not(isnan(mcwv(j, p)))
         cw(j, p, k)  = mcwv(j, p);
    elseif isnan(mcwv(j, p))
         cw(j, p, k)  = 0;
    end
end
    %  This extracts the connection weights from the
appropriate role matrix mcwa or mcwv;
    %  if none of them gives a value, then default value 0
is assigned
for m = 1:1:size(mcfwa,2)
    if not(isnan(mcfwa(j, m)))
         cfw(j, m, k)  = X(mcfwa(j, m), k);
```

```
      elseif not(isnan(mcfwv(j, m)))
              cfw(j, m, k)  = mcfwv(j, m);
      elseif isnan(mcfwv(j, m))
              cfw(j, m, k)  = 0;
      end
end
      %  This extracts the combination function weights from the
appropriate role
      %  matrix mcfwa or mcfwv;
      %  if none of them gives a value, then default value 0
is assigned
for p = 1:1:nocfp
      for m = 1:1:nocf
                if not(isnan(mcfpa(j, p, m)))
                        cfp(j, p, m, k)  = X(mcfpa(j, p, m), k);
                elseif not(isnan(mcfpv(j, p, m)))
                        cfp(j, p, m, k)  = mcfpv(j, p, m);
      elseif isnan(mcfpv(j, p, m))
                        cfp(j, p, m, k)  = 1;
                end
      end
end
      %  This extracts the combination function parameter values
from the appropriate
      %  role matrix mcfpa or mcfpv;
      %  if none of them gives a value, then default value 1
is assigned
```

9.4.3 The Iteration Step from t *to* t + Δt

Then, as a second part of the computational reified network engine, for the step from k to $k + 1$ the following is applied for each j; here `X(j,k)` denotes $X_j(t)$ for t = `t(k) = k*dt`; see Box 9.4.

Box 9.4 Second part of the Reified Temporal-Causal Network Engine developed in Matlab; see also Treur (2019b)

```
for m = 1:1:nocf
cfv(j,m,k) =
   bcf(mcf(m), squeeze(cfp(j, :, m, k)), squeeze(cw(j, :, k)).
*squeeze(b(j, :, k)));
end
       %  This calculates the combination function values cfv(j,
m,k) for each
       %  combination function mcf(m) for state j at k
aggimpact(j, k) = dot(cfw(j, :, k), cfv(j, :, k))/sum(cfw
(j, :, k));
       %  The aggregated impact for state j at k as the inproduct
of combination
       %  function weights and combination function values,
scaled by the sum of these
       %  weights
X(j,k + 1) = X(j,k) + s(j,k)*(aggimpact(j,k) - X(j,k))*dt;
       %  The iteration step from k to k + 1 for state j
t(k + 1) = t(k) + dt;
       %  Keeping track of time
```

As can be seen in Box 9.4, the structure of the code of this computational reified network engine is quite compact; the essential computational core is only 5 lines of code! This is possible because (in contrast to the hybrid approach discussed in Chap. 1, Sect. 1.4.1) a unified approach is applied to all levels of reification. This is enabled by the universal difference equation discussed in Chap. 10, and resembles the formulae in Chap. 4, Table 4.1: structure-preserving implementation.

Note that functions with multiple groups of arguments in Matlab get vector arguments where groups of arguments become vectors of variable length. For example, the basic combination function expression $\text{bcf}_i(P_{1,i}, P_{2,i}, W_1V_1, \ldots, W_kV_k)$ as part of the universal difference equation defined in Chap. 4, Sect. 4.3.2 becomes `bcf(i, p, v)` in Matlab with vectors p = $[P_{1,i}, P_{2,i}]$ for function parameters and v = $[W_1V_1, \ldots, W_kV_k]$ for the values of the function arguments. This format `bcf(i, p, v)` is also used as the basis of the combination function library developed (currently numbered by i = 1 to 35); for an overview of this basic combination function library, see Treur (2019b).

9.5 Discussion

The dedicated modeling environment as described in this chapter includes a dedicated specification format for reified networks and comes with an implemented dedicated computational reified network engine, which can simply run such specifications. Moreover, a library consisting currently of 35 combination functions is offered, which can be extended easily. Using this software environment, the design process of a network model can be focused in a declarative manner on the reified network specification that serves as a design description. All kinds of complex (higher order) adaptive dynamics are covered without being bothered by implementation details. This provides a very powerful modeling environment, which is easy to use and really compact. It uses the reified network architecture, and handles all orders of adaptation in a unified manner, using the universal difference equation.

Note that in the software as described at every iteration step from t to $t + \Delta t$ retrieval of information from the role matrices takes place. An alternative option is to apply a form of compilation before running a simulation so that retrieval of information from the role matrices takes place once and not all the time. Such an alternative implementation will be described in Chap. 10, Sect. 10.7. In that case, during the compilation step, the universal difference equation is instantiated for each state by the information retrieved from the role matrices. This results in a collection of different instantiated difference equations for all states that can be run by any general purpose difference equation simulator. It has not been tested yet, but this perhaps could provide higher efficiency at simulation time, although also compilation time has to be added in that case; this is beyond the scope of the current book. More details can be found in Chap. 10, Sect. 10.

References

Treur, J.: Network-Oriented Modeling: Addressing Complexity of Cognitive, Affective and Social Interactions. Springer Publishers (2016)

Treur, J.: Network reification as a unified approach to represent network adaptation principles within a network. In: Proceedings of the 7th International Conference on Natural Computing. Lecture Notes in Computer Science, vol 11324, pp. 344–358. Springer Publishers (2018a)

Treur, J.: Multilevel network reification: representing higher-order adaptivity in a network. In: Proceedings of the 7th International Conference on Complex Networks and their Applications, Complex Networks' 18, vol. 1. Studies in Computational Intelligence, vol. 812, pp. 635–651. Springer (2018b)

Treur, J.: The ins and outs of network-oriented modeling: from biological networks and mental networks to social networks and beyond. In: Transactions on Computational Collective Intelligence, vol. 32, pp. 120–139. Springer Publishers, Contents of Keynote Lecture at ICCCI'18 (2019a)

Treur, J.: Design of a Software Architecture for Multilevel Reified Temporal-Causal Networks (2019b). Doi: https://doi.org/10.13140/rg.2.2.23492.07045. Url: https://www.researchgate.net/publication/333662169

Chapter 10
On the Universal Combination Function and the Universal Difference Equation for Reified Temporal-Causal Network Models

Abstract The universal differential and difference equation form an important basis for reified temporal-causal networks and their implementation. In this chapter, a more in depth analysis is presented of the universal differential and difference equation. It is shown how these equations can be derived in a direct manner and they are illustrated by some examples. Due to the existence of these universal difference and differential equation, the class of temporal-causal networks is closed under reification: by them it can be guaranteed that any reification of a temporal-causal network is itself also a temporal-causal network. That means that dedicated modeling and analysis methods for temporal-causal networks can also be applied to reified temporal-causal networks. In particular, it guarantees that reification can be done iteratively in order to obtain multilevel reified network models that are very useful to model multiple orders of adaptation. Moreover, as shown in Chap. 9, the universal difference equation enables that software of a very compact form can be developed, as all reification levels are handled by one computational reified network engine in the same manner. Alternatively, it is shown how the universal difference or differential equation can be used for compilation by multiple substitution for all states, which leads to another form of implementation. The background of these issues is discussed in the current chapter.

10.1 Introduction

Modeling dynamic processes by simulating and analysing differential or difference equations has a long tradition in almost all scientific disciplines; e.g., Ashby (1960), Brauer and Nohel (1969), Lotka (1956), Port and van Gelder (1995). The Network-Oriented Modeling approach (Treur 2016, 2019a, b) based on temporal-causal networks also has an underlying differential equation format to model the dynamics. The recent extension of this approach to reified temporal-causal networks as addressed in the current book, has extended and generalised the format of the underlying difference and differential equations in order to enable modeling of networks for processes that are adaptive of any order; see Chaps. 3–8 or (Treur 2018a, b).

J. Treur, *Network-Oriented Modeling for Adaptive Networks: Designing Higher-Order Adaptive Biological, Mental and Social Network Models*, Studies in Systems, Decision and Control 251, https://doi.org/10.1007/978-3-030-31445-3_10

In Chap. 3 it was shown that when network reification is applied to a temporal-causal network, it results in a reified network that itself is again a temporal-causal network. In more abstract mathematical terms this can be formulated by the class of temporal-causal networks being closed under the reification operator. This is a very convenient closure property for this class of networks, as because of that all methods developed for temporal-causal networks can also be applied to reified temporal-causal networks. One important example of this is that it enables that reification can also be applied to reified networks, so that the reification can be iterated easily. Thus multilevel reified networks are obtained that in Chaps. 4–8 turned out very useful to model higher order adaptive networks, for example to model plasticity and metaplasticity. Another important example is that mathematical analysis of emerging behaviour as known for temporal-causal networks can also be applied to reified temporal-causal networks, in particular both to the base states and the reification states in a reified network. Chapters 13 and 14 are examples of such mathematical analyses, where mathematical properties of the combination function of a reification state are related to emerging behaviour for the adaptation process.

To prove that the class of temporal-causal networks is closed under reification there is a central role for the universal combination function and the universal difference equation that can be used to describe the base states in the reified network. In Chap. 3, Sect. 3.5 this universal combination function was introduced more or less out of the blue, and the universal difference equation is just derived from this combination function; that function turned out correct, which provides a form of verification afterwards. However, this can be done in a better way. Strictly spoken, it would be possible to derive valid statements from invalid statements, so confirmative verification of a derived statement is not a strict proof in a logical sense. Therefore in this chapter the derivation of the universal combination function is addressed in some more depth and it is also illustrated for specific cases. However, first a short route is described in Sect. 10.2. After this, the longer route is described. In Sect. 10.3 each of the different roles is analysed separately and for each a combination function and difference equation are derived for the base states. Then, in Sect. 10.4 the universal combination function and universal difference equation are derived for all roles at the same time. Section 10.5 shows that the criterion for equilibria for temporal-causal networks also applies to the universal differential equation in reified temporal-causal networks. In Sect. 10.6 it is shown how it can be derived from the role matrices. In Sect. 10.7 it is shown how this universal difference equation can be used for a compilation process by for each state Y substituting the data from the role matrices in them before simulation time and not during simulation time. This may be a useful method to simulate very large reified networks in an efficient manner.

10.2 A Short Route to the Universal Difference and Differential Equation

Recall this expression (7) from Chap. 3, Sect. 3.5 for the combination function and (8) for the difference equation:

$$\begin{aligned}&\mathbf{c}_Y(t, \boldsymbol{\pi}_{1,1}(t), \boldsymbol{\pi}_{1,2}(t), \ldots, \boldsymbol{\pi}_{1,m}(t), \boldsymbol{\pi}_{1,m}(t), V_1, \ldots, V_k)\\ &= \frac{\boldsymbol{\gamma}_{1,Y}(t)\,\mathrm{bcf}_1(\boldsymbol{\pi}_{1,1,Y}(t), \boldsymbol{\pi}_{2,1,Y}(t), V_1, \ldots, V_k) + \cdots + \boldsymbol{\gamma}_{m,Y}(t)\mathrm{bcf}_m(\boldsymbol{\pi}_{1,m,Y}(t)\boldsymbol{\pi}_{2,m,Y}(t), V_1, \ldots, V_k)}{\boldsymbol{\gamma}_{1,Y}(t) + \cdots + \boldsymbol{\gamma}_{m,Y}(t)}\end{aligned} \tag{1}$$

$$Y(t+\Delta t) = Y(t) + \boldsymbol{\eta}_Y(t)[\mathbf{c}_Y(t, \boldsymbol{\omega}_{X_1,Y}(t)X_1(t), \ldots, \boldsymbol{\omega}_{X_k,Y}(t)X_k(t)) - Y(t)]\Delta t \tag{2}$$

Substituting the former expression (1) in the latter (2), the difference equation becomes

$$\begin{aligned}&Y(t+\Delta t) = Y(t)\\ &+ \boldsymbol{\eta}_Y(t)[\frac{\boldsymbol{\gamma}_{1,Y}(t)\,\mathrm{bcf}_1(\boldsymbol{\pi}_{1,1,Y}(t), \boldsymbol{\pi}_{2,1,Y}(t), \boldsymbol{\omega}_{X_1,Y}(t)X_1(t), \ldots, \boldsymbol{\omega}_{X_k,Y}(t)X_k(t)) + \cdots + \boldsymbol{\gamma}_{m,Y}(t)\,\mathrm{bcf}_m(\boldsymbol{\pi}_{1,m,Y}(t), \boldsymbol{\pi}_{2,m,Y}(t), \boldsymbol{\gamma}_{X_1,Y}(t)X_1(t), \ldots, \boldsymbol{\omega}_{X_k,Y}(t)X_k(t))}{\boldsymbol{\gamma}_{1,Y}(t) + \cdots + \boldsymbol{\gamma}_{m,Y}(t)} - Y(t)]\Delta t\end{aligned} \tag{3}$$

Within the reified network the adaptive values of **η**, **ω**, γ and π are represented by their reification states **H**, **W**, **C** and **P**. By substituting these in (3), a difference equation for the reified network is obtained:

$$\begin{aligned}&Y(t+\Delta t) = Y(t)\\ &+ \mathbf{H}_Y(t)[\frac{\mathbf{C}_{1,Y}(t)\,\mathrm{bcf}_1(\mathbf{P}_{1,1,Y}(t), \mathbf{P}_{2,1,Y}(t), \mathbf{W}_{X_1,Y}(t)X_1(t), \ldots, \mathbf{W}_{X_k,Y}(t)X_k(t)) + \cdots + \mathbf{C}_{m,Y}(t)\,\mathrm{bcf}_m(\mathbf{P}_{1,m,Y}(t), \mathbf{P}_{2,m,Y}(t), \mathbf{W}_{X_1,Y}(t)X_1(t), \ldots, \mathbf{W}_{X_k,Y}(t)X_k(t))}{\mathbf{C}_{1,Y}(t) + \cdots + \mathbf{C}_{m,Y}(t)} - Y(t)]\Delta t\end{aligned} \tag{4}$$

In differential equation format (leaving out references to t), this is

$$\mathbf{d}Y/\mathbf{d}t = \mathbf{H}_Y[\frac{\mathbf{C}_{1,Y}\,\mathrm{bcf}_1(\mathbf{P}_{1,1,Y}, \mathbf{P}_{2,1,Y}, \mathbf{W}_{X_1,Y}X_1, \ldots, \mathbf{W}_{X_k,Y}X_k) + \cdots + \mathbf{C}_{m,Y}\mathrm{bcf}_m(\mathbf{P}_{1,m,Y}, \mathbf{P}_{2,m,Y}, \mathbf{W}_{X_1,Y}X_1, \ldots, \mathbf{W}_{X_k,Y}X_k)}{\mathbf{C}_{1,Y} + \cdots + \mathbf{C}_{m,Y}} - Y] \tag{5}$$

Note that this difference and differential equation is not yet in the standard format of a temporal-causal network, as $\mathbf{H}_Y$ is not a constant speed factor. However, it can be rewritten into the temporal-causal network format when a suitable *universal combination function* $\mathbf{c}^*_Y(..)$ is defined. It can be verified by rewriting that when using the combination function defined by (6) below, this universal difference equation (4) (a) becomes in the standard temporal-causal format, and (b) indeed is equivalent to the above difference equation in (3). So, define

$$\mathbf{c}^*_Y(H, C_1, \ldots, C_m, P_{1,1}, P_{2,1}, \ldots, P_{1,m}, P_{2,m}, W_1, \ldots, W_k, V_1, \ldots, V_k, V)$$
$$= H\frac{C_1\text{bcf}_1(P_{1,1}, P_{2,1}, W_1V_1, \ldots, W_kV_k) + \cdots + C_m\text{bcf}_m(P_{1,m}, P_{2,m}, W_1V_1, \ldots, W_kV_k)}{C_1 + \cdots + C_m} + (1-H)V \tag{6}$$

Then this goes as follows. For more explanation and background on this, see Sect. 10.3 and further. Consider the following *universal differential equation* variant, which is (leaving out the reference to *t*), and assuming speed factor 1:

$$\mathbf{d}Y/\mathbf{d}t = \mathbf{c}^*_Y(\mathbf{H}_Y, \mathbf{C}_{1,Y}, \ldots, \mathbf{C}_{m,Y}, \mathbf{P}_{1,1,Y}, \mathbf{P}_{2,1,Y}, \ldots, \mathbf{P}_{1,m,Y}, \mathbf{P}_{2,m,Y}, \mathbf{W}_{X_1,Y}, \ldots, \mathbf{W}_{X_k,Y}, X_1, \ldots, X_k, Y) - Y \tag{7}$$

This is indeed in temporal-causal format. Using (6) it can be rewritten as

$$\mathbf{d}Y/\mathbf{d}t = \mathbf{H}_Y[\frac{\mathbf{C}_{1,Y}\text{bcf}_1(\mathbf{P}_{1,1,Y}, \mathbf{P}_{2,1,Y}, \mathbf{W}_{X_1,Y}X_1, \ldots, \mathbf{W}_{X_k,Y}X_k) + \cdots + \mathbf{C}_{m,Y}\text{bcf}_m(\mathbf{P}_{1,m,Y}, \mathbf{P}_{2,m,Y}, , \mathbf{W}_{X_1,Y}X_1, \ldots, \mathbf{W}_{X_k,Y}X_k)}{\mathbf{C}_{1,Y} + \cdots + \mathbf{C}_{m,Y}}]$$
$$+ (1-\mathbf{H}_Y)Y - Y$$

Now note that this last part $(1 - \mathbf{H}_Y)\, Y - Y$ is just $-\mathbf{H}_Y\, Y$. Then this easily can be rewritten into:

$$\mathbf{d}Y/\mathbf{d}t = \mathbf{H}_Y[\frac{\mathbf{C}_{1,Y}\text{bcf}_1(\mathbf{P}_{1,1,Y}, \mathbf{P}_{2,1,Y}, \mathbf{W}_{X_1,Y}X_1, \ldots, \mathbf{W}_{X_k,Y}X_k) + \cdots + \mathbf{C}_{m,Y}\text{bcf}_m(\mathbf{P}_{1,m,Y}, \mathbf{P}_{2,m,Y}, \mathbf{W}_{X_1,Y}X_1, \ldots, \mathbf{W}_{X_k,Y}X_k)}{\mathbf{C}_{1,Y} + \cdots + \mathbf{C}_{m,Y}} - Y] \tag{8}$$

This (8) is exactly differential equation (5) earlier above; this confirms that the chosen universal combination function $\mathbf{c}^*_Y(..)$ in (6) to get the reified network in temporal-causal network format is right. So far this short route. Here the definition of the combination function (6) may seem to come out of the blue. In the next sections it is shown (via the longer route) how that can be motivated and derived.

10.3 Downward Causal Connections Defining the Special Effect of Reification States

The added reification states have to be integrated to obtain a well-connected overall network. In the first place outward connections from the reification states to the states in the base network are needed, in order to model how they have their special effect on the dynamics in the network. More specifically, it has to be defined how the reification states contribute causally to an aggregated impact on the base network state. In addition to a downward connection, also the combination function for the base state has to be defined for the aggregated impact. Both these downward causal relations and the combination functions will be defined in a generic manner, related to how a specific network characteristic functions in the overall dynamics as part of the intended semantics of a temporal-causal network. That will be discussed in the current section.

In addition, other connections of the reification states are added in order to model specific network adaptation principles. These may concern upward connections from the states of the base network to the reification states, or horizontal mutual connections between reification states within the upper plain, or both, depending on the specific network adaptation principles addressed. These connections are not generic as they are an essential part of the specification of a particular adaptation principle; they have been illustrated for a number of well-known adaptation principles in Chap. 3; see Figs. 3.4–3.10.

10.3.1 The Overall Picture

For the downward connections the general pattern is that each of the reification states $\mathbf{W}_{X_i,Y}$, $\mathbf{H}_Y$ and $\mathbf{C}_Y$ for the reified network characteristics, connection weights, speed factors and combination functions, has a specific causal connection to state Y in the base network, as they all affect Y. These are the (pink) downward arrows from the reification plane to the base plane in Chap. 3, Fig. 3.3; see also Fig. 10.1. Actually $\mathbf{C}_Y$ is a vector of states ($\mathbf{C}_{1,Y}$, $\mathbf{C}_{2,Y}$, ...) with a (small) number of different components $\mathbf{C}_{1,Y}$, $\mathbf{C}_{2,Y}$, ... for different basic combination functions that will be explained below. Note that combination functions may contain some parameters, for example, for the scaled sum combination function the scaling factor $\boldsymbol{\lambda}$, and for the advanced logistic sum combination function the steepness $\boldsymbol{\sigma}$ and the threshold $\boldsymbol{\tau}$. For these parameters also reification states $\mathbf{P}_{i,j,Y}$ can be added, with the possibility to make them adaptive as well. More specifically, for each basic combination function represented by $\mathbf{C}_{j,Y}$ there are two parameters $\pi_{1,i}$ and $\pi_{2,i}$ that are reified by parameter reification states $\mathbf{P}_{1,j,Y}$ and $\mathbf{P}_{2,j,Y}$. All depicted (downward and horizontal)

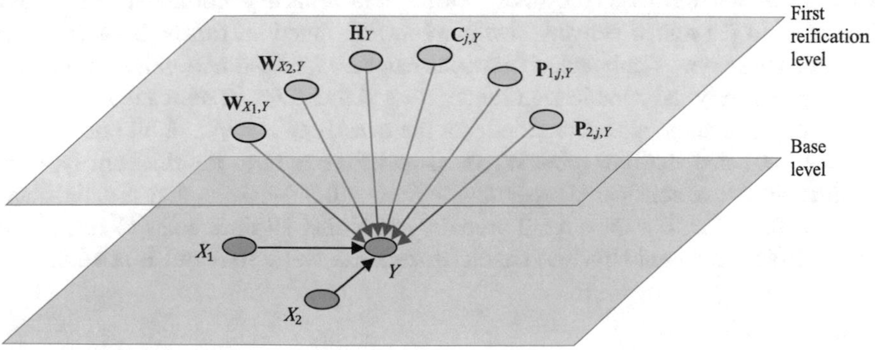

Fig. 10.1 Network reification for temporal-causal networks: downward causal connections from reification states to base network states

connections in Fig. 10.1 get weight 1. Note that this is also a way in which a weighted network can be transformed into an equivalent non-weighted network. In the extended network the speed factors of the base states are set at 1 too.

Note that the 3D layout of these figures and the depicted planes are just for understanding; in a mathematical or computational sense they are not part of the network specification. However, for each of the reification states it is crucial to know what it is that they are reifying and for what base state. Therefore the names of the reification states are chosen in such a way that this information is visible. For example, in the name $\mathbf{H}_Y$ the **H** indicates that it concerns speed factor (indicated by $\boldsymbol{\eta}$) reification and the subscript Y that it is for base state Y. So, in general the bold capital letter **R** in $\mathbf{R}_{subscript}$ indicates the type of reification and the subscript the concerning base state Y, or (for **W**) the pair of states X, Y. This **R** defines the *role* that is played by this reification state. This role corresponds one to one to the characteristics of the base network structure that is reified: connection weight $\boldsymbol{\omega}$, speed factor $\boldsymbol{\eta}$, basic combination function **c(..)**, parameter π. In other words, there are four roles for reification states:

- the role of connection weight reification states $\mathbf{W}_{X_i,Y}$ reifying connection weights $\boldsymbol{\omega}_{X_i,Y}$
- the role of speed factor reification state $\mathbf{H}_Y$ reifying speed factor $\boldsymbol{\eta}_Y$
- the role of combination function reification states $\mathbf{C}_{j,Y}$ reifying combination function $\mathbf{c}_Y(..)$
- the role of parameter reification state $\mathbf{P}_{i,j,Y}$ reifying combination function parameter $\boldsymbol{\pi}_{i,j,Y}$

In accordance with this encoded role information, in principle each reification state has exactly one downward causal connection, which goes to the specified base state Y. In the reified network this downward connection is incorporated according to its role **R** in the aggregation of the causal impacts on Y by a new, dedicated universal combination function for that role. How this is done is explained in more detail in this section.

The general picture is that the base states have more incoming connections now, some of which have specific roles, with special effects according to their role. Therefore in the reified network new combination functions for the base states are needed. These new combination functions can be expressed in a universal manner based on the original combination functions, and the different reification states, but to define them some work is needed. As the overall approach is a bit complex, to get the idea, first the four roles **W**, **H**, **C** and **P** relating to the different types of characteristics are considered separately in Sects. 10.3.2–10.3.4; they are illustrated in Box 10.1–10.3. For the overall process, combining all three roles **W**, **H**, **C** and **P** for all base network structure characteristics, see Sect. 10.3 and Box 10.4.

10.3.2 Downward Causal Connections for Role W for Connection Weight Reification

First, consider only connection weight reification indicated by role **W**. The original difference equation for base state *Y* based on the original combination function $\mathbf{c}_Y(..)$ is

$$Y(t+\Delta t) = Y(t) + \boldsymbol{\eta}_Y[\mathbf{c}_Y(\boldsymbol{\omega}_{X_1,Y}(t)X_1(t), \ldots, \boldsymbol{\omega}_{X_k,Y}(t)X_k(t)) - Y(t)]\Delta t \quad (9)$$

The new combination function $\mathbf{c}^*_Y(..)$ has to aggregate two types of values:

- the base state values $X_1(t), \ldots, X_k(t)$ for the base states from which state *Y* gets its incoming connections
- the reification state values $\mathbf{W}_{X_1,Y}(t) \ldots, \mathbf{W}_{X_k,Y}(t)$ for connection weights

Therefore it has to have arguments for all of these values:

$$\mathbf{c}^*_Y(\mathbf{W}_{X_1,Y}(t), \ldots, \mathbf{W}_{X_k,Y}(t), X_1(t), \ldots, X_k(t))$$

so $\mathbf{c}^*_Y(..)$ has to have this format:

$$\mathbf{c}^*_Y(W_1, \ldots, W_k, V_1, \ldots, V_k)$$

A requirement for this new combination function $\mathbf{c}^*_Y(..)$ in the reified network is

$$Y(t+\Delta t) = Y(t) + \boldsymbol{\eta}_Y\left[\mathbf{c}^*_Y(\mathbf{W}_{X_1,Y}(t), \ldots, \mathbf{W}_{X_k,Y}(t), X_1(t), \ldots, X_k(t)) - Y(t)\right]\Delta t \quad (10)$$

As these two difference Eqs. (9) and (10) must have the same result for $Y(t)$ and $Y(t + \Delta t)$, the requirement for $\mathbf{c}^*_Y(..)$ is that (when $\mathbf{W}_{X_i,Y}(t) = \boldsymbol{\omega}_{X_i,Y}(t)$) it holds

$$\mathbf{c}^*_Y(\mathbf{W}_{X_1,Y}(t), \ldots, \mathbf{W}_{X_k,Y}(t), X_1(t), \ldots, X_k(t)) = \mathbf{c}_Y(\boldsymbol{\omega}_{X_1,Y}(t)X_1(t), \ldots, \boldsymbol{\omega}_{X_k,Y}(t)X_k(t))$$

So the new combination function $\mathbf{c}^*_Y(..)$ for this role has to be defined by

$$\mathbf{c}^*_Y(W_1, \ldots, W_k, V_1, \ldots, V_k) = \mathbf{c}_Y(W_1V_1, \ldots, W_kV_k) \quad (11)$$

where

- W_i stands for $\mathbf{W}_{X_i,Y}(t)$
- V_i stands for $X_i(t)$

In Box 10.1 an example of this combination function relating to Fig. 10.1 is shown. Indeed the requirement is fulfilled when $\mathbf{W}_{X_i,Y}(t) = \boldsymbol{\omega}_{X_i,Y}(t)$:

$$\mathbf{c}^*_Y(\mathbf{W}_{X_1,Y}(t), \ldots, \mathbf{W}_{X_k,Y}(t), X_1(t), \ldots, X_k(t)) = \mathbf{c}_Y(\boldsymbol{\omega}_{X_1,Y}(t)X_1(t), \ldots, \boldsymbol{\omega}_{X_k,Y}(t)X_k(t))$$

Box 10.1 Example of the derived combination function for connection weight reification role $\mathbf{W}_{X_1,Y}$ and $\mathbf{W}_{X_2,Y}$ in the reified network for base state Y from Fig. 10.1.

In this box an example relating to Fig. 10.1 where $m = 2$, $\text{bcf}_1(..) = \mathbf{eucl}_{n,\lambda}(..)$ for $n = 1$, $\text{bcf}_2(..) = \mathbf{alogistic}_{\sigma,\tau}(..)$, where $\mathbf{eucl}_{1,\lambda}(..)$ is assumed for Y.

For connection weight reification the new combination function $\mathbf{c}^*_Y(..)$ for Y is

$$\begin{aligned}\mathbf{c}^*_Y(W_1, W_2, V_1, V_2) &= \mathbf{c}_Y(W_1V_1, W_2V_2) = \mathbf{eucl}(1, \lambda, W_1V_1, W_2V_2)\\ &= (W_1V_1 + W_2V_2)/\lambda\end{aligned}$$

where

$$\begin{aligned}W_1 &= \mathbf{W}_{X_1,Y}(t)\\ W_2 &= \mathbf{W}_{X_2,Y}(t)\\ V_1 &= X_1(t)\\ V_2 &= X_2(t)\end{aligned}$$

10.3.3 Downward Causal Connections for Role H for Speed Factor Reification

Second, reification of speed factors in terms of role **H** is addressed separately; in the new situation in the reified network the combination function needs an extra argument for $\mathbf{H}_Y(t)$. It turns out that to make it work also an extra argument for the current value $Y(t)$ is needed, for the timing modeled by the speed factor:

$$\mathbf{c}^*_Y(\mathbf{H}_Y(t), \boldsymbol{\omega}_{X_1,Y}X_1(t), \ldots, \boldsymbol{\omega}_{X_k,Y}X_k(t), Y(t))$$

So, the format for $\mathbf{c}^*_Y(.)$ becomes:

$$\mathbf{c}^*_Y(H, V_1, \ldots, V_k, V)$$

The requirement for this new function is that (when $\mathbf{H}_Y(t) = \boldsymbol{\eta}_Y(t)$) it holds

$$Y(t+\Delta t) = Y(t) + \boldsymbol{\eta}^*_Y[\mathbf{c}^*_Y(\mathbf{H}_Y(t), \boldsymbol{\omega}_{X_1,Y}X_1(t), \ldots, \boldsymbol{\omega}_{X_k,Y}X_k(t),\ Y(t)) - Y(t)]\Delta t \tag{12}$$

It is assumed that the new speed factor $\boldsymbol{\eta}^*_Y$ is 1; then since (9) and (12) should describe the same values for Y the requirement becomes:

$$\begin{aligned}&\mathbf{c}^*_Y(\mathbf{H}_Y(t), \boldsymbol{\omega}_{X_1,Y}X_1(t), \ldots, \boldsymbol{\omega}_{X_k,Y}X_k(t),\ Y(t)) - Y(t)\\ &\quad = \boldsymbol{\eta}_Y(t)[\mathbf{c}_Y(\boldsymbol{\omega}_{X_1,Y}X_1(t), \ldots, \boldsymbol{\omega}_{X_k,Y}X_k(t)) - Y(t)]\end{aligned}$$

This can be rewritten into

$$\begin{aligned}&\mathbf{c}^*_Y(\mathbf{H}_Y(t), \boldsymbol{\omega}_{X_1,Y}X_1(t), \ldots, \boldsymbol{\omega}_{X_k,Y}X_k(t),\ Y(t))\\ &\quad = \boldsymbol{\eta}_Y(t)\mathbf{c}_Y(\boldsymbol{\omega}_{X_1,Y}X_1(t), \ldots, \boldsymbol{\omega}_{X_k,Y}X_k(t)) + (\mathbf{1} - \boldsymbol{\eta}_Y(t))Y(t)\end{aligned}$$

Now define the combination function $\mathbf{c}^*_Y(..)$ by

$$\mathbf{c}^*_Y(H, V_1, \ldots, V_k,\ V) = H\mathbf{c}_Y(V_1,\ \ldots, V_k) + (1 - H)V \tag{13}$$

where

- H stands for $\mathbf{H}_Y(t)$
- V_i stands for $\boldsymbol{\omega}_{X_i,Y}\ X_i(t)$
- V stands for $Y(t)$

This is a weighted average (with weights speed factor H and $1 - H$) of $\mathbf{c}_Y(V_1, \ldots, V_k)$ and V. Again, in Box 10.2 an example of this combination function relating to Fig. 10.1 is shown. Also here the requirement is fulfilled for $\mathbf{H}_Y(t) = \boldsymbol{\eta}_Y(t)$.

Box 10.2 Example of the derived combination function for speed factor reification role **H** in the reified network for base state Y from Fig. 10.1.
In this box an example relating to Fig. 10.1 where $m = 2$, $\text{bcf}_1(..) = \mathbf{eucl}_{n,\lambda}(..)$ for $n = 1$, $\text{bcf}_2(..) = \mathbf{alogistic}_{\sigma,\tau}(..)$, where $\mathbf{eucl}_{1,\lambda}(..)$ is assumed for Y.
For speed factor reification the new combination function $\mathbf{c}^*_Y(..)$ for Y is

$$\begin{aligned}&\mathbf{c}^*_Y(H, V_1, V_2,\ V) = H\mathbf{c}_Y(V_1, V_2) + (1 - H)V\\ &= H\,\mathbf{eucl}(1, \lambda, V_1, V_2) + (1 - H)V\\ &= H(V_1 + V_2)/\lambda + (1 - H)V\end{aligned}$$

where

- H stands for $\mathbf{H}_Y(t)$
- V_i stands for $\boldsymbol{\omega}_{X_i,Y}X_i(t)$
- V stands for $Y(t)$

10.3.4 Downward Causal Connections for Roles C and P for Combination Function Weight and Parameter Reification

To make reification of combination functions more practical, for the base network a countable number of basic combination functions bcf(..) is assumed. From this sequence of basic combination functions for any arbitrary m a finite subsequence $\text{bcf}_1(..)$,.., $\text{bcf}_m(..)$ of m basic combination functions can be chosen to be used in a specific application. For example with $m = 3$:

$$\text{bcf}_1(..) = \mathbf{id}(..),\quad \text{bcf}_2(..) = \mathbf{ssum}_\lambda(..),\quad \text{bcf}_3(..) = \mathbf{alogistic}_{\sigma,\tau}(..)$$

Note that when more than one argument is used in **id(..)**, the outcome is the sum of these arguments (only one of them will be nonzero when Y has only one incoming connection). For each state Y in the base network *combination function weights* $\gamma_{j,Y}$ are assumed: numbers $\gamma_{1,Y}$, $\gamma_{2,Y}$,… ≥ 0 that change over time. Moreover, combination function parameters $\pi_{1,i,Y}$, $\pi_{2,i,Y}$ are assumed for each basic combination function $\text{bcf}_i(..)$ for Y. The actual combination function $\mathbf{c}_Y(.)$ at time t is expressed as a weighted average by:

$$\mathbf{c}_Y(t, \pi_{1,1,Y}, \pi_{2,1,Y,\ldots,}\pi_{1,m,Y}, \pi_{2,m,Y,,}V_1, \ldots, V_k) = \frac{\gamma_{1,Y}(t)\,\text{bcf}_1\left(\pi_{1,1,Y}, \pi_{2,1,Y}, V_1, \ldots, V_k\right) + \cdots + \gamma_{m,Y}(t)\text{bcf}_m\left(\pi_{1,m,Y}, \pi_{2,m,Y}, V_1, \ldots, V_k\right)}{\gamma_{1,Y}(t) + \cdots + \gamma_{m,Y}(t)} \tag{14}$$

In this way it can be expressed that for Y at each time point t a weighted average of the indicated basic combination functions is applied. This involves multiple basic combination functions if more than one of $\gamma_{j,Y}(t)$ has a nonzero value; just one basic combination function is selected for $\mathbf{c}_Y(.)$, if exactly one of the $\gamma_{j,Y}(t)$ is nonzero. This approach makes it possible, for example, to smoothly switch to another combination function over time by decreasing the value of $\gamma_{j,Y}(t)$ for the earlier chosen basic combination function and increasing the value of $\gamma_{j,Y}(t)$ for the new choice of combination function; see Chap. 3, Sect. 3.7 for an example.

For each basic combination function weight $\gamma_{j,Y}$ a different reification state $\mathbf{C}_{j,Y}$ is added. The value of that state represents the extent to which that basic combination function $\text{bcf}_j(..)$ is applied for state Y. Moreover, the combination function parameters are reified by $\mathbf{P}_{1,1,Y}(t)$, $\mathbf{P}_{2,1,Y}(t)$,…, $\mathbf{P}_{1,m,Y}(t)$, $\mathbf{P}_{2,m,Y}(t)$. The new combination function $\mathbf{c}^*_Y(..)$ needs additional arguments for them, so it gets this format:

$$\mathbf{c}^*_Y(\mathbf{C}_{1,Y}(t),\ldots\mathbf{C}_{m,Y}(t),\mathbf{P}_{1,1,Y}(t),\mathbf{P}_{2,1,Y}(t),\ldots,\mathbf{P}_{1,m,Y}(t),\mathbf{P}_{2,m,Y}(t),\boldsymbol{\omega}_{X_1,Y}X_1(t),\ldots,\boldsymbol{\omega}_{X_k,Y}X_k(t))$$

By using variables C_j, and $P_{i,j}$ for the reified weights $\mathbf{C}_{j,Y}(t)$ and reified parameter values $\mathbf{P}_{i,j,Y}(t)$, the combination function format for $\mathbf{c}^*_Y(..)$ becomes

$$\mathbf{c}^*_Y(C_1,\ldots,C_m,P_{1,1},P_{2,1,\ldots,}P_{1,m},P_{2,m,}V_1,\ldots,V_k)$$

Now the following two difference equations should make the same values for Y:

$$Y(t+\Delta t)=Y(t)+\boldsymbol{\eta}_Y[\mathbf{c}_Y(t,\boldsymbol{\pi}_{1,1,Y},\boldsymbol{\pi}_{2,1,Y,\ldots,}\boldsymbol{\pi}_{1,m,Y},\boldsymbol{\pi}_{2,m,Y,,}\boldsymbol{\omega}_{X_1,Y}X_1(t),\ \ldots,\boldsymbol{\omega}_{X_k,Y}X_k(t))-Y(t)]\Delta t$$
$$Y(t+\Delta t)=Y(t)+\boldsymbol{\eta}_Y[\mathbf{c}^*_Y(\mathbf{C}_{1,Y}(t),\ldots\mathbf{C}_{m,Y}(t),\mathbf{P}_{1,1,Y},\mathbf{P}_{2,1,Y,\ldots,}\mathbf{P}_{1,m,Y},\mathbf{P}_{2,m,Y,,}\boldsymbol{\omega}_{X_1,Y}X_1(t),\ldots,\boldsymbol{\omega}_{X_k,Y}X_k(t),\ Y(t))-Y(t)]\Delta t$$

Therefore the following requirement for the combination function $\mathbf{c}^*_Y(C_1, \ldots, C_m, P_{1,1}, P_{2,1}, \ldots, P_{1,m}, P_{2,m}, V_1, \ldots, V_k)$ is obtained:

$$\begin{aligned}&\mathbf{c}^*_Y(\mathbf{C}_{1,Y}(t),\ldots\mathbf{C}_{m,Y}(t),\mathbf{P}_{1,1,Y}(t),\mathbf{P}_{2,1,Y}(t),\ldots,\mathbf{P}_{1,m,Y}(t),\mathbf{P}_{2,m,Y}(t),\boldsymbol{\omega}_{X_1,Y}X_1(t),\ldots,\boldsymbol{\omega}_{X_k,Y}X_k(t))\\&=\mathbf{c}_Y(t,\boldsymbol{\pi}_{1,1,Y}(t),\boldsymbol{\pi}_{2,1,Y}(t),\ldots,\boldsymbol{\pi}_{1,m,Y}(t),\boldsymbol{\pi}_{2,m,Y}(t),\boldsymbol{\omega}_{X_1,Y}X_1(t),\ldots,\boldsymbol{\omega}_{X_k,Y}X_k(t))\end{aligned}$$

which is

$$\begin{aligned}&\mathbf{c}^*_Y(\mathbf{C}_{1,Y}(t),\ldots\mathbf{C}_{m,Y}(t),\mathbf{P}_{1,1,Y}(t),\mathbf{P}_{2,1,Y}(t),\ldots,\mathbf{P}_{1,m,Y}(t),\mathbf{P}_{2,m,Y}(t),\boldsymbol{\omega}_{X_1,Y}X_1(t),\ldots,\boldsymbol{\omega}_{X_k,Y}X_k(t))\\&=\frac{\gamma_{1,Y}(t)\text{bcf}_1(\pi_{1,1,Y}(t),\pi_{2,1,Y}(t),\boldsymbol{\omega}_{X_1,Y}X_1(t),\ldots,\boldsymbol{\omega}_{X_k,Y}X_k(t))+\ldots+\gamma_{m,Y}(t)\,\text{bcf}_m(\pi_{1,1,Y(t)},\pi_{2,1,Y}(t),\boldsymbol{\omega}_{X_1,Y}X_1(t),\ldots,\boldsymbol{\omega}_{X_k,Y}X_k(t))}{\gamma_{1,Y}(t)+\ldots+\gamma_{m,Y}(t)}\end{aligned}$$

To fullfill this requirement the combination function $\mathbf{c}^*_Y(C_1, \ldots, C_m, P_{1,1}, P_{2,1}, \ldots, P_{1,m}, P_{2,m}, V_1, \ldots, V_k)$ has to be defined by

$$\begin{aligned}&\mathbf{c}^*_Y(C_1,\ldots,C_m,P_{1,1},P_{2,1},\ldots,P_{1,m},P_{2,m},V_1,\ldots,V_k)\\&=\frac{C_1\text{bcf}_1(P_{1,1,Y},P_{2,1,Y},V_1,\cdots,V_k)+\cdots+C_m\,\text{bcf}_m(P_{1,m,Y},\,P_{2,m,Y},\,V_1,\ldots,V_k)}{C_1+\ldots+C_m}\end{aligned} \quad (15)$$

where

- C_j stands for the combination function weight reification $\mathbf{C}_{j,Y}(t)$
- $P_{i,j}$ for the combination function parameter reification $\mathbf{P}_{i,j,Y}(t)$
- V_i for the value $\boldsymbol{\omega}_{X_i,Y}\ X_i(t)$ for base state X_i.

Box 10.3 Example of a derived combination function in the reified network for base states Y from Fig. 10.1 for combination function reification roles **C** and **P**

In this box an example relating to Fig. 10.1 where $m = 2$, $\text{bcf}_1(..) = \mathbf{eucl}_{n,\lambda}(..)$ for $n = 1$, $\text{bcf}_2(..) = \mathbf{alogistic}_{\sigma,\tau}(..)$, where first $\mathbf{eucl}_{1,\lambda}(..)$ is assumed for Y

For combination function reification, assuming $C_{1,Y}(0) = 1$, $C_{2,Y}(0) = 0$, the new combination function $\mathbf{c}^*_Y(..)$ for Y is

$$\mathbf{c}^*_Y(C_1, C_2, P_{1,1}, P_{2,1}, P_{1,2}, P_{2,2}, V_1, V_2)$$
$$= \frac{C_1 \,\text{bcf}_1(P_{1,1,Y}, P_{2,1,Y}, V_1, V_2) + C_2 \,\text{bcf}_2(P_{1,2,Y}, P_{2,2,Y}, V_1, V_2)}{C_1 + C_2}$$
$$= \frac{C_1 \,\mathbf{eucl}(1, \lambda, V_1, V_2) + C_2 \mathbf{alogistic}(\sigma, \tau, V_1, V_2)}{C_1 + C_2}$$
$$= \frac{C_1 \frac{V_1 + V_2}{\lambda} + C_2 \mathbf{alogistic}(\sigma, \tau, V_1, V_2)}{C_1 + C_2}$$

where

- C_j stands for the combination function weight reification $\mathbf{C}_{j,Y}(t)$
- $P_{i,j}$ for the combination function parameter reification $\mathbf{P}_{i,j,Y}(t)$
- V_i for the state value $X_i(t)$ of base state X_i.

This enables over time change from combination function $\mathbf{eucl}_{n,\lambda}(..)$ to combination function $\mathbf{alogistic}_{\sigma,\tau}(..)$ where first $C_1 = 1$ and $C_2 = 0$, and later C_2 becomes 1 and C_1 becomes 0.

Using this combination function, by substitution for the variables it can easily be verified that the requirement is indeed fulfilled. Note that it has to be guaranteed that the case that all C_j become 0 does not occur. For a given combination function adaptation principle, this easily can be achieved by normalising the C_j for each adaptation step so that their sum always stays 1. In Box 10.3 an example of this combination function relating to Fig. 10.1 is shown.

10.4 Deriving the Universal Combination Function and Difference Equation for Reified Networks

Based on the preparation in the previous section, in the current section the universal combination function and universal difference equation for reified networks which apply to all roles at once are presented.

10.4.1 Deriving the Universal Combination Function for Reified Networks

It has been discussed above how in the reified network the causal relations for the base network states can be defined separately for each of the three types of network characteristics. By combining these three in one it can be found that this *universal combination function* for base states Y does all at once:

$$\begin{aligned}
&\mathbf{c}^*_Y(H, C_1, \ldots, C_m, P_{1,1}, P_{2,1}, \ldots, P_{1,m}, P_{2,m}, W_1, \ldots, W_k, V_1, \ldots, V_k, V) \\
&= H\frac{C_1\mathrm{bcf}_1(P_{1,1,Y}, P_{2,1,Y}, W_1V_1, \ldots, W_kV_k) + \cdots + C_m\mathrm{bcf}_m(P_{1,m,Y}, P_{2,m,Y}, W_1V_1, \ldots, W_kV_k)}{C_1 + \cdots + C_m} + (1-H)V \\
&= H[\frac{C_1\mathrm{bcf}_1(P_{1,1,Y}, P_{2,1,Y}, W_1V_1, \ldots, W_kV_k) + \cdots + C_m\mathrm{bcf}_m(P_{1,m,Y}, P_{2,m,Y}, W_1V_1, \ldots, W_kV_k)}{C_1 + \cdots + C_m} - V] + V
\end{aligned} \tag{16}$$

where

- H stands for the speed factor reification $\mathbf{H}_Y(t)$
- C_j for the combination function weight reification $\mathbf{C}_{j,Y}(t)$
- $P_{i,j}$ for the combination function parameter reification $\mathbf{P}_{i,j,Y}(t)$
- W_i for the connection weight reification $\mathbf{W}_{X_i,Y}(t)$
- V_i for the state value $X_i(t)$ of base state X_i
- V for the state value $Y(t)$ of base state Y

See Box 10.4 for a general derivation of this universal combination function and Box 10.5 for an example of its use.

Box 10.4 Deriving the universal combination function and universal difference equation in the reified network for base states.
Here the overall situation is addressed in which all base network structure characteristics $\boldsymbol{\omega}$, $\boldsymbol{\eta}$, $\boldsymbol{\gamma}$, $\boldsymbol{\pi}$ are reified together by reification states **W**, **H**, **C**, and **P**, respectively. The format for the new combination function $\mathbf{c}^*_Y(..)$ needs arguments for all states in the following manner:

$$\mathbf{c}^*_Y(\mathbf{H}_Y(t), \mathbf{C}_{1,Y}(t), \ldots, \mathbf{C}_{m,Y}(t), \mathbf{P}_{1,1,Y}(t), \mathbf{P}_{2,1,Y}(t), \ldots, \mathbf{P}_{1,m,Y}(t), \mathbf{P}_{2,m,Y}(t), \mathbf{W}_{X_1,Y}(t), \ldots, \mathbf{W}_{X_k,Y}(t), X_1(t), \ldots, X_k(t), Y(t))$$

Assuming speed factor $\boldsymbol{\eta}^*_Y = 1$, and connection weights are 1 for the reified network, a new combination function $\mathbf{c}^*_Y(..)$ is needed such that

$$\begin{aligned}
&Y(t+\Delta t) = Y(t) + \boldsymbol{\eta}_Y(t)[\mathbf{c}_Y(t, \boldsymbol{\omega}_{X_1,Y}(t)X_1(t), \ldots, \boldsymbol{\omega}_{X_k,Y}(t)X_k(t)) - Y(t)]\Delta t \\
&Y(t+\Delta t) = Y(t) + [\mathbf{c}^*_Y(\mathbf{H}_Y(t), \mathbf{C}_{1,Y}(t), \ldots, \mathbf{C}_{m,Y}(t), \mathbf{P}_{1,1,Y}(t), \mathbf{P}_{2,1,Y}(t), \ldots, \mathbf{P}_{1,m,Y}(t), \mathbf{P}_{2,m,Y}(t), \mathbf{W}_{X_1,Y}(t), \ldots, \mathbf{W}_{X_k,Y}(t), X_1(t), \ldots, \\
&X_k(t), Y(t)) - Y(t)]\Delta t
\end{aligned}$$

So, the requirement for $\mathbf{c}^*_Y(..)$ is:

$$\begin{aligned}&\mathbf{c}^*_Y(\mathbf{H}_Y(t),\, \mathbf{C}_{1,Y}(t), \ldots, \mathbf{C}_{m,Y}(t), \mathbf{P}_{1,1,Y}(t), \mathbf{P}_{2,1,Y}(t), \ldots, \mathbf{P}_{1,m,Y}(t), \mathbf{P}_{2,m,Y}(t), \mathbf{W}_{X_1,Y}(t), \ldots, \mathbf{W}_{X_k,Y}(t),\, X_1(t), \ldots, X_k(t),\, Y(t))\\ &= Y(t) + \boldsymbol{\eta}_Y(t)[\mathbf{c}_Y(t, \boldsymbol{\omega}_{X_1,Y}(t)X_1(t), \ldots, \boldsymbol{\omega}_{X_k,Y}(t)X_k(t)) - Y(t)]\end{aligned}$$

Assume

$$\begin{aligned}&\mathbf{c}_Y(t, \gamma_{1,Y}(t), \ldots, \gamma_{m,Y}(t), \pi_{1,1,Y}(t), \pi_{2,1,Y}(t), \ldots, \pi_{1,m,Y}(t), \pi_{2,m,Y}(t), V_1, \ldots, V_k)\\ &= \frac{\gamma_{1,Y}(t)\mathrm{bcf}_1\left(\pi_{1,1,Y}(t), \pi_{2,1,Y}(t), V_1, \ldots, V_k\right) + \cdots + \gamma_{m,Y}(t)\mathrm{bcf}_m\left(\pi_{1,m,Y}(t), \pi_{2,m,Y}(t), V_1, \ldots, V_k\right)}{\gamma_{1,Y}(t) + \cdots + \gamma_{m,Y}(t)}\end{aligned}$$

and $\mathbf{C}_{j,Y}(t) = \gamma_{j,Y}(t)$, $\mathbf{H}_Y(t) = \boldsymbol{\eta}_Y(t)$, $\mathbf{W}_{X_i,Y}(t) = \boldsymbol{\omega}_{X_i,Y}(t)$, and $\mathbf{P}_{i,jY}(t) = \pi_{i,j,Y}(t)$ for all i and j.

Now given the above expression the new universal combination function $\mathbf{c}^*_Y(...)$ has to be defined by:

$$\begin{aligned}&\mathbf{c}^*_Y(H, C_1, \ldots, C_m, P_{1,1}, P_{2,1}, \ldots, P_{1,m}, P_{2,m}, W_1, \ldots, W_k, V_1, \ldots, V_k, V)\\ &= H\frac{C_1\mathrm{bcf}_1\left(P_{1,1}, P_{2,1}, W_1V_1, .., W_kV_k\right) + \cdots + C_m\mathrm{bcf}_m\left(P_{1,m}, P_{2,m}, W_1V_1, .., W_kV_k\right)}{C_1 + \cdots + C_m} + (1 - H)V\end{aligned}$$

where

- H stands for the speed factor reification $\mathbf{H}_Y(t)$
- C_j for the combination function weight reification $\mathbf{C}_{j,Y}(t)$
- $P_{i,j}$ for the combination function weight reification $\mathbf{P}_{i,j,Y}(t)$
- W_i for the connection weight reification $\mathbf{W}_{X_i,Y}(t)$
- V_i for the state value of base state X_i
- V for the state value $Y(t)$

Then

$$\begin{aligned}&\mathbf{c}^*_Y(\mathbf{H}_Y(t), \mathbf{C}_{1,Y}(t), \ldots, \mathbf{C}_{m,Y}(t), \mathbf{P}_{1,1,Y}(t), \mathbf{P}_{2,1,Y}(t), \ldots, \mathbf{P}_{1,m,Y}(t), \mathbf{P}_{2,m,Y}(t), \mathbf{W}_{X_1,Y}(t), \ldots, \mathbf{W}_{X_k,Y}(t),\, X_1(t), \ldots, X_k(t),\, Y(t))\\ &= \boldsymbol{\eta}_Y(t)\frac{\gamma_{1,Y}(t)\mathrm{bcf}_1\left(\boldsymbol{\omega}_{X_1,Y}(t)X_1(t), \ldots, \boldsymbol{\omega}_{X_k,Y}(t)X_k(t)\right) + \cdots + \gamma_{m,Y}(t)\mathrm{bcf}_m\left(\boldsymbol{\omega}_{X_1,Y}(t)X_1(t), \ldots, \boldsymbol{\omega}_{X_k,Y}(t)X_k(t)\right)}{\gamma_{1,Y}(t) + \cdots + \gamma_{m,Y}(t)}\\ &\quad + (1 - \boldsymbol{\eta}_Y(t))Y(t))\\ &= Y(t) + \boldsymbol{\eta}_Y(t)[\\ &\frac{\gamma_{1,Y}(t)\mathrm{bcf}_1\left(\pi_{1,1,Y}, \pi_{2,1,Y}, \boldsymbol{\omega}_{X_1,Y}(t)X_1(t), \ldots, \boldsymbol{\omega}_{X_k,Y}(t)X_k(t)\right) + \cdots + \gamma_{m,Y}(t)\mathrm{bcf}_m\left(\pi_{1,m,Y}, \pi_{2,m,Y}, \boldsymbol{\omega}_{X_1,Y}(t)X_1(t), \ldots, \pi_{X_k,Y}(t)X_k(t)\right)}{\gamma_{1,Y}(t) + \cdots + \gamma_{m,Y}(t)} - Y(t)]\\ &= Y(t) + \boldsymbol{\eta}_Y(t)[\mathbf{c}_Y\left(t, \boldsymbol{\omega}_{X_1,Y}(t)X_1(t), \ldots, \boldsymbol{\omega}_{X_k,Y}(t)X_k(t)\right) - Y(t)]\end{aligned}$$

So, this universal combination function $\mathbf{c}^*_Y(..)$ indeed fulfills the requirement.

Box 10.5 An example of the use of the universal combination function for all roles **H**, **C**, **P** and **W**.
Example for Fig. 10.1. For reification of connection weights, speed factors and combination functions and their parameters together, and $\text{bcf}_1(..)$ is the euclidean function **eucl(..)** with order $n = 1$ and $\text{bcf}_2(..)$ the logistic function **alogistic(..)**, and $C_{1,Y}(0) = 1$, $C_{2,Y}(0) = 0$ (so first $\mathbf{eucl}_{1,\lambda}(..)$ is assumed for Y), the new combination function $\mathbf{c}^*_Y(..)$ for Y is (where $P_{1,1}$ for the order n of $\mathbf{eucl}_{n,\lambda}(..)$ is assumed 1):

$$\begin{aligned}
&\mathbf{c}^*_Y(H, C_1, C_2, P_{1,1}, P_{2,1}, P_{1,2}, P_{2,2}, W_1, W_2, V_1, V_2, V) \\
&= H\frac{C_1\text{bcf}_1(P_{1,1}, P_{2,1}, W_1V_1, W_2V_2) + C_2\text{bcf}_2(P_{1,2}, P_{2,2}, W_1V_1, W_2V_2)}{C_1 + C_2} + (1 - H)V \\
&= H\frac{C_1\mathbf{eucl}(P_{1,1}, P_{2,1}, W_1V_1, W_2V_2) + C_2\,\mathbf{alogistic}(P_{1,2}, P_{2,2}, W_1V_1, W_2V_2)}{C_1 + C_2} + (1 - H)V \\
&= H\frac{C_1\frac{W_1V_1 + W_2V_2}{P_{2,1}} + C_2\,\mathbf{alogistic}(P_{1,2}, P_{2,2}, W_1V_1, W_2V_2)}{C_1 + C_2} + (1 - H)V
\end{aligned}$$

10.4.2 The Universal Difference Equation for Reified Networks

In summary, the universal combination function found above in (8) is

$$\begin{aligned}
&\mathbf{c}^*_Y(H,\, C_1, \ldots, C_m, P_{1,1}, P_{2,1}, \ldots, P_{1,m}, P_{2,m}, W_1, \ldots, W_k, V_1, \ldots, V_k, V) \\
&= H\frac{C_1\text{bcf}_1(P_{1,1,Y}, P_{2,1,Y}, W_1V_1, .., W_kV_k) + \cdots + C_m\text{bcf}_m(P_{1,m,Y}, P_{2,m,Y}, W_1V_1, \ldots, W_kV_k)}{C_1 + \cdots + C_m} + (1 - H)V \\
&= H[\frac{C_1\text{bcf}_1(P_{1,1,Y}, P_{2,1,Y}, W_1V_1, \ldots, W_kV_k) + \cdots + C_m\text{bcf}_m(P_{1,m,Y}, P_{2,m,Y}, W_1V_1, \ldots, W_kV_k)}{C_1 + \cdots + C_m} - V] + V
\end{aligned}$$

Based on this, the following *universal difference equation* describes the dynamics of each base state Y within the reified network; in cases of full reification it has no state-specific parameters for network structure characteristics, only variables; therefore it is the same for all states Y:

$$
\begin{aligned}
&Y(t+\Delta t) = Y(t)\\
&\quad + [\mathbf{c}_Y^*(\mathbf{H}_Y(t), \mathbf{C}_{1,Y}(t), \ldots, \mathbf{C}_{m,Y}(t), \mathbf{P}_{1,1,Y}(t), \mathbf{P}_{2,1,Y}(t), \ldots, \mathbf{P}_{1,m,Y}(t), \mathbf{P}_{2,m,Y}(t), \mathbf{W}_{X_1,Y}(t), \ldots, \mathbf{W}_{X_k,Y}(t), X_1(t), \ldots, X_k(t), Y(t)) - Y(t)]\Delta t\\
&= Y(t)\\
&\quad + [\mathbf{H}_Y(t)\frac{\mathbf{C}_{1,Y}(t)\mathrm{bcf}_1(\mathbf{P}_{1,1,Y}(t), \mathbf{P}_{2,1,Y}(t), \mathbf{W}_{X_1,Y}(t)X_1(t), \ldots, \mathbf{W}_{X_k,Y}(t)X_k(t)) + \cdots + \mathbf{C}_{m,Y}(t)\mathrm{bcf}_m(\mathbf{P}_{1,m,Y}(t), \mathbf{P}_{2,m,Y}(t), \mathbf{W}_{X_1,Y}(t)X_1(t), \ldots, \mathbf{W}_{X_k,Y}(t)X_k(t))}{\mathbf{C}_{1,Y}(t) + \cdots + \mathbf{C}_{m,Y}(t)}\\
&\quad + (1 - \mathbf{H}_Y(t))Y(t) - Y(t)]\Delta t\\
&= Y(t)\\
&\quad + [\mathbf{H}_Y(t)\frac{\mathbf{C}_{1,Y}(t)\mathrm{bcf}_1(\mathbf{P}_{1,1,Y}(t), \mathbf{P}_{2,1,Y}(t), \mathbf{W}_{X_1,Y}(t)X_1(t), \ldots, \mathbf{W}_{X_k,Y}(t)X_k(t)) + \cdots + \mathbf{C}_{m,Y}(t)\mathrm{bcf}_m(\mathbf{P}_{1,m,Y}(t), \mathbf{P}_{2,m,Y}(t), \mathbf{W}_{X_1,Y}(t)X_1(t), \ldots, \mathbf{W}_{X_k,Y}(t)X_k(t))}{\mathbf{C}_{1,Y}(t) + \cdots + \mathbf{C}_{m,Y}(t)}\\
&\quad - \mathbf{H}_Y(t)Y(t)]\Delta t\\
&= Y(t)\\
&\quad + \mathbf{H}_Y(t)[\frac{\mathbf{C}_{1,Y}(t)\mathrm{bcf}_1(\mathbf{P}_{1,1,Y}(t), \mathbf{P}_{2,1,Y}(t), \mathbf{W}_{X_1,Y}(t)X_1(t), \ldots, \mathbf{W}_{X_k,Y}(t)X_k(t)) + \cdots + \mathbf{C}_{m,Y}(t)\mathrm{bcf}_m(\mathbf{P}_{1,m,Y}(t), \mathbf{P}_{2,m,Y}(t), \mathbf{W}_{X_1,Y}(t)X_1(t), \ldots, \mathbf{W}_{X_k,Y}(t)X_k(t))}{\mathbf{C}_{1,Y}(t) + \cdots + \mathbf{C}_{m,Y}(t)} - Y(t)]\Delta t
\end{aligned}
\tag{17}
$$

So, this universal difference equation is what defines the dynamics of the whole base network within the reified network. Its differential equation variant is

$$
\begin{aligned}
&\mathbf{d}Y(t)/\mathbf{d}t\\
&= \mathbf{H}_Y(t)[\frac{\mathbf{C}_{1,Y}(t)\mathrm{bcf}_1(\mathbf{P}_{1,1,Y}(t), \mathbf{P}_{2,1,Y}(t), \mathbf{W}_{X_1,Y}(t)X_1(t), \ldots, \mathbf{W}_{X_k,Y}(t)X_k(t)) + \cdots + \mathbf{C}_{m,Y}(t)\mathrm{bcf}_m(\mathbf{P}_{1,m,Y}(t), \mathbf{P}_{2,m,Y}(t), \mathbf{W}_{X_1,Y}(t)X_1(t), \ldots, \mathbf{W}_{X_k,Y}(t)X_k(t))}{\mathbf{C}_{1,Y}(t) + \cdots + \mathbf{C}_{m,Y}(t)} - Y(t)]
\end{aligned}
$$

or by leaving out t:

$$
\begin{aligned}
&\mathbf{d}Y/\mathbf{d}t\\
&= \mathbf{H}_Y[\frac{\mathbf{C}_{1,Y}\mathrm{bcf}_1(\mathbf{P}_{1,1,Y}, \mathbf{P}_{2,1,Y}, \mathbf{W}_{X_1,Y}X_1, \ldots, \mathbf{W}_{X_k,Y}X_k) + \cdots + \mathbf{C}_{m,Y}\mathrm{bcf}_m(\mathbf{P}_{1,m,Y}, \mathbf{P}_{2,m,Y}, \mathbf{W}_{X_1,Y}X_1, \ldots, \mathbf{W}_{X_k,Y}X_k)}{\mathbf{C}_{1,Y} + \cdots + \mathbf{C}_{m,Y}} - Y]
\end{aligned}
\tag{18}
$$

By structure-preserving implementation based on the above universal difference equation, the software environment as described in Chap. 9 has been developed. As can be seen there in Boxes 9.3 and 9.4, the above universal difference equation just occurs in Matlab format in that software environment. However, starting from the above universal difference equation, a different path to implementation can be followed as well. This will be discussed in next section.

10.5 The Criterion for a Stationary Point for the Universal Difference Equation

Recall the criterion for a stationary point:

Criterion for stationary points and equilibria in temporal-causal network models

A state Y in an adaptive temporal-causal network model has a stationary point at t if and only if

$$\boldsymbol{\eta}_Y = 0 \quad \text{or} \quad \mathbf{c}_Y(\boldsymbol{\omega}_{X_1,Y}(t)X_1(t), \ldots, \boldsymbol{\omega}_{X_k,Y}(t)X_k(t)) = Y(t)$$

where X_1, …, X_k are the states with outgoing connections to Y.

An adaptive temporal-causal network model is in an equilibrium state at t if and only if for all states the above criteria hold at t.

Now suppose that some or all of the characteristics $\boldsymbol{\eta}_Y$, $\boldsymbol{\omega}_{X_k,Y}$, $\mathbf{c}_Y(..)$ are reified. Then the above equation becomes the universal differential equation. What is the criterion then? The format shown in (17) above can be rewritten into the format of (18) above; when is the right hand side 0?

$$\mathbf{H}_Y[\frac{\mathbf{C}_{1,Y}\mathrm{bcf}_1(\mathbf{P}_{1,1,Y}, \mathbf{P}_{2,1,Y}, \mathbf{W}_{X_1,Y}X_1, \ldots, \mathbf{W}_{X_k,Y}X_k) + \cdots + \mathbf{C}_{m,Y}\mathrm{bcf}_m(\mathbf{P}_{1,m,Y}, \mathbf{P}_{2,m,Y}, \mathbf{W}_{X_1,Y}X_1, \ldots, \mathbf{W}_{X_k,Y}X_k)}{\mathbf{C}_{1,Y} + \cdots + \mathbf{C}_{m,Y}} - Y] = 0$$

This right hand side of this is 0 if and only if

$$\mathbf{H}_Y = 0 \quad \text{or}$$

$$\frac{\mathbf{C}_{1,Y}\mathrm{bcf}_1(\mathbf{P}_{1,1,Y}, \mathbf{P}_{2,1,Y}, \mathbf{W}_{X_1,Y}X_1, \ldots, \mathbf{W}_{X_k,Y}X_k) + \cdots + \mathbf{C}_{m,Y}\mathrm{bcf}_m(\mathbf{P}_{1,m,Y}, \mathbf{P}_{2,m,Y}, \mathbf{W}_{X_1,Y}X_1, \ldots, \mathbf{W}_{X_k,Y}X_k)}{\mathbf{C}_{1,Y} + \cdots + \mathbf{C}_{m,Y}} = Y$$

Now notice that the part

$$\frac{\mathbf{C}_{1,Y}\mathrm{bcf}_1(\mathbf{P}_{1,1,Y}, \mathbf{P}_{2,1,Y}, \mathbf{W}_{X_1,Y}X_1, \ldots, \mathbf{W}_{X_k,Y}X_k) + \cdots + \mathbf{C}_{m,Y}\mathrm{bcf}_m(\mathbf{P}_{1,m,Y}, \mathbf{P}_{2,m,Y}, \mathbf{W}_{X_1,Y}X_1, \ldots, \mathbf{W}_{X_k,Y}X_k)}{\mathbf{C}_{1,Y} + \cdots + \mathbf{C}_{m,Y}}$$

is precisely $\mathbf{c}_Y(\boldsymbol{\omega}_{X_1,Y}(t)\, X_1(t), \ldots, \boldsymbol{\omega}_{X_k,Y}(t)\, X_k(t))$ in the old criterion above, so this has exactly the same form as the old criterion. Therefore the above criterion also can be used when some or all of the characteristics $\boldsymbol{\eta}_Y$, $\boldsymbol{\omega}_{X_k,Y}(t)$, $\mathbf{c}_Y(..)$ are adaptive.

10.6 Deriving the Difference and Differential Equation from the Role Matrices

In the role matrices all information is available to determine the difference or differential equations. In Box 10.6 it is shown how that can be done. Here it is assumed that the indicated matrix cell provides the static value from the matrix, or, if not a static value, the indicated state name X_k for the adaptive value.

Box 10.6 Derivation of the basic differential equation for the network's dynamics from the role matrices.
Substitute every characteristic by the reference to the cell in the role matrix where this is indicated. So, for the combination function of X_j, in Eq. (2) (use the parameters $\boldsymbol{\pi}_{i,1,X_j}$ and $\boldsymbol{\pi}_{i,2,X_j}$ as first two arguments of a basic combination function):

- for the combination function weight γ_{i,X_j} substitute $\mathbf{mcfw}(j, i)$
- for the parameter $\pi_{i,1,X_j}$ or $\pi_{i,2,X_j}$ substitute $\mathbf{mcfp}(j, 1, i)$ resp. $\mathbf{mcfp}(j, 2, i)$

Then from (2) the following expression in terms of the role matrices results:

$$\mathbf{c}_Y(V_1, .., V_k) = \left[\frac{\begin{array}{c}\mathbf{mcfw}(j,1)\mathbf{bcf}_1(\mathbf{mcfp}(j,1,1), \mathbf{mcfp}(j,2,1), V_1, \ldots, V_k) \\ + \cdots + \mathbf{mcfw}(j,m)\mathbf{bcf}_m(\mathbf{mcfp}(j,1,m), \mathbf{mcfp}(j,2,m), V_1, \ldots, V_k)\end{array}}{\mathbf{mcfw}(j,1) + \cdots + \mathbf{mcfw}(j,m)} \right] \tag{19}$$

Suppose in the role base connectivity matrix **mb** the states specified in the row for X_j are the states X_{i_1}, …, X_{i_k}; these also can be denoted by **mb**(j, 1), …, **mb**(j, k). To get the basic differential equation in terms of the role matrices, as a next step:

- in (3) substitute the single impact $\omega_{\mathbf{mb}(j,i),X_j}$ **mb**(j, i) for V_i
- for connection weight $\omega_{\mathbf{mb}(j,i),X_j}$ substitute $\mathbf{mcw}(j, i)$

Then the following is obtained:

$$\mathbf{c}_Y(\ldots) = \left[\frac{\begin{array}{c}\mathbf{mcfw}(j,1)\mathbf{bcf}_1(\mathbf{mcfp}(j,1,1), \mathbf{mcfp}(j,2,1), \mathbf{mcw}(j,1)\mathbf{mb}(j,1), \ldots, \mathbf{mcw}(j,k)\mathbf{mb}(j,k)) \\ + \cdots + \mathbf{mcfw}(j,m)\mathbf{bcf}_m(\mathbf{mcfp}(j,1,m), \mathbf{mcfp}(j,2,m), \mathbf{mcw}(j,1)\mathbf{mb}(j.1), \ldots, \mathbf{mcw}(j,k)\mathbf{mb}(j,k))\end{array}}{\mathbf{mcfw}(j,1) + \cdots + \mathbf{mcfw}(j,m)} \right] \tag{20}$$

Now to get the differential equation, as a final step

- for the speed factor η_{X_j} substitute $\mathbf{ms}(j, 1)$

Then the differential equation expression becomes:

$$\mathbf{d}X_j/\mathbf{d}t = \mathbf{ms}(j,1) \left[\frac{\begin{array}{c}\mathbf{mcfw}(j,1)\mathbf{bcf}_1(\mathbf{mcfp}(j,1,1), \mathbf{mcfp}(j,2,1), \mathbf{mcw}(j,1)\mathbf{mb}(j,1), \ldots, \mathbf{mcw}(j,k)\mathbf{mb}(j,k)) \\ + \cdots + \mathbf{mcfw}(j,m)\mathbf{bcf}_m(\mathbf{mcfp}(j,1,m), \mathbf{mcfp}(j,2,m), \mathbf{mcw}(j,1)\mathbf{mb}(j.1), \ldots, \mathbf{mcw}(j,k)\mathbf{mb}(j,k))\end{array}}{\mathbf{mcfw}(j,1) + \cdots + \mathbf{mcfw}(j,m)} - X_j \right] \tag{21}$$

Note that for states often only one combination function is selected and has nonzero weight. Then expression (21) in Box 10.6 simplifies to (e.g., for $\mathbf{bcf}_i(..)$):

$$\begin{aligned}&\mathbf{d}X_j/\mathbf{d}t \\ &= \mathbf{ms}(j,1)[\mathbf{bcf}_i(\mathbf{mcfp}(j,1,i), \mathbf{mcfp}(j,2,i), \mathbf{mcw}(j,1)\mathbf{mb}(j,1), \ldots, \mathbf{mcw}(j,k)\mathbf{mb}(j,k)) - X_j]\end{aligned} \tag{22}$$

As a form of verification, this can be filled for state X_2 in the example Social Network from Chap. 2, Box 2.2, so $j = 2$, and $i = 1$ as can be seen in **mcfw**. It can be found in **mb** that $k = 9$ for X_2.

$$\begin{aligned}&\mathbf{d}X_2/\mathbf{d}t\\&= \mathbf{ms}(2,1)[\mathbf{bcf}_1(\mathbf{mcfp}(2,1,1), \mathbf{mcfp}(2,2,1), \mathbf{mcw}(2,1)\mathbf{mb}(2,1), \ldots, \mathbf{mcw}(2,9)\mathbf{mb}(2,9)) - X_2]\end{aligned} \tag{23}$$

To get the idea, from the role matrices in Chap. 2, Box 2.2, all values can be found, for example:

$$\begin{aligned}&\mathbf{ms}(2,1) = 0.5\\&\mathbf{mcfp}(2,1,1) = 1\\&\mathbf{mcfp}(2,2,1) = 1.55\\&\mathbf{mcw}(2,1) = 0.1\\&\mathbf{mb}(2,1) = X_1\\&\text{et cetera}\end{aligned}$$

This leads to:

$$\begin{aligned}&\mathbf{d}X_2/\mathbf{d}t\\&= 0.5[\mathbf{bcf}_1(1, 1.55, 0.1X_1, 0.25X_3, 0.15X_4, 0.2X_5, 0.1X_6, 0.1X_7, 0.25X_8, 0.15X_9, 0.25X_{10}) - X_2]\end{aligned}$$

Finally, after also incorporating the combination function weights represented in **mcfw** it provides:

$$\begin{aligned}&\mathbf{d}X_2/\mathbf{d}t\\&= 0.5[\mathbf{eucl}(1, 1.55, 0.1X_1, 0.25X_3, 0.15X_4, 0.2X_5, 0.1X_6, 0.1X_7, 0.25X_8, 0.15X_9, 0.25X_{10}) - X_2]\\&= 0.5[\frac{0.1X_1 + 0.25X_3 + 0.15X_4 + 0.2X_5 + 0.1X_6 + 0.1X_7 + 0.25X_8 + 0.15X_9 + 0.25X_{10}}{1.55} - X_2]\end{aligned} \tag{25}$$

10.7 Compilation of the Universal Differential Equation by Substitution

In the software as described in Chap. 9, the role matrices defining the model are inspected at every simulation step. There is a second option for implementation by separating this work from simulation time, in the form of compiling. Doing so, the one universal difference equation as shown above is instantiated for each of the states with the entries from the role matrices for that state, so it is replaced by n specific difference equations with n the number of states. The resulting set of specific difference (or differential) equations can be run by any general purpose software environment for differential equation simulation. As there are many quite efficient software environments for this, for large-scale reified networks of thousands or even millions of states such environments can be used for successful simulation. This compilation process will be illustrated for the plasticity and metaplasticity example network from Chap. 4 (see Fig. 4.3 and Box 4.1).

Suppose in role base connectivity matrix **mb** the states specified in the row for X_j are the states $X_{i_1}, \ldots, X_{i_k}$, which also can be denoted by **mb**(j, 1), …, **mb**(j, k). So consider again the universal differential equation

$$\mathbf{d}X_j/\mathbf{d}t = \mathbf{H}_Y \left[\frac{\mathbf{C}_{1,Y}\,\mathrm{bcf}_1(\mathbf{P}_{1,1,Y}, \mathbf{P}_{2,1,Y}, \mathbf{W}_{X_1,Y}X_{i_1}, \ldots, \mathbf{W}_{X_k,Y}X_{i_k}) + \ldots + \mathbf{C}_{m,Y}\,\mathrm{bcf}_m(\mathbf{P}_{1,m,Y}, \mathbf{P}_{2,m,Y}, \mathbf{W}_{X_1,Y}X_{i_1}, \ldots, \mathbf{W}_{X_k,Y}X_{i_k})}{\mathbf{C}_{1,Y} + \ldots + \mathbf{C}_{m,Y}} - X_j \right]$$

Here the parts that need substitution have been highlighted, and the role matrix where the entries to be substituted can be found are indicated as follows:

Yellow $\mathbf{H}_Y$ from role matrix **ms** for speed factors
Green $\mathbf{C}_{j,Y}$ from role matrix **mcfw** for combination function weights
Blue $\mathbf{P}_{1,1,Y}$ from role matrix **mcfp** for combination function parameters
Purple $\mathbf{W}_{X_j Y}$ from role matrix **mcw** for connection weights

Here it is assumed that the indicated matrix cell provides the static value from the matrix, or, if not a static value, the indicated state name X_k for the adaptive value. For example, in the role matrices in Chap. 4, Box 4.1 it can be seen that X_1, X_3, X_6, X_7, X_8, X_9 have the standard difference equation with values for the characteristics. So for these states there are just constant values substituted in the universal difference equation. In other cases, such as X_2, there are entries in role matrices that are just names X_j of states; in these cases just that name X_j has to be substituted. Note first the number m of combination functions has to be read from the role matrices **mcfw**, and per state Y, the number k of incoming base connections from role matrix **mb**. For example, from Chap. 4, Box 4.1 it is seen in role matrix **mcfw** that $m = 3$ and for state X_4 from the fourth row in role matrix **mb** that $k = 2$, and the states with incoming connections (from **mb**) are X_2 and X_3.

That makes the following format in particular for X_4

$$\begin{aligned}\mathbf{d}X_4/\mathbf{d}t &= \mathbf{H}_{X_4}\left[\frac{\mathbf{C}_{1,X_4}\mathrm{bcf}_1(\mathbf{P}_{1,1,X_4}, \mathbf{P}_{2,1,X_4}, \mathbf{W}_{X_2,X_4}X_2, \mathbf{W}_{X_3,X_4}X_3) + \mathbf{C}_{2,X_4}\mathrm{bcf}_2(\mathbf{P}_{1,2,X_4}, \mathbf{P}_{2,2,X_4}, \mathbf{W}_{X_2X_4}X_2, \mathbf{W}_{X_3,X_4}X_3) + \mathbf{C}_{3,X_4}\mathrm{bcf}_3(\mathbf{P}_{1,3,X_4}, \mathbf{P}_{2,3,Y}, \mathbf{W}_{X_2,X_4}X_2, \mathbf{W}_{X_3,X_4}X_3)}{\mathbf{C}_{1,X_4} + \mathbf{C}_{2,X_4} + \mathbf{C}_{3,X_4}} - X_4\right]\end{aligned}$$

So consider this state X_4 further. In role matrix **cfw** for connection function weights it can be seen that the weights $\mathbf{C}_{j,X_4}$ are values 0, 1 and 0, respectively. Substituting these values makes the equation much simpler:

$$\mathbf{d}X_4/\mathbf{d}t = \mathbf{H}_{X_4}[\mathrm{bcf}_2(\mathbf{P}_{1,2,X_4}, \mathbf{P}_{2,2,X_4}, \mathbf{W}_{X_2X_4}X_2, \mathbf{W}_{X_3,X_4}X_3) - X_4]$$

In role matrix **mcw** for connection weights it can be seen that state name X_5 is indicated for the connection from X_2. So that name has to be substituted for $\mathbf{W}_{X_2,X_4}$. The other weight has just constant value 1 in role matrix **mcw**, so then 1 can be substituted for $\mathbf{W}_{X_3,X_4}$. Then this is obtained with the remaining spots for further substitution highlighted:

$$\mathbf{d}X_4/\mathbf{d}t = \mathbf{H}_{X_4}\left[\mathrm{bcf}_2\left(\mathbf{P}_{1,2,X_4}, \mathbf{P}_{2,2,X_4}, X_5X_2, X_3\right) - X_4\right]$$

Also in role matrix **mcfp** for the parameters for X_4 there is an adaptive one, namely the second parameter or the second combination function indicates X_7 as an adaptive value, so this has to be substituted for $\mathbf{P}_{2,2,X_4}$; for $\mathbf{P}_{1,2,X_4}$ the value 5 is indicated. These substitutions make

$$\mathbf{d}X_4/\mathbf{d}t = \mathbf{H}_{X_4}\left[\mathrm{bcf}_2(5, X_7, X_5X_2, X_3) - X_4\right]$$

The speed factor 0.5 from role matrix **ms** can be substituted, and the function $\mathbf{alogistic}_{\sigma,\tau}(..)$ can be substituted for $\mathrm{bcf}_2(..)$:

$$\mathbf{d}X_4/\mathbf{d}t = 0.5\left[\mathbf{alogistic}_{5,X_7}(X_5X_2, X_3) - X_4\right] \tag{26}$$

Similarly the following instantiated difference equations can be found

$$\begin{aligned} \mathbf{d}X_2/\mathbf{d}t &= 0.5\left[\mathbf{alogistic}_{5,X_6}(X_1) - X_2\right] \\ \mathbf{d}X_5/\mathbf{d}t &= X_8\left[\mathbf{hebb}_{X_9}(X_2, X_4, X_5) - X_5\right] \end{aligned} \tag{27}$$

For these functions their detailed formulae can be substituted. For example, for Hebbian learning

$$\mathbf{hebb}_{X_9}(X_2, X_4, X_5) = X_2X_4(1 - X_5) + X_9X_5$$

this makes it

$$\mathbf{d}X_5/\mathbf{d}t = X_8\left[X_2X_4(1 - X_5) + X_9X_5 - X_5\right] \tag{28}$$

The equations for the other states do not involve adaptive characteristics, so then just values found in the role matrices are substituted. See Box 10.6 for the complete outcome of the compilation.

This illustrates how the universal differential equation can be compiled by replacing it by a set of specific differential equations for each of the states that can be entered in a general purpose differential equation solver. This may imply a gain in efficiency during simulation, which may be beneficial when large scale reified networks are simulated, for example, with thousands of states. The compilation process itself can be time consuming if done by hand, but in future that could also be automated.

Box 10.6 The result of complete compilation for the reified network for plasticity and metaplasticity

$\mathbf{d}X_1/\mathbf{d}t = 0$

$\mathbf{d}X_2/\mathbf{d}t = 0.5\ [\mathbf{alogistic}_{5,X_6}(X_1) - X_2]$

$\mathbf{d}X_3/\mathbf{d}t = 0.2\ [\mathbf{alogistic}_{5,0.2}(X_2) - X_3]$

$\mathbf{d}X_4/\mathbf{d}t = 0.5\ [\mathbf{alogistic}_{5,X_7}(X_5X_2, X_3) - X_4]$

$\mathbf{d}X_5/\mathbf{d}t = X_8\ [X_2X_4\ (1 - X_5) + X_9X_5 - X_5]$

$\mathbf{d}X_6/\mathbf{d}t = 0.3\ [\mathbf{alogistic}_{5,0.7}(-0.4X_2, -0.4X_4, X_6) - X_6]$

$\mathbf{d}X_7/\mathbf{d}t = 0.3\ [\mathbf{alogistic}_{5,0.7}(-0.4X_2, -0.4X_4, X_6) - X_7]$

$\mathbf{d}X_8/\mathbf{d}t = 0.5\ [\mathbf{alogistic}_{5,1}(X_2, X_4, -0.4\ X_5, X_8) - X_8]$

$\mathbf{d}X_9/\mathbf{d}t = 0.1\ [\mathbf{alogistic}_{5,1}(X_2, X_4, X_5, X_9) - X_9]$

There is a generic way to write the compiled differential equations down in a symbolic manner, in terms of the cell references in the role matrices as follows. Substitute every reification state by the reference to the cell in the role matrix where this is indicated. So, for the equation of X_j:

- for $\mathbf{H}_{X_j}$ substitute $\mathbf{ms}(j, 1)$
- for $\mathbf{C}_{i,X_j}$ substitute $\mathbf{mcfw}(j, i)$
- for $\mathbf{P}_{i,1,X_j}$ or $\mathbf{P}_{i,2,X_j}$ substitute $\mathbf{mcfp}(j, 1, i)$ or $\mathbf{mcfp}(j, 2, i)$
- for $\mathbf{W}_{X_i,X_j}$ substitute $\mathbf{mcw}(j, i)$

Then the following equation results

$$\mathbf{d}X_j/\mathbf{d}t = \mathbf{ms}(j,1)[\frac{\begin{array}{c}\mathbf{mcfw}(j,1)\mathbf{bcf}_1(\mathbf{mcfp}(j,1,1), \mathbf{mcfp}(j,2,1), \mathbf{mcw}(j,1)\mathbf{mb}(j,1), \ldots, \mathbf{mcw}(j,k)\mathbf{mb}(j,k))\\ + \cdots + \mathbf{mcfw}(j,m)\mathbf{bcf}_m(\mathbf{mcfp}(j,1,m), \mathbf{mcfp}(j,2,m), \mathbf{mcw}(j,1)\mathbf{mb}(j.1), \ldots, \mathbf{mcw}(j,k)\mathbf{mb}(j,k))\end{array}}{\mathbf{mcfw}(j,1) + \cdots + \mathbf{mcfw}(j,m)} - X_j] \tag{29}$$

When for states often only one combination function is selected and has nonzero weight, (5) simplifies to (e.g., for $\mathbf{bcf}_i$(..)):

$$\mathbf{d}X_j/\mathbf{d}t = \mathbf{ms}(j,1)[\mathbf{bcf}_i(\mathbf{mcfp}(j,1,i), \mathbf{mcfp}(j,2,i), \mathbf{mcw}(j,1)\mathbf{mb}(j,1), \ldots, \mathbf{mcw}(j,k)\mathbf{mb}(j,k)) - X_j] \tag{30}$$

Here, to evaluate this expression, the references in the cells of the role matrices are interpreted as strings; so, for example, if in that cell it is written X_4, then that is substituted in the above expression to get the resulting differential equation.

10.8 Discussion

In this chapter a more in depth analysis was presented for the universal differential or difference equation that is an important basis for reified temporal-causal networks. It was shown how this equation can be derived and it was illustrated by some examples. Due to the existence of this specific universal difference or differential equation, it can be guaranteed that any reification of a temporal-causal network is itself also a temporal-causal network: the class of temporal-causal networks is closed under reification. That means that dedicated modeling and analysis methods for temporal-causal networks can also be applied to reified temporal-causal networks. In particular, reification can be done iteratively so that multilevel reified network models are obtained that are very useful to model multiple orders of adaptation. In addition, the fact that the universal difference or differential equation is the same for all states, and has not a number of instantiations for different states, makes that it indeed is universal. This supports structure preserving implementation where the core of the program code for the computational reified temporal-causal network engine has the same simple universal structure expressed in only a few lines of code, as can be seen in Chap. 9, Sect. 9.4.3, Box 9.4.

References

Ashby, W.R.: Design for a Brain. Chapman and Hall, London (second extended edition). First edition, 1952 (1960)

Brauer, F., Nohel, J.A.: Qualitative Theory of Ordinary Differential Equations. Benjamin (1969)

Lotka, A.J.: Elements of Physical Biology. Williams and Wilkins Co. (1924), Dover Publications (1956)

Port, R.F., van Gelder, T.: Mind as Motion: Explorations in the Dynamics of Cognition. MIT Press, Cambridge, MA (1995)

Treur, J.: Network-Oriented Modeling: Addressing Complexity of Cognitive, Affective and Social Interactions. Springer Publishers, Berlin (2016)

Treur, J.: Network reification as a unified approach to represent network adaptation principles within a network. In: Proceedings of the 7th International Conference on Natural Computing. Lecture Notes in Computer Science, vol. 11324, pp. 344–358. Springer Publishers, Berlin (2018a)

Treur, J.: Multilevel network reification: representing higher order adaptivity in a network. In: Proceedings of the 7th International Conference on Complex Networks and their Applications, ComplexNetworks'18, vol. 1. Studies in Computational Intelligence, vol. 812, pp. 635–651. Springer, Berlin (2018b)

Treur, J.: The ins and outs of network-oriented modeling: from biological networks and mental networks to social networks and beyond. In: Transactions on Computational Collective Intelligence. Contents of Keynote Lecture at ICCCI'18, vol. 32, pp. 120–139. Springer Publishers, Berlin (2019a)

Treur, J.: Design of a software architecture for multilevel reified temporal-causal networks (2019b). https://doi.org/10.13140/rg.2.2.23492.07045. URL: https://www.researchgate.net/publication/333662169

Part V
Mathematical Analysis of How Emerging Network Behaviour Relates to Base Network Structure

Chapter 11
Relating Emerging Network Behaviour to Network Structure

Abstract Emerging behaviour of a network is a consequence of the network's structure. However, it may often not be easy to find out how this relation between structure and behaviour exactly is. In this chapter, results are presented on how certain properties of network structure determine network behaviour. The network structure characteristics considered include both connectivity characteristics in terms of being strongly connected, and aggregation characteristics in terms of properties of combination functions to aggregate multiple impacts on a state. In particular, results are found for networks that are strongly connected and combination functions that are strictly monotonically increasing and scalar-free. This class of combination functions includes linear combination functions such as scaled sum functions but also nonlinear ones such as Euclidean combination functions of any order *n* and scaled geometric mean combination functions. In addition, some results are found on how timing characteristics affect final outcomes of the network behaviour.

Keywords Network structure · Social contagion · Asymptotic network behavior · Social convergence · Mathematical analysis

11.1 Introduction

The emerging behaviour of networks is often considered an interesting and sometimes fascinating consequence of the network's structure. Although the emerging behaviour is entailed by the network structure, finding the relation between network structure and network behaviour may be a real challenge. Often simulations under varying settings of the structure characteristics are used to reveal just a glimpse of this relation. But sometimes it is possible to find out how certain properties of the emerging behaviour can be derived in a mathematical manner from certain characteristics of the network structure. In this chapter such cases are shown, using the Network-Oriented Modeling approach from Chap. 2 and (Treur 2016b, 2019) as a vehicle. This approach enables to derive theoretical results that predict emerging behavior that is observed in specific cases of simulations.

J. Treur, *Network-Oriented Modeling for Adaptive Networks: Designing Higher-Order Adaptive Biological, Mental and Social Network Models*, Studies in Systems, Decision and Control 251, https://doi.org/10.1007/978-3-030-31445-3_11

Network structure is in principle described by a number of characteristics that, as considered here, concern: (1) *connectivity* characteristics, describing how different parts of the network connect, (2) *aggregation* characteristics, describing how multiple connections to the same node are handled, and (3) *timing* characteristics, describing how fast network states change over time. For temporal-causal networks, more specifically, such characteristics relate to connection weights defining the connectivity, combination functions defining aggregation of the impacts of multiple states on a given state, and speed factors defining the speed of change of a state. The challenge then is to find out how properties of connection weights, combination functions and speed factors relate to emerging behavior.

In particular, in this chapter it will be addressed what behaviour emerges concerning equilibrium states that are reached; for example:

- What are bounds between which in the end equilibrium values of states occur?
- How much variation occurs for these equilibrium values?
- Under which conditions a common equilibrium value for the different states occurs?

Answers for such questions can be relevant, for example, to predict the spread of information or opinions, or social contagion of emotions; e.g. (Bosse et al. 2015; Castellano et al. 2009).

As a result of the mathematical analysis performed, a number of properties of a network structure have been identified such that any network with a structure satisfying these properties show similar emerging behavior. These structure properties include connectivity characteristics of the network and aggregation characteristics in terms of properties of combination functions used to aggregate the impact of multiple incoming connections to a node. The identified properties of the combination functions define a class of functions most of which are nonlinear, although linear functions are still included. Among the nonlinear ones are Euclidean combination functions of any order n, and scaled geometric mean combination functions. Examples of combination functions that do not belong to this class are minimum and maximum combination functions and logistic sum combination functions. Also some results are presented on how timing characteristics of the network affect the outcomes of the network's behaviour.

In this chapter, in Sect. 11.2 basic concepts are introduced. Section 11.3 shows simulation examples of the emerging behaviour phenomena that can be observed. In Sect. 11.4 properties of network structure are defined that are relevant for the considered types of emerging behaviour. Section 11.5 discusses a number of results for the relation between network structure and network behaviour addressing the questions above. These have been proven mathematically; proofs are included in Chap. 15, Sect. 15.6. Section 11.6 examines a further set of simulations, this time with focus on Euclidean, scaled geometric mean and scaled maximum combination functions. A result is found relating Euclidean combination functions of very high order n to scaled maximum combination functions. Section 11.7 is a final discussion.

11.2 Conceptual and Numerical Representation of a Network

The modeling perspective (Treur 2016b, 2019) used in this chapter, interprets connections in a network in terms of causality and dynamics; see Chap. 2. In the *temporal-causal networks* used, nodes in a network are interpreted as states that vary over time, and the connections are interpreted as causal relations that define how each state can affect other states over time. To define such a network structure, three main elements have to be addressed:

(a) connectivity (the connections in the network)
(b) aggregation (how multiple connections to one node are aggregated)
(c) timing (how timing of the different states takes place).

These three notions determine the characteristics of the network structure. For temporal-causal networks they are modeled by connection weights, combination functions, and speed factors, respectively, which is summarized as:

(a) **Connectivity**

- connection weights from a state X to a state Y, denoted by $\boldsymbol{\omega}_{X,Y}$

(b) **Aggregation**

- a combination function for each state Y, denoted by $\mathbf{c}_Y(..)$

(c) **Timing**

- a speed factor for each state Y, denoted by $\boldsymbol{\eta}_Y$.

Based on these, a conceptual representation of a temporal-causal network model includes labels for connection weights, combination functions, and speed factors; see the upper part (first 5 rows) of Table 11.1. Note that in the current chapter only networks with nonnegative connection weights are considered.

Combination functions are similar to the functions used in a static manner in the (deterministic) Structural Causal Model perspective described, for example, in (Pearl 2000). However, here they are used in a dynamic manner. For example, (Pearl 2000), p. 203, denotes nodes by V_i and the functions corresponding to combination functions by f_i. Pearl (2000) also points at the problem of underspecification for aggregation of multiple connections, as in the often used graph representations the role of combination functions f_i for nodes V_i, is lacking, and they are therefore not a full specification of the network structure.

To provide sufficient flexibility, for each state a specific combination function can be chosen to specify how multiple causal impacts on this state are aggregated. A number of standard combination functions are available as options in the *combination function library* currently including up to tens of functions, but also new functions can be added to the library.

Table 11.1 Conceptual and numerical representations of a temporal-causal network model)

Concept	Conceptual representation	Explanation
States and connections	X, Y, $X \rightarrow Y$	Describes the nodes and links of a network structure (e.g., in graphical or matrix format)
Connection weight	$\boldsymbol{\omega}_{X,Y}$	The *connection weight* $\boldsymbol{\omega}_{X,Y} \in [-1, 1]$ represents the strength of the causal impact of state X on state Y through connection $X \rightarrow Y$
Aggregating multiple impacts on a state	$\mathbf{c}_Y(..)$	For each state Y (a reference to) a *combination function* $\mathbf{c}_Y(..)$ is chosen to combine the causal impacts of other states on state Y
Timing of the effect of causal impact	$\boldsymbol{\eta}_Y$	For each state Y a *speed factor* $\boldsymbol{\eta}_Y \geq 0$ is used to represent how fast a state is changing upon causal impact

Concept	**Numerical representation**	**Explanation**
State values over time t	$Y(t)$	At each time point t each state Y in the model has a real number value, usually in [0, 1]
Single causal impact	$\mathbf{Impact}_{X,Y}(t) = \boldsymbol{\omega}_{X,Y}\, X(t)$	At t state X with a connection to Y has an impact on Y, using connection weight $\boldsymbol{\omega}_{X,Y}$
Aggregating multiple causal impacts	$\mathbf{aggimpact}_Y(t)$ $= \mathbf{c}_Y(\mathbf{impact}_{X_1,Y}(t), \ldots, \mathbf{impact}_{X_k,Y}(t))$ $= \mathbf{c}_Y(\boldsymbol{\omega}_{X_1,Y}X_1(t), \ldots, \boldsymbol{\omega}_{X_k,Y}X_k(t))$	The aggregated causal impact of $k \geq 1$ states $X_1, \ldots, X_k$ on Y at t, is determined using combination function $\mathbf{c}_Y(..)$
Timing of the causal effect	$Y(t+\Delta t) = Y(t) +$ $\boldsymbol{\eta}_Y[\mathbf{aggimpact}_Y(t) - Y(t)]\Delta t$ $= Y(t) +$ $\boldsymbol{\eta}_Y[\mathbf{c}_Y(\boldsymbol{\omega}_{X_1,Y}X_1(t), \ldots, \boldsymbol{\omega}_{X_k,Y}X_k(t)) - Y(t)]\Delta t$	The causal impact on Y is exerted over time gradually, using speed factor $\boldsymbol{\eta}_Y$; here the X_i are all $k \geq 1$ states with outgoing connections to state Y

The lower part of Table 11.1 shows the numerical representation describing the dynamics of a temporal-causal network by difference equations, defined on the basis of the network structure as described by the upper part of that table. Thus dynamic semantics is associated in a numerical-mathematically defined manner to any conceptual temporal-causal network specification. This provides a well-defined relation between network structure and network dynamics at the base level. The difference equations in Table 11.1 last row can be used both for simulation and for mathematical analysis. In Fig. 11.1 the basic relation between structure and dynamics is indicated by the horizontal arrow in the lower part representing the base level; see also in Chap. 2, Fig. 2.6. The upper part will be addressed in Sects. 11.4, 11.5 and 11.6.

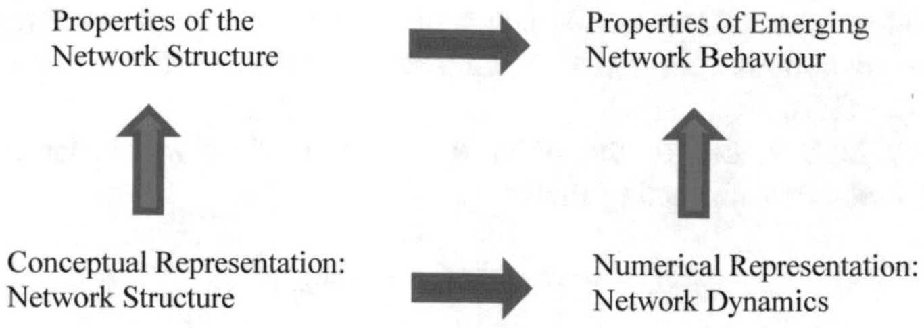

Fig. 11.1 Bottom layer: the conceptual representation defines the numerical representation. Top layer: properties of network structure entail properties of emerging network behaviour

Often used examples of combination functions are the *identity* function **id**(.) for states with impact from only one other state, the scaled maximum and minimum function $\mathbf{smax}_\lambda$(..) and $\mathbf{smin}_\lambda$(..), the *scaled sum* function $\mathbf{ssum}_\lambda$(.), and the scaled geometric mean function $\mathbf{sgeomean}_\lambda$(.), all with scaling factor $\boldsymbol{\lambda}$, and the *advanced logistic sum* combination function $\mathbf{alogistic}_{\sigma,\tau}$(..) with steepness $\boldsymbol{\sigma}$ and threshold $\boldsymbol{\tau}$:

$$\begin{aligned}
&\mathbf{id}(V) = V\\
&\mathbf{smax}_\lambda(V_1, \ldots, V_k) = \mathbf{max}(V_1, \ldots, V_k)/\boldsymbol{\lambda}\\
&\mathbf{smin}_\lambda(V_1, \ldots, V_k) = \mathbf{min}(V_1, \ldots, V_k)/\boldsymbol{\lambda}\\
&\mathbf{ssum}_\lambda(V_1, \ldots, V_k) = (V_1 + \cdots + V_k)/\boldsymbol{\lambda}\\
&\mathbf{sgeomean}_\lambda(V_1, \ldots, V_k) = \sqrt[k]{\frac{V_1 * \ldots * V_k}{\boldsymbol{\lambda}}}\\
&\mathbf{alogistic}_{\sigma,\tau}(V_1, \ldots, V_k) = \left[\frac{1}{1+\mathrm{e}^{-\boldsymbol{\sigma}(V_1+\cdots+V_k-\boldsymbol{\tau})}} - \frac{1}{1+\mathrm{e}^{\boldsymbol{\sigma\tau}}}\right](1+\mathrm{e}^{-\boldsymbol{\sigma\tau}})
\end{aligned} \tag{11.1}$$

In addition to the above functions, generalising the scaled sum function, a *Euclidean combination function* is defined as

$$\mathbf{eucl}_{n,\lambda}(V_1, \ldots, V_k) = \sqrt[n]{\frac{V_1^n + \cdots + V_k^n}{\boldsymbol{\lambda}}} \tag{11.2}$$

where n is the *order* (which can be any positive natural number but also any positive real number), and $\boldsymbol{\lambda}$ is again a scaling factor. Note that indeed for n = 1 (first order) we get the scaled sum function

$$\mathbf{eucl}_{1,\lambda}(V_1, \ldots, V_k) = \mathbf{ssum}_\lambda(V_1, \ldots, V_k) \tag{11.3}$$

For n = 2 it is the second-order Euclidean combination function defined by

$$\mathbf{eucl}_{2,\lambda}(V_1, \ldots, V_k) = \sqrt{\frac{V_1^2 + \cdots + V_k^2}{\boldsymbol{\lambda}}} \tag{11.4}$$

This second-order Euclidean combination function often occurs in aggregating the error value in optimisation and in parameter tuning using the root-mean-square deviation (RMSD).

For very high values of the order n the limit of an nth order normalised Euclidean function with scaling factor

$$\lambda(n) = \omega_{X_1,Y}{}^n + \cdots + \omega_{X_k,Y}{}^n \tag{11.5}$$

is a normalised scaled maximum function with scaling factor $\max\left(\omega_{X_1,Y}, \ldots, \omega_{X_k,Y}\right)$:

$$\lim_{n\to\infty} \mathbf{eucl}_{n,\lambda(n)}(V_1, \ldots, V_k) = \mathbf{smax}_{\max\left(\omega_{X_1,Y},\ldots,\omega_{X_k,Y}\right)}(V_1, \ldots, V_k) \tag{11.6}$$

This will be shown both by simulation and by mathematical analysis later in Sect. 11.6.

11.3 Examples of a Network's Emerging Behaviour

In this section a few examples of a Social Network for social contagion are discussed. They all concern a fully connected network.

11.3.1 The Example Social Network

The example network is shown in Fig. 11.2 and has connection weights and speed factors as shown in the role matrices in Box 11.1, with initial values shown in Table 11.2. This is actually the same example as used in Chap. 2 for a first analysis.

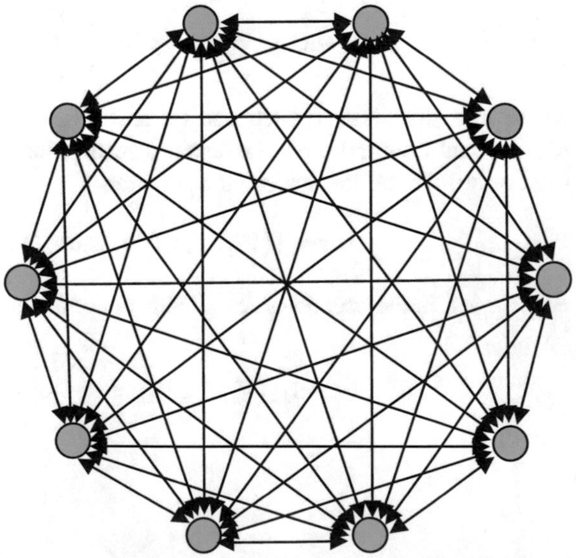

Fig. 11.2 The example social network

Table 11.2 Initial values for the example

Initial values

X_1	X_2	X_3	X_4	X_5	X_6	X_7	X_8	X_9	X_{10}
0.1	0.3	0.9	0.8	0.5	0.6	0.85	0.05	0.25	0.4

Box 11.1 Role matrices for the example network

mb **base connectivity**	1	2	3	4	5	6	7	8	9
X_1	X_2	X_3	X_4	X_5	X_6	X_7	X_8	X_9	X_{10}
X_2	X_1	X_3	X_4	X_5	X_6	X_7	X_8	X_9	X_{10}
X_3	X_1	X_2	X_4	X_5	X_6	X_7	X_8	X_9	X_{10}
X_4	X_1	X_2	X_3	X_5	X_6	X_7	X_8	X_9	X_{10}
X_5	X_1	X_2	X_3	X_4	X_6	X_7	X_8	X_9	X_{10}
X_6	X_1	X_2	X_3	X_4	X_5	X_7	X_8	X_9	X_{10}
X_7	X_1	X_2	X_3	X_4	X_5	X_6	X_8	X_9	X_{10}
X_8	X_1	X_2	X_3	X_4	X_5	X_6	X_7	X_9	X_{10}
X_9	X_1	X_2	X_3	X_4	X_5	X_6	X_7	X_8	X_{10}
X_{10}	X_1	X_2	X_3	X_4	X_5	X_6	X_7	X_8	X_9

mcw **connection weights**	1	2	3	4	5	6	7	8	9
X_1	0.25	0.1	0.25	0.25	0.25	0.2	0.1	0.25	0.2
X_2	0.1	0.25	0.15	0.2	0.1	0.1	0.25	0.15	0.25
X_3	0.2	0.25	0.25	0.1	0.25	0.2	0.1	0.25	0.2
X_4	0.1	0.2	0.1	0.2	0.25	0.15	0.25	0.15	0.2
X_5	0.2	0.1	0.2	0.15	0.25	0.2	0.05	0.2	0.1
X_6	0.15	0.2	0.15	0.8	0.25	0.2	0.15	0.1	0.2
X_7	0.1	0.15	0.1	0.25	0.2	0.1	0.25	0.2	0.15
X_8	0.25	0.25	0.25	0.15	0.1	0.25	0.2	0.15	0.8
X_9	0.25	0.25	0.1	0.25	0.2	0.25	0.15	0.1	0.2
X_{10}	0.1	0.25	0.15	0.25	0.15	0.1	0.25	0.25	0.15

mcfw **combination function weights**	1 **eucl**	2 **alogistic**
X_1	1	
X_2	1	
X_3	1	
X_4	1	
X_5	1	
X_6	1	
X_7	1	
X_8	1	
X_9	1	
X_{10}	1	

mcfp **function**	1 eucl		2 **alogistic**	
parameter	1	2	1 σ	2 τ
X_1	1	1.85		
X_2	1	1.55		
X_3	1	1.8		
X_4	1	1.6		
X_5	1	1.45		
X_6	1	2.2		
X_7	1	1.5		
X_8	1	2.4		
X_9	1	1.75		
X_{10}	1	1.65		

ms **speed factors**	1
X_1	0.8
X_2	0.5
X_3	0.8
X_4	0.5
X_5	0.5
X_6	0.5
X_7	0.8
X_8	0.5
X_9	0.5
X_{10}	0.5

11.3.2 Three Simulations with Different Emerging Behaviour

For this example Social Network simulations have been performed for two types of combination functions: scaled sum and advanced logistic sum combination functions. In Sect. 11.6 similar simulations will be shown for scaled geometric mean, Euclidean, and scaled maximum combination functions. Figure 11.3 shows three different example simulations (all with step size $\Delta t = 0.25$):

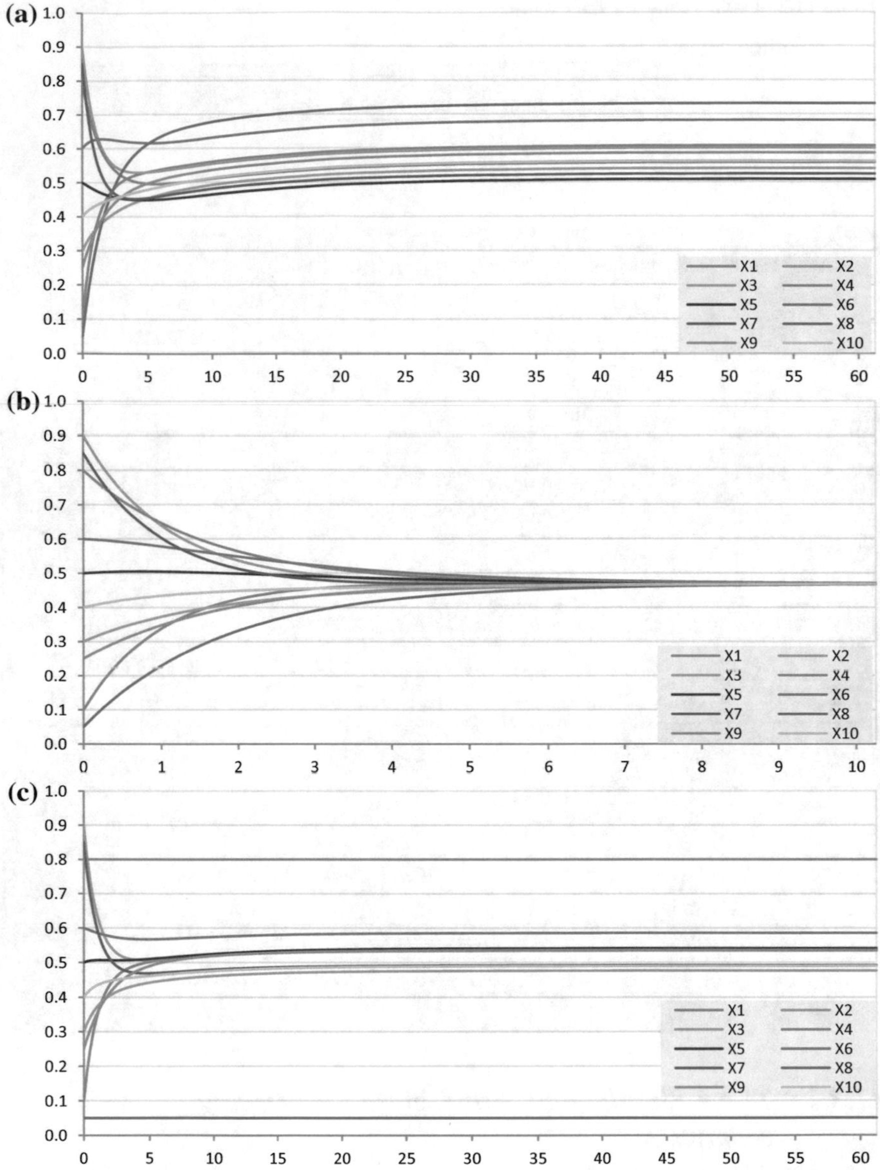

Fig. 11.3 The example network of Fig. 11.2 with **a** upper graph: advanced logistic sum combination functions with steepness $\boldsymbol{\sigma} = 1.5$, threshold $\boldsymbol{\tau} = 0.3$ (no common equilibrium value), **b** middle graph: normalised scaled sum functions (common equilibrium value), **c** lower graph: normalised scaled sum functions with constant X_4 (at 0.8) and X_8 (at 0.05) (no common value)

- in the upper graph advanced logistic sum combination functions are used,
- in the middle graph normalized scaled sum functions, and
- in the lower graph scaled sum functions while two states are independent and remain constant.

It turns out that in one of the three cases convergence to a common equilibrium value takes place, but not in the other two cases; instead some (imperfect) form of clustering seems to take place. How can we explain these differences from the structure of the networks? This question will be answered in Sect. 11.5.

11.4 Relevant Network Structure Characteristics

As explained in Sect. 11.2 and Table 11.1 the basic difference equations describing network dynamics relate to network structure. This covers the lower part of Fig. 11.1. As emerging behaviour shows itself over longer time durations, and the difference equations describe the very small steps in the dynamics, to relate emerging network behaviour to network structure, the gap between these small steps and longer time durations has to be bridged. How that can be done is discussed in the current section and in Sect. 11.5. Here it is discussed which properties of network structure (see Fig. 11.1, left upper corner) underly the behavioural differences (right upper corner in Fig. 11.1) shown in Sect. 11.3 and Fig. 11.3. Proofs can be found in Chap. 15, Sect. 15.6.

Properties of all three main elements of the network's structure (connectivity, aggregation, and timing) have turned out relevant as determining factors for the network's emerging behaviour. Properties of the network's *aggregation* are expressed by combination functions, and properties of the network's *connectivity* are expressed by the connections and their weights. First, in Sect. 11.4.1 the relevant properties of combination functions for aggregation are addressed, and next in Sect. 11.4.2 relevant properties of the network's connectivity. In Sect. 11.5.4 the third main element of a network's structure determining the network's behaviour, namely *timing*, is analysed as well.

11.4.1 Relevant Network Aggregation Characteristics in Terms of Properties of Combination Functions

For the combination functions describing the network's aggregation characteristics, the following properties are relevant. Whether or not they are fulfilled can make differences in emerging behaviour of the type shown in Fig. 11.3.

Definition 1 (Properties of combination functions)

(a) A function c(..) is called *nonnegative* if c(V_1, …, V_k) $\geq$ 0 for all V_1, …, V_k
(b) A function c(..) *respects* 0 if V_1, …, $V_k \geq 0 \Rightarrow$ [c(V_1, …, V_k) = 0 $\Leftrightarrow$ V_1 = ⋯ = V_k = 0]
(c) A function c(..) is called *monotonically increasing* if

$$U_i \leq V_i \text{ for all } i \Rightarrow \mathrm{c}(U_1, \ldots, U_k) \leq \mathrm{c}(V_1, \ldots, V_k)$$

(d) A function c(..) is called *strictly monotonically increasing* if

$$U_i \leq V_i \text{ for all } i, \text{and } U_j < V_j \text{ for at least one } j \Rightarrow \mathrm{c}(U_1, \ldots, U_k) < \mathrm{c}(V_1, \ldots, V_k)$$

(e) A function c(..) is called *scalar-free* if c(αV_1, …, αV_k) = αc(V_1, …, V_k) for all $\alpha > 0$

In Table 11.3 it is shown which functions have which of the properties (c) to (e) from Definition 1. The following propositions are useful to prove that certain combination functions have the above properties.

Proposition 1 Linear combinations with positive coefficients of functions that are (strictly) *monotonic or scalar-free also are* (strictly) monotonic or scalar-free, respectively.

Proposition 2 Any function composed of monotonically increasing or decreasing functions including an even number of monotonically decreasing functions is monotonically increasing. The same holds for strictly monotonically increasing or decreasing.

Proposition 3 For every $n > 0$ a Euclidean combination function of nth degree is strictly monotonic, scalar-free, symmetric and respects 0.

The properties (a) and (b) are basic properties silently assumed to hold for all combination functions considered here. Sometimes combination functions are defined in such a way that (a) automatically holds:

$$\mathrm{c}^*(V_1, \ldots, V_k) = \begin{array}{ll} \mathrm{c}(V_1, \ldots, V_k) & \text{if } \mathrm{c}(V_1, \ldots, V_k) \geq 0 \\ 0 & \text{otherwise} \end{array}$$

Properties (d) and (e) define a specific class of combination functions; this class includes all Euclidean combination functions and geometric mean combination functions, but logistic sum combination functions do not belong to this class, as they are not scalar-free. Also maximum-based combination functions do not belong to this class as they are monotonic but not strict. A number of results on emerging behaviour will be discussed for this class in particular; note that most functions in this class are nonlinear.

See Table 11.3 for which functions have which of the properties (c) to (e) from Definition 1.

Table 11.3 Characteristics of Definition 1 for the example combination functions

	(c)	(d)	(e)
id(.)	+	+	+
$\mathbf{ssum}_{\lambda}$**(..)**	+	+	+
$\mathbf{eucl}_{n,\lambda}$**(..)**	+	+	+
smin(..)	+	–	+
smax(..)	+	–	+
$\mathbf{sgeomean}_{\lambda}$**(..)** for $V_i > 0$	+	+	+
$\mathbf{alogistic}_{\sigma,\tau}$**(..)**	+	+	–

Proposition 4 (Proportional outcomes)
If in a temporal-causal network all combination functions are scalar-free and in some Scenario 1 the initial values for the states are a factor $\boldsymbol{\rho}$ times the initial values in a Scenario 2, then for every t the state values in Scenario 1 are $\boldsymbol{\rho}$ times the corresponding state values in Scenario 2, assuming $\boldsymbol{\eta}_Y\Delta t \leq 1$ for all states Y. This also holds for the equilibrium values when an equilibrium is reached.

Proposition 5 (Order preservation)
If in a temporal-causal network all combination functions are monotonically increasing and in some Scenario 1 the initial values for the states are $\leq$ the initial values in a Scenario 2, then for every t the state values in Scenario 1 are $\leq$ the corresponding state values in Scenario 2, assuming $\boldsymbol{\eta}_Y\Delta t \leq 1$ for all states Y. This also holds for the equilibrium values when an equilibrium is reached. Similarly for decreasing.

Definition 2 (normalised network)
A network is *normalised* or uses normalised combination functions if for each state Y it holds $\mathbf{c}_Y\left(\boldsymbol{\omega}_{X_1,Y},\ldots,\boldsymbol{\omega}_{X_k,Y}\right) = 1$, where $X_1,\ldots, X_k$ are the states from which Y gets its incoming connections.

This normalisation can be achieved in two ways:

(1) **normalisation by adjusting the combination functions**

If any combination function $\mathbf{c}_Y$**(..)** is replaced by $\mathbf{c}'_Y$**(..)** defined as

$$\mathbf{c}'_Y(V_1,\ldots,V_k) = \mathbf{c}_Y(V_1,\ldots,V_k)/\mathbf{c}_Y\left(\boldsymbol{\omega}_{X_1,Y},\ldots,\boldsymbol{\omega}_{X_k,Y}\right) \tag{11.7}$$

then the network is normalised: $\mathbf{c}'_A\left(\boldsymbol{\omega}_{X_1,Y},\ldots,\boldsymbol{\omega}_{X_k,Y}\right) = 1$

(2) **normalisation by adjusting the connection weights**

For scalar-free combination functions also normalisation is possible by adapting the connection weights; define

$$\boldsymbol{\omega}'_{X_i,Y} = \boldsymbol{\omega}_{X_i,Y}/\mathbf{c}_Y\left(\boldsymbol{\omega}_{X_1,Y},\ldots,\boldsymbol{\omega}_{X_k,Y}\right) \tag{11.8}$$

Then the network becomes normalised; indeed $\mathbf{c}_Y\left(\boldsymbol{\omega}'_{X_1,Y},\ldots,\boldsymbol{\omega}'_{X_k,Y}\right) = 1$.

For different example functions, following normalisation (1) above, their normalised variants are given by Table 11.4.

11.4.2 Relevant Network Connectivity Characteristics: Being Strongly Connected

Another important determinant for emerging behaviour is formed by the network's connectivity characteristics, in particular in how far the network has paths connecting any two states:

Definition 3 **(reachable, strongly connected and symmetric network)**

(a) State Y is *reachable* from state X if there is a directed path from X to Y with nonzero connection weights and speed factors.
(b) A network is *connected* if between every two states there is a (nondirected) path with nonzero connection weights and speed factors. It is *strongly connected* if any state Y is reachable from any state X.
(c) A network is *fully connected* if for any states X, Y there is a (direct) connection from X to Y.
(d) A network is called *weakly symmetric* if for all nodes X and Y it holds $\boldsymbol{\omega}_{X,Y} = 0 \Leftrightarrow \boldsymbol{\omega}_{Y,X} = 0$ or, equivalently: $\boldsymbol{\omega}_{X,Y} > 0 \Leftrightarrow \boldsymbol{\omega}_{Y,X} > 0$. The network is called *fully symmetric* if $\boldsymbol{\omega}_{X,Y} = \boldsymbol{\omega}_{Y,X}$ for all nodes X and Y. An adaptive network is called *continually* (*weakly/fully*) *symmetric* if at all time points it is (weakly/fully) symmetric.
(e) A state Y is called *independent* if for any incoming connection with connection weight $\boldsymbol{\omega}_{X,Y} > 0$ the speed factor of Y is 0 (or no incoming connections exist).

Note that an independent state is not reachable from any other state. The term independent means that its behaviour over time is not affected by the other states. Either its value can remain constant (when the speed factor is 0), or it can show any autonomously defined dynamics (see, for example, state X_6 in Fig. 11.4).

Definition 4 **(symmetric combination function)**

A combination function is *symmetric* in a subset S of its arguments if for any $U_1, \ldots, U_k$ is obtained from $V_1, \ldots, V_k$ by a permutation of the arguments in S, it holds $\mathbf{c}(U_1, \ldots, U_k) = \mathbf{c}(V_1, \ldots, V_k)$. It is *fully symmetric* if S is the set of all arguments.

Table 11.4 Normalisation of combination functions

Combination function	Notation	Normalising scaling factor	Normalised combination function
Identity function	**id(.)**	$\omega_{X,Y}$	$V/\omega_{X,Y}$
Scaled sum	$\mathbf{ssum}_\lambda(V_1, \ldots, V_k)$	$\omega_{X_1,Y} + \cdots + \omega_{X_k,Y}$	$(V_1 + \cdots + V_k)/(\omega_{X_1,Y} + \cdots + \omega_{X_k,Y})$
Scaled maximum	$\mathbf{smax}_\lambda(V_1, \ldots, V_k)$	$\max(\omega_{X_1,Y}, \ldots, \omega_{X_k,Y})$	$\max(V_1, \ldots, V_k)/\max(\omega_{X_1,Y}, \ldots, \omega_{X_k,Y})$
Scaled minimum	$\mathbf{smin}_\lambda(V_1, \ldots, V_k)$	$\min(\omega_{X_1,Y}, \ldots, \omega_{X_k,Y})$	$\min(V_1, \ldots, V_k)/\min(\omega_{X_1,Y}, \ldots, \omega_{X_k,Y})$
Scaled geometric mean	$\mathbf{sgeomean}_\lambda(V_1, \ldots, V_k)$	$\omega_{X_1,Y} * \ldots * \omega_{X_k,Y}$	$\sqrt[k]{\frac{V_1 * \ldots * V_k}{\omega_{X_1,Y} * \ldots * \omega_{X_k,Y}}}$
Euclidean	$\mathbf{eucl}_{n,\lambda}(V_1, \ldots, V_k)$	$\omega_{X_1,Y}^n + \cdots + \omega_{X_k,Y}^n$	$\sqrt[n]{\frac{V_1^n + \cdots + V_k^n}{\omega_{X_1,Y}^n + \cdots + \omega_{X_k,Y}^n}}$
Advanced logistic	$\mathbf{alogistic}_{\sigma,\tau}(V_1, \ldots, V_k)$	$\mathbf{alogistic}_{\sigma,\tau}(\omega_{X_1,Y}, \ldots, \omega_{X_k,Y})$	$\frac{\frac{1}{1+e^{-\sigma(V_1 + \cdots + V_k - \tau)}} - \frac{1}{1+e^{\sigma\tau}}}{\frac{1}{1+e^{-\sigma(\omega_{X_1,Y} + \cdots + \omega_{X_k,Y} - \tau)}} - \frac{1}{1+e^{\sigma\tau}}}$

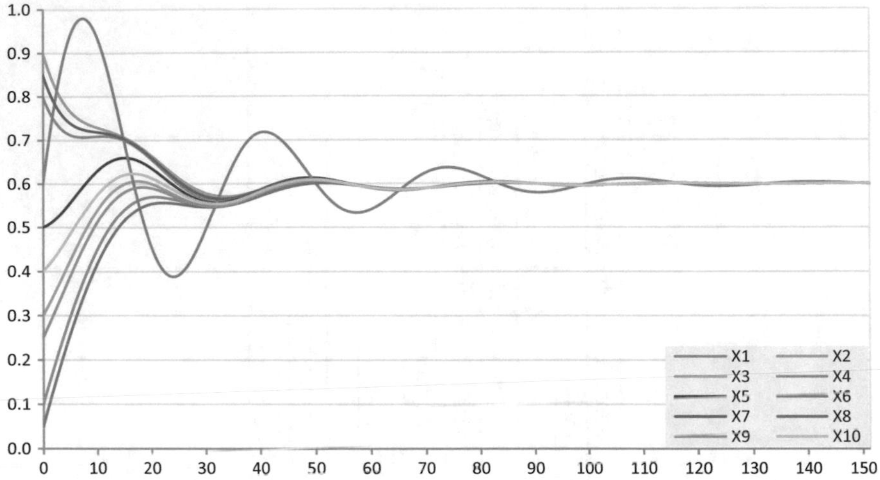

Fig. 11.4 How the dynamics of one independent state X_6 affects all states over time

11.5 Results Relating Emerging Behaviour to Network Structure

This section focuses on the emerging behaviour properties and how they relate to the main structure characteristics connectivity, aggregation, and timing. From these, the first two were discussed in Sect. 11.4, and the third one, timing, will be discussed in Sect. 11.5.4 below. In the current section, it will be shown how properties of these three elements entail properties of emerging behaviour (the horizontal arrow in the upper part of Fig. 11.1). First a few basic definitions and results; see also (Treur 2016a).

11.5.1 Basic Definitions and Results

Definition 5 (stationary point and equilibrium)
A state Y has a *stationary point* at t if $\mathbf{d}Y(t)/\mathbf{d}t = 0$. The network is in *equilibrium* at t if every state Y of the network has a stationary point at t.

Applying this for the specific differential equation format for a temporal-causal network model, a more specific criterion can be formulated in terms of the network structure characteristics $\boldsymbol{\omega}_{X,Y}$, $\mathbf{c}_Y(..)$, $\boldsymbol{\eta}_Y$:

Lemma 1 (Criterion for a stationary point in a temporal-causal network)
Let Y be a state and $X_1, \ldots, X_k$ the states from which state Y gets its incoming connections. Then Y has a stationary point at t if and only if $\boldsymbol{\eta}_Y = 0$ or $\mathbf{c}_Y\left(\boldsymbol{\omega}_{X_1,Y}X_1(t), \ldots, \boldsymbol{\omega}_{X_k,Y}X_k(t)\right) = Y(t)$.

The following proposition and theorem show that for normalised scalar-free combination functions, always when all states have the same value (for example, initially), an equilibrium occurs. For proofs, see Chap. 15, Sect. 15.6.

Proposition 6 Suppose a network with nonnegative connections has normalised scalar-free combination functions.

(a) If X_1, …, X_k are the states from which Y gets its incoming connections, and $X_1(t) = \cdots = X_k(t) = V$ for some common value V, then also $\mathbf{c}_Y(\boldsymbol{\omega}_{X_1,Y}X_1(t), \ldots, \boldsymbol{\omega}_{X_k,Y}X_k(t)) = V$.
(b) If, moreover, the combination functions are monotonic, and $V_1 \leq X_1(t)$, …, $X_k(t) \leq V_2$ for some values V_1 and V_2, then also $V_1 \leq \mathbf{c}_Y(\boldsymbol{\omega}_{X_1,Y}X_1(t), \ldots, \boldsymbol{\omega}_{X_k,Y}X_k(t)) \leq V_2$ and if $\boldsymbol{\eta}_Y\ \Delta t \leq 1$ and $V_1 \leq Y(t) \leq V_2$ then $V_1 \leq Y(t + \Delta t) \leq V_2$.

Theorem 1 (common state values provide equilibria)
Suppose a network with nonnegative connections is based on normalised and scalar-free combination functions. Then the following hold.

(a) Whenever all states have the same value V, the network is in an equilibrium state.
(b) If for every state for its initial value V it holds $V_1 \leq V \leq V_2$, then for all t for every state Y it holds $V_1 \leq Y(t) \leq V_2$. In an achieved equilibrium for every state for its equilibrium value V it holds $V_1 \leq V \leq V_2$.

11.5.2 Common Equilibrium Values for Acyclic and Strongly Connected Networks

In this section the focus is on networks with neat connectivity properties, namely being acyclic or being strongly connected. For these types some results are discussed below. However, also for any network connectivity good results are possible, but these results depend on the network's connectivity structure in terms of its strongly connected components. Those more general results are addressed in Chap. 12.

Theorem 1 does not tell whether other types of equilibria, where the values are not the same, are possible as well, for example, as shown in the first and third graph in Fig. 11.3. In subsequent theorems it is shown that in many cases no other types of equilibria occur. As a first case, consider a network without cycles. Then the following theorem has been proven by applying induction over the acyclic graph connections starting from the independent states and thereby using Proposition 6 and Lemma 1.

Theorem 2 (equilibrium states provide common state values; acyclic case)
Suppose an acyclic network with nonnegative connections is based on normalised and scalar-free combination functions.

(a) If in an equilibrium state the independent states all have the same value V, then all states have the same value V.
(b) If, moreover, the combination functions are monotonic, and in an equilibrium state the independent states all have values V with $V_1 \leq V \leq V_2$, then all states have values V with $V_1 \leq V \leq V_2$.

Next, a basic Lemma for dynamics of normalised networks with combination functions which are monotonically increasing and scalar-free.

Lemma 2 Let a normalised network with nonnegative connections be given with combination functions that are monotonically increasing and scalar-free; then:

(a)

(i) If for some node Y at time t for all nodes X with $\boldsymbol{\omega}_{X,Y} > 0$ it holds $X(t) \leq Y(t)$, then $Y(t)$ is decreasing at t: $\mathbf{d}Y(t)/\mathbf{d}t \leq 0$.
(ii) If the combination functions are strictly increasing and a node X exists with $X(t) < Y(t)$ and $\boldsymbol{\omega}_{X,Y} > 0$, and the speed factor of Y is nonzero, then $Y(t)$ is strictly decreasing at t: $\mathbf{d}Y(t)(t)/\mathbf{d}t < 0$.

(b)

(i) If for some node Y at time t for all nodes X with $\boldsymbol{\omega}_{X,Y} > 0$ it holds $X(t) \geq Y(t)$, then $Y(t)$ is increasing at t: $\mathbf{d}Y(t)/\mathbf{d}t \geq 0$.
(ii) *If*, the combination function is strictly increasing and a node X exists with $X(t) > Y(t)$ and $\boldsymbol{\omega}_{X,Y} > 0$, and the speed factor of Y is nonzero, then $Y(t)$ is strictly increasing at t: $\mathbf{d}Y(t)(t)/\mathbf{d}t > 0$.

Using Lemma 1 and 2 the following proposition has been proven for strongly connected networks with cycles.

Theorem 3 (common equilibrium state values; strongly connected case) Suppose the network has normalised, scalar-free and strictly monotonic combination functions, then:

(a) If the network is strongly connected, then in an equilibrium state all states have the same value.
(b) Suppose the network has one or more independent states and the subnetwork without these independent states is strongly connected. If in an equilibrium state all independent states have values V with $V_1 \leq V \leq V_2$, then all states have values V with $V_1 \leq V \leq V_2$. In particular, when all independent states have the same value V, then all states have this same value V.

Using Lemma 1 and 2 the following slightly more general theorem has been proven for (connected) networks with cycles and possibly with an independent state.

Theorem 4 (equilibrium states provide common state values) Suppose a (possibly cyclic) network with nonnegative connections is based on normalised, strictly monotonically increasing and scalar-free combination functions, then:

(a) If in an equilibrium state, a state Y with nonzero speed factor has highest state value or lowest state value, then all states X from which Y is reachable have the same equilibrium state value as Y.
(b) Suppose except for at most one independent state, every state Y is reachable from all other states X. Then in an equilibrium state all states have the same state value.
(c) Under the conditions of (b) the equilibrium state is attracting, and the common equilibrium state value is between the highest and lowest previous or initial state values.

Theorems 2, 3 and 4 can be applied to many cases and then prove that all states converge to the same value. For example, this explains why for the second simulation in Fig. 11.3 convergence to one common value takes place, but not for the first and third case. For the first case this is because it does not satisfy the scalar-free condition, and the for the third case because it does not satisfy the condition on reachability: one exceptional independent state is allowed but not two, as occurs in the third example in Fig. 11.3.

As an illustration for another function satisfying the above conditions of being scalar-free and strictly monotonically increasing, in Fig. 11.5 a simulation example is shown for a normalised scaled geometric mean function. This function is indeed scalar free, monotonically increasing, and strictly monotonically increasing as long as the values are nonzero. So by Theorem 3 a common equilibrium value may be expected.

As predicted by Theorems 1 and 3 all state values indeed end up in the same value and this value is between the minimal and maximal initial values.

11.5.3 The Effect of Independent States on a Network's Emerging Behaviour

The one exceptional independent state allowed in Theorem 4(b) can have any independently preset constant value, and all other state values converge to this value. But it is also possible to give this state an autonomous pattern over time that converges to some limit value $\underline{\mathbf{V}}$ for $t \to \infty$. Then over time all state values will more or less follow this pattern and end up in the same equilibrium value $\underline{\mathbf{V}}$, all according to Theorem 4(b); see Fig. 11.4. Therefore:

Corollary 1 Assume the conditions of Theorem 4 hold, and one state X is independent, for which its value over time is described by the function $f(t)$, so $X(t) = f(t)$ for all t. If $\lim_{t \to \infty} f(t) = V$, then this V is the common equilibrium value for all states.

Corollary 1 is illustrated by Fig. 11.5 where X_6 is an independent state and has dynamics based on $f(t) = b_2 + b_1 e^{-a_2 t} \sin(2\pi a_1 t)$ with $a_1 = 0.03$, $a_2 = 0.035$,

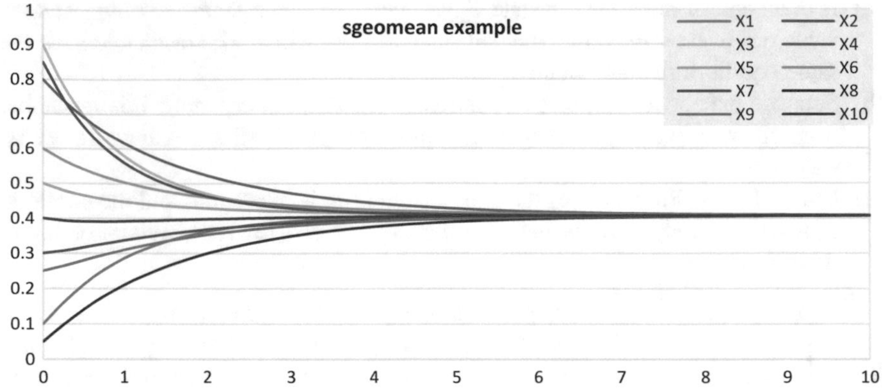

Fig. 11.5 Example simulation for the normalised scaled geometric mean function

$b_1 = 0.5$, $b_2 = 0.6$. Here the outgoing connections of X_6 have been increased according to Table 11.5 to get a stronger effect.

Figure 11.6 shows two cases in which condition (b) of Theorem 4 is not fulfilled: there are two independent states. In the upper graph X_3 and X_7 are constant at values 0.85 and 0.9, respectively, and all other equilibrium values turn out to end up between these values. In the lower graph, X_1 and X_8 are constant at values 0.1 and 0.05, respectively, and also here all other equilibrium values turn out to end up between these values. So, even if these two states both have very low or very high values, still the other state values end up between these values. Note that as in Fig. 11.3c, here the equilibrium values are not equal, although in this case they are close to each other. This is consistent with Theorems 3 and 4.

Theorem 3(b) shows that under the conditions assumed there, all equilibrium value are in between the highest and lowest initial values of independent states, which is illustrated in Fig. 11.6.

More about this can be found when also the role of timing is taken into account as modeled by speed factors.

Table 11.5 Adjusted weights (Compared to Box 11.1) of outgoing connections from X_6

Connection weights	X_1	X_2	X_3	X_4	X_5	X_6	X_7	X_8	X_9	X_{10}
X_6	0.6	0.9	0.7	0.95	0.9		0.8	0.7	0.8	0.9

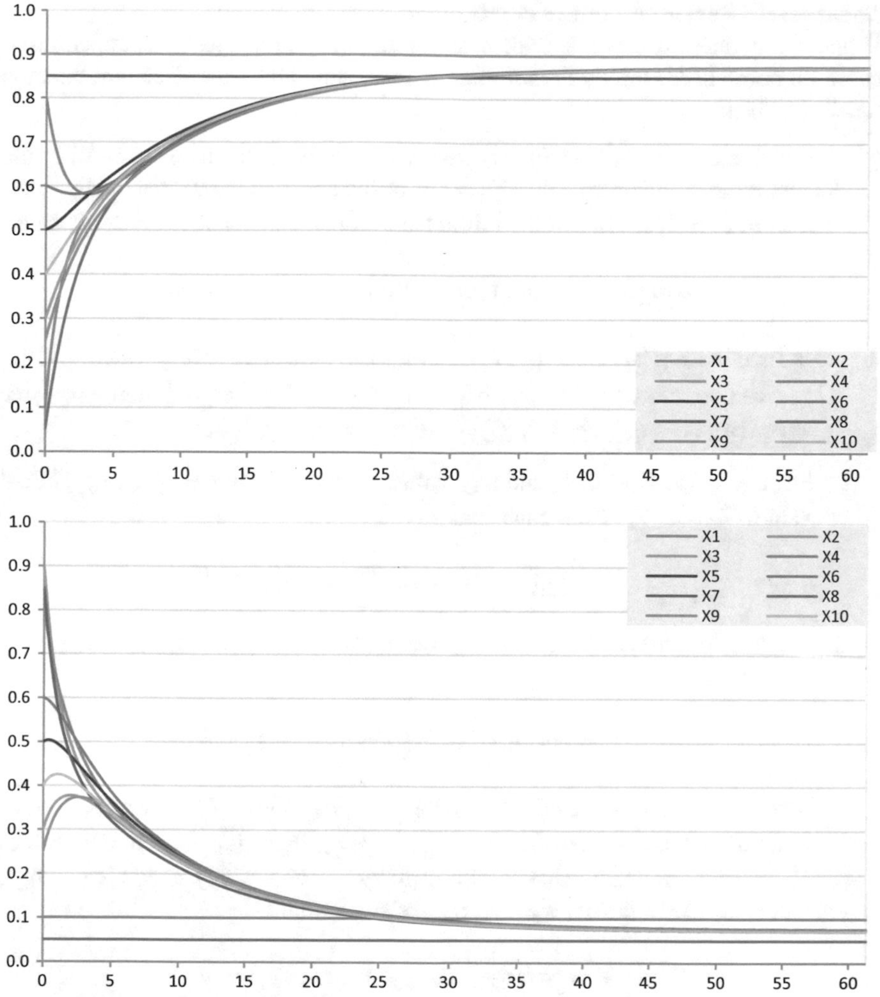

Fig. 11.6 How multiple independent states affect all equilibrium values

11.5.4 How Timing Affects a Common Equilibrium Value

Taking also *timing* into account, as modeled by speed factors, the following Theorem 5 is a further refinement of the above. It shows that under some assumptions any value between the highest and lowest initial value can be the common equilibrium value.

Theorem 5 (Variability of the common equilibrium value)
Suppose a connected network with n states and only nonnegative connections is based on normalised, strictly monotonically increasing and scalar-free combination functions, then:

(a) A function $\underline{\mathbf{eqf}}$: $[0, 1]^{2n} \rightarrow [0, 1]$ exists that assigns in an achieved equilibrium the common equilibrium value $\underline{\mathbf{V}}_{X_i} = \underline{\mathbf{V}}$ of the states to the values of the speed factors $\mathbf{\eta}_{X_i}$, $i = 1, \ldots, n$, and the initial state values V_{X_i}, $i = 1, \ldots, n$ of all states:

$$\underline{\mathbf{eqf}}\left(\mathbf{\eta}_{X_1}, \ldots, \mathbf{\eta}_{X_n}, V_{X_1} \ldots, V_{X_n}\right) = \underline{\mathbf{V}}_{X_i} = \underline{\mathbf{V}} \text{ for all } i$$

(b) This function $\underline{\mathbf{eqf}}\left(\mathbf{\eta}_{X_1}, \ldots, \mathbf{\eta}_{X_n}, V_{X_1} \ldots, V_{X_n}\right)$ is surjective: every value $V \in [0, 1]$ can occur as some achieved equilibrium value $\underline{\mathbf{V}} = \underline{\mathbf{eqf}}\left(\mathbf{\eta}_{X_1}, \ldots, \mathbf{\eta}_{X_n}, V_{X_1} \ldots, V_{X_n}\right)$. More specifically, it holds:

(i) For any value $V \in [0, 1]$ and any values of speed factors $\mathbf{\eta}_{X_1}, \ldots, \mathbf{\eta}_{X_n}$, initial values $V_{X_1} \ldots, V_{X_n}$ exist such that

$$\underline{\mathbf{V}} = \underline{\mathbf{eqf}}(\mathbf{\eta}_{X_1}, \ldots, \mathbf{\eta}_{X_n}, V_{X_1} \ldots, V_{X_n}) = V$$

(ii) For any initial values $V_{X_1} \ldots, V_{X_n}$, values of speed factors $\mathbf{\eta}_{X_1}, \ldots, \mathbf{\eta}_{X_n}$ exist such that

$$\underline{\mathbf{V}} = \underline{\mathbf{eqf}}\left(\mathbf{\eta}_{X_1}, \ldots, \mathbf{\eta}_{X_n}, V_{X_1} \ldots, V_{X_n}\right) = V_{X_i}$$

(iii) Moreover, if it is assumed that $\underline{\mathbf{eqf}}\left(\mathbf{\eta}_{X_1}, \ldots, \mathbf{\eta}_{X_n}, V_{X_1} \ldots, V_{X_n}\right)$ is a continuous function of the speed factors $\mathbf{\eta}_{X_1}, \ldots, \mathbf{\eta}_{X_n}$, then for any initial values $V_{X_1} \ldots, V_{X_n}$, and any value V with $\min(V_{X_1} \ldots, V_{X_n}) \leq V \leq \max(V_{X_1} \ldots, V_{X_n})$, values of speed factors $\mathbf{\eta}_{X_1}, \ldots, \mathbf{\eta}_{X_n}$ exist such that

$$\underline{\mathbf{V}} = \underline{\mathbf{eqf}}\left(\mathbf{\eta}_{X_1}, \ldots, \mathbf{\eta}_{X_n}, V_{X_1} \ldots, V_{X_n}\right) = V$$

Theorem 5 shows that both the initial values and the speed factors affect the common equilibrium value. Note that Theorem 5b) (iii) depends on the assumption that the common equilibrium value is a continuous function of the speed factors. This will depend on characteristics of the combination functions, but it is not clear by which types of combination functions this assumption is satisfied, in addition to being normalised, strictly monotonically increasing and scalar-free combination functions. Should they be continuous? Differentiable? With continuous partial derivatives which are bounded? Or smooth of a certain order, maybe even of infinite order? These are still open questions. However, as illustrated in Fig. 11.7, Theorem 5b) (iii) *at least has been confirmed in simulation experiments*.

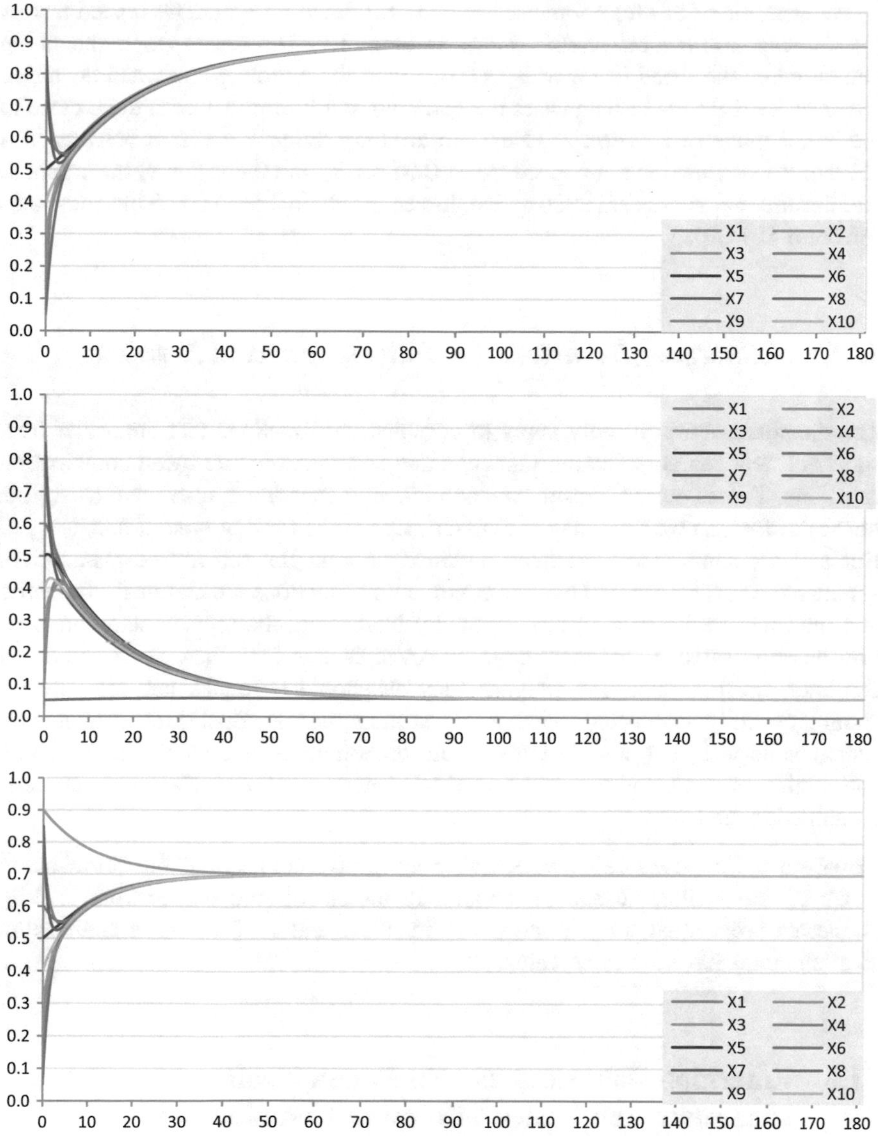

Fig. 11.7 How any value can become a common equilibrium value by using appropriate speed factors (here normalised scaled sum functions were used)

Here for any arbitrary choice for V by adapting values of speed factors in a smooth manner always such values could be determined such that V became the common equilibrium value $\underline{\mathbf{V}}$, thereby clearly showing experimentally that the function **eqf(..)** was continuous.

An example of this for normalised scaled sum functions where two speed factors (of X_3, resp. X_8) *are adapted is shown in* Fig. 11.7. The upper graph shows the pattern when the speed factor of X_3 is 0.001, and the middle graph when the speed factor of X_8 is 0.001. In both cases the common equilibrium value is quite close to the initial value of X_3 *resp.* X_8. Next, an arbitrary value $V = 0.7$ in between was chosen. It was found that for speed factor 0.04 for X_3 and 0.645 for X_8 the common equilibrium value was 0.700001. See lower graph in Fig. 11.7. This illustrates Theorem 5b) (iii).

11.5.5 Emerging Behaviour for Fully Connected Networks

The theorems above all only apply to combination functions that are scalar free, such as Euclidean combination functions and scaled geometric mean combination functions. The advanced logistic sum combination function is also often used; it is not scalar free, so we don't have discussed any results for that type of function yet. But at least some result has been obtained without the scalar free assumption, summarized in Theorem 6. Here the combination functions are assumed symmetric and (not necessarily strictly) monotonically increasing; the logistic sum combination function satisfies that condition. However, in this case there is an additional condition on the connection weights: they all should be equal, and the network should be fully connected. This result still applies to Euclidean combination functions as well, but, also to logistic sum combination functions, and maximum and minimum combination functions, which indeed are symmetric and monotonic combination functions.

Theorem 6 Suppose in a network with nonnegative connections the combination functions are symmetric and monotonically increasing and the network is fully connected with equal weights: $\boldsymbol{\omega}_{X,Y} = \boldsymbol{\omega}$ for all X and Y. Then in an equilibrium state all states have the same value.

11.6 Emerging Behaviour for Euclidean, Scaled Maximum and Scaled Minimum Combination Functions

In Sect. 11.3 example simulations were discussed only for scaled sum, advanced logistic sum and scaled geometric mean combination functions. In the current section also scaled maximum and minimum and Euclidean combination functions of different orders are discussed in some more depth. At the end a relationship between normalised Euclidean combination functions and normalised maximum combination functions is found in simulations and mathematically proven. The example used is the same as in Sect. 11.3 with characteristics and initial values shown in Box 11.1 and Table 11.2. Step size was $\Delta t = 0.25$.

11.6.1 Emerging Behaviour for Euclidean Combination Functions of Different Orders

In this section simulations for the Euclidean combination functions of varying order n are discussed, for higher values n = 2, 4, 10 and 100 (see Figs. 11.8 and 11.9), and for a lower value n = 0.0001 (see Fig. 11.12). The following theorem is for now only a conjecture, as a proof of it is complicated that increasing n leads to increasing outcomes.

Theorem 7 (strictly monotonically increasing trend of eucl(n, V_1, ..., V_k) for n) For equal state values between 0 and 1, the equilibrium values for normalised Euclidean combination functions have a strictly monotonically increasing trend as function of the order n.

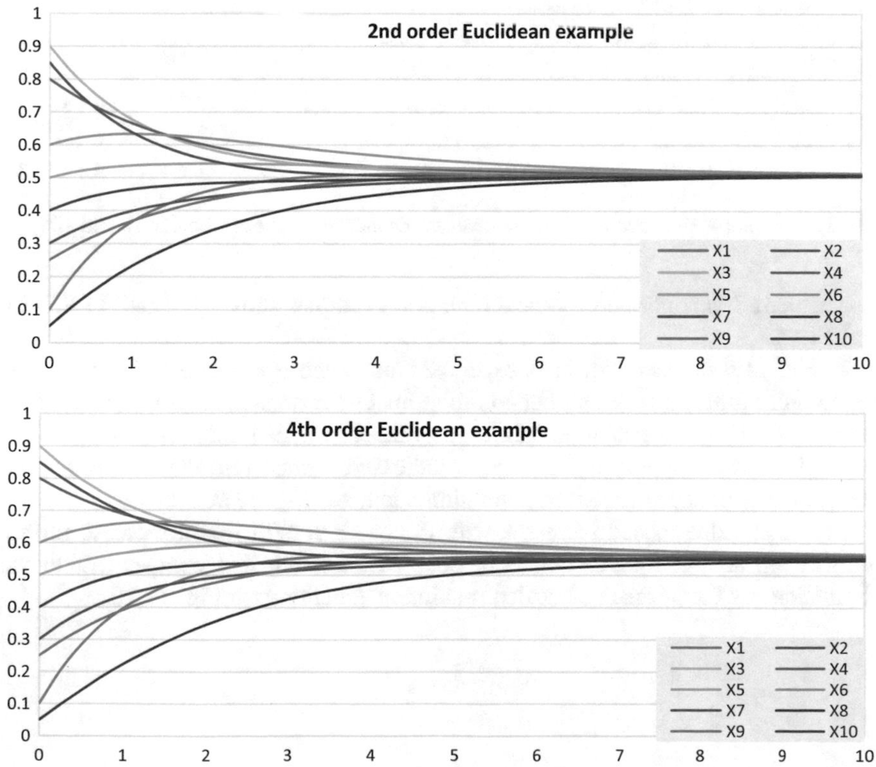

Fig. 11.8 Example simulations for the normalised Euclidean functions of order 2 and 4

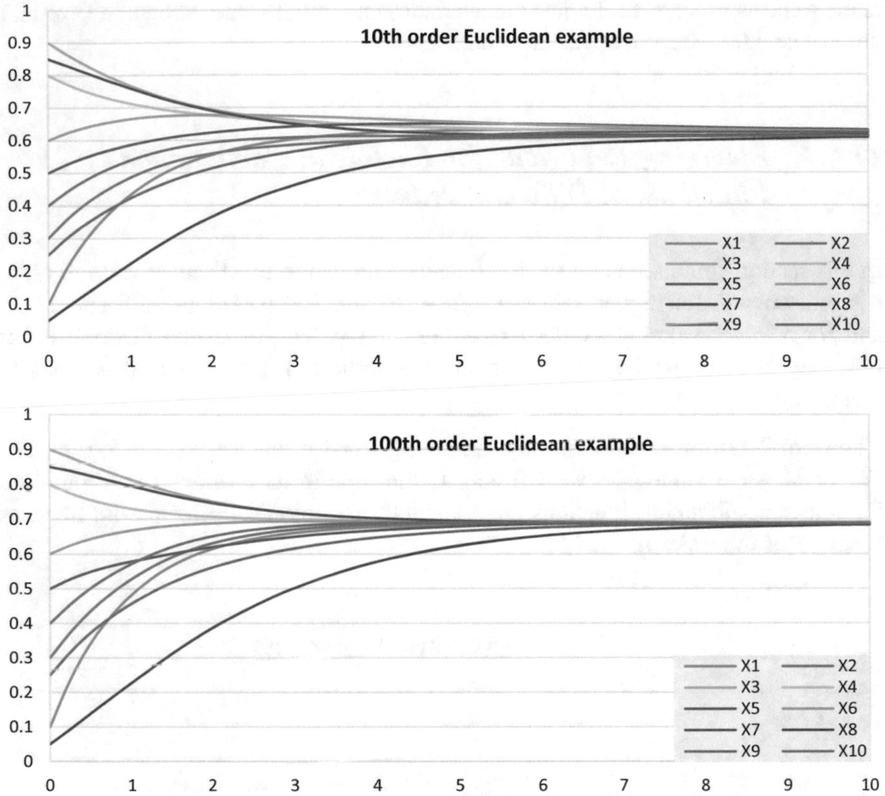

Fig. 11.9 Example simulations for the normalised Euclidean functions of order 10 and 100

Theorem 7 is confirmed by the simulation examples shown in Figs. 11.8, 11.9 and 11.12.

In Fig. 11.8 the behavior is as expected from Theorems 1 and 3. Note that the fourth order function gets a higher common final value than the second order, while the second order one gets almost the same value as the first order one in Fig. 11.3. To explore what happens if the order is increased further, also the 10th order and 100th order Euclidean functions were simulated; see Fig. 11.9.

The final value indeed increases with the order n. Will it still increase further until 1? An answer for this comes in Sect. 11.6.2 when it is compared with the simulation for the normalised scaled maximum function; see Fig. 11.10.

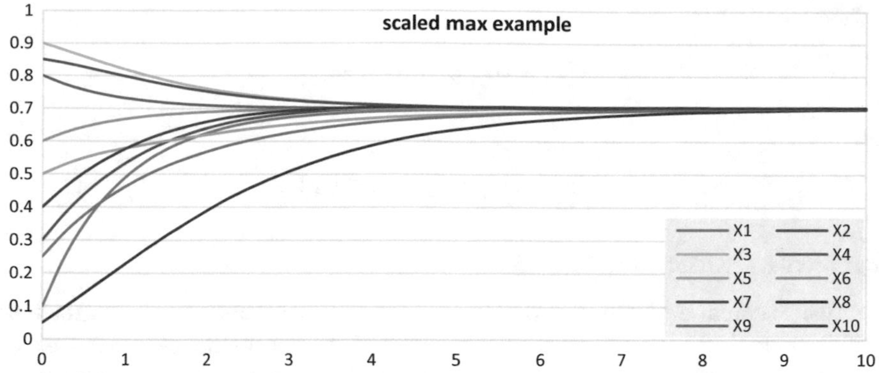

Fig. 11.10 Example simulation for the normalised scaled maximum function

11.6.2 Comparing Equilibrium Values for Euclidean Combination Functions and Scaled Maximum Combination Functions

Note that this pattern shown in Fig. 11.10 is not predicted by Theorem 3 or 4 as the scaled maximum function is not strictly monotonically increasing. But it also does not contradict these theorems as they do not formulate an if and only if relation.

Maybe a bit surprisingly, the pattern of the normalised maximum combination function is very similar to the pattern for the 100th order Euclidean combination function. Indeed, it has been found by mathematical proof that when the order n is increased, the nth order Euclidean function approximates the normalised scaled max function; see Theorem 8 below. To prove this, first in Lemma 3 a general mathematical relation between radical and max expressions is shown.

Lemma 3 (Relating radical and max expressions)
Suppose $a_1, .., a_k$ are any nonnegative real numbers. Then

$$\lim_{n\to\infty} \sqrt[n]{a_1^n + \cdots + a_k^n} = \max(a_1, \ldots, a_k)$$

Theorem 8 Let for each n the normalised Euclidean combination function $\mathbf{eucl}_{n,\lambda(n)}(V_1, \ldots, V_k)$ be given with normalising factor $\lambda(n)$, and let the normalised scaled maximum combination function $\mathbf{smax}_\lambda(V_1, \ldots, V_k)$ be given with scaling factor λ. Then for all $V_1, \ldots, V_k$ it holds

$$\lim_{n\to\infty} \mathbf{eucl}_{n,\lambda(n)}(V_1, \ldots, V_k) = \mathbf{smax}_\lambda(V_1, \ldots, V_k)$$

where

$$\lambda(n) = \omega^n_{X_1,Y} + \cdots + \omega^n_{X_k,Y}$$

and

$$\lambda = \max\left(\omega_{X_1,Y}, \ldots, \omega_{X_k,Y}\right)$$

For proofs, see Chap. 15, Sect. 15.6.

This Theorem 8 gives some kind of hint for why the graphs for the normalised 100th order Euclidean combination function and for the normalised scaled maximum combination function in Figs. 11.9 and 11.10 are very similar: they are almost the same combination function. It also suggests that for higher orders than 100 the final value will not become much higher, although Theorem 7 is no exact proof for this point.

Returning to the issue that the scaled maximum function does not fulfil the requirement of being strictly monotonically increasing, indeed there are cases in which the conclusions of Theorem 3 do not apply. An example is the slightly adjusted network shown in Box 11.2. Here this time the mutual connections between X_1, X_2, and X_3, and the mutual connections between X_8, X_9, and X_{10} all have weight 1. The other weights remain the same.

Box 11.2 Role matrices showing adjusted mutual connection weights for X_1, X_2, and X_3, and for X_8, X_9, and X_{10}

mb base connectivity	1	2	3	4	5	6	7	8	9
X_1	X_2	X_3	X_4	X_5	X_6	X_7	X_8	X_9	X_{10}
X_2	X_1	X_3	X_4	X_5	X_6	X_7	X_8	X_9	X_{10}
X_3	X_1	X_2	X_4	X_5	X_6	X_7	X_8	X_9	X_{10}
X_4	X_1	X_2	X_3	X_5	X_6	X_7	X_8	X_9	X_{10}
X_5	X_1	X_2	X_3	X_4	X_6	X_7	X_8	X_9	X_{10}
X_6	X_1	X_2	X_3	X_4	X_5	X_7	X_8	X_9	X_{10}
X_7	X_1	X_2	X_3	X_4	X_5	X_6	X_8	X_9	X_{10}
X_8	X_1	X_2	X_3	X_4	X_5	X_6	X_7	X_9	X_{10}
X_9	X_1	X_2	X_3	X_4	X_5	X_6	X_7	X_8	X_{10}
X_{10}	X_1	X_2	X_3	X_4	X_5	X_6	X_7	X_8	X_9

mcw connection weights	1	2	3	4	5	6	7	8	9
X_1	1	1	0.25	0.25	0.25	0.2	0.1	0.25	0.2
X_2	1	1	0.15	0.2	0.1	0.1	0.25	0.15	0.25
X_3	1	1	0.25	0.1	0.25	0.2	0.1	0.25	0.2
X_4	0.1	0.2	0.1	0.2	0.25	0.15	0.25	0.15	0.2
X_5	0.2	0.1	0.2	0.15	0.25	0.2	0.05	0.2	0.1
X_6	0.15	0.2	0.15	0.8	0.25	0.2	0.15	0.1	0.2
X_7	0.1	0.15	0.1	0.25	0.2	0.1	0.25	0.2	0.15
X_8	0.25	0.25	0.25	0.15	0.1	0.25	0.2	1	1
X_9	0.25	0.25	0.1	0.25	0.2	0.25	0.15	1	1
X_{10}	0.1	0.25	0.15	0.25	0.15	0.1	0.25	1	1

In Fig. 11.11 simulation results are shown for the normalised scaled maximum combination function. Now clustering takes place with the first group X_1, X_2, and X_3 ending up in value 0.5415, the second group X_8, X_9, and X_{10} in value 0.325, and the remaining group X_4, X_5, X_6 and X_7 in exactly 0.7. This shows that indeed

Theorem 3 is not applicable for scaled maximum combination functions (because that function is not strictly monotonous).

Note that for practical reasons of limited machine precision it is less straightforward to simulate this modified network for a 100th order Euclidean combination function. The reason is that the 100th powers of the single impacts $V_{i,j} = \omega_{Xi,Xj}\, X_j(t)$ from different states X_i on X_j are of very different order of magnitude. They may easily be a factor up to 10^{60} or more different, for example, and when added in $V_{1,j}{}^{100} + \cdots + V_{k,j}{}^{100}$ to form the aggregated impact

$$\sqrt[100]{\frac{V_{1,j}{}^{100} + \cdots + V_{k,j}{}^{100}}{\lambda}}$$

the contributions of the smaller terms are vanishing due to limited machine precision. So, then it is as if for X_1, X_2, and X_3, and for X_8, X_9, and X_{10} a network is simulated with only the connection weights 1 and the rest 0, so then clusters show up in an artificial manner. To simulate this in a correct manner, machine precision of at least 60 digits would be needed.

Returning to the Euclidean combination function and the original network connectivity depicted in Box 11.1, also very small positive real numbers for n can be explored. Figure 11.12 shows a simulation graph for $n = 0.0001$. It turns out that this graph is practically equal to the one for the normalised scaled minimum combination function. There may be a theorem similar to Theorem 8 but then for $n \to 0$ and the minimum operator to explain this. For now, this is left open.

All in all, it turns out that by varying the order n from (close to) 0 to ∞, the common equilibrium value for the states of the example network varies with n from around 0.4 to around 0.7, where these boundaries are the values reached for the normalised scaled minimum and maximum combination functions.

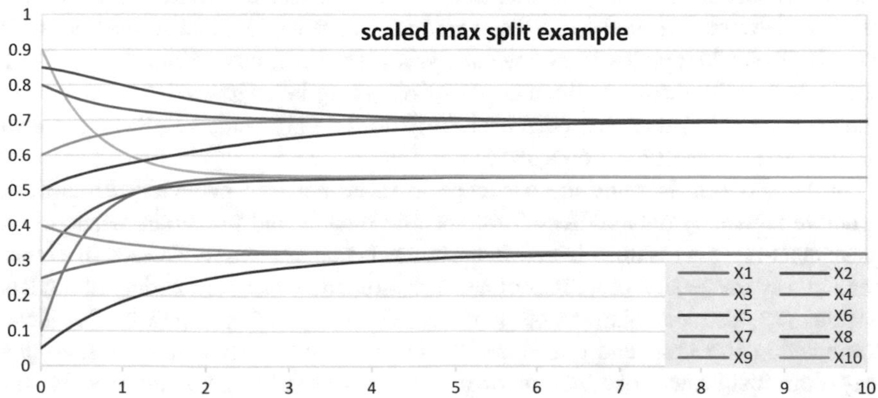

Fig. 11.11 Example simulation for the normalised scaled maximum function for the modified network of Box 11.2

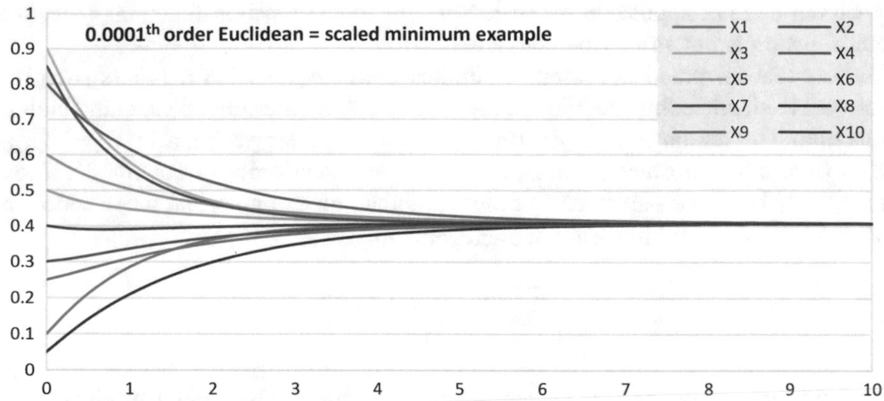

Fig. 11.12 Example simulation for the normalised Euclidean functions of order 0.0001, which turns out to be practically equal to the outcome for the normalised scaled minimum function

11.7 Discussion

Emerging behaviour of a network is a consequence of the characteristics of the network structure, although it can be challenging to find out how this relation between structure and behaviour exactly is. In this chapter a number of theorems and proofs were presented on how certain identified properties of network structure determine network behaviour, in particular concerning equilibrium values reached. Parts of this chapter were adopted from (Treur 2018a). The considered network structure characteristics include the main elements network *connectivity characteristics* (in how far other states of the network are reachable from a given state), network *aggregation characteristics* in terms of properties of combination functions used to aggregate multiple impacts on a single state, and network *timing characteristics* in terms of speed factors for the states. This challenge was addressed using a Network-Oriented Modeling approach based on temporal-causal network models (Treur 2016b, 2019), and the analysis they allow, as a vehicle; e.g., (Treur 2017). Within Mathematics the analysis of emerging behaviour of dynamical systems has a long history that goes back to (Picard 1891; Poincaré 1881); see also (Brauer and Nohel 1969; Lotka 1956).

In the network literature the challenge to relate network behaviour to network structure is usually only addressed for specific models and functions, where often these functions are assumed linear. In the current chapter it was addressed in a more general way for general properties of network structure covering a variety of models or functions, also including various nonlinear functions such as *n*th order Euclidean combination functions and scaled geometric mean combination functions. In this way extra insight was obtained in what properties exactly make that specific network structures lead to certain network behaviour.

As an example, a special case of Theorem 3, for one specific model with one specific combination function and speed factor, and just for fully connected networks appeared in (Bosse et al. 2015). In that paper it does not become clear what it exactly is that makes that the considered model generates that type of emerging behaviour. The much more general formulation presented and proven in Theorem 3 here is new. It shows that the structure-behaviour relation depends on the one hand on *connectivity* of the network, but on the other hand also on *aggregation* (the ways in which multiple impacts on a single state are aggregated), and on *timing* (how fast states respond on impact they receive). It has been found that aggregation structure properties for the combination functions such as strict monotonicity and being scalar-free are crucial. These properties define a relatively wide class of functions including linear (scaled sum) functions but mostly also nonlinear functions such as Euclidean functions and product-based (scaled) geometric mean functions. It also has been shown that networks with examples of combination functions not in this class, such as logistic sum combination functions (not scalar-free) and maximum-based combination functions (not strictly monotonic) do not show the same behaviour.

Theorems 4 and 5 are also new and explore the bounds and range for the equilibrium values reached. Theorem 6 first appeared in (Treur 2017) in the context of an adaptive network to model a specific preferential attachment principle, but actually covers the much more general setting as provided in the current chapter. In (Hendrickx and Tsitsiklis 2013) also analysis of emerging behaviour was addressed, but for a different class of networks.

Concerning connectivity, the work presented in the current chapter mainly addresses strongly connected networks. As a next step, more general networks have been addressed that may or may not be strongly connected. Analysis has been performed based on such a network's strongly connected components. Results have been found for any network, without any condition on strong connectivity; see Chap. 12 or (Treur 2018b).

In other further work, also for adaptive networks a similar challenge has been addressed: how do certain properties of adaptation principles lead to certain types of adaptive network behaviour. For example, adaptation based on a Hebbian learning principle [see Chap. 14 or (Treur 2018d)], or based on a homophily principle [see Chap. 13 or (Treur 2018c)] have been analysed, and properties of the combination functions used have been identified that lead to certain expected behaviour for the adaptive connection weights.

References

Bosse, T., Duell, R., Memon, Z.A., Treur, J., van der Wal, C.N.: Agent-based modeling of emotion contagion in groups. Cogn. Comput. **7**, 111–136 (2015)

Brauer, F., Nohel, J.A.: Qualitative Theory of Ordinary Differential Equations. Benjamin (1969)

Castellano, C., Fortunato, S., Loreto, V.: Statistical physics of social dynamics. Rev. Mod. Phys. **81**, 591–646 (2009)

Hendrickx, J.M., Tsitsiklis, J.N.: Convergence of type-symmetric and cut-balanced consensus seeking systems. IEEE Trans. Autom. Control **58**(1), 214–218. (2013) Extended version: arXiv:1102.2361v2[cs.SY]

Lotka, A.J.: Elements of Physical Biology. Williams and Wilkins Co. (1924), 2nd ed. Dover Publications (1956)

Pearl, J.: Causality. Cambridge University Press, Cambridge (2000)

Picard, E.: Traité d'Analyse, vol. 1 (1891), vol. 2 (1893)

Poincaré, H.: Mémoire sur les courbes défine par une équation différentielle (On curves defined by differential equations) (1881–1882) (1881)

Treur, J.: Verification of temporal-causal network models by mathematical analysis. Vietnam J. Comput. Sci. **3**, 207–221 (2016a)

Treur, J.: Network-Oriented Modeling: Addressing Complexity of Cognitive, Affective and Social Interactions. Springer Publishers (2016b)

Treur, J.: Modelling and analysis of the dynamics of adaptive temporal-causal network models for evolving social interactions. Comput. Soc. Netw. **4**, article 4, 1–20 (2017)

Treur, J.: Relating emerging network behaviour to network structure. In: Proceedings of the 7th International Conference on Complex Networks and their Applications, Complex Networks' 18, vol. 1. Studies in Computational Intelligence, vol. 812, pp. 619–634. Springer Publishers (2018a)

Treur, J.: Mathematical analysis of a network's asymptotic behaviour based on its strongly connected components. In: Proceedings of the 7th International Conference on Complex Networks and their Applications, Complex Networks' 18, vol. 1. Studies in Computational Intelligence, vol. 812, pp. 663–679. Springer Publishers (2018b)

Treur, J.: Relating an adaptive social network's structure to its emerging behaviour based on homophily. In: Proceedings of the 7th International Conference on Complex Networks and their Applications, Complex Networks' 18. Springer, vol. 2. Studies in Computational Intelligence, vol. 813, pp. 341–356. Springer Publishers (2018c)

Treur, J.: Relating an adaptive network's structure to its emerging behaviour for Hebbian learning. In: Martín-Vide, C., Vega-Rodríguez, M.A., Fagan, D., O'Neill, M. (eds.), Theory and Practice of Natural Computing: 7th International Conference, TPNC 2018. Lecture Notes in Computer Science, vol. 11324, pp. 359–373. Springer Publishers (2018d)

Treur, J.: The ins and outs of network-oriented modeling: from biological networks and mental networks to social networks and beyond. Trans. Comput. Collective Intell. **32**, 120–139. Keynote Lecture at ICCCI'18 (2019)

Chapter 12
Analysis of a Network's Emerging Behaviour via Its Structure Involving Its Strongly Connected Components

Abstract In this chapter, it is addressed how network structure can be related to network behaviour. If such a relation is studied, that usually concerns only strongly connected networks and only linear functions describing the aggregation of multiple impacts. In this chapter both conditions are generalised. General theorems are presented that relate emerging behaviour of a network to the network's structure characteristics. The network structure characteristics on the one hand concern network connectivity in terms of the network's strongly connected components and their mutual connections; this generalises the condition of being strongly connected (as addressed in Chap. 11) to a very general condition. On the other hand, the network structure characteristics considered concern aggregation by generalising from linear combination functions to any combination functions that are normalised, monotonic and scalar-free, so that many nonlinear functions are also covered (which also was done in Chap. 11). Thus the contributed theorems generalise existing theorems on the relation between network structure and network behaviour that only address specific cases (such as acyclic networks, fully and strongly connected networks, and theorems addressing only linear functions).

12.1 Introduction

In many cases, the relation between network structure and its emerging behaviour is only studied by performing simulation experiments. In this chapter, it is shown how within a certain context it is also possible to analyse mathematically how certain behaviours relate to certain properties of the network structure. For the network's structure, two types of characteristics are considered: (1) characteristics of the network's connectivity, and (2) characteristics of the aggregation in the network. In Chap. 11, the above question was only addressed for quite specific connectivity characteristics, in particular, for the case of an acyclic network, and the case of a strongly connected network. For these connectivity characteristics, the current chapter uses the general setting based on the strongly connected components of the network to develop a mathematical analysis for the general case. Tools were adopted from the area of Graph Theory, in particular, the manner to identify the

J. Treur, *Network-Oriented Modeling for Adaptive Networks: Designing Higher-Order Adaptive Biological, Mental and Social Network Models*, Studies in Systems, Decision and Control 251, https://doi.org/10.1007/978-3-030-31445-3_12

connectivity structure within a graph by decomposition of the graph according to its (maximal) strongly connected components and the resulting (acyclic) condensation graph (Harary et al. 1965), Chap. 3, and in addition the notion of stratification of an acyclic directed graph; e.g., Chen (2009).

Besides the connectivity characteristics of the network structure, the theorems presented here also take into account the aggregation characteristics of the network structure, by identifying relevant properties of the combination functions by which the impacts from multiple incoming connections are aggregated. It applies not to just one most simple type of (for example, linear) functions, but to a wider class of functions: those combination functions that are characterised as being monotonic, scalar-free and normalised. These properties of combination functions already turned out important in Chap. 11 for acyclic and strongly connected networks, and will also turn out to be important for the general case concerning the connectivity characteristics. This class of functions includes not only the often used linear functions, but also nonlinear functions such as nth order Euclidean combination functions and normalised scaled geometric mean functions.

The theorems explain which are the relevant characteristics that make that these combination functions contribute to certain behaviour when $t \to \infty$. It will be shown how using the above mentioned tools from Graph Theory, together with the aggregation characteristics of combination functions mentioned, enable to address the general case and obtain theorems about it. These theorems apply to arbitrary types of networks, but among the foci of application, in particular, are the types of example network models of which several are described in Treur (2016b):

(1) Mental Networks describing the dynamics of mental processes as the (usually cyclic) interaction of the mental states involved, and behaviour resulting from this,
(2) Social Networks describing social contagion processes for opinions, beliefs, emotions, for example,
(3) Integrative Networks that integrate (1) and (2).

Note that especially Mental Networks are often not strongly connected, although some parts may be. Typically they use sensory input that in general may not be affected by the behaviour, and because of that such input is not on any cycle of the network. Therefore they cannot be treated like strongly connected networks, but the theory developed here based on a decomposition by strongly connected components does apply (for applying the analysis to an example of such a Mental Network, see Sect. 12.7.2 below). Social Networks may often be strongly connected, but also in that case external nodes may be involved that affect them, which makes the whole network not strongly connected. Therefore for applicability on such types of networks the generalisation from strongly connected networks to general types of networks is important.

The foci of applicability on the three types of networks (1)–(3) mentioned above also makes that only addressing linear functions would be too limited. Especially for Mental Networks, often nonlinear functions are used. Therefore the challenge is also to stretch the type of analysis to at least certain types of nonlinear functions.

To apply the theorems introduced in this chapter to any given network, first the decomposition of the network into its strongly connected components is determined. Multiple efficient algorithms are available to determine these strongly connected components; e.g., see Bloem et al. (2006), Fleischer et al. (2000), Gentilini et al. (2003), Li et al. (2014), Tarjan (1972), Wijs et al. (2016), Lacki (2013). The connections between these components are identified, as represented in an acyclic condensation graph, and stratification of this graph is introduced. Based on this acyclic and stratified structure added to the original network, the theorems will show whether and which states within the network will end up in a common equilibrium value, and more in general determine bounds for the equilibrium values of the states.

The research presented here has been initiated from the angle of mathematical analysis and verification of network models in comparison to simulations for these models. For more background on this angle, see, for example, Treur (2016a) or Treur (2016b), Chap. 12. Like verification in Software Engineering is very useful for the quality of developed software e.g., Drechsler (2004), Fisher (2007), verification in network modeling is a useful means to get implementations of network models in accordance with the specifications of the models, and eliminate implementation errors. If a simulation of an implemented network model contradicts one or more of the results presented in the current chapter for the specification of the network model, then this pinpoints that something is wrong: a discrepancy between specification and implementation of the network model that needs to be addressed. Afterwards, it turned out that the contributions presented here also have some relations to research conducted from a different angle, namely on control of networks; e.g., Liu et al. (2011, 2012), Moschoyiannis et al. (2016), Haghighi and Namazi (2015), Karlsen and Moschoyiannis (2018). These relations will be discussed in the Discussion section.

In Sect. 12.2 the basic definition of network used is summarised. Section 12.3 discusses emerging behaviour, illustrated for an example network. Section 12.4 presents the definitions of the Graph Theory tools for the considered network connectivity characteristics; in Sect. 12.5 the identified aggregation characteristics in terms of combination functions are defined. In Sect. 12.6 the main theorems are formulated and it is pointed out how they were proven, thereby referring to Chap. 15, Sect. 15.7 for more complete proofs. In Sect. 12.7 more in-depth analysis is added, and in particular, applicability is illustrated for a type of network which is not a Social Network: a Mental Network describing sharing behaviour based on emotional charge. Section 12.8 is a final discussion.

12.2 Temporal-Causal Networks

This section describes the definition of the concept of network model used: temporal-causal network model. This is a notion of network that covers all types of discrete and smooth continuous dynamical systems, as has been shown in Treur (2017), building further, among others, on Ashby (1960) and Port and van Gelder (1995).

A temporal-causal network model is based on three notions defining the network structure characteristics: connection weight (Connectivity), combination function (Aggregation), and speed factor (Timing); see Table 12.1, upper part. Here the word temporal in temporal-causal refers to the causality. A library with a number (currently 35) of standard combination functions is available as options to choose from; but also own-defined functions can be used.

In Table 12.1, lower part it is shown how a conceptual representation of network structure defines a numerical representation of network dynamics; see also (Treur 2016b), Chap. 2, or (Treur 2019). Here $X_1, \ldots, X_k$ with $k \geq 1$ are the states from which state Y gets its incoming connections. This defines the detailed dynamic semantics of a temporal-causal network. Note that in the current chapter all connection weights are assumed nonnegative.

The difference equations in the last row in Table 12.1 can be used for simulation and mathematical analysis. They can also be written in differential equation format:

$$\mathbf{d}Y(t)/\mathbf{d}t = \boldsymbol{\eta}_Y[\mathbf{c}_Y(\boldsymbol{\omega}_{X_1,Y}X_1(t), \ldots, \boldsymbol{\omega}_{X_k,Y}X_k(t)) - Y(t)] \tag{12.1}$$

Table 12.1 Conceptual and numerical representations of a temporal-causal network

Concepts	Notation	Explanation
States and connections	X, Y, $X \rightarrow Y$	Describes the nodes and links of a network structure (e.g., in graphical or matrix format)
Connection weight	$\boldsymbol{\omega}_{X,Y}$	*Connection weight* $\boldsymbol{\omega}_{X,Y} \in [-1, 1]$ represents the strength of the impact of state X on state Y through connection $X \rightarrow Y$
Aggregating multiple impacts	$\mathbf{c}_Y(..)$	For each state Y a *combination function* $\mathbf{c}_Y(..)$ is chosen to combine the causal impacts of other states on state Y
Concepts	**Numerical representation**	**Explanation**
State values over time t	$Y(t)$	At each time point t each state Y has a real number value, usually in [0, 1]
Single causal impact	$\mathbf{impact}_{X,Y}(t) = \boldsymbol{\omega}_{X,Y}\, X(t)$	At t state X with connection to state Y has an impact on Y, using weight $\boldsymbol{\omega}_{X,Y}$
Aggregating multiple impacts	$\mathbf{aggimpact}_Y(t) = \mathbf{c}_Y(\mathbf{impact}_{X_1,Y}(t), \ldots, \mathbf{impact}_{X_k,Y}(t)) = \mathbf{c}_Y(\boldsymbol{\omega}_{X_1,Y}X_1(t), \ldots, \boldsymbol{\omega}_{X_k,Y}X_k(t))$	The aggregated impact of $k \geq 1$ states $X_1, \ldots, X_k$ on Y at t, is determined using combination function $\mathbf{c}_Y(..)$
Timing of the causal effect	$Y(t + \Delta t) = Y(t) + \boldsymbol{\eta}_Y [\mathbf{aggimpact}_Y(t) - Y(t)] \Delta t = Y(t) + \boldsymbol{\eta}_Y [\mathbf{c}_Y(\boldsymbol{\omega}_{X_1,Y}X_1(t), \ldots, \boldsymbol{\omega}_{X_k,Y}X_k(t)) - Y(t)] \Delta t$	The impact on Y is exerted over time gradually, using speed factor $\boldsymbol{\eta}_Y$

Note that combination functions usually are functions on the 0–1 interval within the real numbers: $[0, 1]^k \rightarrow [0, 1]$. Moreover, note that the condition $k \geq 1$ in Table 12.1 makes that by definition the above general format only applies to states Y with at least one incoming connection. However, in a network there also may be states Y without any incoming connection; for example, such states can serve as external input. Their dynamics can be specified in an independent manner by any mathematical function f: $[0, \infty) \rightarrow [0, 1]$ over time t:

$$Y(t) = f(t) \quad \text{for all} \quad t \tag{12.2}$$

Special cases of this are states Y with constant values over time, where for some constant c ∈ [0, 1] it holds $f(t) = c$ for all t. For such constant states, still the general format can be used as well, as long as the speed factor $\boldsymbol{\eta}_Y$ is set at 0 and the combination function is well-defined for zero arguments: then the general format reduces to $Y(t + \Delta t) = Y(t)$, and therefore the initial value is kept over time. But there are also other possible types of external input, for example, a repeated alternation of values 0 and 1 for some time intervals to model episodes in which a stimulus occurs and episodes in which it does not.

Examples of often used combination functions (see also Treur 2016b, Chap. 2, Table 2.10) are the following:

- The *identity* function **id**(.) for states with only one impact

$$\mathbf{id}(V) = V$$

- the *scaled sum* function $\mathbf{ssum}_\lambda$(..) with scaling factor $\boldsymbol{\lambda}$

$$\mathbf{ssum}_\lambda(V_1, \ldots, V_k) = \frac{V_1 + \cdots + V_k}{\boldsymbol{\lambda}}$$

- the scaled minimum function $\mathbf{smin}_\lambda$**(..)** with scaling factor $\boldsymbol{\lambda}$

$$\mathbf{smin}_\lambda(V_1, \ldots, V_k) = \frac{\min(V_1, \ldots, V_k)}{\lambda}$$

- the scaled maximum function **smax(..)** with scaling factor $\boldsymbol{\lambda}$

$$\mathbf{smax}_\lambda(V_1, \ldots, V_k) = \frac{\max(V_1, \ldots, V_k)}{\boldsymbol{\lambda}}$$

- the *simple logistic sum* combination function $\mathbf{slogistic}_{\sigma,\tau}$(..) with steepness $\boldsymbol{\sigma}$ and threshold $\boldsymbol{\tau}$, defined by

$$\mathbf{slogistic}_{\sigma,\tau}(V_1, \ldots, V_k) = \frac{1}{1 + e^{-\sigma(V_1 + \cdots + V_k - \tau)}}$$

- the *advanced logistic sum* combination function $\mathbf{alogistic}_{\sigma,\tau}(..)$ with steepness σ and threshold τ, defined by

$$\mathbf{alogistic}_{\sigma,\tau}(V_1,\ldots,V_k) = \left[\frac{1}{1+\mathrm{e}^{-\sigma(V_1+\cdots+V_k-\tau)}} - \frac{1}{1+\mathrm{e}^{\sigma\tau})}\right](1+\mathrm{e}^{-\sigma\tau})$$

- the *Euclidean combination function* of nth order with scaling factor λ (generalising the scaled sum $\mathbf{ssum}_{\lambda}(..)$ for $n = 1$) defined by

$$\mathbf{eucl}_{n,l}(V_1,\ldots,V_k) = \sqrt[n]{\frac{V_1^n + \cdots + V_k^n}{\lambda}}$$

  Here n can be any positive integer, or even any positive real number.
- the *scaled geometric mean combination function* with scaling factor λ

$$\mathbf{sgeomean}_{\lambda}(V_1,\ldots,V_k) = \sqrt[k]{\frac{V_1 * \ldots * V_k}{\lambda}}$$

For example, scaled minimum and maximum functions are often used in fuzzy logic inspired modelling and modeling uncertainty in AI, and the logistic sum functions are often used in neural network inspired modeling. The scaled sum functions, which are a special (linear) case of Euclidean functions, are often used in modeling of social networks. Geometric mean combination functions relate to product-based combination rules often used for probability-based approaches.

Recall from Chap. 11 the picture shown in Fig. 12.1. It also applies here. The basic relation between structure and dynamics is indicated by the horizontal arrow in the lower part. The upward arrows point at relevant properties of the structure and of the behaviour of the network. Relevant properties of the network structure are addressed in Sect. 12.4 (properties of the connectivity structure based on the network's strongly connected components) and Sect. 12.5 (properties of the aggregation structure based on combination functions). For behaviour, in particular, the equilibria that occur will be discussed. Section 12.3 presents basic definitions and shows examples of this. In Sect. 12.6, the main results are presented as depicted

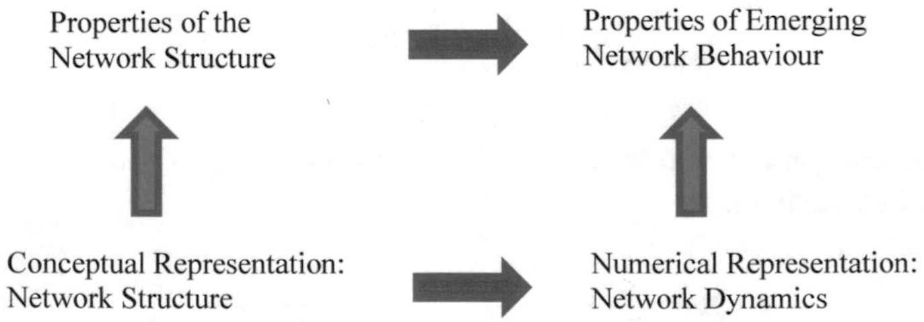

Fig. 12.1 Bottom layer: the conceptual representation defines the numerical representation. Top layer: properties of network structure entail properties of emerging network behaviour

by the upper horizontal arrow in Fig. 12.1. These results mostly have the form that certain network structure properties entail certain network behaviour properties.

12.3 Emerging Behaviour of a Network

Behaviour for $t \rightarrow \infty$ will be explored by analysing possible equilibria that can occur.

12.3.1 Basics on Stationary Points and Equilibria for Temporal-Causal Networks

Stationary points and equilibria are defined as follows.

Definition 1 (stationary point and equilibrium) A state *Y* has a *stationary point* at *t* if $\mathbf{d}Y(t)/\mathbf{d}t = 0$.
The network is in *equilibrium* at *t* if every state *Y* of the model has a stationary point at *t*.

Given the specific differential equation format for a temporal-causal network model the following criterion can be found:

Lemma 1 (Criterion for a stationary point in a temporal-causal network) Let *Y* be a state and X_1, … , X_k the states from which state *Y* gets its incoming connections. Then *Y* has a stationary point at *t* if and only if

$$\boldsymbol{\eta}_Y = 0 \quad \text{or} \quad \mathbf{c}_Y(\boldsymbol{\omega}_{X_1,Y}X_1(t), \ldots, \boldsymbol{\omega}_{X_k,Y}X_k(t)) = Y(t)$$

■

12.3.2 An Example Network

As an illustration the example network shown in Fig. 12.2 is used. The role matrices including the connection weights, speed factors, combination function weights, and combination function parameters, and the initial values used are shown in Box 12.1. The simulation for $\Delta t = 0.5$ is shown in Fig. 12.3.

Note that state X_1 has no incoming connections; in the simulation, it has initial value 0.9 and this stays constant at this level due to having speed factor 0. Also, X_5 has 0.9 as an initial value. The other states have an initial value 0. Note that in Sect. 12.6 theorems are presented from which it follows that the initial values of states X_2 to X_4 and X_8 to X_{10} are irrelevant for the emerging behavior as they do not have any effect on the final behaviour; therefore they were initially set at 0 here.

The speed factor of the states X_2 to X_{10} is 0.5. The combination function used is a normalised scaled sum function (see Sect. 12.5 for more details on normalisation).

12.3.3 Simulations for the Example Network

In the simulation shown in Fig. 12.3 states X_1 to X_4 all end up at value 0.9, states X_5 to X_7 all at value 0.3 and states X_8 to X_{10} at different individual values 0.681, 0.490, and 0.389, respectively. Overall, there is some clustering, but also some states get their unique value. It can be observed that these unique values are in between the cluster values. These observations will be confirmed by the mathematical analysis presented later.

Box 12.1 Role matrices and initial values for the example network shown in Fig. 12.2 as simulated in Fig. 12.3

mb **base connectivity**	1	2
X_1		
X_2	X_1	X_4
X_3	X_2	
X_4	X_1	X_3
X_5	X_6	
X_6	X_7	
X_7	X_5	
X_8	X_3	X_{10}
X_9	X_5	X_8
X_{10}	X_7	X_9

mcw **connection weights**	1	2
X_1		
X_2	0.8	0.6
X_3	1	
X_4	0.5	0.2
X_5	0.7	
X_6	0.8	
X_7	0.6	
X_8	0.8	0.6
X_9	0.8	0.8
X_{10}	0.8	0.7

ms **speed factors**	1
X_1	0
X_2	0.5
X_3	0.5
X_4	0.5
X_5	0.5
X_6	0.5
X_7	0.5
X_8	0.5
X_9	0.5
X_{10}	0.5

mcfw **combination function weights**	1 **eucl**
X_1	1
X_2	1
X_3	1
X_4	1
X_5	1
X_6	1
X_7	1
X_8	1
X_9	1
X_{10}	1

mcfp **function**	1 **eucl**	
parameter	1 n	2 λ
X_1	1	1
X_2	1	1.4
X_3	1	1
X_4	1	0.7
X_5	1	0.7
X_6	1	0.8
X_7	1	0.6
X_8	1	1.4
X_9	1	1.6
X_{10}	1	1.5

iv **initial values**	1
X_1	0.9
X_2	0
X_3	0
X_4	0
X_5	0.9
X_6	0
X_7	0
X_8	0
X_9	0
X_{10}	0

The combination function used for the simulation in Fig. 12.3 is the first order euclidean function (or scaled sum function), which is linear. It might be believed that this pattern depends on the combination function being linear. However, this is not

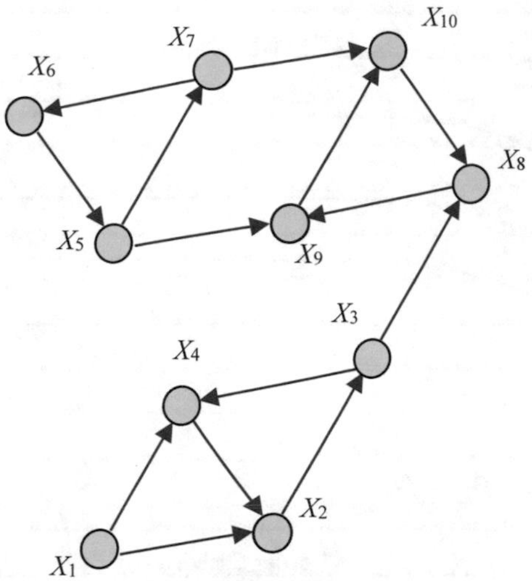

Fig. 12.2 Example network

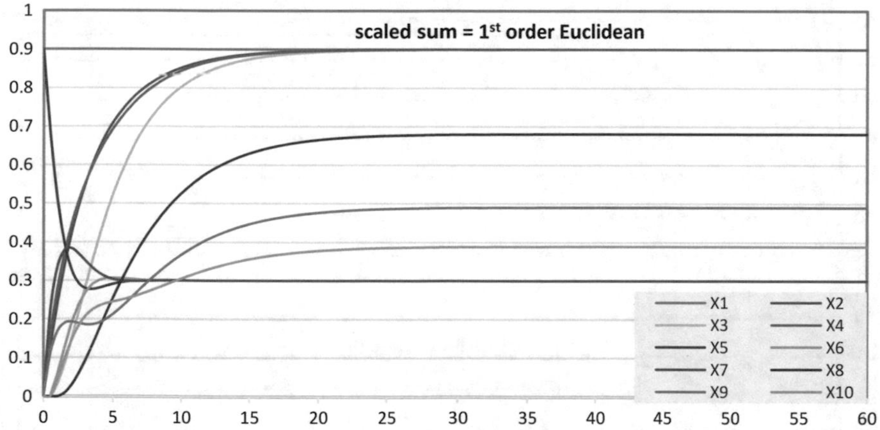

Fig. 12.3 Example simulation for linear scaled sum combination functions

the case. In Fig. 12.4 three simulations are shown for nonlinear combination functions, namely higher order Euclidean combination functions of order 2, 4 and 8, respectively. It is shown that the overall pattern is very similar with the same two groups going for 0.3 and 0.9, and the remaining three states X_8 to X_{10} getting each at different values but between these two values 0.3 and 0.9. The only difference is that the latter three values differ for the four considered combination functions, although they are in the same order. Note that in the graph for the 8th order Euclidean combination function state X_8, in the end, gets a value very close but not equal to 0.9.

The question of how such emerging asymptotic patterns can be explained will be addressed in the next three sections. It will be analysed how the pattern depends on

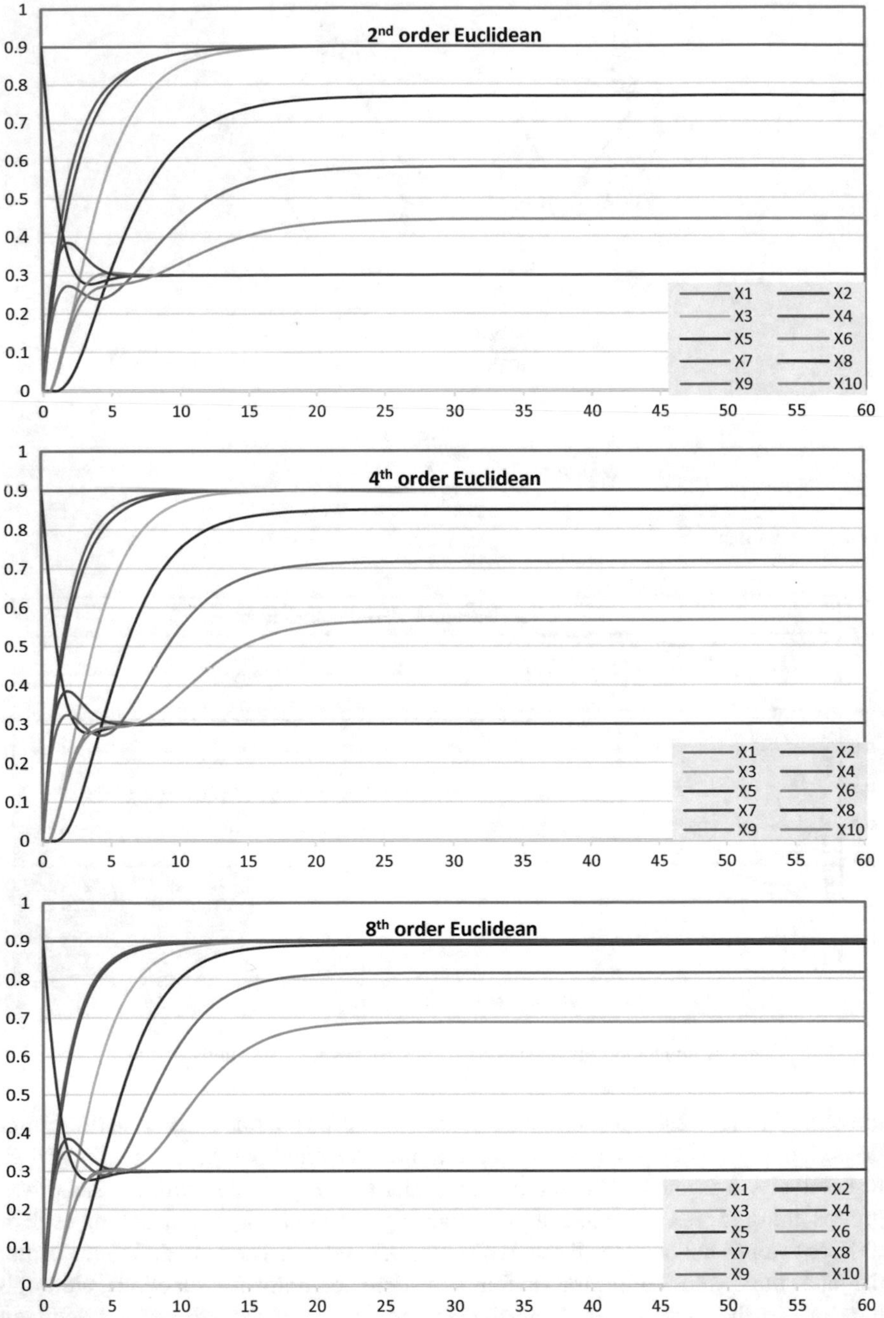

Fig. 12.4 Simulations for nonlinear higher order Euclidean combination functions of order 2, 4, and 8

the network's characteristics, in particular on the connectivity characteristics of the network and the aggregation characteristics modeled by the combination functions. Each of these two factors will be discussed first Sects. 12.4 and 12.5, respectively, after which in Sect. 12.6 they will be related to the emerging behaviour patterns.

12.4 Network Connectivity Characteristics Based on Strongly Connected Components

When broadening the scope of analysis for a wider class of network concerning connectivity characteristics, analysis based on the notion of strongly connected component is useful. Although it had to be rediscovered first, this is known from Graph Theory as turned out afterwards.

12.4.1 A Network's Strongly Connected Components

Most of the following definitions can be found, for example, in (Harary et al. 1965), Chap. 3, or in (Kuich 1970), Sect. 6. Note that here only nonnegative connection weights are considered.

Definition 2 (reachability and strongly connected components)

(a) State Y is *reachable* from state X if there is a directed path from X to Y with nonzero connection weights and speed factors.
(b) A network N is *strongly connected* if every two states are mutually reachable within N.
(c) A state is called *independent* if it is not reachable from any other state.
(d) A *subnetwork* of network N is a network whose states and connections are states and connections of N.
(e) A *strongly connected component* C of a network N is a strongly connected subnetwork of N such that no larger strongly connected subnetwork of N contains it as a subnetwork.

Strongly connected components C can be determined by choosing any node X of N and adding all nodes that are on any cycle through X. When a node X is not on any cycle, then it will form a singleton strongly connected component C by itself; this applies to all nodes of N with indegree or outdegree zero. Efficient algorithms have been developed to determine the strongly connected components of a graph; for example, see Bloem et al. (2006), Fleischer et al. (2000), Gentilini et al. (2003), Li et al. (2014), Tarjan (1972), Wijs et al. (2016). The strongly connected components of the example network from Fig. 12.2 are shown in Fig. 12.5.

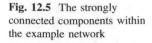

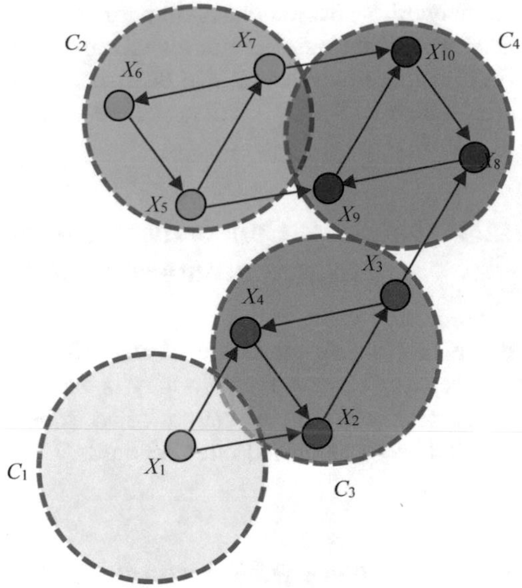

Fig. 12.5 The strongly connected components within the example network

12.4.2 *The Stratified Condensation Graph of a Network*

Based on the strongly connected components, a form of an abstracted picture of the network can be made, called the condensation graph; see Fig. 12.6.

Definition 3 (condensation graph) The *condensation* C(*N*) of a network *N* with respect to its strongly connected components is a graph whose nodes are the strongly connected components of *N* and whose connections are determined as follows: there is a connection from node C_i to node C_j in C(*N*) if and only if in *N* there is at least one connection from a node in the strongly connected component C_i to a node in the strongly connected component C_j.

A condensation graph C(*N*) is always an acyclic graph. The following theorem summarizes this; see also Harary et al. (1965), Chap. 3, Theorems 3.6 and 3.8, or Kuich (1970), Sect. 6.

Theorem 1 (acyclic condensation graph)

(a) For any network *N*, its condensation graph C(*N*) is acyclic and has at least one state of outdegree zero and at least one state of indegree zero.
(b) The network *N* is acyclic itself if and only if it is graph-isomorphic to C(*N*). In this case, the nodes in C(*N*) all are singleton sets {*X*} containing one state *X* from *N*.
(c) The network *N* is strongly connected itself if and only if C(*N*) only has one node; this node is the set of all states of *N*.

■

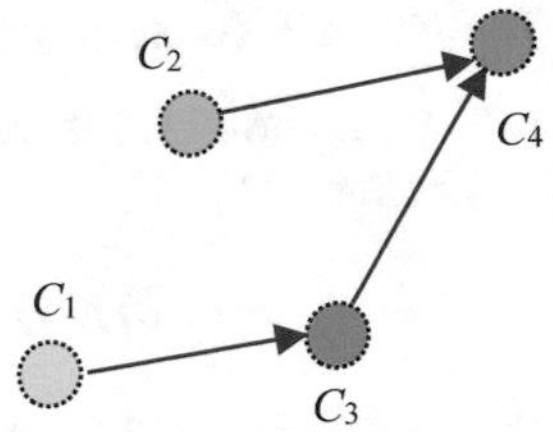

Fig. 12.6 Condensation of the example network by its strongly connected components: the directed acyclic condensation graph C(*N*)

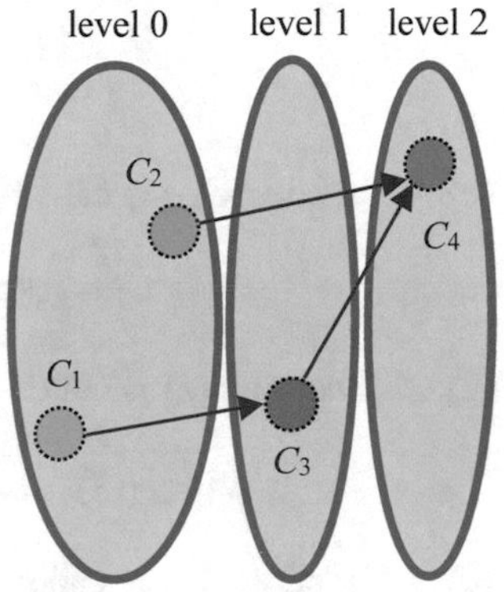

Fig. 12.7 Stratified condensation graph SC(*N*) for the example network

The structure of an acyclic graph is much simpler than the structure of a cyclic graph. For example, for any acyclic directed graph a stratification structure can be defined; e.g., Chen (2009). Here such construction is applied in particular to the condensation graph C(*N*) thus obtaining a stratified condensation graph SC (*N*) which will turn out very useful in Sect. 12.6; see Fig. 12.7.

Definition 4 (stratified condensation graph) The *stratified condensation graph* for network *N*, denoted by SC(*N*), is the condensation graph C(*N*) together with a leveled partition $S_0, \ldots, S_{h-1}$ in strata S_i such that $S_0 \cup \ldots \cup S_{h-1}$ is the set of all nodes of C(*N*) and the S_i are mutually disjoint, which is defined inductively as follows. Here, *h* is the height of C(*N*), i.e., the length of the longest path in C(*N*).

(i) The stratum S_0 is the set of nodes in C(*N*) without incoming connections in C(*N*).
(ii) For each $i > 0$ the stratum S_i is the set of nodes in C(*N*) for which all incoming connections in C(*N*) come only from nodes in $S_0, \ldots, S_{i-1}$.

If node *X* is in stratum S_i, its *level* is *i*.

12.5 Network Aggregation Characteristics Based on Properties of Combination Functions

The following network aggregation characteristics based on properties of combination functions have been found to relate to emerging behaviour as discussed in Sect. 12.3. Note that for combination functions it is (silently) assumed that $c(V_1, \ldots, V_k) = 0$ iff $V_i = 0$ for all *i*.

Definition 5 (monotonic, scalar-free, and additive for a combination function)

(a) A function c(..) is called *monotonically increasing* if for all values U_i, V_i it holds

$$U_i \leq V_i \quad \text{for all} \quad i \Rightarrow \mathrm{c}(U_1, \ldots, U_k) \leq \mathrm{c}(V_1, \ldots, V_k)$$

(b) A function c(..) is called *strictly monotonically increasing* if

$$U_i \leq V_i \quad \text{for all} \quad i, \text{and } U_j < V_j \quad \text{for at least one}$$
$$j \Rightarrow \mathrm{c}(U_1, \ldots, U_k) < \mathrm{c}(V_1, \ldots, V_k)$$

(c) A function c(..) is called *scalar-free* if for all $\alpha > 0$ and all $V_1, \ldots, V_k$ it holds

$$\mathrm{c}(\alpha V_1, \ldots, \alpha V_k) = \alpha\, \mathrm{c}(V_1, \ldots, V_k)$$

(d) A function c(..) is called *additive* if for all $U_1, \ldots, U_k$ and $V_1, \ldots, V_k$ it holds

$$\mathrm{c}(U_1 + V_1, \ldots, U_k + V_k) = \mathrm{c}(U_1, \ldots, U_k) + \mathrm{c}(V_1, \ldots, V_k)$$

(e) A function c(..) is called *linear* if it is both scalar-free and additive.

Note that these characteristics vary over the different examples of combination functions. Table 12.2 shows which of these characteristics apply to which combination functions. In general, the theorems that follow in Sect. 12.6 have the characteristics (a), (b) and (c) as conditions, so as can be seen in Table 12.2 they apply to **id(.)**, **ssum**$_\lambda$**(..)**, **eucl**$_{n,\lambda}$**(..)**, and **sgeomean**$_\lambda$**(..)** (of which only the first two are linear and the last two are nonlinear, assuming $n \neq 1$ for the third one and nonzero values for the fourth one). The theorems do not apply to **smin**$_\lambda$**(..)** and **smax**$_\lambda$**(..)** (not strictly monotonous), and to **slogistic**$_{\sigma,\tau}$**(..)**, and **alogistic**$_{\sigma,\tau}$**(..)** (not scalar-free). Note that different functions satisfying (a), (b) and (c) can also be combined to get more complex functions by using linear combinations with positive coefficients and function composition.

Table 12.2 Characteristics of Definition 5 for the example combination functions

	(a)	(b)	(c)	(d)	(e)
id(.)	+	+	+	+	+
ssum$_\lambda$**(..)** (= **eucl**$_{n,\lambda}$**(..)** for $n = 1$)	+	+	+	+	+
eucl$_{n,\lambda}$**(..)** for $n \neq 1$	+	+	+	−	−
sgeomean$_\lambda$**(..)** for nonzero values	+	+	+	−	−
smin$_\lambda$**(..)**	+	−	+	−	−
smax$_\lambda$**(..)**	+	−	+	−	−
slogistic$_{\sigma,\tau}$**(..)**	+	+	−	−	−
alogistic$_{\sigma,\tau}$**(..)**	+	+	−	−	−

Definition 6 (normalised) A network is *normalised* if for each state Y it holds $\mathbf{c}_Y(\boldsymbol{\omega}_{X_1,Y}, \ldots, \boldsymbol{\omega}_{X_k,Y}) = 1$, where $X_1, \ldots, X_k$ are the states from which Y gets its incoming connections.

As an example, for a Euclidean combination function of nth order the scaling parameter choice $\boldsymbol{\lambda}_Y = \boldsymbol{\omega}_{X_1,Y^n} + \cdots + \boldsymbol{\omega}_{X_k,Y^n}$ will provide a normalised network. This can be done in general as follows:

(1) **normalising a combination function**
If any combination function $\mathbf{c}_Y(..)$ is replaced by $\mathbf{c}'_Y(..)$ defined as

$$\mathbf{c}'_Y(V_1, \ldots, V_k) = \mathbf{c}_Y(V_1, \ldots, V_k)/\mathbf{c}_Y(\boldsymbol{\omega}_{X_1,Y}, \ldots, \boldsymbol{\omega}_{X_k,Y})$$

(note $\mathbf{c}_Y(\boldsymbol{\omega}_{X_1,Y}, \ldots, \boldsymbol{\omega}_{X_k,Y}) > 0$ since $\boldsymbol{\omega}_{X_i,Y} > 0$), then the network becomes normalised.

(2) **normalising the connection weights (for scalar-free combination functions)**
For scalar-free combination functions also normalisation is possible by adapting the connection weights; define $\boldsymbol{\omega}'_{X_i,Y} = \boldsymbol{\omega}_{X_i,Y} / \mathbf{c}_Y(\boldsymbol{\omega}_{X_1,Y}, \ldots, \boldsymbol{\omega}_{X_k,Y})$, then indeed it holds:

$$\mathbf{c}_Y(\boldsymbol{\omega}'_{X_1,Y}, \ldots, \boldsymbol{\omega}'_{X_k,Y}) = \mathbf{c}(\boldsymbol{\omega}_{X_1,Y}/c_Y(\boldsymbol{\omega}_{X_1,Y}, \ldots, \boldsymbol{\omega}_{X_k,Y}), \ldots, \boldsymbol{\omega}_{X_k,Y}/ \\ \mathbf{c}(\boldsymbol{\omega}_{X_1,Y}, \ldots, \boldsymbol{\omega}_{X_k,Y})) = 1$$

Normalisation is a necessary condition for applying the theorems developed in Sect. 12.6. Simulation is still possible when the network is not normalised. But the effect then usually is that activation is lost in an artificial manner (if the function values are lower than normalised) so that all values go to 0, or that activation is amplified in an artificial manner (if the function values are higher than normalised) so that all values go to 1. That makes less interesting behaviour for practical applications and also less interesting analysis.

For different example functions, following normalisation step (1) above, their normalised variants are given by Table 12.3.

Some of the implications of the above-defined characteristics are illustrated in the following proposition. This will be used in Sect. 12.6; for a proof, see Chap. 15, Sect. 15.7.

Proposition 1 Suppose the network is normalised.

(a) If the combination functions are scalar-free and $X_1, \ldots, X_k$ are the states from which Y gets its incoming connections, and $X_1(t) = \cdots = X_k(t) = V$ for some common value V, then also $\mathbf{c}_Y(\boldsymbol{\omega}_{X_1,Y}X_1(t), \ldots, \boldsymbol{\omega}_{X_k,Y}X_k(t)) = V$.

(b) If the combination functions are scalar-free and $X_1, \ldots, X_k$ are the states with outgoing connections to Y, and for $U_1, \ldots, U_k, V_1, \ldots, V_k$ and $\boldsymbol{\alpha} \geq 0$ it holds $V_i = \boldsymbol{\alpha} U_i$, then $\mathbf{c}_Y(\boldsymbol{\omega}_{X_1,Y}V_1, \ldots, \boldsymbol{\omega}_{X_k,Y}V_k) = \boldsymbol{\alpha}\, \mathbf{c}_Y(\boldsymbol{\omega}_{X_1,Y}U_1, \ldots, \boldsymbol{\omega}_{X_k,Y}U_k)$.
If in this situation in two different simulations, state values $X_i(t)$ and $X'_i(t)$ are generated then $X'_i(t) = \boldsymbol{\alpha} X_i(t) \quad \Rightarrow X'_i(t+\Delta t) = \boldsymbol{\alpha} X_i(t+\Delta t)$.

Table 12.3 Normalisation of the different examples of combination functions

Combination function	Notation	Normalising scaling factor	Normalised combination function
Identity function	**id(.)**	$\omega_{X,Y}$	$V/\omega_{X,Y}$
Scaled sum	$\mathbf{ssum}_\lambda(V_1, \ldots, V_k)$	$\omega_{X_1,Y} + \cdots + \omega_{X_k,Y}$	$(V_1 + \cdots + V_k)/(\omega_{X_1,Y} + \cdots + \omega_{X_k,Y})$
Scaled maximum	$\mathbf{smax}_\lambda(V_1, \ldots, V_k)$	$\max(\omega_{X_1,Y}, \ldots, \omega_{X_k,Y})$	$\max(V_1, \ldots, V_k)/\max(\omega_{X_1,Y}, \ldots, \omega_{X_k,Y})$
Scaled minimum	$\mathbf{smin}_\lambda(V_1, \ldots, V_k)$	$\min(\omega_{X_1,Y}, \ldots, \omega_{X_1,Y})$	$\min(V_1, \ldots, V_k)/\min(\omega_{X_1,Y}, \ldots, \omega_{X_k,Y})$
Euclidean	$\mathbf{eucl}_{n,\lambda}(V_1, \ldots, V_k)$	$\omega^n_{X_1,Y} + \cdots + \omega^n_{X_k,Y}$	$\sqrt[n]{\frac{V_1^n + \cdots + V_k^n}{\omega^n_{X_1,Y} + \cdots + \omega^n_{X_k,Y}}}$
Simple logistic	$\mathbf{slogistic}_{\sigma,\tau}(V_1, \ldots, V_k)$	$\mathbf{slogistic}_{\sigma,\tau}(\omega_{X_1,Y}, \ldots, \omega_{X_k,Y})$	$\frac{1+e^{-\sigma\omega(X_1,Y + \cdots + X_k,Y - \tau\omega)}}{1+e^{-\sigma(V_1 + \cdots + V_k - \tau)}}$
Advanced logistic	$\mathbf{alogistic}_{\sigma,\tau}(V_1, \ldots, V_k)$	$\mathbf{alogistic}_{\sigma,\tau}(\omega_{X_1,Y}, \ldots, \omega_{X_k,Y})$	$\frac{\frac{1}{1+e^{-\sigma(V_1 + \cdots + V_k - \tau)}} - \frac{1}{1+e^{\sigma\tau}}}{\frac{1}{1+e^{-\sigma\omega(X_1,Y + \cdots + X_k,Y - \tau\omega)}} - \frac{1}{1+e^{\sigma\tau}}}$

(c) If the combination functions are additive and $X_1, \ldots, X_k$ are the states from which Y gets its incoming connections, then for values $U_1, \ldots, U_k, V_1, \ldots, V_k$ it holds

$$\mathbf{c}_Y(\omega_{X_1,Y}(U_1+V_1), \ldots, \omega_{X_k,Y}(U_k+V_k)) = \mathbf{c}_Y(\omega_{X_1,Y}U_1, \ldots, \omega_{X_k,Y}U_k) + \mathbf{c}_Y(\omega_{X_1,Y}V_1, \ldots, \omega_{X_k,Y}V_k)$$

If in this situation in three different simulations, state values $X_i(t)$, $X'_i(t)$ and $X'_i(t)$ are generated then

$$X''_i(t) = X_i(t) + X'_i(t) \quad \Rightarrow \quad X''_i(t+\Delta t) = X_i(t+\Delta t) + X'_i(t+\Delta t)$$

(d) If the combination functions are scalar-free and monotonically increasing, and $X_1, \ldots, X_k$ are the states from which Y gets its incoming connections, and $V_1 \leq X_1(t), \ldots, X_k(t) \leq V_2$ for some values V_1 and V_2, then also

$$V_1 \leq \mathbf{c}_Y(\omega_{X_1,Y}X_1(t), \ldots, \omega_{X_k,Y}X_k(t)) \leq V_2$$

and if $\boldsymbol{\eta}_Y \Delta t \leq 1$ and $V_1 \leq Y(t) \leq V_2$ then $V_1 \leq Y(t + \Delta t) \leq V_2$.

12.6 Network Behaviour and Network Structure Characteristics

How the network structure characteristics concerning connectivity and aggregation as discussed in Sects. 12.4 and 12.5 relate to emerging network behaviour is discussed in this section.

12.6.1 Network Behaviour for Special Cases

As a first case, a network without cycles is considered. The following theorem has been proven using Lemma 1 from Sect. 12.3 and Proposition 1; see also Chap. 11, Theorem 1, or Treur (2018a).

Theorem 2 (common state values provide equilibria) Suppose a network with nonnegative connections is based on normalised and scalar-free combination functions, and the states without any incoming connection have a constant value. Then the following holds.

(a) Whenever all states have the same value *V*, the network is in an equilibrium state.
(b) If for every state for its initial value *V* it holds $V_1 \leq V \leq V_2$, then for all *t* for every state *Y* it holds $V_1 \leq Y(t) \leq V_2$. In an achieved equilibrium for every state for its equilibrium value *V* it holds $V_1 \leq V \leq V_2$.

■

Also this theorem is adopted from Chap. 11, Theorem 2.

Theorem 3 (Common equilibrium state values; acyclic case) Suppose an acyclic network with nonnegative connections is based on normalised and scalar-free combination functions.

(a) If in an equilibrium state the independent states all have the same value *V*, then all states have the same value *V*.
(b) If, moreover, the combination functions are monotonically increasing, and in an equilibrium state the independent states all have values *V* with $V_1 \leq V \leq V_2$, then all states have values *V* with $V_1 \leq V \leq V_2$.

■

The following is a useful basic lemma for dynamics of normalised networks with combination functions that are (strictly) monotonically increasing and scalar-free.

Lemma 2 Let a normalised network with nonnegative connections be given and its combination functions are monotonically increasing and scalar-free; then the following hold:

(a)

(i) If for some node *Y* at time *t* for all nodes *X* with $\boldsymbol{\omega}_{X,Y} > 0$ it holds $X(t) \leq Y(t)$, then *Y*(*t*) is decreasing at *t*: $\mathbf{d}Y(t)/\mathbf{d}t \leq 0$.
(ii) If the combination functions are strictly increasing and at time *t* for all nodes *X* with $\boldsymbol{\omega}_{X,Y} > 0$ it holds $X(t) \leq Y(t)$, and a node *X* exists with $X(t) < Y(t)$ and $\boldsymbol{\omega}_{X,Y} > 0$, and the speed factor of *Y* is nonzero, then *Y*(*t*) is strictly decreasing at *t*: $\mathbf{d}Y(t)/\mathbf{d}t < 0$.

(b)

(i) If for some node Y at time t for all nodes X with $\omega_{X,Y} > 0$ it holds $X(t) \geq Y(t)$, then $Y(t)$ is increasing at t: $\mathbf{d}Y(t)/\mathbf{d}t \geq 0$.
(ii) If, the combination function is strictly increasing and at time t for all nodes X with $\omega_{X,Y} > 0$ it holds $X(t) \geq Y(t)$, and a node X exists with $X(t) > Y(t)$ and $\omega_{X,Y} > 0$, and the speed factor of Y is nonzero, then $Y(t)$ is strictly increasing at t: $\mathbf{d}Y(t)/\mathbf{d}t > 0$.

■

The following theorem has been proven for strongly connected networks with cycles using Lemma 1 and 2; see Chap. 11 (Lemma 2 and Theorem 3) or Treur (2018a).

Theorem 4 (Common equilibrium state values; strongly connected cyclic case) Suppose the combination functions of the normalised network *N* are scalar-free and strictly monotonically increasing. Then the following hold.

(a) If the network is strongly connected itself, then in an equilibrium state all states have the same value.
(b) Suppose the network has one or more independent states and the subnetwork without these independent states is strongly connected. If in an equilibrium state all independent states have values V with $V_1 \leq V \leq V_2$, then all states have values V with $V_1 \leq V \leq V_2$. In particular, when all independent states have the same value V, then all states have this same equilibrium value V.

■

12.6.2 *Network Behaviour for the General Case*

The first general, main theorem is formulated by Theorems 5 and 6.

Theorem 5 (main theorem on equilibrium state values, part I) Suppose the network N is normalised and its combination functions are scalar-free and strictly monotonic. Let SC(*N*) be the stratified condensation graph of *N*. *Then in an equilibrium state of N the following hold.*

(a) Suppose $C \in$ SC(*N*) is a strongly connected component of *N* of level 0, and in case it consists of a single state without any incoming connection, this state has a constant value. Then the following hold:

   (i) All states in *N* belonging to *C* have the same equilibrium value *V*.
   (ii) If for the initial values *V* of all states in *N* belonging to *C* it holds $V_1 \leq V \leq V_2$, then also for the equilibrium values *V* of all states in *C* it holds $V_1 \leq V \leq V_2$.

(iii) In particular, when all initial values of states in N belonging to C are equal to one value V, then the equilibrium value of all states in C is also V.

(b) Let $C \in \text{SC}(N)$ be a strongly connected component of N of level i > 0. Let $C_1, \ldots, C_k \in \text{SC}(N)$ be the strongly connected components of N from which C gets an incoming connection within the condensation graph SC(N). Then the following hold.

(i) If for the equilibrium values V of all states in N belonging to $C_1 \cup \cdots \cup C_k$ it holds $V_1 \leq V \leq V_2$, then for all states in N belonging to C for their equilibrium value V it holds $V_1 \leq V \leq V_2$.

(ii) In particular, when all equilibrium values of all states in N belonging to $C_1 \cup \cdots \cup C_k$ are equal to one value V, then also the equilibrium values of all states in N belonging to C are equal to the same V.

Proof

(a)

(i) follows from Theorem 3(a).
(ii) follows from Proposition 1(b).
(iii) This follows from (ii) with $V_1 = V_2 = V$.

(b)

(i) This follows from Theorem 3(b) applied to C augmented with (as independent states) the states in $C_1 \cup \ldots \cup C_k$ with outgoing connections to states in C, with their values and these connections.
(ii) follows from (i) with $V_1 = V_2 = V$.

■

Theorem 6 (main theorem on equilibrium state values, part II) Suppose the network N is normalised and its combination functions are scalar-free and strictly monotonic. Let SC(N) be the stratified condensation graph of N. Then in an equilibrium state of N the following hold.

(a) If the equilibrium values of all states in every strongly connected component of level 0 in SC(N) are equal to one value V, then the equilibrium state values of all states in N are equal to the same value V.
(b) If for the equilibrium values V of all states in every strongly connected component of level 0 in SC(N) it holds $V_1 \leq V \leq V_2$, then for the equilibrium state values V of all states in N it holds $V_1 \leq V \leq V_2$.
(c) Suppose the states without any incoming connection have a constant value. If the initial values of all states in every strongly connected component of level 0 in SC(N) are equal to one value V, then for the equilibrium state values of all states in N are equal to the same value V.

(d) Suppose the states without any incoming connection have a constant value. If for the initial values V of all states in every strongly connected component of level 0 in SC(N) it holds $V_1 \leq V \leq V_2$, then for the equilibrium state values V of all states in N it holds $V_1 \leq V \leq V_2$.

Proof This follows by using induction over the number of strata in SC(N) and applying Theorem 4(a) for the level 0 stratum and Theorem 4(b) for the induction step from the strata of level $j < i$ to the stratum of level $i > 0$.

■

As an illustration, for the example simulation, the following implications of these theorems can be found.

- **Level 0 components**
  The strongly connected components of level 0 are the subnetworks based on $\{X_1\}$ and $\{X_5, X_6, X_7\}$ (see Figs. 12.5 and 12.6). As shown in Box 12.1, the initial values of X_1 and X_5 are 0.9, and the initial values for all other states are 0. From Theorem 5(a)(i) and 4(a)(ii), it follows that the equilibrium value of X_1 is 0.9, which indeed is the case, and those of X_5, X_6, X_7 are the same and $\leq$ 0.9; this is indeed confirmed in Fig. 12.3, as these three equilibrium values of X_5, X_6, X_7 are all 0.3. This value 0.3 depends on the initial values of the states and the connection weights, which are not taken into account in the theorems; however, see also Theorem 7 below.
- **Level 1 component**
  For the level 1 component C_3, based on $\{X_2, X_3, X_4\}$, it goes as follows. The only incoming connection for C_3 is from X_1, which has equilibrium value 0.9 (implied by Theorem 5(a)(ii)). By Theorem 5(b)(ii) it follows that X_2, X_3, X_4 all have the same equilibrium value 0.9; this is indeed confirmed in Fig. 12.3.
- **Level 2 component**
  The level 2 component C_4 is based on $\{X_8, X_9, X_{10}\}$. It has two incoming connections, one from X_3 in C_3 and one from X_5 in C_2. Their equilibrium values are 0.9 and 0.3, respectively, so they are not equal. Therefore the above theorems do not imply that the equilibrium values of X_8, X_9, X_{10} are the same; indeed in Fig. 12.3 they are different: 0.681, 0.490, and 0.389, respectively. But there is still an implication from Theorem 5(b)(i), namely, that these equilibrium values should be $\geq$ 0.3 and $\leq$ 0.9. This is indeed confirmed in Fig. 12.3.

This illustrates how the above theorems have implications for simulations. Note that the specific equilibrium values 0.681, 0.490, and 0.389 are not predicted here. They also depend on the connection weights for the states X_8, X_9, X_{10} within component C_4, and these are not taken into account in the theorems; however see also below, in the last part of this section.

Consider a variation, by setting the initial value of X_1 at 0.3 instead of 0.9. Then all equilibrium values turn out to become the same 0.3; see Fig. 12.8. Now the values of all states in the level 0 components C_1 and C_2 have the same value 0.3. As above, also the states in C_3 have the equilibrium value 0.3 because they are only

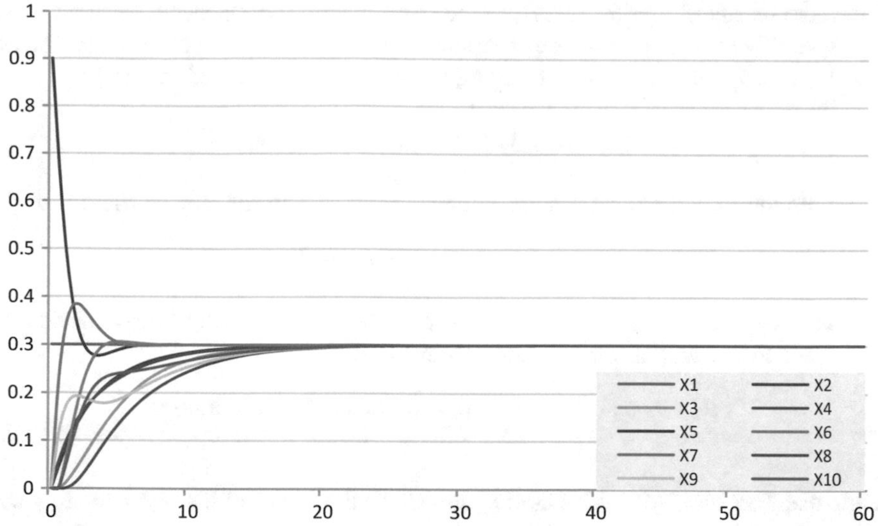

Fig. 12.8 Variation of the example simulation for initial value 0.3 of X_1

affected by X_1 which has value 0.3. But now the equilibrium values of both X_3 in C_3 and X_5 in C_2 are the same 0.3, so this time Theorem 5(b)(ii) can be applied to derive that all states in C_4 also have that same equilibrium value 0.3.

This predicts that all states of the network have value 0.3 in the equilibrium. Alternatively, Theorem 6(a) can be applied for this case. By that theorem from the equal equilibrium values in the level 0 components C_1 and C_2 it immediately follows that all states in all components in the network have that same equilibrium value.

As seen above, in the theorems the level 0 components play a central role, as initial nodes in the stratified condensation graph SC(N). Therefore it can be useful to know more about them, for example, how their initial values determine all equilibrium values in the network. This is addressed for the case of a linear combination function in the following theorem. For a proof, see Chap. 15, Sect. 15.7.

Theorem 7 (equilibrium state values in relation to level 0 components in the linear case) Suppose the network N is normalised and the combination functions are strictly monotonically increasing and linear. Assume that the states at level 0 that form a singleton component on their own are constant.
Then the following hold:

(a) For each state Y its equilibrium value is independent of the initial values of all states at some level $i > 0$. It is only dependent on the initial values for the states at level 0.

(b) More specifically, let $B_1, \ldots, B_p$ be the states in level 0 components. Then for each state Y its equilibrium value eq_Y is described by a linear function of the initial values $V_1, \ldots, V_p$ for $B_1, \ldots, B_p$, according to the following weighted average:

$$\text{eq}_Y(V_1, \ldots, V_p) = d_{B_1,Y}V_1 + \cdots + d_{B_p,Y}V_p$$

Here the $d_{B_i,Y}$ are real numbers between 0 and 1 and the sum of them is 1:

$$d_{B_1,Y} + \cdots + d_{B_p,Y} = 1$$

(c) Each $d_{B_i,Y}$ is the equilibrium value for Y when the following initial values are used: $V_i = 1$ and all other initial values are 0:

$$d_{B_i,Y} = \text{eq}_Y(0, \ldots, 0, 1, 0, \ldots, 0) \text{ with 1 as } i\text{th argument.}$$

Note that Theorem 7(c) can be used to determine the values of the numbers $d_{B_i,Y}$ by simulation for each of these p initial value settings. However, in Sect. 12.7 it will also be shown how they can be determined by symbolically solving the equilibrium equations. Based on Theorem 7, for the case of linear combination functions, for level 0 components after each value $d_{B_i,Y}$ is determined, any equilibrium value can be predicted from the initial values by the identified linear expression.

Note that for the case of linear combination functions the equilibrium equations are linear and could be solved algebraically. But this does not provide additional information for nonsingleton level 0 components. They have an infinite number of solutions as every common value V is a solution; apparently, the linear equations always have a mutual dependency in this case. However, for components of level $i > 0$, solving the linear equations can provide specific values, due to the specific input values they get from one or more lower level components. In Sect. 12.7 such implications of the theorems for some example networks are shown. The next theorems show some variations on Theorem 7.

Theorem 8 (equilibrium state values for level 0 components) Suppose the network N with states $X_1, \ldots, X_n$ is normalised and strongly connected. Then the following hold.

(a) If the combination functions of the network N are scalar-free, then for given connection weights and speed factors, for any value $V \in [0, 1]$ there are initial values such that V is the common state value in an equilibrium achieved from these initial values.

(b) For given connection weights and speed factors, let eq: $[0, 1]^n \rightarrow [0, 1]$ be the function such that $\text{eq}(V_1, \ldots, V_n)$ is the common state value for an equilibrium achieved from initial values $X_i(0) = V_i$ for all i. Then $\text{eq}(0, \ldots, 0) = 0$, $\text{eq}(1, \ldots, 1) = 1$, and the following hold:

   (i) If the combination functions of the network are scalar-free, then eq is scalar-free

(ii) If the combination functions of the network are additive, then eq is additive.

(c) Suppose the combination functions of the network N are linear. For given connection weights and speed factors for each i let e_i be the achieved common equilibrium value for initial values $X_i(0) = 1$ and $X_j(0) = 0$ for all $j \neq i$, *i.e.*, $e_i = \text{eq}(0, \ldots, 0, 1, 0, \ldots, 0)$ with 1 as ith argument. Then the sum of the e_i *is* 1, *i.e.*, $e_1 + \cdots + e_n = 1$ and in the general case for these given connection weights and speed factors, the common equilibrium value eq(…) is a linear, monotonically increasing, continuous and differentiable function of the initial values $V_1, \ldots, V_n$ satisfying the following linear relation:

$$\text{eq}(V_1, \ldots, V_n) = e_1 V_1 + \ldots + e_n V_n$$

If the combination functions of N are strictly increasing, then $e_i > 0$ for all i, and eq is also strictly increasing.

Proof (a) This follows from Proposition 1(a) or (d) with $V_1 = V_2 = V$.
(b) and (c) This follows from Proposition 1(b) and (c), and Lemma 2. ■

For a proof of the following theorem, see Chap. 15, Sect. 15.7.

Theorem 9 (equilibrium state values for components of level $i > 0$**)** Suppose the network is normalised, and consists of a strongly connected component plus a number of independent states $A_1, \ldots, A_p$ with outgoing connections to this strongly connected component. Then the following hold

(a) Suppose the combination functions are scalar-free and $X_1, \ldots, X_k$ are the states from which Y gets its incoming connections. If for $U_1, \ldots, U_k, V_1, \ldots, V_k$ and $\alpha \geq 0$ it holds $V_i = \alpha U_i$ for all i, then $c_Y(\omega_{X_1,Y}V_1, \ldots, \omega_{X_k,Y}V_k) = \alpha\, c_Y(\omega_{X_1,Y}U_1, \ldots, \omega_{X_k,Y}U_k)$

(b) Suppose the combination functions are additive and $X_1, \ldots, X_k$ are the states from which Y gets its incoming connections. Then if for values $U_1, \ldots, U_k, V_1, \ldots, V_k, W_1, \ldots, W_k$ it holds $W_i = U_i + V_i$ for all i, then

$$\mathbf{c}_Y(\omega_{X_1,Y}W_1, \ldots, \omega_{X_k,Y}W_k) = \mathbf{c}_Y(\omega_{X_1,Y}U_1, \ldots, \omega_{X_k,Y}U_k) + \mathbf{c}_Y(\omega_{X_1,Y}V_1, \ldots, \omega_{X_k,Y}V_k)$$

(c) Suppose all combination functions of the network N are linear. Then for given connection weights and speed factors, for each state Y the achieved equilibrium value for Y only depends on the equilibrium values $V_1, \ldots, V_p$ of states $A_1, \ldots, A_p$; the function $\text{eq}_Y(V_1, \ldots, V_p)$ denotes this achieved equilibrium value for Y.

(d) Suppose the combination functions of the network N are linear. For the given connection weights and speed factors for each i let $d_{i,Y}$ be the achieved equilibrium value for state Y in a situation with equilibrium values $A_i = 1$ and $A_j = 0$ for all $j \neq i$, i.e., $d_{i,Y} = \text{eq}_Y(0, \ldots, 0, 1, 0, \ldots, 0)$ with 1 as ith argument.

Then in the general case for these given connection weights and speed factors, for each Y in the strongly connected component its equilibrium value is a linear, monotonically increasing, continuous and differentiable function $\text{eq}_Y(\ldots)$ of the equilibrium values $V_1, \ldots, V_p$ of $A_1, \ldots, A_p$ satisfying the following linear relation: $\text{eq}_Y(V_1, \ldots, V_p) = d_{1,Y}\, V_1 + \cdots + d_{p,Y}\, V_p$. Here the sum of the $d_{i,Y}$ *is* 1: $d_{1,Y} + \cdots + d_{p,Y} = 1$. In particular, the equilibrium values are independent of the initial values for all states Y different from $A_1, \ldots, A_p$. If the combination functions of N are strictly increasing, then $d_{i,Y} > 0$ for all i, and $\text{eq}_Y(..)$ is also strictly increasing.

Note that by using Theorem 3 instead of Theorem 5(b)(ii) in the above proof a similar theorem is obtained for the case of an acyclic network: then the equilibrium values of all states are linear combinations of the values of the initial states.

12.7 Further Implications for Example Networks

In this section, it is shown what further conclusions can be drawn from the theorems presented in Sect. 12.6 for the example described in Sect. 12.3 and for an example Mental Network described in Schoenmaker et al. (2018). This shows that the applicability goes beyond only Social Networks. First, the earlier example described in Sect. 12.3 is analysed; after that the new example will be addressed.

12.7.1 Further Analysis of the Example Network from Sect. 12.3.2

Theorems 7 to 9 are illustrated by the example network shown in Fig. 12.2 as follows. Here there is only one independent constant state X_1 with singleton component. Moreover, the states in the other level 0 component C_2 are X_5, X_6, X_7 respectively (see Fig. 12.5). So, from Theorem 7 it follows that the equilibrium value of any state Y is

$$\text{eq}_Y(V_1, V_2, V_3, V_4) = d_{X_1,Y}V_1 + d_{X_5,Y}V_2 + d_{X_6,Y}V_3 + d_{X_7,Y}V_4 \tag{12.3}$$

where V_1, V_2, V_3, V_4 are the initial values of the states X_1, X_5, X_6, X_7 in the level 0 components C_1 and C_2. For the example states $Y \in \{X_8, X_9, X_{10}\}$ the coefficients $d_{X1,Y}$, $d_{X5,Y}$, $d_{X6,Y}$, $d_{X7,Y}$ have been determined by simulation for the connection weights shown in Box 12.1 (and using speed factors 0.5), with these results shown in Table 12.4.

So, for example, for $Y = X_8$ the four coefficients are:

$$d_{X_1,X_8} = 0.634921 \quad d_{X_5,X_8} = 0.121693 \quad d_{X_6,X_8} = 0.121693 \quad d_{X_7,X_8} = 0.121693$$

Table 12.4 Coefficients of the linear relations between equilibrium values and initial values

	X_1	X_5	X_6	X_7
X_8	0.634921	0.121693	0.121693	0.121693
X_9	0.31746	0.227513	0.227513	0.227513
X_{10}	0.148148	0.283951	0.283951	0.283951
	$d_{X1,Y}$	$d_{X5,Y}$	$d_{X6,Y}$	$d_{X7,Y}$

Therefore, in a sense, the equilibrium value of X_8 can be considered to be determined for 63.5% by the constant value of X_1 and for 12.2% by each of the initial values of X_5, X_6, X_7. These four values indeed sum up to 1 or 100%. Note that in this case, the last three coefficients happen to be equal, as for the sake of simplicity this component is just one cycle and is therefore highly symmetric; this is not always the case. More specifically, given the above values, for the considered case the equilibrium value for X_8 is given by

$$\text{eq}_{X_8}(V_1, V_2, V_3, V_4) = 0.634921V_1 + 0.121693V_2 + 0.121693V_3 + 0.121693V_4 \tag{12.4}$$

with V_1 the constant value of X_1 and V_2, V_3, V_4 the initial values of X_5, X_6, X_7, respectively. This is indeed confirmed in simulations.

For the considered example, as the scaled sum used is linear, solving the linear equations can also provide specific values. In this way, in line with Theorems 7 to 9 the specific equilibrium values of the states X_8, X_9 and X_{10} in C_4 can be determined algebraically from the values of the states X_3, X_5, and X_7 in the lower level components C_2 and C_3 (repetitive digits in italics). Using a symbolic solver (the online WIMS Linear Solver tool[1] was used), this can be done more in general. The linear equilibrium equations are:

$$\begin{aligned}
(\omega_{X_3,X_8} + \omega_{X_{10},X_8})X_8 &= \omega_{X_3,X_8}X_3 + \omega_{X_{10},X_8}X_{10} \\
(\omega_{X_5,X_9} + \omega_{X_8,X_9})X_9 &= \omega_{X_5,X_9}X_5 + \omega_{X_8,X_9}X_8 \\
(\omega_{X_7,X_{10}} + \omega_{X_9,X_{10}})X_{10} &= \omega_{X_7,X_{10}}X_7 + \omega_{X_9,X_{10}}X_9
\end{aligned} \tag{12.5}$$

These general equations have the following unique symbolic solution when X_3, X_5, X_7 are assumed given from the lower level components:

[1]http://wims.unice.fr/wims/wims.cgi?session=DH1DFC9A6E.3&+lang=en&+module=tool%2Flinear%2Flinsolver.en.

$$\begin{aligned}
X_8 = {} & (\omega_{X_{10},X_8}(\omega_{X_7,X_{10}}\omega_{X_8,X_9}X_7 + \omega_{X_5,X_9}\omega_{X_7,X_{10}}X_7 + \omega_{X_5,X_9}\omega_{X_9,X_{10}}X_5) \\
& + \omega_{X_3,X_8}(\omega_{X_8,X_9}(\omega_{X_9,X_{10}} + \omega_{X_7,X_{10}}) + \omega_{X_5,X_9}(\omega_{X_9,X_{10}} + \omega_{X_7,X_{10}}))X_3) \\
& /(\omega_{X_3,X_8}(\omega_{X_8,X_9}(\omega_{X_9,X_{10}} + \omega_{X_7,X_{10}}) + \omega_{X_5,X_9}(\omega_{X_9,X_{10}} + \omega_{X_7,X_{10}})) \\
& + \omega_{X_{10},X_8}(\omega_{X_5,X_9}(\omega_{X_9,X_{10}} + \omega_{X_7,X_{10}}) + \omega_{X_7,X_{10}}\omega_{X_8,X_9}))
\end{aligned}$$

$$\begin{aligned}
X_9 = {} & (\omega_{X_{10},X_8}(\omega_{X_7,X_{10}}\omega_{X_8,X_9}X_7 + \omega_{X_5,X_9}(\omega_{X_9,X_{10}} + \omega_{X_7,X_{10}})X_5) \\
& + \omega_{X_3,X_8}\omega_{X_5,X_9}(\omega_{X_9,X_{10}} + \omega_{X_7,X_{10}})X_5 + \omega_{X_3,X_8}\omega_{X_8,X_9}(\omega_{X_9,X_{10}} + \omega_{X_7,X_{10}})X_3) \\
& /(\omega_{X_3,X_8}(\omega_{X_8,X_9}(\omega_{X_9,X_{10}} + \omega_{X_7,X_{10}}) + \omega_{X_5,X_9}(\omega_{X_9,X_{10}} + \omega_{X_7,X_{10}})) \\
& + \omega_{X_{10},X_8}(\omega_{X_5,X_9}(\omega_{X_9,X_{10}} + \omega_{X_7,X_{10}}) + \omega_{X_7,X_{10}}\omega_{X_8,X_9}))
\end{aligned}$$

$$\begin{aligned}
X_{10} = {} & (\omega_{X_3,X_8}(\omega_{X_7,X_{10}}\omega_{X_8,X_9}X_7 + \omega_{X_5,X_9}\omega_{X_7,X_{10}}X_7 + \omega_{X_5,X_9}\omega_{X_9,X_{10}}X_5) \\
& + \omega_{X_{10},X_8}(\omega_{X_7,X_{10}}w_{X_8,X_9}X_7 + \omega_{X_5,X_9}\omega_{X_7,X_{10}}X_7 + \omega_{X_5,X_9}\omega_{X_9,X_{10}}X_5) \\
& + \omega_{X_3,X_8}\omega_{X_8,X_9}\omega_{X_9,X_{10}}X_3)/(\omega_{X_3,X_8}(\omega_{X_8,X_9}(\omega_{X_9,X_{10}} + \omega_{X_7,X_{10}}) + \omega_{X_5,X_9}(\omega_{X_9,X_{10}} + \omega_{X_7,X_{10}})) \\
& + \omega_{X_{10},X_8}(\omega_{X_5,X_9}(\omega_{X_9,X_{10}} + \omega_{X_7,X_{10}}) + \omega_{X_7,X_{10}}\omega_{X_8,X_9}))
\end{aligned}$$

From this the values of the coefficients $d_{X_i,Y}$ of the linear relation from Theorem 8 can be determined:

$$\begin{aligned}
X_8 = {} & (\omega_{X_{10},X_8}(\omega_{X_7,X_{10}}\omega_{X_8,X_9}X_7 + \omega_{X_5,X_9}\omega_{X_7,X_{10}}X_7 + \omega_{X_5,X_9}\omega_{X_9,X_{10}}X_5) \\
& + \omega_{X_3,X_8}(\omega_{X_8,X_9}(\omega_{X_9,X_{10}} + \omega_{X_7,X_{10}}) + \omega_{X_5,X_9}(\omega_{X_9,X_{10}} + \omega_{X_7,X_{10}}))X_3) \\
& /(\omega_{X_3,X_8}(\omega_{X_8,X_9}(\omega_{X_9,X_{10}} + \omega_{X_7,X_{10}}) + \omega_{X_5,X_9}(\omega_{X_9,X_{10}} + \omega_{X_7,X_{10}})) \\
& + \omega_{X_{10},X_8}(\omega_{X_5,X_9}(\omega_{X_9,X_{10}} + \omega_{X_7,X_{10}}) + \omega_{X_7,X_{10}}\omega_{X_8,X_9}))
\end{aligned}$$

$$\begin{aligned}
d_{X_3,X_8} = {} & \omega_{X_3,X_8}(\omega_{X_8,X_9}(\omega_{X_9,X_{10}} + \omega_{X_7,X_{10}}) + \omega_{X_5,X_9}(\omega_{X_9,X_{10}} + \omega_{X_7,X_{10}})) \\
& /(\omega_{X_3,X_8}(\omega_{X_8,X_9}(\omega_{X_9,X_{10}} + \omega_{X_7,X_{10}}) + \omega_{X_5,X_9}(\omega_{X_9,X_{10}} + \omega_{X_7,X_{10}})) \\
& + \omega_{X_{10},X_8}(\omega_{X_5,X_9}(\omega_{X_9,X_{10}} + \omega_{X_7,X_{10}}) + \omega_{X_7,X_{10}}\omega_{X_8,X_9}))
\end{aligned}$$

$$\begin{aligned}
d_{X_5,X_8} = {} & \omega_{X_{10},X_8}\omega_{X_5,X_9}\omega_{X_9,X_{10}} \\
& /(\omega_{X_3,X_8}(\omega_{X_8,X_9}(\omega_{X_9,X_{10}} + \omega_{X_7,X_{10}}) + \omega_{X_5,X_9}(\omega_{X_9,X_{10}} + \omega_{X_7,X_{10}})) \\
& + \omega_{X_{10},X_8}(\omega_{X_5,X_9}(\omega_{X_9,X_{10}} + \omega_{X_7,X_{10}}) + \omega_{X_7,X_{10}}\omega_{X_8,X_9}))
\end{aligned}$$

$$\begin{aligned}
d_{X_7,X_8} = {} & \omega_{X_{10},X_8}(\omega_{X_7,X_{10}}\omega_{X_8,X_9} + \omega_{X_5,X_9}\omega_{X_7,X_{10}}) \\
& /(\omega_{X_3,X_8}(\omega_{X_8,X_9}(\omega_{X_9,X_{10}} + \omega_{X_7,X_{10}}) + \omega_{X_5,X_9}(\omega_{X_9,X_{10}} + \omega_{X_7,X_{10}})) \\
& + \omega_{X_{10},X_8}(\omega_{X_5,X_9}(\omega_{X_9,X_{10}} + \omega_{X_7,X_{10}}) + \omega_{X_7,X_{10}}\omega_{X_8,X_9}))
\end{aligned}$$

$$\begin{aligned}
X_9 = {} & (\omega_{X_{10},X_8}(\omega_{X_7,X_{10}}\omega_{X_8,X_9}X_7 + \omega_{X_5,X_9}(\omega_{X_9,X_{10}} + w_{X_7,X_{10}})X_5) \\
& + \omega_{X_3,X_8}\omega_{X_5,X_9}(\omega_{X_9,X_{10}} + \omega_{X_7,X_{10}})X_5 + \omega_{X_3,X_8}\omega_{X_8,X_9}(\omega_{X_9,X_{10}} + \omega_{X_7,X_{10}})X_3) \\
& /(\omega_{X_3,X_8}(\omega_{X_8,X_9}(\omega_{X_9,X_{10}} + \omega_{X_7,X_{10}}) + \omega_{X_5,X_9}(\omega_{X_9,X_{10}} + \omega_{X_7,X_{10}})) \\
& + \omega_{X_{10},X_8}(\omega_{X_5,X_9}(\omega_{X_9,X_{10}} + \omega_{X_7,X_{10}}) + \omega_{X_7,X_{10}}\omega_{X_8,X_9}))
\end{aligned}$$

$$
\begin{aligned}
d_{X_3,X_9} = {} & \omega_{X_3,X_8}\omega_{X_8,X_9}(\omega_{X_9,X_{10}}+\omega_{X_7,X_{10}}) \\
& /(w_{X_3,X_8}(w_{X_8,X_9}(w_{X_9,X_{10}}+w_{X_7,X_{10}})+w_{X_5,X_9}(w_{X_9,X_{10}}+w_{X_7,X_{10}})) \\
& +\omega_{X_{10},X_8}(\omega_{X_5,X_9}(\omega_{X_9,X_{10}}+\omega_{X_7,X_{10}})+\omega_{X_7,X_{10}}\omega_{X_8,X_9}))
\end{aligned}
$$

$$
\begin{aligned}
d_{X_5,X_9} = {} & \omega_{X_5,X_9}(\omega_{X_{10},X_8}+\omega_{X_3,X_8})(\omega_{X_9,X_{10}}+\omega_{X_7,X_{10}}) \\
& /(\omega_{X_3,X_8}(\omega_{X_8,X_9}(\omega_{X_9,X_{10}}+\omega_{X_7,X_{10}})+\omega_{X_5,X_9}(\omega_{X_9,X_{10}}+\omega_{X_7,X_{10}})) \\
& +\omega_{X_{10},X_8}(\omega_{X_5,X_9}(\omega_{X_9,X_{10}}+\omega_{X_7,X_{10}})+\omega_{X_7,X_{10}}\omega_{X_8,X_9}))
\end{aligned}
$$

$$
\begin{aligned}
d_{X_7,X_9} = {} & \omega_{X_{10},X_8}\omega_{X_7,X_{10}}\omega_{X_8,X_9} \\
& /(\omega_{X_3,X_8}(\omega_{X_8,X_9}(\omega_{X_9,X_{10}}+\omega_{X_7,X_{10}})+\omega_{X_5,X_9}(\omega_{X_9,X_{10}}+\omega_{X_7,X_{10}})) \\
& +\omega_{X_{10},X_8}(\omega_{X_5,X_9}(\omega_{X_9,X_{10}}+\omega_{X_7,X_{10}})+\omega_{X_7,X_{10}}\omega_{X_8,X_9}))
\end{aligned}
$$

$$
\begin{aligned}
X_{10} = {} & (\omega_{X_3,X_8}(\omega_{X_7,X_{10}}\omega_{X_8,X_9}X_7+\omega_{X_5,X_9}\omega_{X_7,X_{10}}X_7+\omega_{X_5,X_9}\omega_{X_9,X_{10}}X_5) \\
& +\omega_{X_{10},X_8}(\omega_{X_7,X_{10}}\omega_{X_8,X_9}X_7+\omega_{X_5,X_9}X_7+\omega_{X_5,X_9}\omega_{X_7,X_{10}}X_7+\omega_{X_5,X_9}\omega_{X_9,X_{10}}X_5) \\
& +\omega_{X_3,X_8}\omega_{X_8,X_9}\omega_{X_9,X_{10}}X_3)/(\omega_{X_3,X_8}(\omega_{X_8,X_9}(\omega_{X_9,X_{10}}+\omega_{X_7,X_{10}})+\omega_{X_5,X_9}(\omega_{X_9,X_{10}}+\omega_{X_7,X_{10}})) \\
& +\omega_{X_{10},X_8}(\omega_{X_5,X_9}(\omega_{X_9,X_{10}}+\omega_{X_7,X_{10}})+\omega_{X_7,X_{10}}\omega_{X_8,X_9}))
\end{aligned}
$$

$$
\begin{aligned}
d_{X_3,X_{10}} = {} & \omega_{X_3,X_8}\omega_{X_8,X_9}\omega_{X_9,X_{10}}/(\omega_{X_3,X_8}(\omega_{X_8,X_9}(\omega_{X_9,X_{10}}+\omega_{X_7,X_{10}})+\omega_{X_5,X_9}(\omega_{X_9,X_{10}}+\omega_{X_7,X_{10}})) \\
& +\omega_{X_{10},X_8}(\omega_{X_5,X_9}(\omega_{X_9,X_{10}}+\omega_{X_7,X_{10}})+\omega_{X_7,X_{10}}\omega_{X_8,X_9}))
\end{aligned}
$$

$$
\begin{aligned}
d_{X_5,X_{10}} = {} & \omega_{X_5,X_9}\omega_{X_9,X_{10}}(\omega_{X_3,X_8}+\omega_{X_{10},X_8}) \\
& /(\omega_{X_3,X_8}(\omega_{X_8,X_9}(\omega_{X_9,X_{10}}+\omega_{X_7,X_{10}})+\omega_{X_5,X_9}(\omega_{X_9,X_{10}}+\omega_{X_7,X_{10}})) \\
& +\omega_{X_{10},X_8}(\omega_{X_5,X_9}(\omega_{X_9,X_{10}}+\omega_{X_7,X_{10}})+\omega_{X_7,X_{10}}\omega_{X_8,X_9}))
\end{aligned}
$$

$$
\begin{aligned}
d_{X_7,X_{10}} = {} & \omega_{X_7,X_{10}}(\omega_{X_3,X_8}+\omega_{X_{10},X_8})(\omega_{X_8,X_9}+\omega_{X_5,X_9}) \\
& /(\omega_{X_3,X_8}(\omega_{X_8,X_9}(\omega_{X_9,X_{10}}+\omega_{X_7,X_{10}})+\omega_{X_5,X_9}(\omega_{X_9,X_{10}}+\omega_{X_7,X_{10}})) \\
& +\omega_{X_{10},X_8}(\omega_{X_5,X_9}(\omega_{X_9,X_{10}}+\omega_{X_7,X_{10}})+\omega_{X_7,X_{10}}\omega_{X_8,X_9}))
\end{aligned}
$$

As a special case, if all ω's are set equal to one ω, then the denominator becomes $7\omega^3$ and the following values are obtained:

$$
\begin{array}{lll}
d_{X_3,X_8} = 4/7 & d_{X_3,X_9} = 2/7 & d_{X_3,X_{10}} = 1/7 \\
d_{X_5,X_8} = 1/7 & d_{X_5,X_9} = 4/7 & d_{X_5,X_{10}} = 2/7 \\
d_{X_7,X_8} = 2/7 & d_{X_7,X_9} = 1/7 & d_{X_7,X_{10}} = 4/7
\end{array}
$$

Then the linear relations for the equilibrium values become:

$$X_8 = {}^4/_7X_3 + {}^1/_7X_5 + {}^2/_7X_7$$
$$X_9 = {}^2/_7X_3 + {}^4/_7X_5 + {}^1/_7X_7$$
$$X_{10} = {}^1/_7X_3 + {}^2/_7X_5 + {}^4/_7X_7$$

12.8 Analysis of an Example Mental Network

In this section, applicability is illustrated for a type of network which is not a social network. In general Theorems 7 to 9 can be applied for many cases of networks that receive external input. This varies from Mental Networks that get input from external stimuli to Social Networks that are affected by context factors such as broadcasts from external sources that are received by members of the network. As an example of this, for the mental area, the Mental Network model from Schoenmaker et al. (2018) has been analysed. The strongly connected components are as shown in Fig. 12.9, with stratified condensation graph as in Fig. 12.10; for the connection weights and other values, see the role matrices in Box 12.5. The model describes how the emotional charge of a received tweet affects the decision to retweet it. It can be explained by the following scenario considering Mark sending a tweet to Tim in which he expresses that he cannot wait to sing in the Christmas choir next week.

> This tweet contains both information and emotional charge: there is a choir performance next week, and secondly, Mark makes clear that he cannot wait for this event to happen. Tim's interpretation of this message is positively influenced by the fact that Mark and Tim

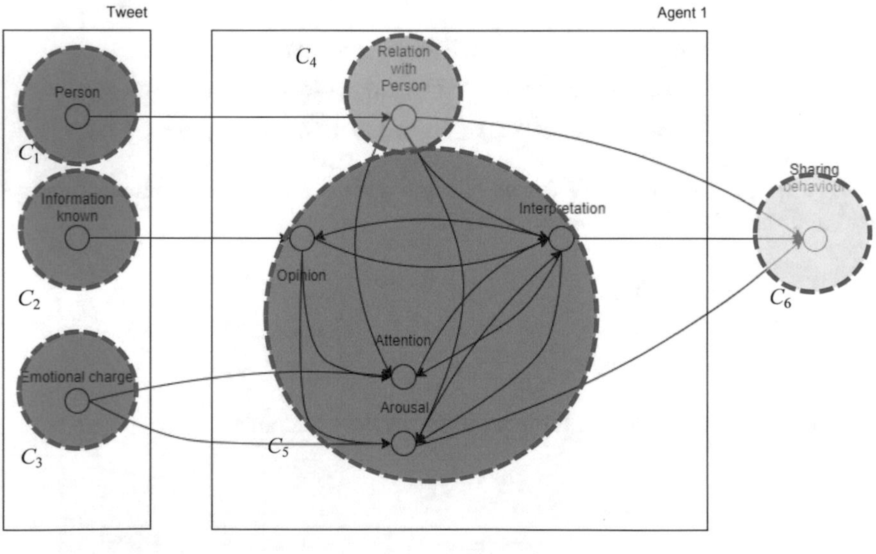

Fig. 12.9 The strongly connected components within the second example network

are friends. Tim does like to visit choir performances; therefore, he already has a positive association on the information that this event will take place. Reading about this Christmas performance, Tim gets slightly aroused and is focusing on the message. Mark's enthusiasm amplifies Tim's attention and arousal, which in turn lead to a positive interpretation of the tweet. Tim's positive interpretation of the message coupled with the fact that he is good friends with Mark and is excited about this performance leads to Tim's decision to retweet Mark's original Tweet. (Schoenmaker et al. 2018), p. 138

The states within the box Agent 1 all have a scaled sum combination function. The final state Sharing has $\mathbf{alogistic}_{\sigma,\tau}$**(..)** as combination function. For a complete overview of the role matrices, see Box 12.5. For the analysis, the above theorems can be applied to the network when the state Sharing is left out of consideration.

The stratified condensation graph for this network is shown in Fig. 12.10. From this stratified condensation graph a number of conclusions can be drawn:

- The level 0 states are the states Person, Information known and Emotional charge in C_1, C_2, and C_3, respectively; therefore these three states are the determining factors for the whole network.
- The level 1 state Relation with person will have the same equilibrium value as the level 0 state Person in C_1.
- When all level 0 states have the same equilibrium value V, then also all level 1 and level 2 states Relation with person, Opinion, Attention, Arousal, and Interpretation will have that same equilibrium value V. For example, when all level 0 states are constant 1, then all states as mentioned will end up in equilibrium value 1.
- When the level 0 states have different equilibrium values, then the level 2 states Opinion, Attention, Arousal and Interpretation are expected to have different equilibrium values too, these values lay between the maximal and minimal values at level 0.

More specifically, in numbers, the following can be concluded. Suppose any given constant values A_1, A_2, A_3 for the level 0 components in C_1, C_2, C_3, respectively. Then:

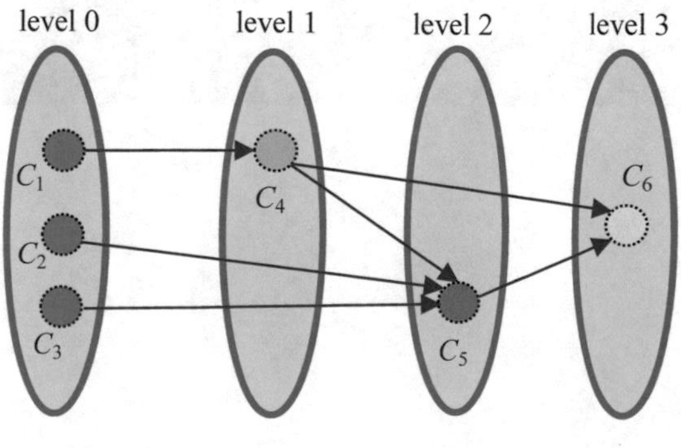

Fig. 12.10 Stratified condensation graph SC(*N*) for the second example network

- at level 1 the equilibrium value in C_4 is A_1
- at level 2 the equilibrium values of all four states in C_5 are between min(A_1, A_2, A_3) and max(A_1, A_2, A_3)
- these equilibrium values of the four states in C_5 are linear functions in the form of weighted sums of A_1, A_2, A_3
- when all A_i = A for one value, then at level 2 the equilibrium values of the states in C_5 are A as well

In Box 12.2 the differential equations are shown for this second example model, and in Box 12.3 the equilibrium equations.

Box 12.2 Overview of the differential equations of the second example network models

$$\begin{aligned}
\mathbf{d}\text{Relation}/\mathbf{dt} &= \eta_{\text{Relation}}\left[\omega_{\text{Person,Relation}}\text{Person} - \text{Relation}\right] \\
\mathbf{d}\text{Opinion}/\mathbf{dt} &= \eta_{\text{Opinion}}\left[(\omega_{\text{Information,Opinion}}\text{Information} \right.\\
&\quad \left. + \omega_{\text{Interpretation,Opinion}}\text{Interpretation})/\lambda_{\text{Opinion}} - \text{Opinion}\right] \\
\mathbf{d}\text{Interpretation}/\mathbf{dt} &= \eta_{\text{Interpretation}}\left[(\omega_{\text{Relation,Interpretation}}\text{Relation} \right.\\
&\quad + \omega_{\text{Opinion,Interpretation}}\text{Opinion} \\
&\quad + \omega_{\text{Attention,Interpretation}}\text{Attention} \\
&\quad \left. + \omega_{\text{Arousal,Interpretation}}\text{Arousal})/\lambda_{\text{Interpretation}} - \text{Interpretation}\right] \\
\mathbf{d}\text{Attention}/\mathbf{dt} &= \eta_{\text{Attention}}\left[(\omega_{\text{Emotion,Attention}}\text{Emotion} + \omega_{\text{Relation,Attention}}\text{Relation} \right.\\
&\quad + \omega_{\text{Opinion,Attention}}\text{Opinion} \\
&\quad \left. + \omega_{\text{Interpretation,Attention}}\text{Interpretation})/\lambda_{\text{Attention}} - \text{Attention}\right] \\
\mathbf{d}\text{Arousal}/\mathbf{dt} &= \eta_{\text{Arousal}}\left[(\omega_{\text{Emotion,Arousal}}\text{Emotion} + \omega_{\text{Relation,Arousal}}\text{Relation} \right.\\
&\quad + \omega_{\text{Opinion,Arousal}}\text{Opinion} \\
&\quad \left. + \omega_{\text{Interpretation,Arousal}}\text{Interpretation})/\lambda_{\text{Arousal}} - \text{Arousal}\right] \\
\mathbf{d}\text{Sharing}/\mathbf{dt} &= \eta_{\text{Sharing}}\left[(\omega_{\text{Relation,Sharing}}\text{Relation} \right.\\
&\quad + \omega_{\text{Interpretation,Sharing}}\text{Interpretation} \\
&\quad \left. + \omega_{\text{Arousal,Sharing}}\text{Arousal})/\lambda_{\text{Sharing}} - \text{Sharing}\right]
\end{aligned}$$

Box 12.3 Overview of the equilibrium equations of the second example network model

$$\begin{aligned}
\text{Relation} &= \omega_{\text{Person,Relation}}\text{Person}\\
\text{Opinion} &= (\omega_{\text{Information,Opinion}}\text{Information}\\
&\quad + \omega_{\text{Interpretation,Opinion}}\text{Interpretation})/\lambda_{\text{Opinion}}\\
\text{Interpretation} &= (\omega_{\text{Relation,Interpretation}}\text{Relation} + \omega_{\text{Opinion,Interpretation}}\text{Opinion}\\
&\quad + \omega_{\text{Attention,Interpretation}}\text{Attention}\\
&\quad + \omega_{\text{Arousal,Interpretation}}\text{Arousal})/\lambda_{\text{Interpretation}}\\
\text{Attention} &= (\omega_{\text{Emotion,Attention}}\text{Emotion} + \omega_{\text{Relation,Attention}}\text{Relation}\\
&\quad + \omega_{\text{Opinion,Attention}}\text{Opinion}\\
&\quad + \omega_{\text{Interpretation,Attention}}\text{Interpretation})/\lambda_{\text{Attention}}\\
\text{Arousal} &= (\omega_{\text{Emotion,Arousal}}\text{Emotion} + \omega_{\text{Relation,Arousal}}\text{Relation}\\
&\quad + \omega_{\text{Opinion,Arousal}}\text{Opinion}\\
&\quad + \omega_{\text{Interpretation,Arousal}}\text{Interpretation})/\lambda_{\text{Arousal}}\\
\text{Sharing} &= (\omega_{\text{Relation,Sharing}}\text{Relation} + \omega_{\text{Interpretation,Sharing}}\text{Interpretation}\\
&\quad + \omega_{\text{Arousal,Sharing}}\text{Arousal})/\lambda_{\text{Sharing}}
\end{aligned}$$

The linear equilibrium equations for the states other than Sharing can be solved in a symbolic manner to obtain explicit algebraic expressions for their equilibrium values (again the online WIMS Linear Solver tool was used); see Box 12.4. Here subscripts are abbreviated for the sake of briefness.

Box 12.4 Explicit algebraic solutions of the equilibrium equations of the second example network model; adopted from Schoenmaker et al. (2018)

$$\textbf{Person} = X_1 = \text{A}_1 \quad \textbf{Information} = X_2 = \text{A}_2 \quad \textbf{Emotion} = X_3 = \text{A}_3$$

$$\textbf{Relation} = X_4 = \omega_{\text{P,R}}\text{A}_1$$

$$\begin{aligned}
\textbf{Opinion} = X_5 = -[&\text{A}_1\omega_{\text{Int,O}}\omega_{\text{P,R}}(\lambda_{\text{Ar}}\lambda_{\text{At}}\omega_{\text{R,Int}}\\
&+ \lambda_{\text{Ar}}\omega_{\text{At,Int}}\omega_{\text{RAt}} + \lambda_{\text{At}}\omega_{\text{Ar,Int}}\omega_{\text{R,Ar}})\\
&+ \text{A}_3(\lambda_{\text{Ar}}\omega_{\text{At,Int}}\omega_{\text{E,At}} + \lambda_{\text{At}}\omega_{\text{Ar,Int}}\omega_{\text{E,Ar}})\omega_{\text{Int,O}}\\
&+ \text{A}_2\omega_{\text{Inf,O}}((-\lambda_{\text{Ar}}\omega_{\text{At,Int}}\omega_{\text{Int,At}}) - \lambda_{\text{At}}\omega_{\text{Ar,Int}}\omega_{\text{Int,Ar}} + \lambda_{\text{Ar}}\lambda_{\text{At}}\lambda_{\text{I}})\\
&/[\omega_{\text{Int,O}}(\lambda_{\text{Ar}}\lambda_{\text{At}}\omega_{\text{O,Int}} + \lambda_{\text{Ar}}\omega_{\text{At,Int}}\omega_{\text{O,At}} + \lambda_{\text{At}}\omega_{\text{Ar,Int}}\omega_{\text{O,Ar}})\\
&+ \lambda_{\text{O}}(\lambda_{\text{Ar}}\omega_{\text{At,Int}}\omega_{\text{Int,At}} + \lambda_{\text{At}}\omega_{\text{Ar,Int}}\omega_{\text{Int,Ar}} - \lambda_{\text{Ar}}\lambda_{\text{At}}\lambda_{\text{I}})]
\end{aligned}$$

$$
\begin{aligned}
\textbf{Interpretation} = X_6 = &-[\mathrm{A}_1\lambda_{\mathrm{O}}\omega_{\mathrm{P,R}}(\lambda_{\mathrm{Ar}}\lambda_{\mathrm{At}}\omega_{\mathrm{R,Int}} + \lambda_{\mathrm{Ar}}\omega_{\mathrm{At,Int}}\omega_{\mathrm{R,At}} + \lambda_{\mathrm{At}}\omega_{\mathrm{Ar,Int}}\omega_{\mathrm{R,Ar}}) \\
&+ \mathrm{A}_2\omega_{\mathrm{Inf,O}}(\lambda_{\mathrm{Ar}}\lambda_{\mathrm{At}}\omega_{\mathrm{O,Int}} + \lambda_{\mathrm{Ar}}\omega_{\mathrm{At,Int}}\omega_{\mathrm{O,At}} + \lambda_{\mathrm{At}}\omega_{\mathrm{Ar,Int}}\omega_{\mathrm{O,Ar}}) \\
&+ \mathrm{A}_3\lambda_{\mathrm{O}}(\lambda_{\mathrm{Ar}}\omega_{\mathrm{At,Int}}\omega_{\mathrm{E,At}} + \lambda_{\mathrm{At}}\omega_{\mathrm{Ar,Int}}\omega_{\mathrm{E,Ar}})] \\
&/[\omega_{\mathrm{Int,O}}(\lambda_{\mathrm{Ar}}\lambda_{\mathrm{At}}\omega_{\mathrm{O,Int}} + \lambda_{\mathrm{Ar}}\omega_{\mathrm{At,Int}}\omega_{\mathrm{O,At}} + \lambda_{\mathrm{At}}\omega_{\mathrm{Ar,Int}}\omega_{\mathrm{O,Ar}}) \\
&+ \lambda_{\mathrm{O}}(\lambda_{\mathrm{Ar}}\omega_{\mathrm{At,Int}}\omega_{\mathrm{Int,At}} + \lambda_{\mathrm{At}}\omega_{\mathrm{Ar,Int}}\omega_{\mathrm{Int,Ar}} - \lambda_{\mathrm{Ar}}\lambda_{\mathrm{At}}\lambda_{\mathrm{I}})]
\end{aligned}
$$

$$
\begin{aligned}
\textbf{Attention} = X_7 = &-[\mathrm{A}_1\omega_{\mathrm{P,R}}(\omega_{\mathrm{Int,O}}(\lambda_{\mathrm{Ar}}\omega_{\mathrm{O,At}}\omega_{\mathrm{R,Int}} \\
&+ \omega_{\mathrm{Ar,Int}}(\omega_{\mathrm{O,At}}\omega_{\mathrm{R,Ar}} - \omega_{\mathrm{O,Ar}}\omega_{\mathrm{R,At}}) - \lambda_{\mathrm{Ar}}\omega_{\mathrm{O,Int}}\omega_{\mathrm{R,At}}) \\
&+ \lambda_{\mathrm{O}}(\lambda_{\mathrm{Ar}}\omega_{\mathrm{Int,At}}\omega_{\mathrm{R,Int}} \\
&+ \omega_{\mathrm{Ar,Int}}(\omega_{\mathrm{Int,At}}\omega_{\mathrm{R,Ar}} - \omega_{\mathrm{Int,Ar}}\omega_{\mathrm{R,At}}) + \lambda_{\mathrm{Ar}}\lambda_{\mathrm{I}}\omega_{\mathrm{R,At}})) \\
&+ \mathrm{A}_3(\omega_{\mathrm{Int,O}}(\omega_{\mathrm{Ar,Int}}(\omega_{\mathrm{E,Ar}}\omega_{\mathrm{O,At}} - \omega_{\mathrm{E,At}}\omega_{\mathrm{O,Ar}}) - \lambda_{\mathrm{Ar}}\omega_{\mathrm{E,At}}\omega_{\mathrm{O,Int}}) \\
&+ \lambda_{\mathrm{O}}(\omega_{\mathrm{Ar,Int}}(\omega_{\mathrm{E,Ar}}\omega_{\mathrm{Int,At}} - \omega_{\mathrm{E,At}}\omega_{\mathrm{Int,At}}) + \lambda_{\mathrm{Ar}}\lambda_{\mathrm{I}}\omega_{\mathrm{E,At}})) \\
&+ \mathrm{A}_2\omega_{\mathrm{Inf,O}}(\lambda_{\mathrm{Ar}}\omega_{\mathrm{Int,At}}\omega_{\mathrm{O,Int}} \\
&+ \omega_{\mathrm{Ar,Int}}(\omega_{\mathrm{Int,At}}\omega_{\mathrm{O,Ar}} - \omega_{\mathrm{Int,Ar}}\omega_{\mathrm{O,At}}) + \lambda_{\mathrm{Ar}}\lambda_{\mathrm{I}}\omega_{\mathrm{O,At}})] \\
&/[\omega_{\mathrm{Int,O}}(\lambda_{\mathrm{Ar}}\lambda_{\mathrm{At}}\omega_{\mathrm{O,Int}} + \lambda_{\mathrm{Ar}}\omega_{\mathrm{At,Int}}\omega_{\mathrm{O,At}} + \lambda_{\mathrm{At}}\omega_{\mathrm{Ar,Int}}\omega_{\mathrm{O,Ar}}) \\
&+ \lambda_{\mathrm{O}}(\lambda_{\mathrm{Ar}}\omega_{\mathrm{At,Int}}\omega_{\mathrm{Int,At}} + \lambda_{\mathrm{At}}\omega_{\mathrm{Ar,Int}}\omega_{\mathrm{Int,Ar}} - \lambda_{\mathrm{Ar}}\lambda_{\mathrm{At}}\lambda_{\mathrm{I}})]
\end{aligned}
$$

$$
\begin{aligned}
\textbf{Arousal} = X_8 = &-[\mathrm{A}_1\omega_{\mathrm{P,R}}(\omega_{\mathrm{Int,O}}(\lambda_{\mathrm{At}}\omega_{\mathrm{O,Ar}}\omega_{\mathrm{R,Int}} \\
&+ \omega_{\mathrm{At,Int}}(\omega_{\mathrm{O,Ar}}\omega_{\mathrm{R,At}} - \omega_{\mathrm{O,At}}\omega_{\mathrm{R,Ar}}) - \lambda_{\mathrm{At}}\omega_{\mathrm{O,Int}}\omega_{\mathrm{R,Ar}}) \\
&+ \lambda_{\mathrm{O}}(\lambda_{\mathrm{At}}\omega_{\mathrm{Int,Ar}}\omega_{\mathrm{R,Int}} \\
&+ \omega_{\mathrm{At,Int}}(\omega_{\mathrm{Int,Ar}}\omega_{\mathrm{R,At}} - \omega_{\mathrm{Int,At}}\omega_{\mathrm{R,Ar}}) + \lambda_{\mathrm{At}}\lambda_{\mathrm{I}}\omega_{\mathrm{R,Ar}})) \\
&+ \mathrm{A}_3(\omega_{\mathrm{Int,O}}(\omega_{\mathrm{At,Int}}(\omega_{\mathrm{E,At}}\omega_{\mathrm{O,Ar}} - \omega_{\mathrm{E,Ar}}\omega_{\mathrm{O,At}}) - \lambda_{\mathrm{At}}\omega_{\mathrm{E,Ar}}\omega_{\mathrm{O,Int}}) \\
&+ \lambda_{\mathrm{O}}(\omega_{\mathrm{At,Int}}(\omega_{\mathrm{E,At}}\omega_{\mathrm{Int,Ar}} - \omega_{\mathrm{E,Ar}}\omega_{\mathrm{Int,At}}) + \lambda_{\mathrm{At}}\lambda_{\mathrm{I}}\omega_{\mathrm{E,Ar}})) \\
&+ \mathrm{A}_2\omega_{\mathrm{Inf,O}}(\lambda_{\mathrm{At}}\omega_{\mathrm{Int,Ar}}\omega_{\mathrm{O,Int}} \\
&+ \omega_{\mathrm{At,Int}}(\omega_{\mathrm{Int,Ar}}\omega_{\mathrm{O,At}} - \omega_{\mathrm{Int,At}}\omega_{\mathrm{O,Ar}}) + \lambda_{\mathrm{At}}\lambda_{\mathrm{I}}\omega_{\mathrm{O,Ar}})] \\
&/[\omega_{\mathrm{Int,O}}(\lambda_{\mathrm{Ar}}\lambda_{\mathrm{At}}\omega_{\mathrm{O,Int}} + \lambda_{\mathrm{Ar}}\omega_{\mathrm{At,Int}}\omega_{\mathrm{O,At}} + \lambda_{\mathrm{At}}\omega_{\mathrm{Ar,Int}}\omega_{\mathrm{O,Ar}}) \\
&+ \lambda_{\mathrm{O}}(\lambda_{\mathrm{Ar}}\omega_{\mathrm{At,Int}}\omega_{\mathrm{Int,At}} + \lambda_{\mathrm{At}}\omega_{\mathrm{Ar,Int}}\omega_{\mathrm{Int,Ar}} - \lambda_{\mathrm{Ar}}\lambda_{\mathrm{At}}\lambda)]
\end{aligned}
$$

As can be seen, each of the equilibrium values is a linear combination of the three values A_1, A_2, A_3 (as predicted by Theorem 8), where the coefficients are expressed in terms of specific connection weights and scaling factors. For example, this means that if all of these values A_1, A_2, A_3 are reduced by 20%, all equilibrium values will be reduced by 20%. This indeed is the case in simulation examples. If the values of the connection weights and scaling factors are assigned as in the role matrices in Box 12.5, then the outcomes of the equilibrium values are (here the italic digits are repetitive):

$$\text{Person} = X_1 = \text{A}_1 \quad \text{Information} = X_2 = \text{A}_2 \quad \text{Emotion} = X_3 = \text{A}_3$$
$$\text{Relation} = X_4 = \text{A}_1$$
$$\text{Opinion} = X_5 = 0.17307692\text{A}_3 + 0.682692307\text{A}_2 + 0.1442307692\text{A}_1$$
$$\text{Interpretation} = X_6 = 0.40384615\text{A}_3 + 0.259615384\text{A}_2 + 0.336538461\text{A}_1$$
$$\text{Attention} = X_7 = 0.65384615\text{A}_3 + 0.13461538\text{A}_2 + 0.21153846\text{A}_1$$
$$\text{Arousal} = X_8 = 0.65384615\text{A}_3 + 0.13461538\text{A}_2 + 0.21153846\text{A}_1$$

Box 12.5 Example values for the connection weights, adopted from Schoenmaker et al. (2018)

mb base connectivity	1	2	3	4
X_1				
X_2				
X_3				
X_4	X_1			
X_5	X_2	X_6		
X_6	X_4	X_5	X_7	X_8
X_7	X_3	X_4	X_5	X_6
X_8	X_3	X_4	X_5	X_6
X_9	X_4	X_6	X_8	

mcw connection weights	1	2	3	4
X_1				
X_2				
X_3				
X_4	1			
X_5	1	0.75		
X_6	0.5	0.75	0.75	0.75
X_7	1	0.25	0.25	0.25
X_8	1	0.25	0.25	0.25
X_9	0.5	1	1	

mcfw combination function weights	1 eucl	2 alogistic
X_1	1	
X_2	1	
X_3	1	
X_4	1	
X_5	1	
X_6	1	
X_7	1	
X_8	1	
X_9		1

mcfp function	1 eucl		2 alogistic	
parameter	1 n	2 λ	1 σ	2 τ
X_1	1	1		
X_2	1	1		
X_3	1	1		
X_4	1	1		
X_5	1	1.75		
X_6	1	2.75		
X_7	1	1.75		
X_8	1	1.75		
X_9			2.5	1.25

ms speed factors	1
X_1	0
X_2	0
X_3	0
X_4	0.5
X_5	0.5
X_6	0.5
X_7	0.5
X_8	0.5
X_9	0.5

iv initial values	1
X_1	A_1
X_2	A_2
X_3	A_3
X_4	0
X_5	0
X_6	0
X_7	0
X_8	0
X_9	0

It can be seen that each of these equilibrium state values is a weighted average of A_1, A_2, and A_3 (for each the sum of these weights is 1, as predicted by Theorem 8). Therefore, in particular, when all A_i are 1, all of these outcomes are 1. If only A_1 and A_2 are 1, then the outcomes depend just on the emotional charge A_3:

$$
\begin{aligned}
\text{Person} = X_1 &= 1.0\,\text{Information} = X_2 = 1.0\,\text{Emotion} = X_3 = A_3 \\
\text{Relation} = X_4 &= 1.0 \\
\text{Opinion} = X_5 &= 0.17307692A_3 + 0.82692307 \\
\text{Interpretation} = X_6 &= 0.40384615A_3 + 0.59615384 \\
\text{Attention} = X_7 &= 0.6538461A_3 + 0.3461538 \\
\text{Arousal} = X_8 &= 0.6538461A_3 + 0.3461538 \\
\text{Sharing} = X_9 &= \mathbf{alogistic}_{\sigma,\tau}(0.5, 0.40384615A_3 \\
&\quad + 0.59615384, 0.6538461A_3 + 0.3461538)
\end{aligned}
$$

It can be seen from this analysis that the equilibrium values of Attention and Arousal depend for about 65% on the emotional charge level and as a consequence, the impact of the emotional charge on the equilibrium value of Interpretation is about 40%. The effect of emotional charge on Sharing works through two causal pathways: via Interpretation and via Arousal. This leads to the function

$$
\begin{aligned}
\mathbf{Sharing} = \mathbf{alogistic}_{\sigma,\tau}&(0.5, 0.40384615A_3 \\
&+ 0.59615384, 0.6538461A_3 + 0.3461538)
\end{aligned} \tag{12.6}
$$

of A_3, which is a monotonically increasing function of A_3.

12.9 Discussion

To analyse and predict from its structure what behaviour a given network model will eventually show is in general a challenging issue. For example, do all states in the network eventually converge to the same value? Some results are available for the case of acyclic, fully connected or strongly connected networks and for linear combination functions only; e.g., Bosse et al. (2015). It is often believed that when nonlinear functions are used, such results become impossible. Also, networks that are not strongly connected are often not addressed as they are more difficult to handle. This chapter shows what is still possible beyond the case of linear combination functions and also beyond the case of strongly connected networks. Parts of this chapter were adopted from Treur (2018b).

In this chapter general theorems were presented that relate network behaviour to the network structure characteristics. The relevant network structure characteristics concern two types of them:

- Network connectivity characteristics in terms of the network's strongly connected components and their mutual connections as shown in the network's condensation graph
- Network aggregation characteristics in terms of the combination functions used to aggregate the effects of multiple incoming connections (in particular, monotonicity, scalar-freeness, and normalisation).

The first item makes the approach applicable to any type of network connectivity, thus going beyond the limitation to strongly connected networks. The second item makes the approach applicable to a wider class of combination functions (most of which are nonlinear) going beyond the limitation to linear functions. However, there are also nonlinear functions that are not covered by this class. Some examples not covered are logistic functions, discrete threshold functions, and boolean functions, for example, as used in Karlsen and Moschoyiannis (2018), Watts (2002). The current chapter provides a first step to cover certain types of nonlinear functions. Nonlinear functions not covered yet form a next challenge that has been left open for now. In future research also other types of nonlinear functions will be explored further. Note that the notion of temporal-causal network itself is not a limitation as it is a very general notion which covers all types of discrete or smooth dynamical systems, and all systems of first-order differential equations. For these results, see Treur (2017), building further, among others, on Ashby (1960) and Port and van Gelder (1995).

The presented theorems subsume and generalise existing theorems for specific cases such as similar theorems for acyclic networks, fully connected networks and strongly connected networks (e.g., Theorems 3 and 4 in Sect. 12.6), and theorems addressing only linear combination functions as one fixed type of combination function; e.g., Theorem 3 at p. 120 of Bosse et al. (2015).

The theorems can be applied to predict behaviour of a given network, or to determine initial values in order to get some expected behaviour. In particular, they can be used as a method of verification to check the correctness of the implementation of a network. If simulation outcomes contradict the implications of the theorems, then some debugging of the implementation may be needed.

As already indicated in the Introduction section, after having developed the theorems presented here, it has turned out that these contributions also have some relations to research conducted from a different angle, namely on control of networks; e.g., Liu et al. (2011, 2012), Moschoyiannis et al. (2016), Haghighi and Namazi (2015), Karlsen and Moschoyiannis (2018). In that area, e.g., Liu et al. (2011, 2012), usually a system of linear differential equations is used for the dynamics of the considered network with N nodes $x_1, \ldots, x_N$ represented over time t by states $\mathbf{x}(t) = (x_1(t), \ldots, x_N(t))$. The dynamics is based on the connections with weights a_{ij} from x_j to x_i, overall represented by a matrix $\mathbf{A} = (a_{ij})$. For the control M additional nodes $u_1, \ldots, u_M$ are added, which are numerically represented over time t by states $\mathbf{u}(t) = (u_1(t), \ldots, u_M(t))$. These are meant to provide input at all time points in order to affect some of the network states (called drivers) over time. The latter nodes have connections to these driver nodes represented by an $N \times M$ input matrix $\mathbf{B} = (b_{ij})$ where b_{ij} represents the weight of the connection from node u_j to node x_i. Then overall the dynamics of the extended network can be represented as

$$d\mathbf{x}(t)/dt = \mathbf{Ax}(t) + \mathbf{Bu}(t)$$

Reachability within the network relates to the powers of matrix **A** and controllability of the network from the states $u_1, \ldots, u_M$ relates to the combined $N \times NM$ matrix $\mathrm{C} = (\mathbf{B}, \mathbf{AB}, \ldots, \mathbf{A}^{N-1}\mathbf{B}) \in \mathrm{R}^{N \times NM}$. Although precise mathematical criteria e.g., Kalman (1963) exist for this matrix **C** characterizing controllability of the network, such criteria often cannot be applied in practice as they depend on the precise values of the connection weights a_{ij} and in practical contexts usually these are not known. Therefore in literature such as (Liu et al. 2011, 2012) these weights are considered parameters, which introduces some complications: criteria for certain slightly different forms of (called structural) controllability are expressed in relation to these parameters e.g., Lin (1974); however, such criteria apply to (by far) most but not exactly all of such linear systems.

In contrast to the network control approach sketched in the previous paragraph, in the approach presented in the current chapter the lack of knowledge of specific weight values is not an issue, as these specific values are not used. Moreover, the theorems and their proofs do not make use of linearity assumptions, but instead of identified properties of a wider class of functions also including (a subset of the class of) nonlinear functions. Another difference is that the angle of controlling a network was not addressed in the current chapter, as the focus was on an angle of verification of a network model. However, some of the theorems still can be used for controlling a network. For example, Theorem 6 can be applied when the states $u_1, \ldots, u_M$ of the vector **u** in the above formalisation get outgoing connections (represented in matrix **B**) to the states within the level 0 components in the original network. Then the states within the level 0 components in the original network are used as drivers. More specifically, this theorem provides the following results for the considered class of nonlinear functions extending the class of linear functions:

- Theorem 6(a) and (b) show that the whole network can be controlled by only controlling the final equilibrium values of the states within the level 0 components of the network. This actually can be done by extending the network by nodes $\boldsymbol{u}_i$ that are connected to the states in level 0 components of the original network. In the extended network this leads to singleton level 0 components $\{u_i\}$ and the other levels are increased by 1; for example, the level 0 components in the original network now become level 1 components in the extended network. Then from Theorem 6(a) and (b) it follows that the equilibrium values of all states in the network depend on the equilibrium values of the states u_i in the level 0 components $\{u_i\}$, and these equilibrium values are $\lim_{t\to\infty} u_i(t)$, $i = 1, \ldots, M$.
- Note that if the u_i are kept constant over time, these limit values of the u_i are just the initial values $u_i(0)$; in this case Theorem 6(c) and (d) apply. For example, for

this case Theorem 6(c) shows that if these initial values $u_i(0)$ are all set at 1, then after some time all states of the network will get equilibrium value 1.

This illustrates how all states of the network can be controlled by only controlling the states within the level 0 components. Note that this has a partial overlap with what is found in Liu et al. (2012) for the linear case, where also a decomposition based on the network's strongly connected components is used. In Theorems 7 to 9 above it is described that some more can be said about how exactly the equilibrium value of each of the network's nodes depends on the initial or final values of the states in the level 0 components. In particular for the linear case, this equilibrium value of each state of the network is a linear function of the initial or equilibrium values of the states in the level 0 components.

References

Ashby, W.R.: Design for a Brain: The Origin of Adaptive Behaviour. Chapman and Hall, London, second extended edition (first edition, 1952) (1960)

Bloem, R., Gabow, H.N., Somenzi, F.: An algorithm for strongly connected component analysis in n log n symbolic steps. Form. Meth. Syst. Des. **28**, 37–56 (2006)

Bosse, T., Duell, R., Memon, Z.A., Treur, J., van der Wal, C.N.: Agent-based modelling of emotion contagion in groups. Cogn. Comp. **7**(1), 111–136 (2015)

Chen, Y.: General spanning trees and reachability query evaluation. In: Desai, B.C. (ed.) Proceedings of the 2nd Canadian Conference on Computer Science and Software Engineering, C3S2E'09, pp. 243–252. ACM Press, New York (2009)

Drechsler, R.: Advanced Formal Verification. Kluwer Academic Publishers, Dordrecht (2004)

Fisher, M.S.: Software Verification and Validation: An Engineering and Scientific Approach. Springer Science + Business Media, New York, NY (2007)

Fleischer, L.K., Hendrickson, B., Pınar, A.: On identifying strongly connected components in parallel. In: Rolim J. (ed.) Parallel and Distributed Processing. IPDPS 2000. Lecture Notes in Computer Science, vol. 1800, pp. 505–511. Springer, Berlin (2000)

Gentilini, R., Piazza, C., Policriti, A.: Computing strongly connected components in a linear number of symbolic steps. In: Proceedings of SODA'03, pp. 573–582 (2003)

Haghighi, R., Namazi, H.: Algorithm for identifying minimum driver nodes based on structural controllability. In: Mathematical Problems in Engineering, vol. 2015, Article ID 192307. http://dx.doi.org/10.1155/2015/192307 (2015)

Harary, F., Norman, R.Z., Cartwright, D.: Structural Models: An Introduction to the Theory of Directed Graphs. Wiley, New York (1965)

Kalman, R.E.: Mathematical description of linear dynamical systems. J. Soc. Indus. Appl. Math. Ser. A **1**, 152 (1963)

Karlsen, M., Moschoyiannis, S.: Evolution of control with learning classifier systems. Appl. Netw. Sci. **3**, 30 (2018)

Kuich, W.: On the entropy of context-free languages. Inf. Contr. **16**, 173–200 (1970)

Łacki, J.: Improved deterministic algorithms for decremental reachability and strongly connected components. ACM Trans. Algorithms **9**(3), Article 27 (2013)

Li, G., Zhu, Z., Cong, Z., Yang, F.: Efficient decomposition of strongly connected components on GPUs. J. Syst. Architect. **60**(1), 1–10 (2014)

Lin, C.-T.: Structural controllability. IEEE Trans. Automat. Contr. **19**, 201–208 (1974)

Liu, Y.Y., Slotine, J.J., Barabasi, A.L.: Controllability of complex networks. Nature **473**, 167–173 (2011)

Liu, Y.Y., Slotine, J.J., Barabasi, A.L.: Control centrality and hierarchical structure in complex networks. PLOS One **7**(9), e44459 (2012). https://journals.plos.org/plosone/article?id=10.1371/journal.pone.0044459#s4

Moschoyiannis, S., Elia, N., Penn, A.S., Lloyd, D.J.B., Knight, C.: A web-based tool for identifying strategic intervention points in complex systems. In: Brihaye, T., Delahaye, B., Jezequel, L., Markey, N., Srba, J. (eds.) Casting Workshop on Games for the Synthesis of Complex Systems and 3rd International Workshop on Synthesis of Complex Parameters (Cassting'16/SynCoP'16). EPTCS, vol. 220, pp. 39–52 (2016)

Port, R.F., van Gelder, T.: Mind as Motion: Explorations in the Dynamics of Cognition. MIT Press, Cambridge, MA (1995)

Schoenmaker, R., Treur, J., Vetter, B.: A temporal-causal network model for the effect of emotional charge on information sharing. Biol. Inspired Cogn. Arch. **26**, 136–144 (2018)

Tarjan, R.: Depth-first search and linear graph algorithms. SIAM J. Comput. **1**(2), 146–160 (1972)

Treur, J.: Verification of temporal-causal network models by mathematical analysis. Vietnam. J. Comput. Sci. **3**, 207–221 (2016a)

Treur, J.: Network-Oriented Modeling: Addressing the Complexity of Cognitive, Affective and Social Interactions. Springer Publishers, Berlin (2016b)

Treur, J.: On the applicability of network-oriented modeling based on temporal-causal networks: why network models do not just model networks. J. Inf. Telecommun. **1**(1), 23–40 (2017)

Treur, J.: Relating emerging network behaviour to network structure. In: Proceedings of the 7th International Conference on Complex Networks and their Applications, ComplexNetworks'18, vol. 1. Studies in Computational Intelligence, vol. 812, pp. 619–634. Springer Publishers, Berlin (2018a)

Treur, J.: Mathematical analysis of a network's asymptotic behaviour based on its strongly connected components. In: Proceedings of the 7th International Conference on Complex Networks and their Applications, ComplexNetworks'18, vol. 1. Studies in Computational Intelligence, vol. 812, pp. 663–679. Springer Publishers, Berlin (2018b)

Treur, J.: The ins and outs of network-oriented modeling: from biological networks and mental networks to social networks and beyond. Trans. Comput. Collect. Intell. **32**, 120–139. Paper for Keynote Lecture at ICCCI'18 (2019)

Watts, D.J.: A simple model of global cascades on random networks. Proc. Natl. Acad. Sci. U. S. A. **99**(9), 5766–5771 (2002)

Wijs, A., Katoen, J.P., Bošnacki, D.: Efficient GPU algorithms for parallel decomposition of graphs into strongly connected and maximal end components. Form. Methods Syst. Des. **48**, 274–300 (2016)

Part VI
Mathematical Analysis of How Emerging Network Behaviour of Adaptive Networks Relates to Reified Network Structure

Chapter 13
Relating a Reified Adaptive Network's Structure to Its Emerging Behaviour for Bonding by Homophily

Abstract In this chapter it is analysed how emerging behaviour in an adaptive social network for bonding can be related to characteristics of the adaptive network's structure, which includes the structure of the adaptation principles incorporated. In particular, this is addressed for adaptive social networks for bonding based on homophily and for community formation in such adaptive social networks. To this end, relevant characteristics of the reified network structure (including the adaptation principle) have been identified, such as a tipping point for similarity as used for homophily. Applying network reification, the adaptive network characteristics are represented by reification states in the extended network, and adaptation principles are described by characteristics of these reification states, in particular their connectivity characteristics (their connections) and their aggregation characteristics (in terms of their combination functions). According to this network reification approach, as one of the results it has been found how the emergence of communities strongly depends on the value of this similarity tipping point. Moreover, it is shown that some characteristics entail that the connection weights all converge to 0 (for persons in different communities) or 1 (for persons within one community).

13.1 Introduction

In this chapter, the emerging behaviour of the coevolution of social contagion (Levy and Nail 1993) and bonding by homophily (McPherson et al. 2001; Pearson et al. 2006) is analysed. In particular, it is analysed how emerging communities based on the coevolution of social contagion and bonding by homophily can be related to characteristics of the adaptive network's structure. For this adaptive case, this network structure includes the structure of the homophily adaptation principle that is incorporated. The homophily adaptation principle expresses how 'being alike' strengthens the connection between two persons (McPherson et al. 2001; Pearson et al. 2006). Social contagion implies that the stronger two persons are connected, the more they will become alike (Levy and Nail 1993). Thus a reciprocal

J. Treur, *Network-Oriented Modeling for Adaptive Networks: Designing Higher-Order Adaptive Biological, Mental and Social Network Models*, Studies in Systems, Decision and Control 251, https://doi.org/10.1007/978-3-030-31445-3_13

causal relation between the two processes occurs. It is known from simulations (Axelrod 1997; Holme, and Newman 2006; Sharpanskykh and Treur 2014; Vazquez 2013; Vazquez et al. 2007) that the emerging behaviour of adaptive network models combining these two processes as a form of coevolution often shows community formation. In the resulting network structure within a community persons have high mutual connection weights and a high extent of 'being alike', and persons from different communities have low mutual connection weights and a low extent of 'being alike'.

Relevant characteristics of the network and the homophily adaptation principle have been identified, such as a tipping point for homophily. As one of the results, it has been found how the emergence of communities strongly depends on the value of this tipping point. Moreover, it is shown that some characteristics of the reified network structure entail that the connection weights all converge to 0 (for states in different emerging communities) or 1 (for states within one emerging community).

In general, it is a challenging issue for dynamic models to predict what patterns of behaviour will emerge, and how their emergence depends on the structure of the model. The latter includes chosen values used for characteristics of the model's structure (and settings). This also applies to network models, where behaviour depends in some way on the network structure, defined by network characteristics such as connectivity (connections and their weights); e.g., Turnbull et al. (2018). In this context, the issue is how emerging network behaviour relates to network structure; for example, see (Treur 2018a). It can be an even more challenging issue when coevolution that occurs in adaptive networks is considered, in which case by a mutual causal interaction both the states and the network characteristics change over time. In this case the connections in the network change according to certain adaptation principles which depend on certain adaptation characteristics.

In the current chapter, the emergence of communities based on the coevolution of social contagion (Levy and Nail 1993) and bonding by homophily (McPherson et al. 2001; Pearson et al. 2006) is analysed in some depth. By mathematical analysis, it is found out how emerging communities relate to the characteristics of the network and of the specific homophily adaptation principle used in combination with a social contagion principle.

The issue was addressed using the Network-Oriented Modeling approach based on temporal-causal networks (Treur 2016, 2019) as a vehicle. For temporal-causal networks, characteristics of the network structure are Connectivity, Aggregation, and Timing, represented by connection weights, combination functions, and speed factors, respectively. For the type of adaptive networks considered, the connection weights are dynamic based on the homophily principle. When network reification, see Chap. 3 or (Treur 2018d) is applied, these adaptation principles are represented as an extended part of the network using certain reification states in the network extension. These reification states have their own (reified) network structure characteristics defining their Connectivity, Aggregation, and Timing by their connection weights, speed factors and combination functions. So, the challenge then is how the emergent behaviour of the reified adaptive network depends on these characteristics and, in particular, on Aggregation characteristics of reification states in terms of properties of the combination functions used by them for bonding by homophily.

In Fig. 13.1 the basic relation between structure and dynamics of a reified network model is indicated by the horizontal arrow in the lower part representing the base level. For a reified network these also apply to the reification states. The network structure characteristics (addressed in Sect. 13.5) cover properties of the adaptation principle based on bonding by homophily represented by the reification states at the reification level (in Sect. 13.5.1) and properties of the social contagion in the base network (in Sect. 13.5.2). Properties of the behaviour are discussed in Sect. 13.4 first by a number of simulation experiments, and they are formalised and related to the network structure properties (the horizontal arrow in the upper part of Fig. 13.1) in Sect. 13.6.

In the research reported here characteristics of the combination function describing the aggregation used by the homophily adaptation principle have been identified that play an important role in community formation, among which the tipping point for the similarity between two persons. This is the point where 'being alike' turns into 'not being alike', or conversely. In this chapter, results are discussed that have been proven mathematically for this relationship between network structure and network behavior for the coevolution process. In particular, for emerging communities what the connections become between persons from one community or persons from different communities, and how different persons from one community or from different communities become. Note that these results have not been proven for one specific model or combination function, but for whole classes of models with a variety of combination functions that fulfill certain properties.

In this chapter, in Sect. 13.2 the Network-Oriented Modeling approach used is briefly outlined. In Sect. 13.3 adaptive networks based on homophily are described and a number of functions that can be used to model them. Section 13.4 shows some example simulations. In Sect. 13.5 relevant reified network structure characteristics are defined that are used in Sect. 13.6 to prove results on the relation between network structure and behaviour. In Sect. 13.7 it is shown how the obtained results relate to the strongly connected components of a network. In Sect. 13.8 an overview is presented of various simulation experiments concerning bonding by homophily, most of which in relation to empirical data. Finally, Sect. 13.9 is a discussion.

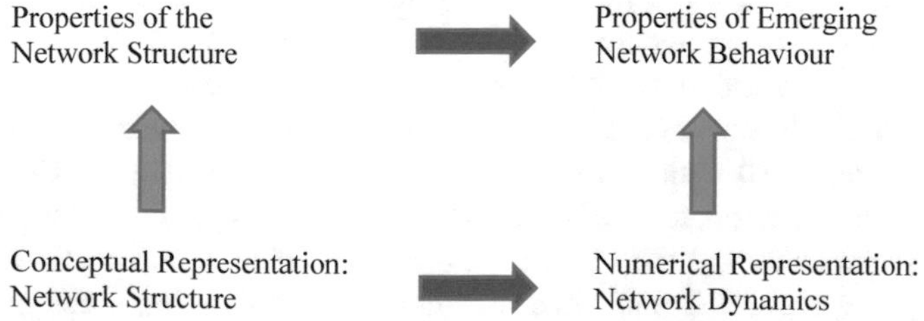

Fig. 13.1 Bottom layer: the conceptual representation of a reified network model defines the numerical representation. Top layer: properties of reified network structure entail properties of emerging reified network behaviour

13.2 Network-Oriented Modeling by Temporal-Causal Networks

In order to undertake any mathematical analysis of networks, in the first place, a solid definition of the concept of network is needed, based on well-defined semantics. In the current chapter, the interpretation of connections based on causality and dynamics forms a basis of the structure and semantics of the considered networks. It is a deterministic dynamic modeling approach, for example in the line of (Ashby 1960), in contrast to, stochastic modeling approaches as, for example, used in Axelrod (1997). It can be positioned as a branch in the causal modelling area which has a long tradition in AI; e.g., see Kuipers and Kassirer (1983), Kuipers (1984), Pearl (2000). It distinguishes itself by a dynamic perspective on causal relations, according to which causal relations exert causal effects over time, and these causal relations themselves can also change over time.

More specifically, the nodes in a network are interpreted here as states (or state variables) that vary over time, and the connections are interpreted as causal relations that define how each state can affect other states over time. This type of network has been called a *temporal-causal network* (Treur 2016b); note that the word temporal here refers to the causality, not to the network. Temporal-causal networks that themselves change over time as well are called adaptive temporal-causal networks; e.g., Gross and Sayama (2009). So, in cases of adaptive temporal-causal networks, in addition to the node states also the connections are assumed to change over time and are therefore treated like states as well.

A conceptual representation of a temporal-causal network model by a *labeled graph* provides a fundamental basis. Such a conceptual representation includes representing in a declarative manner states (also called nodes) and connections between them that represent (causal) impacts of states on each other. This part of a conceptual representation is often depicted in a *conceptual picture* by a graph with nodes and directed connections. However, a *complete conceptual representation* of a temporal-causal network model also includes a number of labels for such a graph. A notion of *strength of a connection* is used as a label for connections, some way to *aggregate multiple causal impacts* on a state is used, and a notion of *speed of change* of a state is used for timing of the processes. Note that states have one value; they can relate one by one to states of persons or agents, for example, the strength of their opinion states. It is also possible to model each person by more than one state, for example, an opinion and an emotion state per person. In such a case a person does not relate to a single state but to a subnetwork consisting of multiple states.

The three described notions for network structure characteristics Connectivity, Aggregation, and Timing, are called connection weight $\boldsymbol{\omega}_{X,Y}$, combination function $\mathbf{c}_Y(..)$, and speed factor $\boldsymbol{\eta}_Y$. They make the graph of states and connections a labeled graph (e.g., see Fig. 13.2), and form the defining structure of a temporal-causal network model in the form of a conceptual representation; for a summary, see also Table 13.1, top half: rows 1–5. Note that although in general that is not always required, for the current chapter all connection weights are assumed nonnegative: $\boldsymbol{\omega}_{X,Y} \in [0, 1]$ for all X and Y.

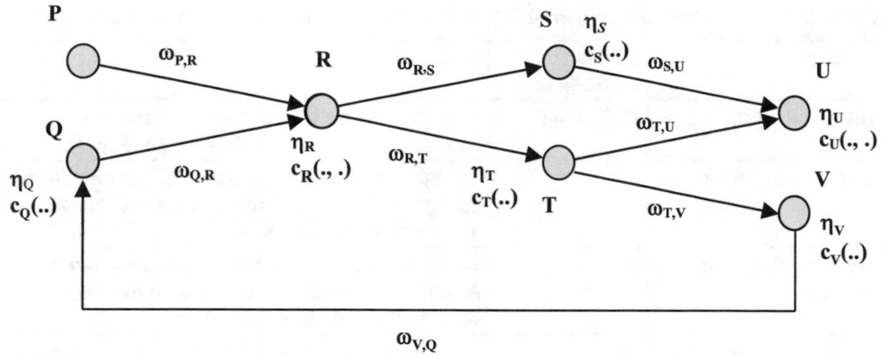

Fig. 13.2 Conceptual representation of an example temporal-causal network model: adopted from Treur (2016b)

The interpretation of a network based on causality and dynamics can be expressed in a formal-numerical way, thus associating semantics to any conceptual temporal-causal network representation in a detailed numerical-mathematically defined manner. For a summary, see Table 13.1, bottom half: rows 6 to 10. This shows how a conceptual representation based on states and connections enriched with labels for connection weights, combination functions, and speed factors, can be transformed into a numerical representation (Treur 2016b, Chap. 2). A more detailed explanation of the difference equation format, taken from Treur (2016b, Chap. 2, pp. 60–61), is as follows; see also Fig. 13.3. The aggregated impact value $\mathbf{aggimpact}_Y(t)$ at time t pushes the value of Y up or down, depending on how it compares to the current value of Y. So, $\mathbf{aggimpact}_Y(t)$ is compared to the current value $Y(t)$ of Y at t by taking the difference between them (also see Fig. 13.3): $\mathbf{aggimpact}_Y(t) - Y(t)$. If this difference is positive, which means that $\mathbf{aggimpact}_Y(t)$ at time t is higher than the current value of Y at t, in the time step from t to $t + \Delta t$ (for some small Δt) the value $Y(t)$ will increase in the direction of the higher value $\mathbf{aggimpact}_Y(t)$. This increase is done proportional to the difference, with proportion factor $\boldsymbol{\eta}_Y \Delta t$: the increase is $\boldsymbol{\eta}_Y\, [\mathbf{aggimpact}_Y(t) - Y(t)]\, \Delta t$; see Fig. 13.3. By this format, the parameter $\boldsymbol{\eta}_Y$ indeed acts as a speed factor by which it can be specified how fast state Y should change upon causal impact.

There are many different approaches possible to address the issue of combining multiple causal impacts. To provide sufficient flexibility, the Network-Oriented Modelling approach based on temporal-causal networks incorporates for each state a way to specify how multiple causal impacts on this state are aggregated by a combination function. For this aggregation, a library with a number of standard combination functions are available as options (currently 35), but also own-defined functions can be added. The difference equations in Fig. 13.3 and in the last row in Table 13.1 constitute the overall numerical representation of the temporal-causal network model and can be used for simulation and mathematical analysis; it can also be written in differential equation format:

Table 13.1 Conceptual and numerical representations of a temporal-causal network model

Concept	Conceptual representation	Explanation
States and connections	X, Y, $X \rightarrow Y$	Describes the nodes and links of a network structure (e.g., in graphical or matrix format)
Connection weight	$\boldsymbol{\omega}_{X,Y}$	The *connection weight* $\boldsymbol{\omega}_{X,Y} \in [-1, 1]$ represents the strength of the causal impact of state X on state Y through connection $X \rightarrow Y$
Aggregating multiple impacts on a state	$\mathbf{c}_Y(..)$	For each state Y (a reference to) a *combination function* $\mathbf{c}_Y(..)$ is chosen to combine the causal impacts of other states on state Y
Timing of the effect of causal impact	$\boldsymbol{\eta}_Y$	For each state Y a *speed factor* $\boldsymbol{\eta}_Y \geq 0$ is used to represent how fast a state is changing upon causal impact

Concept	Numerical representation	Explanation
State values over time t	$Y(t)$	At each time point t each state Y in the model has a real number value, usually in [0, 1]
Single causal impact	$\mathbf{impact}_{X,Y}(t) = \boldsymbol{\omega}_{X,Y}\, X(t)$	At t state X with a connection to state Y has an impact on Y, using connection weight $\boldsymbol{\omega}_{X,Y}$
Aggregating multiple causal impacts	$\mathbf{aggimpact}_Y(t) = \mathbf{c}_Y(\mathbf{impact}_{X_1,Y}(t), \ldots, \mathbf{impact}_{X_k,Y}(t)) = \mathbf{c}_Y(\boldsymbol{\omega}_{X_1,Y} X_1(t), \ldots, \boldsymbol{\omega}_{X_k,Y} X_k(t))$	The aggregated causal impact of multiple states X_i on Y at t, is determind using combination function $\mathbf{c}_Y(..)$
Timing of the causal effect	$Y(t+\Delta t) = Y(t) + \boldsymbol{\eta}_Y[\mathbf{aggimpact}_Y(t) - Y(t)]\Delta t = Y(t) + \boldsymbol{\eta}_Y[\mathbf{c}_Y(\boldsymbol{\omega}_{X_1,Y} X_1(t), \ldots, \boldsymbol{\omega}_{X_k,Y} X_k(t)) - Y(t)]\Delta t$	The causal impact on Y is exerted over time gradually, using speed factor $\boldsymbol{\eta}_Y$; here the X_i are all states with outgoing connections to state Y

$$Y(t+\Delta t) = Y(t) + \boldsymbol{\eta}_Y\left[\mathbf{c}_Y\left(\boldsymbol{\omega}_{X_1,Y} X_1(t), \ldots, \boldsymbol{\omega}_{X_k,Y} X_k(t)\right) - Y(t)\right]\Delta t \tag{13.1}$$

$$\mathbf{d}Y(t)/\mathbf{d}t = \boldsymbol{\eta}_Y\left[\mathbf{c}_Y\left(\boldsymbol{\omega}_{X_1,Y} X_1(t), \ldots, \boldsymbol{\omega}_{X_k,Y} X_k(t)\right) - Y(t)\right]$$

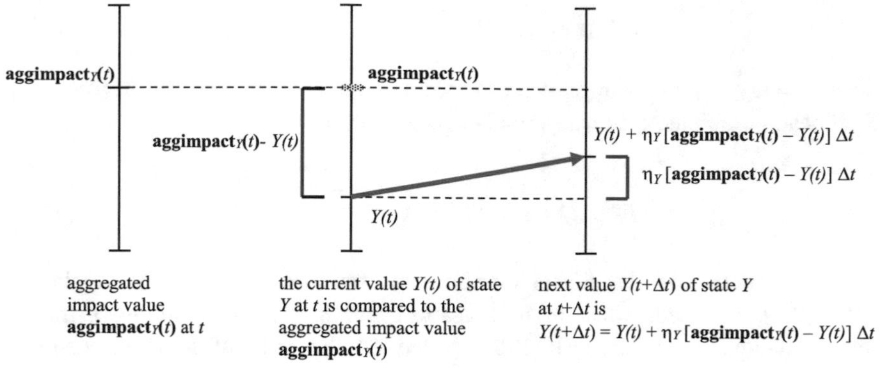
aggimpact$_Y$(t)
aggimpact$_Y$(t)
aggimpact$_Y$(t)- Y(t)
Y(t) + η$_Y$ [aggimpact$_Y$(t) – Y(t)] Δt
η$_Y$ [aggimpact$_Y$(t) – Y(t)] Δt
Y(t)
aggregated impact value aggimpact$_Y$(t) at t
the current value Y(t) of state Y at t is compared to the aggregated impact value aggimpact$_Y$(t)
next value Y(t+Δt) of state Y at t+Δt is
Y(t+Δt) = Y(t) + η$_Y$ [aggimpact$_Y$(t) – Y(t)] Δt

Fig. 13.3 How **aggimpact**$_Y(t)$ makes a difference for state $Y(t)$ in the time step from t to $t + \Delta t$

For reified adaptive networks, connection weights are dynamic and therefore handled by reification states for them, with their own connectivity and aggregation characteristics in terms of connections and combination functions. This will be shown in more detail in Sects. 13.3–13.5.

Often used examples of combination functions are the *identity* function **id**(.) for states with impact from only one other state, the *scaled sum* function $\mathbf{ssum}_\lambda$(.) with scaling factor $\boldsymbol{\lambda}$, and the *advanced logistic sum* combination function $\mathbf{alogistic}_{\sigma,\tau}$(..) with steepness $\boldsymbol{\sigma}$ and threshold $\boldsymbol{\tau}$:

$$
\begin{aligned}
&\mathbf{id}(V) = V \\
&\mathbf{ssum}_\lambda(V_1, \ldots, V_k) = \frac{V_1 + \cdots + V_k}{\lambda} \\
&\mathbf{alogistic}_{\sigma,\tau}(V_1, \ldots, V_k) = \left[\frac{1}{1 + \mathrm{e}^{-\sigma(V_1 + \cdots + V_k - \tau)}} - \frac{1}{1 + \mathrm{e}^{\sigma\tau}}\right](1 + \mathrm{e}^{-\sigma\tau})
\end{aligned}
\tag{13.2}
$$

Note that for $\lambda = 1$, the scaled sum function is just the sum function **sum(..)**, and this sum function can also be used as the identity function in case of one incoming connection. In addition to the above functions, a *Euclidean combination function* is defined as

$$
\mathbf{c}(V_1, \ldots, V_k) = \mathbf{eucl}_{n,\lambda}(V_1, \ldots, V_k) = \sqrt[n]{\frac{V_1^n + \cdots + V_k^n}{\lambda}} \tag{13.3}
$$

where $\boldsymbol{n}$ is the *order* (which can be any positive natural number but also any positive real number), and $\boldsymbol{\lambda}$ is a scaling factor. This can be used when all connection weights are non-negative, but in the specific case that $\boldsymbol{n}$ is an odd natural number, also negative connection weights can be allowed. A Euclidean combination function is called *normalised* if the scaling factor $\boldsymbol{\lambda}$ is chosen in such a way that $\mathbf{c}(\boldsymbol{\omega}_{X_1,Y}, \ldots, \boldsymbol{\omega}_{X_k,Y}) = 1$; this is achieved for $\boldsymbol{\lambda} = \boldsymbol{\omega}_{X_1,Y}^n + \cdots + \boldsymbol{\omega}_{X_k,Y}^n$. Note that for $\boldsymbol{n} = 1$ (first order) the scaled sum function is obtained:

$$\mathbf{eucl}_{1,\lambda}(V_1, \ldots, V_k) = \mathbf{ssum}_{\lambda}(V_1, \ldots, V_k) \tag{13.4}$$

Then $\boldsymbol{\lambda} = \boldsymbol{\omega}_{X_1,Y} + \cdots + \boldsymbol{\omega}_{X_k,Y}$ makes it normalised. For $\boldsymbol{n}$ = 2 it is the second-order Euclidean combination function $\mathbf{eucl}_{2,\lambda}(..)$ defined by:

$$\mathbf{eucl}_{2,\lambda}(V_1, \ldots, V_k) = \sqrt{\frac{V_1^2 + \cdots + V_k^2}{\lambda}} \tag{13.5}$$

Such a second-order Euclidean combination function is also often applied in aggregating the error value in optimisation and in parameter tuning using the root-mean-square deviation (RMSD), based on the Sum of Squared Residuals (SSR).

13.3 Reified Adaptive Networks for Bonding Based on Homophily

The homophily principle addresses bonding between persons. It describes how connections between two persons are strengthened or weakened depending on the extent of similarity between them: more 'being alike' will make the persons more 'like each other' (McPherson et al. 2001; Pearson et al. 2006). This is modelled by how the states $X(t)$ and $Y(t)$ the persons X and Y have at time t compare to each other, for example, indicating whether at time t they enjoy being physically active (high value) or not (low value), or indicating the extent to which they agree with a certain opinion. According to the homophily principle the weight $\boldsymbol{\omega}_{X,Y}$ of the connection from X to Y is changing over time dynamically, depending on how the state levels $X(t)$ and $Y(t)$ differ.

13.3.1 Modeling the Bonding by Homophily Adaptation Principle by Reification

As the connection weight $\boldsymbol{\omega}_{X,Y}$ is dynamic, using network reification it is handled as a reification state $\mathbf{W}_{X,Y}$ with its own combination function $\mathbf{c}_{\mathbf{W}_{X,Y}}(V_1, V_2, W)$, called in this chapter a *homophily combination function*. Here V_1 and V_2 refer to the values for $X(t)$ and $Y(t)$ of the states X and Y involved and W to the value $\mathbf{W}_{X,Y}(t)$ of the reified connection weight $\mathbf{W}_{X,Y}$. See Fig. 13.4, where the homophily functions $\mathbf{c}_{\mathbf{W}_{X_1,Y}}(V_1, V_2, W)$ and $\mathbf{c}_{\mathbf{W}_{X_2,Y}}(V_1, V_2, W)$ are indicated as labels (like the labels in Fig. 13.2) for the reification states $\mathbf{W}_{X_1,Y}$ and $\mathbf{W}_{X_2,Y}$. Similarly labels for adaptation speed can be added, and labels for the incoming connections for the reification states.

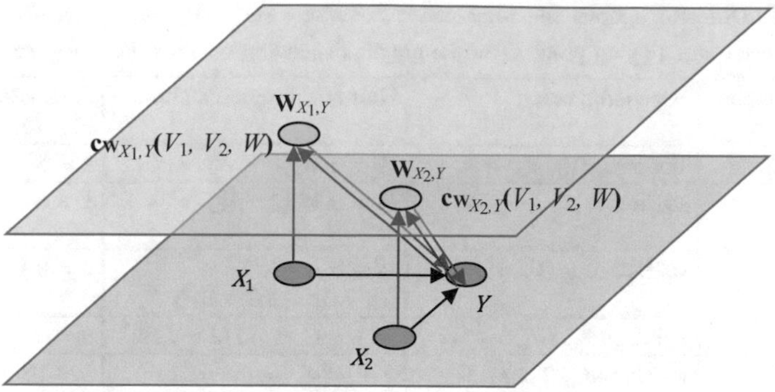

Fig. 13.4 Homophily combination functions as labels in the reified network

Then the standard difference and differential equation format for temporal-causal networks as shown in Sect. 13.2 is applied:

$$\begin{aligned} \mathbf{W}_{X,Y}(t+\Delta t) &= \mathbf{W}_{X,Y}(t) + \boldsymbol{\eta}_{\mathbf{W}_{X,Y}}[\mathbf{cw}_{X,Y}(X(t), Y(t), \mathbf{W}_{X,Y}(t)) - \mathbf{W}_{X,Y}(t)]\Delta t \\ \mathbf{dW}_{X,Y}/\mathbf{d}t &= \boldsymbol{\eta}_{\mathbf{W}_{X,Y}}[\mathbf{cw}_{X,Y}(X, Y, \mathbf{W}_{X,Y}) - \mathbf{W}_{X,Y}] \end{aligned} \quad (13.6)$$

Here $\boldsymbol{\eta}_{\mathbf{W}_{X,Y}}$ is the speed factor of connection weight reification state $\mathbf{W}_{X,Y}$. It determines how fast adaptation takes place. Note that

- $\mathbf{W}_{X,Y}(t)$ increases at t if and only if $\mathbf{cw}_{X,Y}$ $(X(t), Y(t), \mathbf{W}_{X,Y}(t)) > \mathbf{W}_{X,Y}(t)$
- $\mathbf{W}_{X,Y}(t)$ decreases if and only if $\mathbf{cw}_{X,Y}(X(t), Y(t), \mathbf{W}_{X,Y}(t)) < \mathbf{W}_{X,Y}(t)$
- $\mathbf{W}_{X,Y}(t)$ is stationary if and only if $\mathbf{cw}_{X,Y}(X(t), Y(t), \mathbf{W}_{X,Y}(t)) = \mathbf{W}_{X,Y}(t)$

Specific adaptation principles formalising bonding by homophily can be obtained by picking specific functions for the homophily combination function $\mathbf{cw}_{X,Y}(V_1, V_2, W)$ for a given application domain, or even different combination functions $\mathbf{cw}_{X,Y}(V_1, V_2, W)$ for different persons within a given domain. There are many options for such choices; see, for example, Table 13.2 and Fig. 13.5. Therefore it may be more useful to define classes of such homophily combination functions characterised by certain properties of them. Then as will be shown in Sect. 13.6, results can be proven for such a class instead of for each homophily combination function separately. Formal definitions of such properties of the function $\mathbf{cw}_{X,Y}(V_1, V_2, W)$ will be given in Sect. 13.5, but here some examples of homophily combination functions are shown.

13.3.2 *Various Examples of Homophily Combination Functions*

A very simple example of a homophily combination function $\mathbf{cw}_{X,Y}(V_1, V_2, W)$ is a linear function in $D = |V_1 - V_2|$ defined as follows:

Table 13.2 Different options for combination functions $\mathbf{cw}_{\omega_{X,Y}}(V_1, V_2, W)$ for the homophily principle based on a tipping point τ; for the graphs depending on $D = |V_1 - V_2|$, see Fig. 13.5

Function type	Function name	Numerical representation	Parameters in Fig. 13.5
Simple Linear	$\mathbf{slhomo}_{\tau,\alpha}(V_1, V_2, W)$	$W + \alpha\, W\,(1 - W)\,(\tau - D)$	$\alpha = 6$
Simple quadratic 1	$\mathbf{sq1homo}_{\tau,\alpha}(V_1, V_2, W)$	$W + \alpha\, W\,(1 - W)\,(\tau^2 - D^2)$	$\alpha = 6$
Simple quadratic 2	$\mathbf{sq2homo}_{\tau,\alpha,\delta}(V_1, V_2, W)$	$W + \alpha\,((\tau + \delta)^2 - (D + \delta)^2)$	$\delta = 0.15$, $\alpha = 10$
Cubic	$\mathbf{cubehomo}_{\tau,\alpha}(V_1, V_2, W)$	$W + \alpha\,(1 - W)\,(1 - D/\tau)^3$	$\alpha = 0.9$
Logistic 1	$\mathbf{log1homo}_{\tau,\sigma}(V_1, V_2, W)$	$\frac{W}{W+(1-W)e^{(D-\tau)}}$	$\sigma = 10$
Logistic 2	$\mathbf{slog2homo}_{\tau,\sigma,\alpha}(V_1, V_2, W)$	$W + \alpha \frac{W(1-W)}{1+e^{-(D-\tau)}}$	$\sigma = 4$, $\alpha = 5$
Sine-based	$\mathbf{sinhomo}_{\tau,\alpha}(V_1, V_2, W)$	$W - \alpha\,(1 - W)\,\sin(\pi\,(D - \tau)/2)$	$\alpha = 2$
Tangent-based	$\mathbf{tanhomo}_{\tau,\alpha}(V_1, V_2, W)$	$W - \alpha\,(1 - W)\,\tan(\pi\,(D - \tau)/2)$	$\alpha = 2$
Exponential	$\mathbf{exphomo}_{\tau,\sigma}(V_1, V_2, W)$	$1 - (1 - W)\,e^{\sigma(D-\tau)}$	$\sigma = 10$

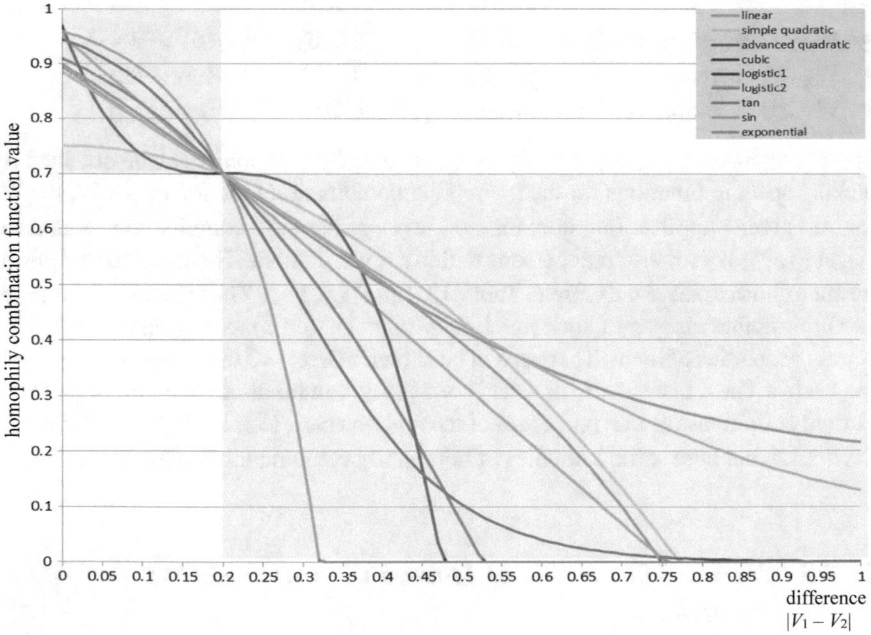

Fig. 13.5 Graphs for different options for homophily combination functions $\mathbf{cw}_{\omega_{X,Y}}(V_1, V_2, W)$ with tipping point $\tau = 0.2$ and $W = 0.7$ with difference $D = |V_1 - V_2|$ on the horizontal axis

$$\mathbf{c}_{\mathbf{W}_{X,Y}}(V_1, V_2, W) = W + \boldsymbol{\beta}(\tau - D) \tag{13.7}$$

Here $\boldsymbol{\beta}$ is a modulation factor that still can be chosen, and τ is a tipping point (or threshold) parameter. This function may have the disadvantage that when $W = 0$ it may be negative (when $D > \tau$) or when $W = 1$ it may be higher than 1 (when $D < \tau$), so that it has to be cut off to avoid that the reified connection weight value $\mathbf{W}_{X,Y}$ goes outside the interval [0, 1]. This can also be remedied in a smooth manner by choosing $\boldsymbol{\beta}$ as a function $\boldsymbol{\beta}(W) = \boldsymbol{\alpha}\, W(1 - W)$ of W which can suppress the term $\tau - D$ when W comes closer to 0 or 1. This function makes that $\mathbf{W}_{X,Y}$ will not cross the boundaries 0 and 1:

$$\mathbf{c}_{\mathbf{W}_{X,Y}}(V_1, V_2, W) = W + \boldsymbol{\alpha}\, W(1 - W)(\tau - |V_1 - V_2|) \tag{13.8}$$

Here $\boldsymbol{\alpha}$ is a modulation or amplification parameter; the higher its value, the stronger the effect (either positive or negative) of homophily on the connection. Using this homophily combination function, the dynamic relations for $\mathbf{W}_{X,Y}$ are:

$$\begin{aligned}
&\mathbf{dW}_{X,Y}/\mathbf{d}t = \boldsymbol{\eta}_{\mathbf{W}_{X,Y}} \boldsymbol{\alpha}\, \mathbf{W}_{X,Y}(1 - \mathbf{W}_{X,Y})(\tau - |X - Y|) \\
&\mathbf{W}_{X,Y}(t + \Delta t) = \mathbf{W}_{X,Y}(t) + \boldsymbol{\eta}_{\mathbf{W}_{X,Y}} \boldsymbol{\alpha}\, \mathbf{W}_{X,Y}(t)(1 - \mathbf{W}_{X,Y}(t))(\tau - |X(t) - Y(t)|)\Delta t
\end{aligned} \tag{13.9}$$

As a variant of this homophily combination function $\mathbf{c}_{\mathbf{W}_{X,Y}}(V_1, V_2, W)$ that is linear in D, the following function can be obtained as a quadratic function of $D = |V_1 - V_2|$:

$$\mathbf{c}_{\mathbf{W}_{X,Y}}(V_1, V_2, W) = W + \boldsymbol{\alpha}\, W(1 - W)\left(\tau^2 - (V_1 - V_2)^2\right) \tag{13.10}$$

Using this combination function, the dynamic relations for the reification state $\mathbf{W}_{X,Y}$ are:

$$\begin{aligned}
&\mathbf{dW}_{X,Y}/\mathbf{d}t = \boldsymbol{\eta}_{\mathbf{W}_{X,Y}} \boldsymbol{\alpha}\, \mathbf{W}_{X,Y}(1 - \mathbf{W}_{X,Y})\left(\tau - (X - Y)^2\right) \\
&\mathbf{W}_{X,Y}(t + \Delta t) = \mathbf{W}_{X,Y}(t) + \boldsymbol{\eta}_{\mathbf{W}_{X,Y}} \boldsymbol{\alpha}\, \mathbf{W}_{X,Y}(t)(1 - \mathbf{W}_{X,Y}(t))\left(\tau - (X(t) - Y(t))^2\right)\Delta t
\end{aligned} \tag{13.11}$$

In Table 13.2 and Fig. 13.5 these linear and quadratic combination function and a number of other examples of homophily combination functions are depicted, with $D = |V_1 - V_2|$ on the horizontal axis and $\mathbf{c}_{\mathbf{W}_{X,Y}}(V_1, V_2, W)$ on the vertical axis for $\tau = 0.2$ and $W = 0.7$. In Sect. 13.6 the results that have been proven are general in the sense that they apply to all of such functions, not just one specific function. To this end, in Sect. 13.5 relevant properties shared by these functions are defined. The simple linear combination function $\mathbf{slhomo}_{\tau,\alpha}(V_1, V_2, W)$ was used in Blankendaal et al. (2016), and the simple quadratic combination function $\mathbf{sqhom}_{\tau,\alpha}(V_1, V_2, W)$ in van Beukel et al. (2017). A more advanced logistic combination function based on $\mathbf{log2homo}_{\tau,\sigma,\alpha}(V_1, V_2, W)$ was explored in Boomgaard et al. (2018).

13.4 Example Simulations for the Coevolution of Social Contagion and Bonding

In this section, a few example simulations for the coevolution of social contagion and bonding based on homophily are described. It will be shown how communities emerge and more specifically how their emergence depends on properties of the chosen functions for homophily. These examples will be used in Sect. 13.6 to illustrate the results that have been proven. The examples concern a fully connected social network of 10 states with speed factors $\mathbf{\eta}_Y = 0.1$ for states Y and initial values $\mathbf{W}_{X,Y}(0)$ for the reified connection weight from state X to state Y as shown in Table 13.3.

As combination functions for social contagion for the states, the normalised scaled sum functions were used, and as combination functions for the connections for each $\mathbf{W}_{X,Y}$ the simple linear homophily function $\mathbf{slhomo}_{\tau,\alpha}(V_1, V_2, W)$ with tipping point $\tau = 0.1$, and speed factor $\mathbf{\eta}_{\mathbf{W}_{X,Y}} = 0.4$. The graphs in the left-hand side of Figs. 13.6 and 13.7 show time on the horizontal axis and activation values of states at the vertical axis until it ends up in an equilibrium state, i.e., all values become constant. In the same figures on the right-hand side matrices with the final connection weights are shown.

13.4.1 Simulations for Varying Homophily Modulation Factor

The homophily modulation factor α is varying from 1 to 10 in Fig. 13.6, and from 11 to 15 in Fig. 13.7. More specifically, at the left-hand side of Figs. 13.6 and 13.7 graphs of simulations of the state values are shown up to time point 100

Table 13.3 Connection matrix for the initial connection weights and state speed factor for the example network

Connections	X_1	X_2	X_3	X_4	X_5	X_6	X_7	X_8	X_9	X_{10}
X_1		0.1	0.2	0.1	0.2	0.15	0.1	0.25	0.25	0.1
X_2	0.25		0.25	0.2	0.1	0.2	0.15	0.25	0.25	0.25
X_3	0.1	0.25		0.1	0.2	0.15	0.1	0.25	0.1	0.15
X_4	0.25	0.15	0.25		0.15	0.8	0.25	0.15	0.25	0.25
X_5	0.25	0.2	0.1	0.2		0.25	0.2	0.1	0.2	0.15
X_6	0.25	0.1	0.25	0.25	0.25		0.1	0.25	0.25	0.1
X_7	0.2	0.1	0.2	0.15	0.2	0.2		0.2	0.15	0.25
X_8	0.1	0.25	0.1	0.25	0.05	0.15	0.25		0.1	0.25
X_9	0.25	0.15	0.25	0.15	0.2	0.1	0.2	0.15		0.15
X_{10}	0.2	0.25	0.2	0.2	0.1	0.2	0.15	0.8	0.2	

α = 1

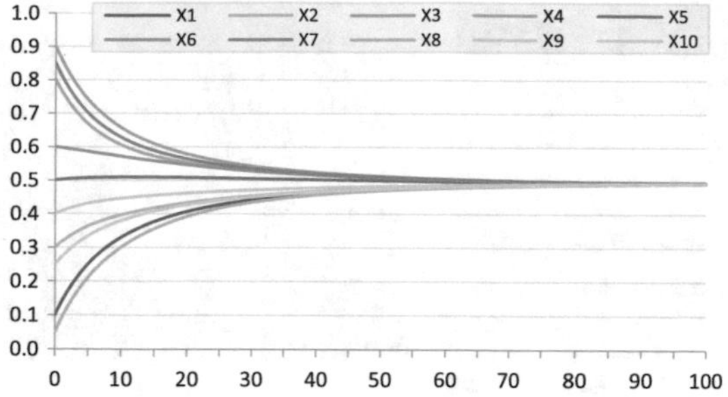

	X_1	X_2	X_3	X_4	X_5	X_6	X_7	X_8	X_9	X_{10}
X_1	0	1	1	1	1	1	1	1	1	1
X_2	1	0	1	1	1	1	1	1	1	1
X_3	1	1	0	1	1	1	1	1	1	1
X_4	1	1	1	0	1	1	1	1	1	1
X_5	1	1	1	1	0	1	1	1	1	1
X_6	1	1	1	1	1	0	1	1	1	1
X_7	1	1	1	1	1	1	0	1	1	1
X_8	1	1	1	1	1	1	1	0	1	1
X_9	1	1	1	1	1	1	1	1	0	1
X_{10}	1	1	1	1	1	1	1	1	1	0

α = 5

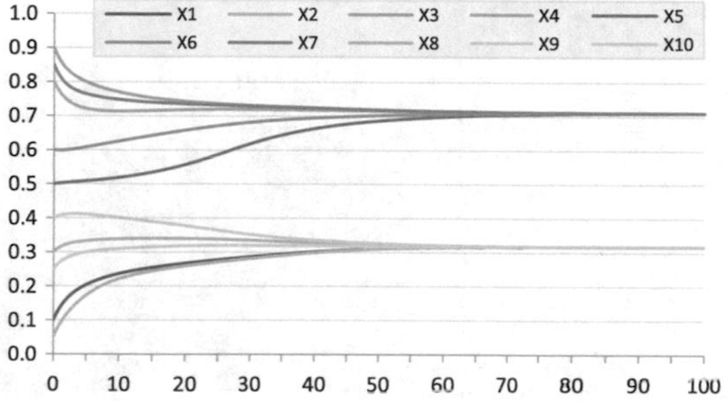

	X_1	X_2	X_3	X_4	X_5	X_6	X_7	X_8	X_9	X_{10}
X_1	0	1	0	0	0	0	0	1	1	1
X_2	1	0	0	0	0	0	0	1	1	1
X_3	0	0	0	1	1	1	1	0	0	0
X_4	0	0	1	0	1	1	1	0	0	0
X_5	0	0	1	1	0	1	1	0	0	0
X_6	0	0	1	1	1	0	1	0	0	0
X_7	0	0	1	1	1	1	0	0	0	0
X_8	1	1	0	0	0	0	0	0	1	1
X_9	1	1	0	0	0	0	0	1	0	1
X_{10}	1	1	0	0	0	0	0	1	1	0

α = 10

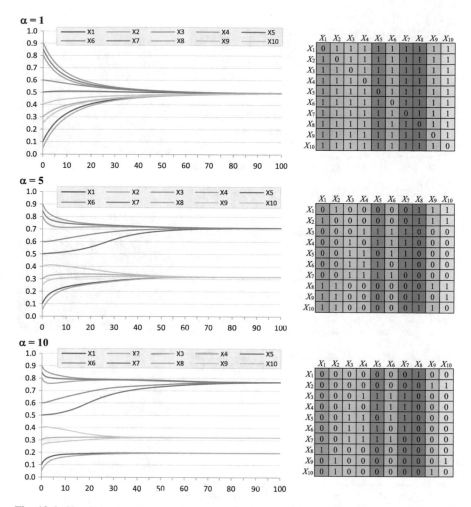

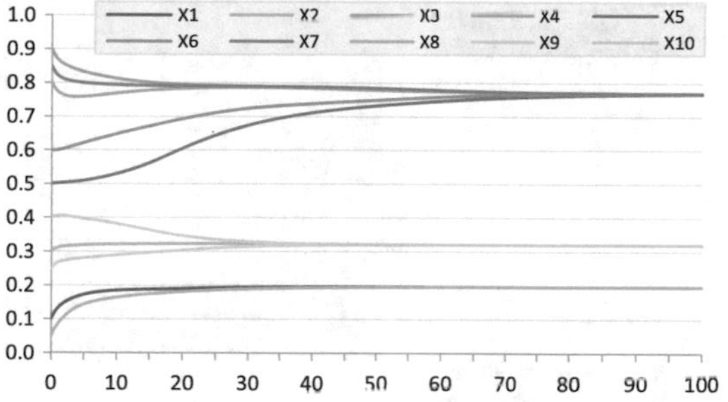

	X_1	X_2	X_3	X_4	X_5	X_6	X_7	X_8	X_9	X_{10}
X_1	0	0	0	0	0	0	0	1	0	0
X_2	0	0	0	0	0	0	0	0	1	1
X_3	0	0	0	1	1	1	1	0	0	0
X_4	0	0	1	0	1	1	1	0	0	0
X_5	0	0	1	1	0	1	1	0	0	0
X_6	0	0	1	1	1	0	1	0	0	0
X_7	0	0	1	1	1	1	0	0	0	0
X_8	1	0	0	0	0	0	0	0	0	0
X_9	0	1	0	0	0	0	0	0	0	1
X_{10}	0	1	0	0	0	0	0	0	1	0

Fig. 13.6 Simulations of the example network for homophily modulation factor $\alpha = 1$, 5 and 10

(with $\Delta t = 0.05$); at the right-hand side the connection matrices are shown at time point 500 (with $\Delta t = 0.25$), with an accuracy of 3 digits (0 means < 0.001, 1 means > 0.999). It turns out that all connection weights converge to 0 or 1.

Note that the modulation factor α models the strength of the effect of homophily. A higher α makes that connections change earlier in the coevolution process (compared to the pace of the social contagion) and due to that more clusters occur, as can be seen in Fig. 13.6. For $\alpha = 1$ all states end up in one cluster (all become connected by weights 1), for $\alpha = 5$ in two clusters (two subgroups of states $\{X_1, X_2, X_8, X_9, X_{10}\}$ and $\{X_3, X_4, X_5, X_6, X_7\}$ get connections with weight 1 and form clusters in this way; between these clusters the weights are 0), and for $\alpha = 10$ in three clusters: $\{X_1, X_8\}$, $\{X_2, X_9, X_{10}\}$ and $\{X_3, X_4, X_5, X_6, X_7\}$). Moreover, it can be observed that no clusters emerge with state values at a distance less than tipping point $\tau = 0.1$.

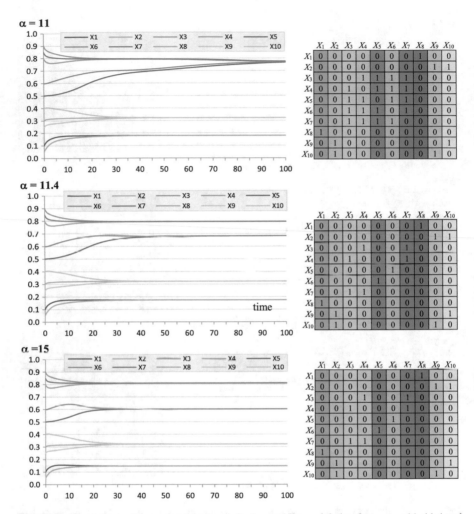

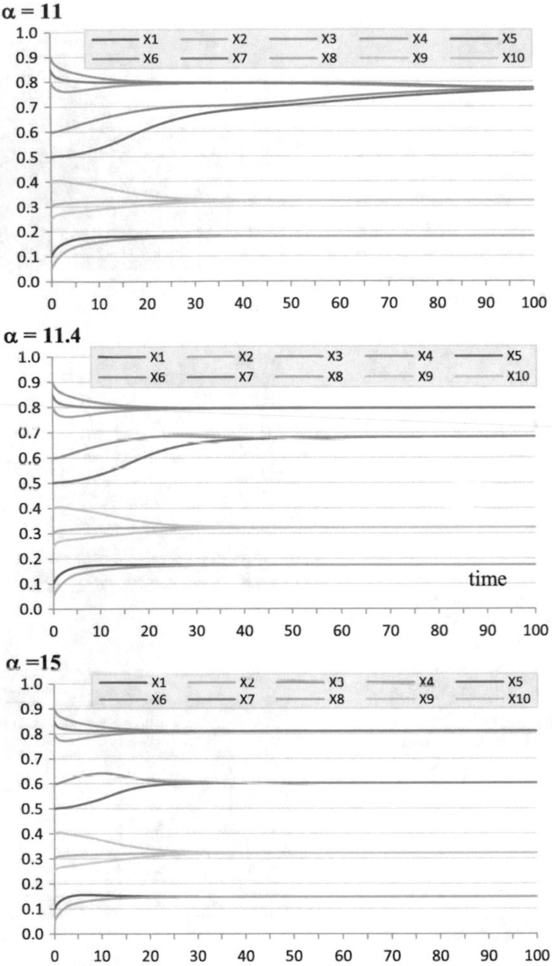

	X_1	X_2	X_3	X_4	X_5	X_6	X_7	X_8	X_9	X_{10}
X_1	0	0	0	0	0	0	0	1	0	0
X_2	0	0	0	0	0	0	0	0	1	1
X_3	0	0	0	1	1	1	1	0	0	0
X_4	0	0	1	0	1	1	1	0	0	0
X_5	0	0	1	1	0	1	1	0	0	0
X_6	0	0	1	1	1	0	1	0	0	0
X_7	0	0	1	1	1	1	0	0	0	0
X_8	1	0	0	0	0	0	0	0	0	0
X_9	0	1	0	0	0	0	0	0	0	1
X_{10}	0	1	0	0	0	0	0	0	1	0

	X_1	X_2	X_3	X_4	X_5	X_6	X_7	X_8	X_9	X_{10}
X_1	0	0	0	0	0	0	0	1	0	0
X_2	0	0	0	0	0	0	0	0	1	1
X_3	0	0	0	1	0	0	1	0	0	0
X_4	0	0	1	0	0	0	1	0	0	0
X_5	0	0	0	0	0	1	0	0	0	0
X_6	0	0	0	0	1	0	0	0	0	0
X_7	0	0	1	1	0	0	0	0	0	0
X_8	1	0	0	0	0	0	0	0	0	0
X_9	0	1	0	0	0	0	0	0	0	1
X_{10}	0	1	0	0	0	0	0	0	1	0

	X_1	X_2	X_3	X_4	X_5	X_6	X_7	X_8	X_9	X_{10}
X_1	0	0	0	0	0	0	0	1	0	0
X_2	0	0	0	0	0	0	0	0	1	1
X_3	0	0	0	1	0	0	1	0	0	0
X_4	0	0	1	0	0	0	1	0	0	0
X_5	0	0	0	0	0	1	0	0	0	0
X_6	0	0	0	0	1	0	0	0	0	0
X_7	0	0	1	1	0	0	0	0	0	0
X_8	1	0	0	0	0	0	0	0	0	0
X_9	0	1	0	0	0	0	0	0	0	1
X_{10}	0	1	0	0	0	0	0	0	1	0

Fig. 13.7 Simulations of the example network for homophily modulation factor α = 11, 11.4 and 15

13.4.2 Exploring the Birth of an Extra Cluster

Still increasing the modulation factor α, Fig. 13.7 zooms in at the birth of a fourth cluster. For $\alpha = 11$ it can be seen that for the cluster $\{X_3, X_4, X_5, X_6, X_7\}$ the states X_5 and X_6 join X_3, X_4, and X_7 in a very late stage, and (compared to the case $\alpha = 10$ in Fig. 13.6) even a slight hesitation for that may be observed from time point 20 to 40 while the values of X_5 and X_6 (around 0.7) have distance of about the tipping point $\tau = 0.1$ from the values (around 0.8) for X_3, X_4, and X_7. This hesitation can be confirmed, as increasing the modulation factor α by just 0.4 to 11.4 shows how a fourth separate cluster emerges for $\{X_5, X_6\}$. This fourth cluster converges to state value 0.683 whereas the other cluster $\{X_3, X_4, X_7\}$ converges to state value 0.796, which is a difference of 0.113: just above the tipping point $\tau = 0.1$. This may suggest that the model keeps the clusters at a distance of at least τ; this is one of the issues that will be analysed further in Sect. 13.6. For higher values of $\boldsymbol{\alpha}$, such as $\boldsymbol{\alpha} = 15$ no more than these four clusters emerge.

13.5 Relevant Aggregation Characteristics for Bonding by Homophily and for Social Contagion

This section addresses definitions for the characteristics for the network structure and in particular also for the homophily adaptation principle that have been identified as relevant for the adaptive network's behavior (following the upward arrow in the left side of Fig. 13.1).

13.5.1 Relevant Aggregation Characteristics for Bonding by Homophily

As adaptation principles are specified by the network characteristics of their reification, in particular the combination functions of the reification states for the adaptive connection weights, the properties of the aggregation are described as properties of such homophily combination functions. The following are considered plausible assumptions for a homophily combination function $\mathbf{c}(V_1, V_2, W)$ for reification state $\mathbf{W}_{X,Y}$ for $\boldsymbol{\omega}_{X,Y}$; here $D = |V_1 - V_2|$:

- $\mathbf{c}(V_1, V_2, W)$ is a function: $[0, 1] \times [0, 1] \times [0, 1] \rightarrow [0, 1]$
- $\mathbf{c}(V_1, V_2, W)$ is a monotonically decreasing function of D
- For D close to 0 and $W < 1$ it holds $\mathbf{c}(V_1, V_2, W) > W$ (i.e., $\mathbf{W}_{\boldsymbol{\omega}_{X,Y}}$ will increase)
- For D close to 1 and $W > 0$ it holds $\mathbf{c}(V_1, V_2, W) < W$ (i.e., $\mathbf{W}_{\boldsymbol{\omega}_{X,Y}}$ will decrease)

Note that when the first condition is not fulfilled, usually the function is cut off at 1 (when its value would be above 1) or at 0 (when its value would be below 0); see also in Fig. 13.5. Relatively simple functions $\mathbf{c}(V_1, V_2, W)$ that satisfy these requirements are obtained when a *tipping point* τ (a fixed number between 0 and 1) is assumed such that for $0 < \mathbf{W}_{X,Y} < 1$ it holds

- upward change of $\mathbf{W}_{X,Y}$ when $D < \tau$ $\quad$ $\mathbf{c}(V_1, V_2, W) > W$ when $D < \tau$
- no change of $\mathbf{W}_{X,Y}$ when $D = \tau$ $\quad$ $\mathbf{c}(V_1, V_2, W) = W$ when $D = \tau$
- downward change of $\mathbf{W}_{X,Y}$ when $D > \tau$ $\quad$ $\mathbf{c}(V_1, V_2, W) < W$ when $D > \tau$

These criteria are formalised in the following definition of properties of any function $c(V_1, V_2, W)$, but considered here as properties for homophily combination function in particular. So for the function $c(V_1, V_2, W)$, keep in mind that this is applied to a homophily combination function for reification state $\mathbf{W}_{X,Y}$ of $\boldsymbol{\omega}_{X,Y}$. It will be shown in Sect. 13.6 how these properties relate to the equilibrium behaviour of the network.

Definition 1 (Tipping point and strict tipping point)

(a) The function $c(V_1, V_2, W)$: $[0, 1] \times [0, 1] \times [0, 1] \rightarrow [0, 1]$ has *tipping point* τ for V_1 and V_2 if for all W with $0 < W < 1$ and all V_1, V_2 it holds

(i) $c(V_1, V_2, W) > W \Leftrightarrow |V_1 - V_2| < \tau$
(ii) $c(V_1, V_2, W) = W \Leftrightarrow |V_1 - V_2| = \tau$
(iii) $c(V_1, V_2, W) < W \Leftrightarrow |V_1 - V_2| > \tau$

(b) The function $c(V_1, V_2, W)$ has a *strict tipping point* τ if it has tipping point τ and in addition it holds:

(i) If $|V_1 - V_2| < \tau$ then $c(V_1, V_2, 0) > 0$
(ii) If $|V_1 - V_2| > \tau$ then $c(V_1, V_2, 1) < 1$

Note Definition 1(a) can be reformulated in the sense that any function $c(V_1, V_2, W)$ that is monotonically decreasing in $D = |V_1 - V_2|$ and goes for any W with $0 < W < 1$ through the point with $D = |V_1 - V_2| = \tau$ and $c(V_1, V_2, W) = W$ has tipping point τ. This is illustrated in Fig. 13.5 by the yellow areas. In particular, this applies to all example functions shown in Table 13.2 as Fig. 13.5 shows that they all are monotonically decreasing and go through the point (0.2, 0.7) for which $D = 0.2$ and $c(V_1, V_2, W) = 0.7$.

Note that condition (b) of this Definition 1 is not fulfilled when $|V_1 - V_2| < \tau$ and $c(V_1, V_2, 0) = 0$ or when $|V_1 - V_2| > \tau$ and $c(V_1, V_2, 1) = 1$. For example, for $\mathbf{slhomo}_{\tau,\alpha}(V_1, V_2, W)$ and $\mathbf{sqhomo}_{\tau,\alpha}(V_1, V_2, W)$ these cases do occur as always $\mathbf{slhomo}_{\tau,\alpha}(V_1, V_2, 0) = 0$ and $\mathbf{slhomo}_{\tau,\alpha}(V_1, V_2, 1) = 1$ due to the factor $W(1 - W)$, and the same for $\mathbf{sqhomo}_{\tau,\alpha}(V_1, V_2, W)$. Therefore they do have a tipping point but they do not have a strict tipping point. In the third paragraph in Sect. 13.8 it is discussed how in a practical application having a tipping point but not having a strict tipping point is not a big problem, but some care is needed in not assigning initial values 0 or 1 to connection weights, as these values will never change in such cases. Note that when a function with strict tipping point τ is used, any two nodes X, Y with $\boldsymbol{\omega}_{X,Y}(t) = 0$ (which usually is interpreted as not being connected) can become connected over time: when $|X(t) - Y(t)| < \tau$, then by Definition 1(b)(i) it holds $\mathbf{W}_{X,Y}(t') > 0$ for $t' > t$.

Some examples of combination functions having a strict tipping point τ have been explored in more detail in Sharpanskykh and Treur (2014) and Boomgaard et al. (2018). In the former the advanced quadratic combination function $\mathbf{aqhomo}_{\tau,\sigma}(V_1, V_2, W)$ and in the latter the advanced logistic function $\mathbf{alog2homo}_{\tau,\sigma,\alpha}(V_1, V_2, W)$ has been explored; see also Treur (2016, Chap. 11, p. 309):

$$\begin{aligned}\mathbf{aqhomo}_{\tau,\sigma}(V_1, V_2, W) &= W + \mathrm{Pos}\left(\alpha\left(\tau^2 - (V_1 - V_2)^2\right)\right)(1 - W) \\ &\quad - \mathrm{Pos}(-\alpha(\tau^2 - (V_1 - V_2)^2))W \\ \mathbf{alog2homo}_{\tau,\sigma,\alpha}(V_1, V_2, W) &= W + \mathrm{Pos}\left(\alpha\left(0.5 - \frac{1 - W}{1 + \mathrm{e}^{-\sigma(|V_1 - V_2| - \tau)}}\right)\right) \\ &\quad - \mathrm{Pos}\left(-\alpha\left(0.5 - \frac{W}{1 + \mathrm{e}^{-\sigma(|V_1 - V_2| - \tau)}}\right)\right)\end{aligned} \tag{13.12}$$

Here Pos(x) = (|x| + x)/2. The following lemma shows some properties of this operator.

Lemma 1

(a) Pos(x) = x when x is positive, else Pos(x) = 0.
So always Pos(x) $\geq$ 0, and when Pos(x) > 0, then Pos(−x) = 0.

(b) For any numbers α and β the following are equivalent:

(i) $\alpha\,\mathrm{Pos}(x) + \beta\,\mathrm{Pos}(-x) = 0$
(ii) $\alpha\,\mathrm{Pos}(x) = 0$ and $\beta\,\mathrm{Pos}(-x) = 0$
(iii) $x = 0$ or $x > 0$ and $\alpha = 0$ or $x < 0$ and $\beta = 0$.

This format using the double Pos(..) function can often be used to create a combination function satisfying the strict tipping point condition as a variation on a combination function satisfying only the tipping point condition. This is shown in the following proposition that assumes any close distance function denoted by d(τ, D). More specifically, the following proposition makes it quite easy to obtain functions with tipping point τ or with strict tipping point τ. For a proof of these Propositions, see Chap. 15, Sect. 15.8.

Proposition 1 Suppose for any function d(τ, D) it holds

$$\begin{aligned}\mathrm{d}(\tau, D) > 0 \quad &\text{iff } D < \tau \\ \mathrm{d}(\tau, D) < 0 \quad &\text{iff } D > \tau\end{aligned}$$

Then the following hold:

(a) For any $\alpha > 0$ the function

$$\mathrm{c}(V_1, V_2, W) = W + \alpha\, W(1 - W)\,\mathrm{d}(\tau, |V_1 - V_2|)$$

satisfies the tipping point condition, but not strict

(b) For any $\alpha > 0$ the function

$$c'(V_1, V_2, W) = W + \alpha\,\text{Pos}(d(\tau, |V_1 - V_2|))(1 - W) - \alpha\,\text{Pos}(-d(\tau, |V_1 - V_2|))W$$

satisfies the strict tipping point condition.

Proposition 1 can easily be applied to simple linear or quadratic functions $d(\tau, D)$ such as:

$$d(\tau, D) = \tau - D$$

$$d(\tau, D) = \tau^2 - D^2$$

but also to functions such as

$$d(\tau, D) = 0.5 - \frac{1}{1 + e^{-\sigma(D-\tau)}}$$

as is shown in Proposition 2(b) and (c).

The following proposition shows for four cases how it can be proven that some homophily combination function satisfies the tipping point or strict tipping point conditions. For proofs, see Chap. 15, Sect. 15.8.

Proposition 2

(a) **log1hom**$_{\tau,\alpha}(V_1, V_2, W)$ has tipping point τ, and is not strict
(b) **slog2hom**$_{\tau,\alpha}(V_1, V_2, W)$ has tipping point τ, and is not strict
(c) **alog2hom**$_{\tau,\alpha}(V_1, V_2, W)$ has a strict tipping point τ
(d) **exphomo**$_{\tau,\sigma}(V_1, V_2, W)$ has a tipping point τ and is not strict

The following proposition shows that weighted averages of functions with tipping point τ also have a tipping point τ, and the same for having a strict tipping point.

Proposition 3 A weighted average (with positive weights) of homophily combination functions with tipping point τ also has tipping point τ, and with strict tipping point τ, also has strict tipping point τ.

13.5.2 Relevant Aggregation Characteristics for Social Contagion

When more characteristics on network structure and adaptation combination functions are assumed, more refined results can be found, as will be shown in Sect. 13.6. Next, consider the combination functions used for social contagion of the states.

Definition 2 (Properties of combination functions for social contagion)

(a) A function c(..) is called *monotonically increasing* if

$$U_i \leq V_i \text{ for all } i \Rightarrow \mathrm{c}(U_1, \ldots, U_k) \leq \mathrm{c}(V_1, \ldots, V_k)$$

(b) A function c(..) is called *strictly monotonically increasing* if

$$U_i \leq V_i \text{ for all } i, \text{and } U_j < V_j \text{ for at least one } j \Rightarrow \mathrm{c}(U_1, \ldots, U_k) < \mathrm{c}(V_1, \ldots, V_k)$$

(c) A function c(..) is called *scalar-free* if $\mathrm{c}(\alpha V_1, \ldots, \alpha V_k) = \alpha\, \mathrm{c}(V_1, \ldots, V_k)$ for all $\alpha > 0$

Definition 3 (Normalised network) A network is *normalised* or uses normalised combination functions if for each state Y it holds $\mathrm{c}_Y(\omega_{X_1,Y}, \ldots, \omega_{X_k,Y}) = 1$, where $X_1, \ldots, X_k$ are the states from which Y gets incoming connections.

Note that $\mathrm{c}_Y(\omega_{X_1,Y}, \ldots, \omega_{X_k,Y})$ is an expression in terms of the parameter(s) of the combination function and $\omega_{X_1,Y}, \ldots, \omega_{X_k,Y}$. To require this expression to be equal to 1 provides a constraint on these parameters: an equation relating the parameter value(s) of the combination functions to the parameters $\omega_{X_1,Y}, \ldots, \omega_{X_k,Y}$. To satisfy this property, often the parameter(s) can be given suitable values. For example, for a Euclidean combination function $\lambda_Y = \omega^n_{X_1,Y} + \ldots + \omega^n_{X_k,Y}$ will provide a normalised network. This can be done in general:

(1) normalisation by adjusting the combination functions. If any combination function $\mathbf{c}_Y(..)$ is replaced by $\mathbf{c}'_Y(..)$ defined as

$$\mathbf{c}'_Y(V_1, \ldots, V_k) = \mathbf{c}_Y(V_1, \ldots, V_k)/\mathbf{c}_Y(\omega_{X_1,Y}, \ldots, \omega_{X_k,Y})$$

then the network becomes normalised: indeed $\mathbf{c}'_A(\omega_{X_1,Y}, \ldots, \omega_{X_k,Y}) = 1$

(2) normalisation by adjusting the connection weights (for scalar-free combination functions). For scalar-free combination functions also normalisation is possible by adapting the connection weights; define:

$$\omega'_{X_i,Y} = \omega_{X_i,Y}/\mathbf{c}_Y(\omega_{X_1,Y}, \ldots, \omega_{X_k,Y})$$

Then the network becomes normalised; indeed it holds:

$$\mathrm{c}_Y(\omega'_{X_1,Y}, .., \omega'_{X_k,Y}) = \mathbf{c}(\omega_{X_1,Y}/\mathrm{c}_Y(\omega_{X_1,Y}, .., \omega_{X_k,Y}), .., \omega_{X_k,Y}/\mathrm{c}(\omega_{X_1,Y}, .., \omega_{X_k,Y})) = 1$$

Definition 4 (Symmetric network and symmetric combination function)

a) A network is called *weakly symmetric* if for all states X, Y it holds $\omega_{X,Y} > 0 \Leftrightarrow \omega_{Y,X} > 0$. It is *fully symmetric* if for all states X, Y it holds $\omega_{X,Y} = \omega_{Y,X}$.

b) A homophily combination function $c(V_1, V_2, W)$ is called *symmetric* if $c(V_1, V_2, W) = c(V_2, V_1, W)$.

If the homophily combination function $c(V_1, V_2, W)$ is symmetric, the network is fully symmetric if the initial values for $\boldsymbol{\omega}_{X,Y}$ and $\boldsymbol{\omega}_{Y,X}$ are equal. For a proof of the following proposition, again see Chap. 15, Sect. 15.8.

Proposition 4

(a) When the homophily combination function $c(V_1, V_2, W)$ is symmetric, and initially the network is fully symmetric, then the network is continually fully symmetric.
(b) For every $\boldsymbol{n} > 0$ a Euclidean combination function of $\boldsymbol{n}$th degree is strictly monotonically increasing, scalar-free, and symmetric.

13.6 Relating Adaptive Network Structure to Emerging Bonding Behaviour

In this section, an analysis is presented of the behaviour of an adaptive network based on the coevolution of social contagion and bonding by homophily. Here it is found out how the network and adaptation characteristics defined in Sect. 13.5 (such as a tipping point τ), affect the emerging behaviour of the adaptive network. This relates to horizontal arrow in the upper part of Fig. 13.1. In particular, it is addressed how these properties imply that over time the network ends up in certain states (for both the nodes and the connections) that represent community formation. It will turn out that under certain conditions in terms of the properties defined in Sect. 13.5, in the end, a number of communities emerge such that:

- Nodes within one community have similar state values
- Connections between nodes within a community are very strong
- Connections between nodes from different communities are very weak

Note that these phenomena were already observed in the simulation examples in Sect. 13.4, but now it will be proved why under certain conditions they always have to occur. To formalise this, the following general notions are important; e.g., Brauer and Nohel (1969), Hirsch (1984), Lotka (1956).

Definition 5 (**Stationary point and equilibrium**) A state Y has a *stationary point* at t if $\mathbf{d}Y(t)/\mathbf{d}t = 0$, is increasing at t if $\mathbf{d}Y(t)/\mathbf{d}t > 0$, and is decreasing at t if $\mathbf{d}Y(t)/\mathbf{d}t < 0$, and similarly this applies to reification states for adaptive connection weights $\boldsymbol{\omega}$. The network is in *equilibrium* a t if every state Y and every connection weight $\boldsymbol{\omega}$ in the model has a stationary point at t. The equilibrium is *attracting* if any small perturbations of its values lead to convergence to the equilibrium values.

Considering the specific differential equation format for a temporal-causal network model shown in Sect. 13.2, and assuming nonzero speed factors the following more specific criteria for stationary points, and for increasing and decreasing trends, in terms of the combination functions and connection weights are easily found:

Lemma 2 (Criteria for a stationary, increasing and decreasing) Let Y be a state and X_1, ..., X_k the states from which state Y gets its incoming connections. Then

(i) Y has a stationary point at $t \Leftrightarrow \mathbf{c}_Y(\boldsymbol{\omega}_{X_1,Y}X_1(t), \ldots, \boldsymbol{\omega}_{X_k,Y}X_k(t)) = Y(t)$
(ii) Y is strictly increasing at $t \Leftrightarrow \mathbf{c}_Y(\boldsymbol{\omega}_{X_1,Y}X_1(t), \ldots, \boldsymbol{\omega}_{X_k,Y}X_k(t)) > Y(t)$
(iii) Y is strictly decreasing at $t \Leftrightarrow \mathbf{c}_Y(\boldsymbol{\omega}_{X_1,Y}X_1(t), \ldots, \boldsymbol{\omega}_{X_k,Y}X_k(t)) < Y(t)$

Similarly this applies to reification states **W** representing connection weights $\boldsymbol{\omega}$.

13.6.1 Relating Structure and Behaviour Independent of Social Contagion Characteristics

Theorem 1 presents some of the results found for the relation between the emerging equilibrium values for states and for connection weights. It is shown how the distance of the equilibrium values of two states relates to the equilibrium value of their connections. Note that for now no specific assumption is made on the aggregation characteristics for social contagion in terms of properties of the combination functions for it. To prove Theorem 1, the following lemma is a useful means. It shows that when the homophily combination function c(V_1, V_2, W) satisfies the tipping point criterion, a connection weight 0 can only be reached for states X and Y when $|X(t) - Y(t)| > \tau$ and a connection weight 1 can only be reached for states X and Y when $|X(t) - Y(t)| < \tau$. For proofs, again see Chap. 15, Sect. 15.8.

Lemma 3 Suppose the function c(V_1, V_2, W) has tipping point τ for V_1 and V_2. Then

(i) The value 0 for $\mathbf{W}_{X,Y}$ can only be reached from $\mathbf{W}_{X,Y}(t)$ with $0 < \mathbf{W}_{X,Y}(t) < 1$ if $|X(t) - Y(t)| > \tau$
(ii) The value 1 for $\mathbf{W}_{X,Y}$ can only be reached from $\mathbf{W}_{X,Y}(t)$ with $0 < \mathbf{W}_{X,Y}(t) < 1$ if $|X(t) - Y(t)| < \tau$.

This lemma already reveals that connection weights $\boldsymbol{\omega}_{X,Y}$ converging to 0 have some relation to the state values of X and Y having distance more than τ, and connection weights $\boldsymbol{\omega}_{X,Y}$ converging to 1 have some relation to the state values of X and Y having a distance less than τ. These relations between connection weights and state values during a convergence process are made more precise for reaching an equilibrium state in Theorem 1.

Theorem 1 (Relations between equilibrium values for states and for connection weights) Suppose the function c(V_1, V_2, W) has tipping point τ for V_1 and V_2

and an attracting equilibrium state is given with values $\underline{\mathbf{X}}$ for the states X and $\underline{\mathbf{W}}_{X,Y}$ for the connection weight reification states $\mathbf{W}_{X,Y}$. Then the following hold:

(a) If $|\underline{\mathbf{X}} - \underline{\mathbf{Y}}| < \tau$, then the equilibrium value $\underline{\mathbf{W}}_{X,Y}$ is 1; in particular this holds when $\underline{\mathbf{X}} = \underline{\mathbf{Y}}$. Therefore, if $\underline{\mathbf{W}}_{X,Y} < 1$, then $|\underline{\mathbf{X}} - \underline{\mathbf{Y}}| \geq \tau$, and, in particular, $\underline{\mathbf{X}} \neq \underline{\mathbf{Y}}$.
(b) If $|\underline{\mathbf{X}} - \underline{\mathbf{Y}}| > \tau$, then the equilibrium value $\underline{\mathbf{W}}_{X,Y}$ is 0. Therefore, if $\underline{\mathbf{W}}_{X,Y} > 0$, then $|\underline{\mathbf{X}} - \underline{\mathbf{Y}}| \leq \tau$.
(c) $0 < \underline{\mathbf{W}}_{X,Y} < 1$ implies $|\underline{\mathbf{X}} - \underline{\mathbf{Y}}| = \tau$.

This Theorem 1 explains some of the observations made in Sect. 13.4, in particular, that in all of these simulations the connection weights all end up either in value 0 (the state equilibrium values differ at least τ) or in value 1 (the state equilibrium values are equal).

When also some assumptions are made for social contagion, more refined results can be found, for example, as explored in Treur (2018a). The following is a basic Lemma for normalised networks with social contagion combination functions that are monotonically increasing and scalar-free.

Lemma 4 Let a normalised network with nonnegative connections be given with combination functions that are monotonically increasing and scalar-free; then the following hold:

(a)

(i) If for some state Y at time t for all nodes X with $\boldsymbol{\omega}_{X,Y} > 0$ it holds $X(t) \leq Y(t)$, then $Y(t)$ is decreasing at t: $\mathbf{d}Y(t)/\mathbf{d}t \leq 0$.
(ii) If, moreover, the combination function is strictly increasing and a state X exists with $X(t) < Y(t)$ and $\boldsymbol{\omega}_{X,Y} > 0$, then $Y(t)$ is strictly decreasing at t: $\mathbf{d}Y(t)(t)/\mathbf{d}t < 0$.

(b)

(i) If for some state Y at time t for all nodes X with $\boldsymbol{\omega}_{X,Y} > 0$ it holds $X(t) \geq Y(t)$, then $Y(t)$ is increasing at t: $\mathbf{d}Y(t)/\mathbf{d}t \geq 0$.
(ii) If, moreover, the combination function is strictly increasing and a state X exists with $\mathrm{X}(t) > Y(t)$ and $\boldsymbol{\omega}_{X,Y} > 0$, then $Y(t)$ is strictly increasing at t: $\mathbf{d}Y(t)(t)/\mathbf{d}t > 0$.

13.6.2 Relating Structure and Behaviour for Some Social Contagion Characteristics

Now the more specific theorem is obtained for the connection weights if the combination functions used for aggregation within social contagion for the states are assumed strictly monotonically increasing and scalar-free. It states that the connection weights in an attracting equilibrium are all 0 or 1, and when τ is a strict tipping point, that depends just on whether they connect states with the same or a different equilibrium state value.

Theorem 2 (Equilibrium values $\underline{\mathbf{W}}_{X,Y}$ all 0 or 1) Suppose the network is weakly symmetric and normalised, and the combination functions for the social contagion for the base states are strictly monotonically increasing and scalar-free. Suppose that the combination functions c(V_1, V_2, W) for the reification states for the connection weights have a tipping point τ. Then

(a) In an attracting equilibrium state for any states X, Y from $\underline{\mathbf{X}} \neq \underline{\mathbf{Y}}$ it follows $\underline{\mathbf{W}}_{X,Y} = 0$.
(b) In an attracting equilibrium state for any states X, Y with $\underline{\mathbf{X}} = \underline{\mathbf{Y}}$ it holds $\underline{\mathbf{W}}_{X,Y} = 0$ or $\underline{\mathbf{W}}_{X,Y} = 1$.
(c) If c(V_1, V_2, W) has a strict tipping point τ, then in an equilibrium state for any X, Y with $\underline{\mathbf{X}} = \underline{\mathbf{Y}}$ it holds $\underline{\mathbf{W}}_{X,Y} = 1$.

For proofs, see Chap. 15, Sect. 15.8. The following theorem shows the implications for emerging communities or clusters and distances between different equilibrium values for nodes.

Theorem 3 (Partition and equilibrium values of nodes) Suppose the network is weakly symmetric and normalised, the combination functions for the social contagion for the base states are strictly monotonically increasing and scalar-free, and the combination functions for the reification states for the connection weights use tipping point τ and is strict and symmetric. Then in any attracting equilibrium state a partition of the set of states into disjoint subsets $C_1, \ldots, C_p$ occurs such that:

(i) For each C_i the equilibrium values for all the states in C_i are equal: $\underline{\mathbf{X}} = \underline{\mathbf{Y}}$ for all $X, Y \in C_i$.
(ii) Every C_i forms a fully connected network with weights 1: $\underline{\mathbf{W}}_{X,Y} = 1$ for all X, $Y \in C_i$.
(iii) Every two nodes in different C_i have connection weight 0: when $i \neq j$, then $X \in C_i$ and $Y \in C_j$ implies $\underline{\mathbf{W}}_{X,Y} = 0$.
(iv) Any two distinct equilibrium values of states $\underline{\mathbf{X}} \neq \underline{\mathbf{Y}}$ have distance $\geq \tau$. Therefore there are at most $p \leq 1 + 1/\tau$ communities C_i and equilibrium values $\underline{\mathbf{X}}$.

13.7 Characterising Behaviour in Terms of Strongly Connected Components

To analyse connectivity, within Graph Theory the notion of a strongly connected component has been identified. The main parts of the following definitions can be found, for example, in Harary et al. (1965, Chap. 3), or Kuich (1970, Sect. 6). See also Chap. 12 of this volume.

Note that in the current chapter only nonnegative connection weights are considered.

Definition 6 (**reachability and strongly connected components**)

(a) State Y is *reachable* from state X if there is a directed path from X to Y with nonzero connection weights and speed factors.
(b) A network N is *strongly connected* if every two states are mutually reachable within N.
(c) A state is called *independent* if it is not reachable from any other state.
(d) A *subnetwork* of a network N is a network whose states and connections are states and connections of N.
(e) A *strongly connected component* C of a network N is a strongly connected subnetwork of N such that it is maximal: no larger strongly connected subnetwork of N contains it as a subnetwork.

Strongly connected components C can be identified by choosing any node X of N and adding all nodes that are on any cycle through X. Note also that when a node X is not on any cycle, then it will form a singleton strongly connected component C by itself; this applies in particular to all nodes of N with indegree or outdegree zero. There are efficient algorithms available to determine the strongly connected components of a network or graph; for example, see Bloem et al. (2006), Fleischer et al. (2000), Gentilini et al. (2003), Łacki (2013), Li et al. (2014), Tarjan (1972), Wijs et al. (2016).

The strongly connected components form a partition of the nodes of the graph or network. By this partition a more abstract, simpler view of the network can be created, called condensation graph, in which each component becomes one abstract node; this is defined as follows.

Definition 7 (**condensation graph**) The *condensation graph* C(N) of a network N with respect to its strongly connected components is a graph whose nodes are the strongly connected components of N and whose connections are determined as follows: there is a connection from node C_i to node C_j in C(N) if and only if in N there is at least one connection from a node in the strongly connected component C_i to a node in the strongly connected component C_j.

An important result is that a condensation graph C(N) is always an acyclic graph. In a sense, all cycles are locked up inside the components, and the connections between these components do not contain any cycles. The following theorem summarizes this; it was adopted from Harary et al. (1965, Chap. 3, Theorems 3.6 and 3.8), or Kuich (1970, Sect. 6).

Theorem 4 (**Acyclic condensation graph**)

(a) For any network *N* its condensation graph C(*N*) is acyclic, and has at least one state of outdegree zero and at least one state of indegree zero.
(b) The network *N* is acyclic itself if and only if it is graph-isomorphic to C(*N*). In this case, the nodes in C(*N*) all are singleton sets $\{X\}$ containing one state *X* from *N*.
(c) The network *N* is strongly connected itself if and only if C(*N*) only has one node; this node is the set of all states of *N*.

Note that in an adaptive network, the strongly connected components and the condensation graph usually change over time. For example, connection weights that were 0 can become nonzero, or nonzero connection weights can converge to 0. From Theorem 3 it immediately follows that in the equilibrium state the C_i defined there are the strongly connected components of the network. Then the results from Theorem 3 can be rephrased in the above terms in the following way.

Theorem 5 (**Strongly connected components in an attracting equilibrium**) Suppose the network is weakly symmetric and normalised, the combination functions for social contagion between the base nodes are strictly monotonically increasing and scalar-free, and the homophily combination functions for the connection weight reification states use tipping point τ and are strict and symmetric. Then in any attracting equilibrium state the following hold:

(i) There are at most $p \leq 1 + 1/\tau$ strongly connected components.
(ii) Each strongly connected component is fully connected and all states in it have a common equilibrium state value.
(iii) There are no nonzero connections between states from different strongly connected components, and the equilibrium values of these states have distance $\geq \tau$.
(iv) The condensation graph C(*N*) is totally disconnected: it has no connections at all.

The following converse holds as well; this shows how equilibrium states can be characterised by specific properties of the strongly connected components. For a proof: Chap. 15, Sect. 15.8.

Theorem 6 (**Strongly connected components characterisation**) Suppose the network is weakly symmetric, the combination functions for social contagion between the base nodes are strictly monotonically increasing, normalised and scalar-free, and the homophily combination function for the connection weight reification states use tipping point τ and are strict and symmetric. Suppose at some time point t the following hold:

(i) Each strongly connected component *C* is fully connected and all states in *C* have a common state value.
(ii) All connections between states from different strongly connected components have weight 0 and the equilibrium values of these states have distance $> \tau$.

Then the network is in an equilibrium state.

Note that the situation as characterised is a quite specific situation with trivial condensation graph, if compared, for example, to a general situation as addressed in Chap. 12 and Treur (2018c), where usually nontrivial condensation graphs occur.

13.8 Overview of Other Simulation Experiments

In recent years for a number of the functions from Table 13.2 simulation experiments have been performed for real-world domains and related to empirical data. An overview of them is shown in Table 13.4. Some of the experiences are the following.

In general, it was possible to get similar patterns as in the empirical data by using dedicated Parameter Tuning methods such as Simulated Annealing. As may be

Table 13.4 Overview of simulation experiments for different combination functions

Combination functions		Domain and data	Reference and venue
Homophily	Contagion		
$\mathbf{slhomo}_{\sigma,\tau}(..)$	$\mathbf{ssum}_{\lambda}(..)$	Friendships in a classroom Glasgow data, 2016	Blankendaal et al. (2016) ECAI'16
		Social media: blogger appreciation Instagram data	Kozyreva et al. (2018) SocInfo'18
		Social media: music appreciation Twitter data	Gerwen et al. (2019) DCAI'18
	$\mathbf{ssum}_{\lambda}(..)$ $\mathbf{alogistic}_{\sigma,\tau}(..)$	Segregation of queer community Questionnaire data	Heijmans et al. (2019) ICICT'19
	$\mathbf{alogistic}_{\sigma,\tau}(..)$	Social media: Zwarte Piet debate Twitter data	Roller et al. (2017) COMPLEXNETWORKS'17
$\mathbf{sqhomo}_{\sigma,\tau}(..)$	$\mathbf{ssum}_{\lambda}(..)$	Segregation of immigrants Literature data	Kappert et al. (2018) ICCCI'18
	$\mathbf{alogistic}_{\sigma,\tau}(..)$	Friendships in a classroom Glasgow data, 2016	Beukel et al. (2017, 2019) PAAMS'17, Neurocomputing 2019

(continued)

Table 13.4 (continued)

Combination functions		Domain and data	Reference and venue
Homophily	Contagion		
aqhomo$_{\sigma,\tau}$**(..)**	**ssum**$_{\lambda}$**(..)**	Scalefree versus random networks Literature data	Sharpanskykh and Treur (2013, 2014) ICCCI'13, Neurocomputing 2014
log2homo$_{\sigma,\tau}$**(..)**	**ssum**$_{\lambda}$**(..)**	Social media: physical activity Twitter data	Dijk and Treur (2018) ICCCI'18
	alogistic$_{\sigma,\tau}$**(..)**	Friendships in a classroom Knecht data, 2008	Boomgaard et al. (2018) SocInfo'18

expected the remaining Root Mean Square error differed with the different studies and domains, and also depends on how fine-grained the scoring scales for the data were; for example when a 3 points scale was used for empirical data, already because of that in the [0, 1] interval a variation of 0.15 should be expected within the empirical data themselves, and for a 5 points scale 0.1. Then an average deviation between model and empirical data will be at least in that order of magnitude, not smaller than, say 0.15–0.25. This indeed was usually the case in the examples discussed in Table 13.4.

Using the simple linear or simple quadratic functions **slhomo**$_{\sigma,\tau}$**(..)** or **sqhomo**$_{\sigma,\tau}$(..) the change of connection weights is very slow when the weights are close to 0 or 1 due to the factor $W\,(1 - W)$ in the function. The advantage of this slowing down effect is that the connection weight values stay within the [0, 1] interval in a natural manner. But the downside of this factor is that when they are exactly 0 or 1 they will even freeze and not be able to change anyway, and the same when 0 or 1 are used as initial values. This is because **slhomo**$_{\sigma,\tau}$**(..)** or **sqhomo**$_{\sigma,\tau}$**(..)** have a tipping point, but no strict tipping point; also see the remark after Definition 1. The same holds for **log2homo**$_{\sigma,\tau}$**(..)**. So, when these functions are used, it is better to initialise all connection weight values between 0.1 and 0.9 instead of in the full [0, 1] interval. This is different for the advanced linear and quadratic homophily functions **alhomo**$_{\sigma,\tau}$**(..)** or **aqhomo**$_{\sigma,\tau}$**(..)**. For them approaching 0 or 1 is still slow, but leaving 0 or 1 can be fast: they do have a strict tipping point. Note, however, that this is not necessarily always a good property. Maybe when a very strong connection has been formed, even some differences that occur may not affect the connection immediately. So, in some cases, or for some types of persons the slow change close to 0 or 1 as shown by **slhomo**$_{\sigma,\tau}$**(..)** or **sqhomo**$_{\sigma,\tau}$**(..)** may even be more plausible.

Note that in Kozyreva et al. (2018) a slightly different multicriteria variant of the simple linear homophily function was used that takes into account multiple states in the similarity measure: instead of $|V_1 - V_2|$ for 1 criterion, for k criteria the following Euclidean distance formula is used to measure similarity:

$$\sqrt{(V_{1,1} - V_{1,2})^2 + \cdots + (V_{k,1} - V_{k,2})^2} \qquad (13.13)$$

In Blankendaal et al. (2016) and Beukel et al. (2017) for bonding not only a homophily principle was used, but also a 'more becomes more' principle which can be considered a variant of what sometimes is called 'the rich get richer' (Simon 1955; Bornholdt and Ebel 2001), 'cumulative advantage' (de Solla Price 1976), 'the Matthew effect' (Merton 1968), or 'preferential attachment' (Barabási and Albert 1999; Newman 2003). For this 'more becomes more' principle, in Blankendaal et al. (2016) a scaled sum combination function was used for the (reified) weights of the connections to a given other person Y with scaling factor the number k of such connections (resulting in the average of these connection weights), and in Beukel et al. (2017, 2019) and advanced logistic sum:

$$\begin{array}{l} \mathbf{ssum}_k(W_1, \ldots, W_k) \\ \mathbf{alogistic}_{\sigma,\tau}(W_1, \ldots, W_k) \end{array} \qquad (13.14)$$

where W_1, …, W_k refer to the values of the reification states for the weights $\boldsymbol{\omega}_{X_1,Y}, \ldots, \boldsymbol{\omega}_{X_k,Y}$ of all connections of the other person Y. Note that the first scaled sum option $\mathbf{ssum}_k(.)$ adapts to the other person's average connection weight independent of the number of connections, whereas the second $\mathbf{alogistic}_{\sigma,\tau}(\ldots)$ option is more additive as more connections of the other person provide higher values. In both approaches, the two combination functions for both the homophily and the more becomes more principle were combined as a weighted sum.

13.9 Discussion

In this chapter, it was analysed how emerging network behaviour (in particular, community formation) can be related to characteristics of the adaptive network's structure for a reified adaptive network modeling bonding based on similarity. Parts of this chapter were adopted from Treur (2018b). Here the reified network structure characteristics include the aggregation characteristics in terms of the combination function specifying the adaptation principle incorporated. This has been addressed for adaptive social networks for bonding based on homophily (McPherson et al. 2001; Pearson et al. 2006) combined with social contagion for the base states. Relevant characteristics of the network and the adaptation principle have been identified, such as a tipping point for similarity for the combination function for the bonding. As one of the results, it has been found how the emergence of communities strongly depends on the value of this tipping point. It has been shown, for example, that some properties of the structure of the base network and the adaptation principle modeled by the reification entail that the connection weights all converge to 0 (for persons in different communities) or 1 (for persons within one community). More specifically, it has been found how the formation of communities depends on the value τ of this tipping point: there can be no more communities than $1 + 1/\tau$, assuming state values within the interval [0, 1].

The presented results do not concern results for just one type of network or combination function, as more often is found. Instead, they were formulated and proven at a more general level and therefore can be applied not just to specific networks but to classes of networks satisfying the identified relevant properties of network structure and adaptation characteristics. Note, however, that the focus in the current chapter is on deterministic behaviour; therefore stochastic models such as, for example, the one reported in a nontechnical manner in Axelrod (1997), are not covered by this analysis.

It may be an interesting research focus for the future to explore whether and how the analysis results found here have counterparts for stochastic network models. Besides, in Axelrod (1997) also regional differences are addressed. In a more extensive application of the models discussed in the current chapter, that may be an interesting ingredient to address as well.

The results found also have been related to the strongly connected components of the network (Harary et al. 1965, Chap. 3), or Kuich (1970, Sect. 6). A characterisation in terms of properties of the strongly connected components and their relations was found for states of an adaptive network which are attracting equilibrium states.

References

Ashby, W.R.: Design for a Brain. Chapman and Hall, London (second extended edition) (1960) (First edition, 1952)

Axelrod, R.: The dissemination of culture: a model with local convergence and global polarization. J. Conflict Resolut. **41**(2), 203–226 (1997)

Barabási, A.L., Albert, R.: Emergence of scaling in random networks. Science **286**, 509–512 (1999)

Blankendaal, R., Parinussa, S., Treur, J.: A temporal-causal modelling approach to integrated contagion and network change in social networks. In: Proceedings of the 22nd European Conference on Artificial Intelligence, ECAI'16. Frontiers in Artificial Intelligence and Applications, vol. 285, pp. 1388–1396. IOS Press (2016)

Bloem, R., Gabow, H.N., Somenzi, F.: An algorithm for strongly connected component analysis in n log n symbolic steps. Formal Method Syst. Des. **28**, 37–56 (2006)

Boomgaard, G., Lavitt, F., Treur, J.: Computational analysis of social contagion and homophily based on an adaptive social network model. In: Proceedings of the 10th International Conference on Social Informatics, SocInfo'18. Lecture Notes in Computer Science, vol. 11185, pp. 86–101. Springer Publishers (2018)

Bornholdt, S., Ebel, H.: World Wide Webscaling exponent from Simon's 1955 model. Phys. Rev. E 64, art. no. 035104 (2001)

Brauer, F., Nohel, J.A.: Qualitative Theory of Ordinary Differential Equations. Benjamin (1969)
de Solla Price, D.J.: A general theory of bibliometric and other cumulative advantage processes. J. Am. Soc. Inform. Sci. **27**, 292–306 (1976)
Fleischer, L.K., Hendrickson, B., Pınar, A.: On identifying strongly connected components in parallel. In: Rolim, J. (ed.) Parallel and Distributed Processing. IPDPS 2000. Lecture Notes in Computer Science, vol. 1800, pp. 505–511. Springer (2000)
Gentilini, R., Piazza, C., Policriti, A.: Computing strongly connected components in a linear number of symbolic steps. In: Proceedings of the SODA'03, pp. 573–582 (2003)
Glasgow Empirical Data: https://www.stats.ox.ac.uk/~snijders/siena/Glasgow_data.htm (2016)
Gross, T., Sayama, H. (eds.): Adaptive Networks: Theory, Models and Applications. Springer (2009)
Harary, F., Norman, R.Z., Cartwright, D.: Structural Models: An Introduction to the Theory of Directed Graphs. Wiley, New York (1965)
Heijmans, P., van Stijn, J., Treur, J.: Modeling cultural segregation of the queer community through an adaptive social network model. In: Proceedings of the Fourth International Congress on Information and Communication Technology, ICICT'19. Advances in Intelligent Systems and Computing. Springer (2019)
Hirsch, M.W.: The dynamical systems approach to differential equations. Bull. (New Ser.) Am. Math. Soc. 11, 1–64 (1984)
Holme, P., Newman, M.E.J.: Nonequilibrium phase transition in the coevolution of networks and opinions. Phys. Rev. E **74**(5), 056108 (2006)
Kappert, C., Rus, R., Treur, J.: On the emergence of segregation in society: network-oriented analysis of the effect of evolving friendships. In: Nguyen, N.T., Pimenidis, E., Khan, Z., Trawinski, B. (eds.) Computational Collective Intelligence: 10th International Conference, ICCCI 2018, Proceedings, vol. 1. Lecture Notes in Artificial Intelligence, vol. 11055, pp. 178–191. Springer (2018)
Knecht, A.: Empirical data: collected by Andrea Knecht (2008). https://www.stats.ox.ac.uk/*snijders/siena/tutorial2010_data.htm
Kozyreva, O., Pechina, A., Treur, J.: Network-oriented modeling of multi-criteria homophily and opinion dynamics in social media. In: Koltsova, O., Ignatov, D.I., Staab, S. (eds.) Social Informatics: Proceedings of the 10th International Conference on Social Informatics, SocInfo'18, vol. 1. Lecture Notes in AI, vol. 11185, pp. 322–335. Springer (2018)
Kuich, W.: On the Entropy of Context-Free Languages. Information and Control **16**, 173–200 (1970)
Kuipers, B.J.: Commonsense reasoning about causality: deriving behavior from structure. Artif. Intell. **24**, 169–203 (1984)
Kuipers, B.J., Kassirer, J.P.: How to discover a knowledge representation for causal reasoning by studying an expert physician. In: Proceedings of the Eighth International Joint Conference on Artificial Intelligence, IJCAI'83. William Kaufman, Los Altos, CA (1983)
Łacki, J.: Improved deterministic algorithms for decremental reachability and strongly connected components. ACM Trans. Algorithms 9(3), Article 27 (2013)
Levy, D.A., Nail, P.R.: Contagion: a theoretical and empirical review and reconceptualization. Genet. Soc. Gen. Psychol. Monogr. **119**(2), 233–284 (1993)
Li, G., Zhu, Z., Cong, Z., Yang, F.: Efficient decomposition of strongly connected components on GPUs. J. Syst. Architect. **60**(1), 1–10 (2014)
Lotka, A.J.: Elements of Physical Biology. Williams and Wilkins Co. (1924). Dover Publications, 2nd edn (1956)
McPherson, M., Smith-Lovin, L., Cook, J.M.: Birds of a feather: homophily in social networks. Ann. Rev. Sociol. **27**(1), 415–444 (2001)
Merton, R.K.: The Matthew effect in science. Science **159**(1968), 56–63 (1968)
Newman, M.E.J.: The structure and function of complex networks. SIAM Rev. **45**, 167–256 (2003)

Pearl, J.: Causality. Cambridge University Press (2000)
Pearson, M., Steglich, C., Snijders, T.: Homophily and assimilation among sport-active adolescent substance users. Connections **27**(1), 47–63 (2006)
Roller, R., Blommestijn, S.Q., Treur, J.: An adaptive computational network model for multi-emotional social interaction. In: Proceedings of the 6th International Conference on Complex Networks and their Applications, COMPLEXNETWORKS'17. Studies in Computational Intelligence. Springer (2017)
Sharpanskykh, A., Treur, J.: Modelling and analysis of social contagion processes with dynamic networks. In: Bădică, C., Nguyen, N.T., Brezovan, M. (eds.) Computational Collective Intelligence. Technologies and Applications. ICCCI 2013. Lecture Notes in Computer Science, vol. 8083, pp. 40–50. Springer, Berlin (2013)
Sharpanskykh, A., Treur, J.: Modelling and analysis of social contagion in dynamic networks. Neurocomputing **146**(2014), 140–150 (2014)
Simon, H.A.: On a class of skew distribution functions. Biometrika **42**(1955), 425–440 (1955)
Tarjan, R.: Depth-first search and linear graph algorithms. SIAM J. Comput. **1**(2), 146–160 (1972)
Treur, J.: Verification of temporal-causal network models by mathematical analysis. Vietnam J. Comput. Sci. **3**, 207–221 (2016a)
Treur, J.: Network-Oriented Modeling: Addressing Complexity of Cognitive, Affective and Social Interactions. Springer Publishers (2016b)
Treur, J.: Relating emerging network behaviour to network structure. In: Aiello, L.M., Cherifi, C., Cherifi, H., Lambiotte, R., Lió, P., Rocha, L.M. (eds.) Proceedings of the 7th International Conference on Complex Networks and their Applications, ComplexNetworks'18. Studies in Computational Intelligence, vol. 812, pp. 619–634. Springer Publishers (2018a)
Treur, J.: Relating an adaptive social network's structure to its emerging behaviour based on homophily. In: Aiello, L.M., Cherifi, C., Cherifi, H., Lambiotte, R., Lió, P., Rocha, L.M. (eds.) Proceedings of the 7th International Conference on Complex Networks and their Applications, Complex Networks'18. Studies in Computational Intelligence, vol. 812, pp. 635–651. Springer Publishers (2018b)
Treur, J.: Mathematical analysis of a network's asymptotic behaviour based on its strongly connected components. In: Proc. of the 7th International Conference on Complex Networks and their Applications, ComplexNetworks'18, vol. 1. Studies in Computational Intelligence, vol. 812, pp. 663–679. Springer Publishers (2018c)
Treur, J.: Multilevel network reification: representing higher order adaptivity in a network. In: Proceedings of the 7th International Conference on Complex Networks and their Applications, ComplexNetworks'18, vol. 1. Studies in Computational Intelligence, vol. 812, pp. 635–651. Springer (2018d)
Treur, J.: The ins and outs of network-oriented modeling: from biological networks and mental networks to social networks and beyond. Trans. Comput. Collect. Intell. 32, 120–139 (2019). Text of Keynote Lecture at the 10th International Conference on Computational Collective Intelligence, ICCCI'18 (2018)
Turnbull, L., Hütt, M.-T., Ioannides, A.A., Kininmonth, S., Poeppl, R., Tockner, K., Bracken, L.J., Keesstra, S., Liu, L., Masselink, R., Parsons, A.J.: Connectivity and complex systems: Learning from a multi-disciplinary perspective. Appl. Netw. Sci. 3(47) (2018). https://doi.org/10.1007/s41109-018-0067-2
van Beukel, S., Goos, S., Treur, J.: Understanding homophily and more-becomes-more through adaptive temporal-causal network models. In: De la Prieta, F. (ed.) Trends in Cyber-Physical Multi-Agent Systems. The PAAMS Collection—15th International Conference PAAMS'17. Advances in Intelligent Systems and Computing, vol. 619, pp. 16–29. Springer (2017)
van Beukel, S., Goos, S., Treur, J.: An adaptive temporal-causal network model for social networks based on the homophily and more-becomes-more principle. Neurocomputing **338**, 361–371 (2019)

van Dijk, M., Treur, J.: Physical activity contagion and homophily in an adaptive social network model. In: Nguyen, N.T., Pimenidis, E., Khan, Z., Trawinski, B. (eds.) Computational Collective Intelligence: 10th International Conference, ICCCI 2018, Proceedings. vol. 1. Lecture Notes in AI, vol. 11055, pp. 87–98. Springer (2018)

van Gerwen, S., van Meurs, A., Treur, J.: An adaptive temporal-causal network for representing changing opinions on music releases. In: De La Prieta F., Omatu S., Fernández-Caballero, A. (eds.) Distributed Computing and Artificial Intelligence, 15th International Conference. DCAI 2018. Advances in Intelligent Systems and Computing, vol. 800, pp. 357–367. Springer, Cham (2019)

Vazquez, F.: Opinion dynamics on coevolving networks. In: Mukherjee, A., Choudhury, M., Peruani, F., Ganguly, N., Mitra, B. (eds.) Dynamics On and Of Complex Networks, Volume 2, Modeling and Simulation in Science, Engineering and Technology, pp. 89–107. Springer, New York, (2013)

Vazquez, F., Gonzalez-Avella, J.C., Eguíluz, V.M., San Miguel, M.: Time-scale competition leading to fragmentation and recombination transitions in the coevolution of network and states. Phys. Rev. E **76**, 046120 (2007)

Wijs, A., Katoen, J.P., Bošnacki, D.: Efficient GPU algorithms for parallel decomposition of graphs into strongly connected and maximal end components. Formal Methods Syst. Des. **48**, 274–300 (2016)

Chapter 14
Relating a Reified Adaptive Network's Emerging Behaviour Based on Hebbian Learning to Its Reified Network Structure

Abstract In this chapter another challenge is analysed for how emerging behaviour of an adaptive network can be related to characteristics of the adaptive network's structure. By applying network reification, the adaptation structure is modeled itself as a network too: as a subnetwork of the reified network extending the base network. In particular, this time the challenge is addressed for mental networks with adaptive connection weights based on Hebbian learning. To this end relevant properties of the network and the adaptation principle that have been identified are discussed. Using network reification for modeling of the adaptation principle, a central role is played by the combination function specifying the aggregation for the reification states of the connection weights, and in particular, identified mathematical properties of this combination function. As one of the results it has been found that under some conditions in an achieved equilibrium state the value of a connection weight has a functional relation to the values of the connected states that can be identified.

Keywords Reified adaptive network · Hebbian learning · Analysis of behaviour

14.1 Introduction

As for Chap. 13 the challenging issue addressed here is to predict what patterns of behaviour will emerge, and how their emergence depends on the structure of the model, in particular for adaptive network models. Here adaptive behaviour depends in some way on the reified network structure, defined by network characteristics such as connections and their weights, and the aggregation of multiple connections to one node. When adaptive networks are considered, where the network characteristics also change over time, according to certain adaptation principles this poses extra challenges. Such adaptation principles themselves depend on certain adaptation characteristics. It is this latter issue what is the topic of the current chapter: how does emerging behaviour of adaptive networks relate to the characteristics of the network and of the adaptation principles used. More in particular, the

J. Treur, *Network-Oriented Modeling for Adaptive Networks: Designing Higher-Order Adaptive Biological, Mental and Social Network Models*, Studies in Systems, Decision and Control 251, https://doi.org/10.1007/978-3-030-31445-3_14

focus is on adaptive Mental Networks based on Hebbian learning (Bi and Poo 2001; Gerstner and Kistler 2002; Hebb 1949; Keysers and Perrett 2004; Keysers and Gazzola 2014; Kuriscak et al. 2015). Hebbian learning is, roughly stated, based on the principle 'neurons that fire together, wire together' from Neuroscience.

To address the issue, as a vehicle the Network-Oriented Modeling approach based on temporal-causal networks (Treur 2016) is used together with the notion of network reification from Chap. 3 and (Treur 2018a). For temporal-causal networks, characteristics of the network structure Connectivity, Aggregation, and Timing are represented by connection weights, combination functions and speed factors. For the type of adaptive networks considered, the connection weights are dynamic based on Hebbian learning, so they are actually not part of the characteristics of the (static) network structure anymore. Instead, by applying network reification the base network is extended by reification states that represent the network characteristics such as in this case the Connectivity characteristics indicated by connection weights and their dynamics. In the reified network the dynamics of these reification states is defined by the standard concepts for temporal-causal networks: connection weights for Connectivity, speed factors for Timing, and combination functions for Aggregation of the reification states. In particular, the focus here is on the mathematical properties of these combination functions for the reification states as they play a main role in the specification of an adaptation principle; e.g., see Chap. 3 or (Treur 2018a).

Based on the chosen approach, characteristics of Hebbian learning have been identified that play an important role in the emerging behaviour; these characteristics indeed have been expressed as mathematical properties of combination functions for the reification states for connection weights.

In Fig. 14.1 the basic relation between structure and dynamics of a reified network model is indicated by the horizontal arrow in the lower part representing the base level. For a reified network these also apply to the reification states. The properties of network structure focus on properties of the adaptation principle based on Hebbian learning represented by the reification states at the reification level; they are first discussed in a general setting in Sect. 14.4. Properties of the behaviour are addressed in Sect. 14.4 as well and related to the network structure properties (the horizontal arrow in the upper part of Fig. 14.1). In Sect. 14.5 these results are refined by introducing an extra assumption on variable separation.

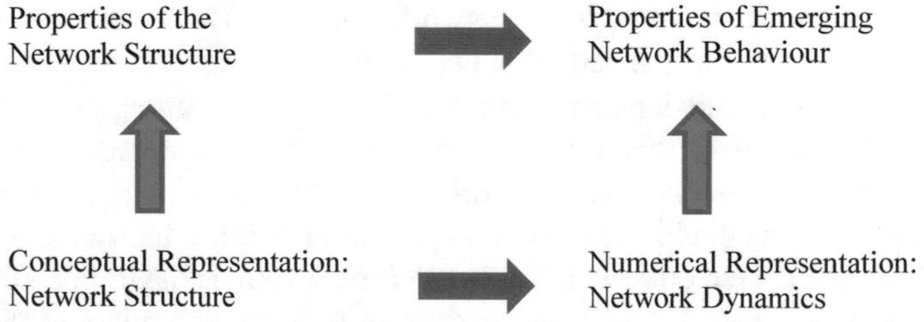

Fig. 14.1 Bottom layer: the conceptual representation of a reified network model defines the numerical representation. Top layer: properties of reified network structure entail properties of emerging reified network behaviour

In this chapter, results will be discussed that have been proven mathematically in this way for this relation between structure and behavior for such reified adaptive network models, in particular, for the equilibrium values of Hebbian learning in relation to equilibrium values of the connected network states. These results have been proven not for one specific model or function, but for classes of combination functions that fulfill certain properties. More specifically, as one of the results it has been found how for the classes of functions considered within an emerging equilibrium state the connection weight and the connected states satisfy a fixed functional relation that can be expressed mathematically.

In this chapter, in Sect. 14.2 the temporal-causal networks that are used as vehicle are briefly introduced. Section 14.3 briefly introduces Hebbian learning and how it can be modeled by a reified network. In Sect. 14.4 the properties of Hebbian learning functions are introduced that define the adaptation principle of the network. Section 14.5 focuses in particular on the class of functions for which a form of variable separation can be applied, In Sect. 14.6 a number of examples are discussed. Finally, Sect. 14.7 is a discussion.

14.2 Temporal-Causal Networks and Network Reification

For the perspective on networks used in the current chapter, the interpretation of connections based on causality and dynamics forms a basis of the structure and semantics of the considered networks. More specifically, the nodes in a network are interpreted here as states (or state variables) that vary over time, and the connections are interpreted as causal relations that define how each state can affect other states over time. This type of network has been called a *temporal-causal network* (Treur 2016). A conceptual representation of a temporal-causal network model by a *labeled graph* provides a fundamental basis. Such a conceptual representation includes representing in a declarative manner states and connections between them that represent (causal) impacts of states on each other. This part of a conceptual representation is often depicted in a *conceptual picture* by a graph with nodes and directed connections. However, a *complete conceptual representation* of a temporal-causal network model also includes a number of labels for such a graph, representing network characteristics such as Connectivity, Aggregation and Timing. A notion of *strength of a connection* is used as a label for Connectivity, some way for *Aggregation of multiple causal impacts* on a state is used, and a notion of *speed of change* of a state is used for Timing of the processes. These three notions, called connection weight, combination function, and speed factor, make the graph of states and connections a labeled graph, and form the defining structure of a temporal-causal network model in the form of a conceptual representation; see Table 14.1, first 5 rows.

There are many different approaches possible to address the issue of combining multiple impacts. To provide sufficient flexibility, the Network-Oriented Modelling approach based on temporal-causal networks incorporates for each state a way to specify how multiple causal impacts on this state are aggregated by a combination

Table 14.1 Concepts of conceptual and numerical representations of a temporal-causal network

Concepts	Notation	Explanation
States and connections	X, Y, $X \rightarrow Y$	Describes the nodes and links of a network structure (e.g., in graphical or matrix format)
Connection weight	$\boldsymbol{\omega}_{X,Y}$	The *connection weight* $\boldsymbol{\omega}_{X,Y} \in [-1, 1]$ represents the strength of the causal impact of state X on state Y through connection $X \rightarrow Y$
Aggregating multiple impacts	$\mathbf{c}_Y(..)$	For each state Y (a reference to) a *combination function* $\mathbf{c}_Y(..)$ is chosen to combine the causal impacts of other states on state Y
Timing of the causal effect	$\boldsymbol{\eta}_Y$	For each state Y a *speed factor* $\boldsymbol{\eta}_Y \geq 0$ is used to represent how fast a state is changing upon causal impact
Concepts	**Numerical representation**	**Explanation**
State values over time t	$Y(t)$	At each time point t each state Y in the model has a real number value, usually in [0, 1]
Single causal impact	$\mathbf{impact}_{X,Y}(t) = \boldsymbol{\omega}_{X,Y}X(t)$	At t state X with connection to state Y has an impact on Y, using weight $\boldsymbol{\omega}_{X,Y}$
Aggregating multiple impacts	$\mathbf{aggimpact}_Y(t)$ $= \mathbf{c}_Y(\mathbf{impact}_{X_1,Y}(t), \ldots, \mathbf{impact}_{X_k,Y}(t))$ $= \mathbf{c}_Y(\boldsymbol{\omega}_{X_1,Y}X_1(t), \ldots, \boldsymbol{\omega}_{X_k,Y}X_k(t))$	The aggregated causal impact of multiple states X_i on Y at t, is determined using combination function $\mathbf{c}_Y(..)$
Timing of the causal effect	$Y(t+\Delta t) = Y(t) + \boldsymbol{\eta}_Y[\mathbf{aggimpact}_Y(t) - Y(t)]\Delta t$ $= Y(t) + \boldsymbol{\eta}_Y[\mathbf{c}_Y(\boldsymbol{\omega}_{X_1,Y}X_1(t), \ldots, \boldsymbol{\omega}_{X_k,Y}X_k(t)) - Y(t)]\Delta t$	The causal impact on Y is exerted over time gradually, using speed factor $\boldsymbol{\eta}_Y$

function. For this aggregation a library with a number of standard combination functions are available as options, but also own-defined functions can be added.

Next, this conceptual interpretation is expressed in a formal-numerical way, thus associating semantics to any temporal-causal network specification in a detailed numerical-mathematically defined manner.

This is done by showing how a conceptual representation based on states and connections enriched with labels for connection weights, combination functions and speed factors, can get an associated numerical representation (Treur 2016), Chap. 2; see Table 14.1, last five rows. The difference equations in the last row in Table 14.1 constitute the numerical representation of the temporal-causal network model and can be used for simulation and mathematical analysis; it can also be written in differential equation format:

$$\begin{aligned} Y(t+\Delta t) &= Y(t) + \boldsymbol{\eta}_Y[\mathbf{c}_Y(\boldsymbol{\omega}_{X_1,Y}X_1(t), \ldots, \boldsymbol{\omega}_{X_k,Y}X_k(t)) - Y(t)]\Delta t \\ \mathbf{d}Y(t)/\mathbf{d}t &= \boldsymbol{\eta}_Y[\mathbf{c}_Y(\boldsymbol{\omega}_{X_1,Y}X_1(t), \ldots, \boldsymbol{\omega}_{X_k,Y}X_k(t)) - Y(t)] \end{aligned} \tag{14.1}$$

In adaptive networks some of the network structure characteristics such as connection weights are dynamic and actually should be treated more in the same way as states. The concept of network reification provides a neat definition for doing this. By introducing additional network states $\mathbf{W}_{X,Y}$ representing them, called reification states, network reification avoids that connection weights $\boldsymbol{\omega}_{X,Y}$ get an ambiguous status. The thus extended network is a reified network. In Sect. 14.3 this will be discussed for Hebbian learning in particular.

14.3 Reified Adaptive Networks for Hebbian Learning

In Sect. 14.3 the basics of Hebbian learning and reification for it are briefly summarized. Next, in Sect. 14.4 relevant properties for Hebbian learning combination functions are discussed and how they imply certain behaviour. Recall from Chap. 1, Fig. 1.4 the way in which Hebbian learning can be modeled by network reification; see also Fig. 14.2 here. Following this reification approach, the adaptation principle and the dynamics it entails gets a specification in the standard form of a temporal-causal network structure, as a subnetwork of the reified network. In particular, the combination function for the reification state and its mathematical properties play a main role.

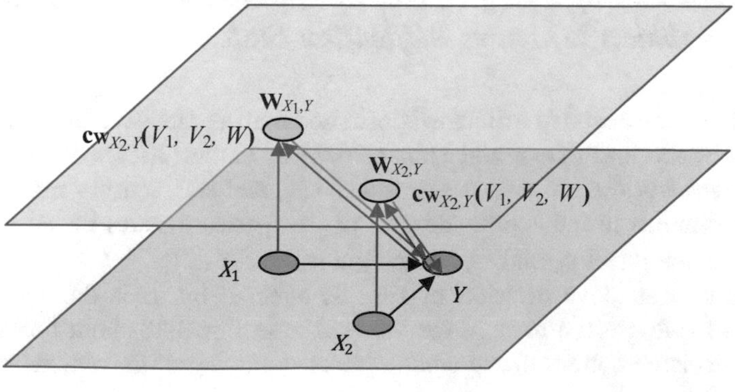

Fig. 14.2 Hebbian learning combination functions as labels in a reified network

14.3.1 Reification States for Hebbian Learning and Their Hebbian Learning Combination Functions

The Hebbian learning principle for the connection between two mental states is sometimes formulated as 'neurons that fire together, wire together'; e.g., (Bi and Poo 2001; Gerstner and Kistler 2002; Hebb 1949; Keysers and Perrett 2004; Keysers and Gazzola 2014; Kuriscak et al. 2015; Zenke et al. 2017). This can be modelled by using the activation values the two mental states $X(t)$ and $Y(t)$ have at time t. Then the reification state $\mathbf{W}_{X,Y}$ for the weight $\boldsymbol{\omega}_{X,Y}$ of the connection from X to Y is changing over time dynamically, depending on these levels $X(t)$ and $Y(t)$, but also on the value of $\mathbf{W}_{X,Y}$ itself. Therefore these *Hebbian learning combination functions* $\mathbf{c}(V_1, V_2, W)$ have suitable arguments refering to the relevant states: V_1 refers to the state value $X(t)$ of X, V_2 to the state value $Y(t)$ of Y and W to the state value $\mathbf{W}_{X,Y}(t)$ of $\mathbf{W}_{X,Y}$. In Fig. 14.2, the Hebbian learning combination functions $\mathbf{c}_{\mathbf{W}_{X_1,Y}}(V_1, V_2, W)$ and $\mathbf{c}_{\mathbf{W}_{X_2,Y}}(V_1, V_2, W)$ are indicated as labels for the reification states $\mathbf{W}_{X_1,Y}$ and $\mathbf{W}_{X_2,Y}$. Similarly, labels for adaptation speed (speed factors indicating the learning rate) can be added, and labels for the incoming connections for the reification states.

Thus in the standard way for temporal-causal networks based on such Hebbian learning combination functions the following difference and differential equations are obtained for the reification states; these define the adaptive dynamics of Hebbian learning:

$$\begin{aligned}\mathbf{W}_{X,Y}(t+\Delta t) &= \mathbf{W}_{X,Y}(t) + \boldsymbol{\eta}_{\mathbf{W}_{X,Y}}\left[\mathbf{c}(X(t), Y(t), \mathbf{W}_{X,Y}(t)) - \mathbf{W}_{X,Y}(t)\right]\Delta t \\ \mathbf{d}\mathbf{W}_{X,Y}(t)/\mathbf{d}t &= \boldsymbol{\eta}_{\mathbf{W}_{X,Y}}\left[\mathbf{c}(X(t), Y(t), \mathbf{W}_{X,Y}(t)) - \mathbf{W}_{X,Y}(t)\right]\end{aligned} \tag{14.2}$$

Here indeed the speed factor $\boldsymbol{\eta}_{\mathbf{W}X,Y}$ now can be interpreted as learning rate for the connection weight.

14.3.2 An Example Reified Network Model with Multiple Hebbian Learning Reification States

An example of an adaptive mental network model using Hebbian learning is shown in Fig. 14.3; see also (Treur and Umair 2011) or (Treur 2016), Chap. 6, p. 163. Here ws_w are world states, ss_w sensor states, srs_w and srs_{e_i} sensory representations states for stimulus w and action effect e_i, ps_{a_i} preparation states for a_i, fs_{e_i} feeling states for action effect e_i, and es_{a_i} execution states for a_i.

It describes adaptive decision making as affected by three different adaptive connections (the green arrows in the base plane in Fig. 14.3) both for direct triggering of decision options a_i and emotion-related valuing of the options by an as-if prediction loop:

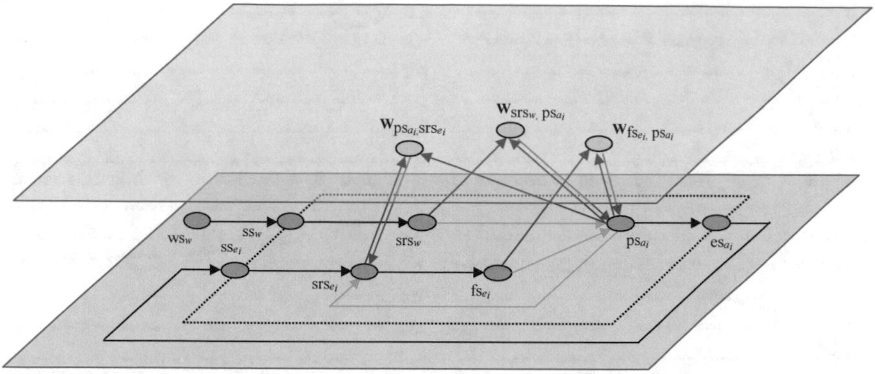

Fig. 14.3 Reified temporal-causal network model for adaptive rational decision making based on emotions

- stimulus-response connection from srs_w to action option preparation ps_{a_i}
- action effect prediction link from ps_{a_i} to effect representation srs_{e_i}
- emotion-related valuing of the action by the connection from feeling state fs_{e_i} to ps_{a_i}

Each of these connections can use Hebbian learning. A relatively simple example, also used in (Treur 2016) in a number of applications (including in Chap. 6 there for the model shown in Fig. 14.3), is the following Hebbian learning combination function:

$$\begin{aligned} &\mathbf{c}_{\mathbf{W}_{X,Y}}(V_1, V_2, W) = V_1V_2(1 - W) + \boldsymbol{\mu}W \\ &\text{or} \\ &\mathbf{c}_{\mathbf{W}_{X,Y}}(X(t), Y(t), \mathbf{W}_{X,Y}(t)) = X(t)Y(t)(1 - \mathbf{W}_{X,Y}(t)) + \boldsymbol{\mu}\mathbf{W}_{X,Y}(t) \end{aligned} \tag{14.3}$$

Here $\boldsymbol{\mu}$ is a persistence parameter. In an emerging equilibrium state it turns out that the equilibrium value for $\mathbf{W}_{X,Y}$ functionally depends on the equilibrium values of X and Y according to a specific formula that has been determined for this case in (Treur 2016), Chap. 12, Sect. 12.5.2. For some example patterns, see Fig. 14.4.

It is shown that when the equilibrium values of X and Y are 1, the equilibrium value for $\mathbf{W}_{X,Y}$ is 0.83 (top row), when the equilibrium values of X and Y are 0.6, the equilibrium value for $\mathbf{W}_{X,Y}$ is 0.64 (middle row), and when the equilibrium values of X and Y are 0, the equilibrium value for $\mathbf{W}_{X,Y}$ is 0 (bottom row). This equilibrium value of $\mathbf{W}_{X,Y}$ is always attracting. The three different rows in Fig. 14.4 illustrate how the equilibrium value of $\mathbf{W}_{X,Y}$ varies with the equilibrium values of X and Y. It is this relation that is analysed in a more general setting in some depth in this chapter. In (Treur 2016), Chap. 12 a mathematical analysis was made for the equilibria of the specific example combination function above (although written in the slightly different but equivalent format as discussed in Chap. 15, Sect. 15.2). In the current chapter a much more general analysis is made which applies to a wide

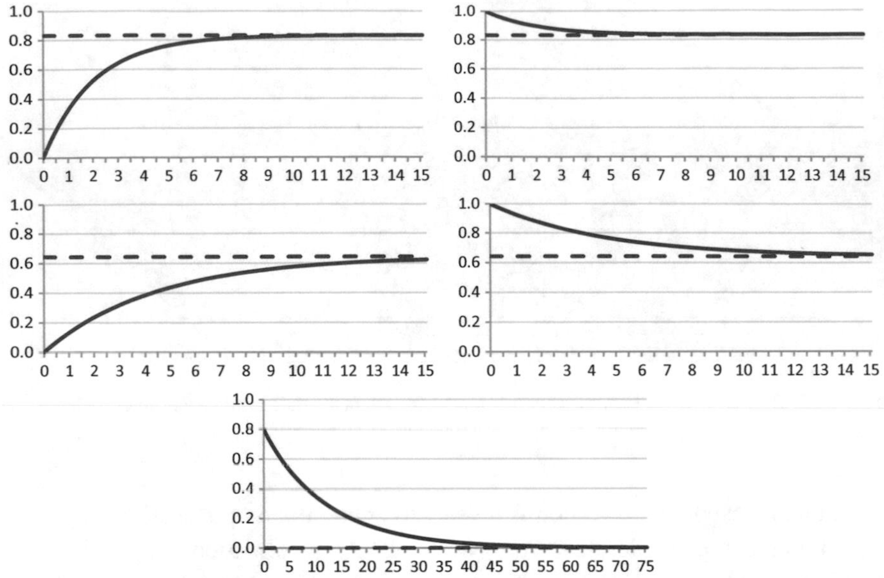

Fig. 14.4 Hebbian learning $\boldsymbol{\eta} = 0.4$, $\boldsymbol{\mu} = 0.8$, $\Delta t = 0.1$; adopted from (Treur 2016), pp. 339–340. **a** Top row: activation levels $X_1 = 1$ and $X_2 = 1$; equilibrium value 0.83. **b** Middle row activation levels $X_1 = 0.6$ and $X_2 = 0.6$, equilibrium value 0.64. **c** Bottom row: activation levels $X_1 = X_2 = 0$; equilibrium value 0 (pure extinction)

class of functions. In Example 1 in Sect. 14.6 below, the above case is obtained as a special case of the more general results found, and more precise numbers will be derived for the equilibrium values.

14.4 Relevant Aggregation Characteristics for Hebbian Learning and the Implied Behaviour

In this section, it is discussed how in a reified network aggregation for the connection weight reification state can be defined by a specific class of combination functions for Hebbian learning, and it will be analysed what equilibrium values can emerge for the learnt connections.

14.4.1 Relevant Aggregation Characteristics for Hebbian Learning

First a basic definition; see also (Brauer and Nohel 1969; Hirsch 1984; Lotka 1956).

Definition 1 (stationary point and equilibrium)
A state Y has a *stationary point* at t if $\mathbf{d}Y(t)/\mathbf{d}t = 0$. The network is in *equilibrium* a t if every state Y of the model has a stationary point at t. A state Y is increasing at t if $\mathbf{d}Y(t)/\mathbf{d}t > 0$; it is decreasing at t if $\mathbf{d}Y(t)/\mathbf{d}t < 0$.

Considering the specific type of differential equation for a temporal-causal network model, and assuming a nonzero speed factor, from (14.1) and (14.2) more specific criteria can be found:

Lemma 1 (Criteria for a stationary, increasing and decreasing)
Let Y be a state and X_1, ..., X_k the states from which state Y gets its incoming connections. Then

$$
\begin{array}{lll}
Y \text{ has a stationary point at } t & \Leftrightarrow & \mathbf{c}_Y(\boldsymbol{\omega}_{X_1,Y}X_1(t), \ldots, \boldsymbol{\omega}_{X_k,Y}X_k(t)) = Y(t) \\
Y \text{ is increasing at } t & \Leftrightarrow & \mathbf{c}_Y(\boldsymbol{\omega}_{X_1,Y}X_1(t), \ldots, \boldsymbol{\omega}_{X_k,Y}X_k(t)) > Y(t) \\
Y \text{ is decreasing at } t & \Leftrightarrow & \mathbf{c}_Y(\boldsymbol{\omega}_{X_1,Y}X_1(t), \ldots, \boldsymbol{\omega}_{X_k,Y}X_k(t)) < Y(t)
\end{array}
$$

These criteria can also be applied to adaptive connection weights based on Hebbian learning combination functions $\mathbf{c}(V_1, V_2, W)$ for the reification states $\mathbf{W}_{X,Y}$:

$$
\begin{array}{lll}
\mathbf{W}_{X,Y}(t) \text{ has a stationary point at } t & \Leftrightarrow & \mathbf{c}(V_1, V_2, W) = W \\
\mathbf{W}_{X,Y}(t) \text{ is increasing at } t & \Leftrightarrow & \mathbf{c}(V_1, V_2, W) > W \\
\mathbf{W}_{X,Y}(t) \text{ is decreasing at } t & \Leftrightarrow & \mathbf{c}(V_1, V_2, W) < W
\end{array}
$$

The following plausible assumptions are made for the Hebbian learning functions used as combination function to specify aggregation of the reification state in the reified network: one set for fully persistent Hebbian learning and one set for hebbian learning with extinction described by a persistence parameter $\boldsymbol{\mu}$; here again V_1 is the argument of the function $\mathbf{c}_Y(..)$ used for $X(t)$, V_2 for $Y(t)$, and W for $\mathbf{W}_{X,Y}(t)$.

Definition 2 (Hebbian Learning Function)
A function c: $[0, 1] \times [0, 1] \times [0, 1] \rightarrow [0, 1]$ is called a *fully persistent Hebbian learning function* if the following hold:

(a) $c(V_1, V_2, W)$ is a monotonically increasing function of V_1 and V_2
(b) $c(V_1, V_2, W) - W$ is a monotonically decreasing function of W
(c) $c(V_1, V_2, W) \geq W$
(d) $c(V_1, V_2, W) = W$ if and only if one of V_1 and V_2 is 0 (or both), or $W = 1$

A function c: $[0, 1] \times [0, 1] \times [0, 1] \rightarrow [0, 1]$ is called a *Hebbian learning function* with *persistence parameter* $\boldsymbol{\mu}$ if the following hold:

(a) $c(V_1, V_2, W)$ is a monotonically increasing function of V_1 and V_2
(b) $c(V_1, V_2, W) - \boldsymbol{\mu}W$ is a monotonically decreasing function of W
(c) $c(V_1, V_2, W) \geq \boldsymbol{\mu}W$
(d) $c(V_1, V_2, W) = \boldsymbol{\mu}W$ if and only if one of V_1 and V_2 is 0 (or both), or $W = 1$

Note that for $\boldsymbol{\mu} = 1$ the function is fully persistent.

14.4.2 Functional Relation for the Equilibrium Value of a Hebbian Learning Reification State

The following proposition shows that for any Hebbian learning function with persistence parameter $\boldsymbol{\mu}$ there exists a monotonically increasing function $f_{\boldsymbol{\mu}}(V_1, V_2)$ which is implicitly defined for given V_1, V_2 by the equation $c(V_1, V_2, W) = W$ in W. When applied to an equilibrium state of an adaptive temporal-causal network, the existence of this function $f_{\boldsymbol{\mu}}(V_1, V_2)$ reveals that in equilibrium states there is a direct and monotonically increasing functional relation of the equilibrium value $\underline{\mathbf{W}}_{X,Y}$ of $\mathbf{W}_{X,Y}$ with the equilibrium values $\underline{X}$, $\underline{Y}$ of the states X and Y. This is described in Theorem 1 below. Proposition 1 describes the functional relation needed for that. For proofs of Propositions 1 and 2, see Chap. 15, Sect. 15.9.

Proposition 1 (functional relation for W)
Suppose that $c(V_1, V_2, W)$ is a Hebbian learning function with persistence parameter $\boldsymbol{\mu}$.

(a) Suppose $\boldsymbol{\mu} < 1$. Then the following hold:
(i) The function $W \rightarrow c(V_1, V_2, W) - W$ on $[0, 1]$ is strictly monotonically decreasing
(ii) There is a unique function $f_{\boldsymbol{\mu}}$: $[0, 1] \times [0, 1] \rightarrow [0, 1]$ such for any V_1, V_2 it holds

$$c(V_1, V_2, f_{\boldsymbol{\mu}}(V_1, V_2)) = f_{\boldsymbol{\mu}}(V_1, V_2)$$

This function $f_{\boldsymbol{\mu}}$ is a monotonically increasing function of V_1, V_2, and is implicitly defined by the above equation. Its maximal value is $f_{\boldsymbol{\mu}}(1, 1)$ and minimum $f_{\boldsymbol{\mu}}(0, 0) = 0$.

(b) Suppose $\boldsymbol{\mu} = 1$. Then there is a unique function f_1: $(0, 1] \times (0, 1] \rightarrow [0, 1]$, such for any V_1, V_2 it holds

$$c(V_1, V_2, f_1(V_1, V_2)) = f_1(V_1, V_2)$$

This function f_1 is a constant function of V_1, V_2 with $f_1(V_1, V_2) = 1$ for all V_1, $V_2 > 0$ and is implicitly defined on $(0, 1] \times (0, 1]$ by the above equation.
If one of V_1, V_2 is 0, then any value of W satisfies the equation $c(V_1, V_2, W) = W$, so no unique function value for $f_1(V_1, V_2)$ can be defined then.

When applied to an equilibrium state of a reified adaptive temporal-causal network, this proposition entails the following Theorem 1. For $\boldsymbol{\mu} < 1$ this follows from Proposition 1 (a) applied to the function c(..). From (a)(i) it follows that the equilibrium value $\underline{\mathbf{W}}_{X,Y}$ is attracting: suppose $\mathbf{W}_{X,Y}(t) < \underline{\mathbf{W}}_{X,Y}$, then from $c(V_1, V_2, \underline{\mathbf{W}}_{X,Y}) - \underline{\mathbf{W}}_{X,Y} = 0$ and the decreasing monotonicity of $W \rightarrow c(V_1, V_2, W) - W$ it follows that $c(V_1, V_2, \mathbf{W}_{X,Y}(t)) - \mathbf{W}_{X,Y}(t) > 0$, and therefore by Lemma 1 $\mathbf{W}_{X,Y}(t)$ is increasing. Similarly, when $\mathbf{W}_{X,Y}(t) > \underline{\mathbf{W}}_{X,Y}$, it is decreasing. For $\boldsymbol{\mu} = 1$ the statement follows from Proposition 1 (b) applied to the function c(..).

Theorem 1 (functional relation for equilibrium values of $\mathbf{W}_{X,Y}$) Suppose in a reified temporal-causal network, c(V_1, V_2, W) is the combination function for reification state $\boldsymbol{W}_{X,Y}$ for connection weight $\boldsymbol{\omega}_{X,Y}$ and is a Hebbian learning function with persistence parameter $\boldsymbol{\mu}$, with $f_{\boldsymbol{\mu}}$ the function defined by Proposition 1. In an achieved equilibrium state the following hold.

(a) Suppose $\boldsymbol{\mu} < 1$. For any equilibrium values $\boldsymbol{X}, \boldsymbol{Y} \in [0, 1]$ of states X and Y the value $f_{\boldsymbol{\mu}}(\boldsymbol{X}, \boldsymbol{Y})$ provides the unique equilibrium value $\underline{\boldsymbol{W}}_{X,Y}$ for $\boldsymbol{W}_{X,Y}$. This $\underline{\boldsymbol{W}}_{X,Y}$ monotonically depends on $\boldsymbol{X}, \boldsymbol{Y}$: it is higher when $\boldsymbol{X}, \boldsymbol{Y}$ are higher. The maximal equilibrium value $\underline{\boldsymbol{W}}_{X,Y}$ of $\boldsymbol{W}_{X,Y}$ is $f_{\boldsymbol{\mu}}(1, 1)$ and the minimal equilibrium value is 0. Moreover, the equilibrium value $\underline{\boldsymbol{W}}_{X,Y}$ is attracting.

(b) Suppose $\boldsymbol{\mu} = 1$. If for the equilibrium values $\boldsymbol{X}, \boldsymbol{Y} \in [0, 1]$ of states X and Y it holds $\boldsymbol{X}, \boldsymbol{Y} > 0$, then $\underline{\boldsymbol{W}}_{X,Y} = 1$. If one of $\boldsymbol{X}, \boldsymbol{Y}$ is 0, then $\underline{\boldsymbol{W}}_{X,Y}$ can be any value in [0, 1]: it does not depend on $\boldsymbol{X}, \boldsymbol{Y}$. So, for $\boldsymbol{\mu} = 1$ the maximal value of $\underline{\boldsymbol{W}}_{X,Y}$ in an equilibrium state is 1 and the minimal value is 0.

14.5 Variable Separation for Hebbian Learning Functions and the Implied Behaviour

There is a specific subclass of Hebbian learning functions that is often used. For this subclass the implied behaviour can be determined more explicitly by obtaining certain algebraic formulae for the function $f_{\boldsymbol{\mu}}$ in Theorem 1.

14.5.1 Hebbian Learning Combination Functions with Variable Separation

Relatively simple functions c(V_1, V_2, W) that satisfy the requirements from Definition 2 are obtained when the state arguments V_1 and V_2 and the connection argument W can be separated as follows.

Definition 3 (variable separation)
The Hebbian learning function c(V_1, V_2, W) with persistence parameter $\boldsymbol{\mu}$ *enables variable separation* by functions cs: [0, 1] × [0, 1] → [0, 1] monotonically increasing and cs: [0, 1] → [0, 1] monotonically decreasing if

$$c(V_1, V_2, W) = cs(V_1, V_2)cc(W) + \boldsymbol{\mu} W$$

where $cs(V_1, V_2) = 0$ if and only if one of V_1, V_2 is 0, and $cc(1) = 0$ and $cc(W) > 0$ when $W < 1$.

The function $cs(V_1, V_2)$ is called the *states factor* and the function $cc(W)$ the *connection factor*.

Note that the s in cs stands for states and the second c in cc for connection. When variable separation holds, the following proposition can be obtained. For this type of function the indicated functional relation can be defined.

Proposition 2 (functional relation for *W* based on variable separation) Assume the Hebbian function $c(V_1, V_2, W)$ with persistence parameter μ enables variable separation by the two functions $cs(V_1, V_2)$ monotonically increasing and $cc(W)$ monotonically decreasing:

$$c(V_1, V_2, W) = cs(V_1, V_2)cc(W) + \mu W$$

Let $h_\mu(W)$ be the function defined for $W \in [0, 1)$ by

$$h_\mu(W) = \frac{(1-\mu)W}{cc(W)}$$

Then the following hold.

(a) When $\mu < 1$ the function $h_\mu(W)$ is strictly monotonically increasing, and has a strictly monotonically increasing inverse g_μ on the range $h_\mu([0, 1))$ of h_μ with $W = g_\mu(h_\mu(W))$ for all $W \in [0, 1)$.
(b) When $\mu < 1$ and $c(V_1, V_2, W) = W$, then $g_\mu(cs(V_1, V_2)) < 1$ and $W < 1$, and it holds

$$h_\mu(W) = cs(V_1, V_2)$$
$$W = g_\mu(cs(V_1, V_2))$$

So, in this case the function f_μ from Theorem is the function composition g_μ o cs of cs followed by g_μ; it holds:

$$f_\mu(V_1, V_2) = g_\mu(cs(V_1, V_2))$$

(c) For $\mu = 1$ it holds $c(V_1, V_2, W) = W$ if and only if $V_1 = 0$ or $V_2 = 0$ or $W = 1$.
(d) For $\mu < 1$ the maximal value W with $c(V_1, V_2, W) = W$ is $g_\mu(cs(1, 1))$, and the minimal equilibrium value W is 0. For $\mu = 1$ the maximal value W is 1 (always when $V_1, V_2 > 0$ holds) and the minimal value is 0 (occurs when one of V_1, V_2 is 0).

Note that by Proposition 2 the function $f_\mu(V_1, V_2)$ can be determined by inverting the function $h_\mu(W) = (1-\mu)W/cc(W)$ to find g_μ and composing the inverse with the function $cs(V_1, V_2)$. This will be shown below for some cases.

14.5.2 Functional Relation for the Equilibrium Value of a Hebbian Learning Reification State: The Variable Separation Case

For the case of an equilibrium state of a reified adaptive temporal network model Proposition 2 entails Theorem 2.

Theorem 2 (functional relation for equilibrium values of $\mathbf{W}_{X,Y}$: variable separation)
Assume in a reified temporal-causal network the Hebbian learning combination function c(V_1, V_2, W) with persistence parameter $\boldsymbol{\mu}$ for $\mathbf{W}_{X,Y}$ enables variable separation by the two functions cs(V_1, V_2) monotonically increasing and cc (W) monotonically decreasing, and the functions f_μ and g_μ is defined as in Propositions 1 and 2. Then the following hold.

(a) When $\boldsymbol{\mu} < 1$ in an achieved equilibrium state with equilibrium values **X**, **Y** for states X and Y and $\mathbf{W}_{X,Y}$ for $\mathbf{W}_{X,Y}$ it holds

$$\underline{\mathbf{W}}_{X,Y} = \mathrm{f}_{\boldsymbol{\mu}}(\mathbf{X}, \mathbf{Y}) = g_{\boldsymbol{\mu}}(\mathrm{cs}(\mathbf{X}, \mathbf{Y})) < 1$$

(b) For $\boldsymbol{\mu} = 1$ in an equilibrium state with equilibrium values **X**, **Y** for states X and Y and $\mathbf{W}_{X,Y}$ for $\mathbf{W}_{X,Y}$ it holds $\mathbf{X} = 0$ or $\mathbf{Y} = 0$ or $\underline{\mathbf{W}}_{X,Y} = 1$.

(c) For $\boldsymbol{\mu} < 1$ in an equilibrium state the maximal equilibrium value $\underline{\mathbf{W}}_{X,Y}$ for $\mathbf{W}_{X,Y}$ is g_μ(cs(1, 1)) < 1, and the minimal equilibrium value $\underline{\mathbf{W}}_{X,Y}$ is 0. For $\boldsymbol{\mu} = 1$ the maximal value is 1 (always when **X**, **Y** > 0 holds for the equilibrium values for the states X and Y) and the minimal value is 0 (which occurs when one of **X**, **Y** is 0).

14.6 Implications for Different Classes of Hebbian Learning Functions

In this section some cases are analysed as corollaries of Theorem 2.

14.6.1 Hebbian Learning Functions with Variable Separation and Linear Connection Factor

First the specific class of Hebbian learning functions enabling variable separation with cc(W) = 1 − W is considered. Then

$$h_{\mu}(W) = \frac{(1-\mu)W}{cc(W)} = \frac{(1-\mu)W}{1-W} \tag{14.4}$$

and the inverse $g_{\mu}(W')$ of $h_{\mu}(W)$ can be determined from (14.4) algebraically as shown in Box 14.1.

Box 14.1 Inverting $h_{\mu}(W)$ for linear connection factor $cc(W) = 1-W$

$$W' = h_{\mu}(W) = \frac{(1-\mu)W}{1-W}$$
$$W'(1-W) = (1-\mu)W$$
$$W' - W'W = (1-\mu)W$$
$$W' = (W' + (1-\mu))W$$
$$W - \frac{W'}{W' + (1-\mu)}$$
$$g_{\mu}(W') = \frac{W'}{W' + (1-\mu)}$$

So it has been found that

$$g_{\mu}(W') = \frac{W'}{W' + (1-\mu)} \tag{14.5}$$

Substitute $W' = cs(V_1, V_2)$ in (14.5) and it is obtained:

$$f_{\mu}(V_1, V_2) = g_{\mu}(cs(V_1, V_2)) = \frac{cs(V_1, V_2)}{(1-\mu) + cs(V_1, V_2)} \tag{14.6}$$

and this is less than 1 because $1-\mu > 0$. From this and Theorem 2 (b) and (c) it follows.

Corollary 1 (cases for connection factor $cc(W) = 1 - W$)
Assume in a reified temporal-causal network the Hebbian learning combination function $c(V_1, V_2, W)$ for $\mathbf{W}_{X,Y}$ with persistence parameter μ enables variable separation by the two functions $cs(V_1, V_2)$ monotonically increasing and $cc(W)$ monotonically decreasing, where for the connection factor it hols $cc(W) = 1 - W$. Then the following hold.

(a) When $\mu < 1$ in an equilibrium state with equilibrium values **X**, **Y** for states X and Y and $\underline{\mathbf{W}}_{X,Y}$ for $\mathbf{W}_{X,Y}$ it holds

$$\underline{\mathbf{W}}_{X,Y} = f_{\mu}(\mathbf{X}, \mathbf{Y}) = \frac{cs(\mathbf{X}, \mathbf{Y})}{(1-\mu) + cs(\mathbf{X}, \mathbf{Y})} < 1$$

Table 14.2 Special cases for variable separation

cc(W)	cs(V_1, V_2)	Equilibrium value $\underline{\mathbf{W}}_{X,Y} = f_\mu(\mathbf{X}, \mathbf{Y})$	Maximal equilibrium value $\underline{\mathbf{W}}_{X,Y}$
$1 - W$	V_1V_2	$\mathbf{X}\,\mathbf{Y}/[(1-\mu) + \mathbf{X}\,\mathbf{Y}]$	$\frac{1}{2-\mu}$
	$\sqrt{V_1V_2}$	$\sqrt{\mathbf{XY}}/[(1-\mu) + \sqrt{\mathbf{XY}}]$	$\frac{1}{2-\mu}$
	$V_1V_2(V_1 + V_2)$	$\frac{\mathbf{XY}(\mathbf{X}+\mathbf{Y})}{(1-\mu)+\mathbf{XY}(\mathbf{X}+\mathbf{Y})}$	$\frac{2}{3-\mu}$
$1 - W^2$	$V_1V_2(V_1 + V_2)$	$\frac{-(1-\mu)+\sqrt{(1-\mu)^2+4(\mathbf{XY}(\mathbf{X}+\mathbf{Y}))^2}}{2\mathbf{XY}(\mathbf{X}+\mathbf{Y})}$	$\frac{-(1-\mu)+\sqrt{(1-\mu)^2+16}}{4}$

(b) For $\mu = 1$ in an equilibrium state with equilibrium values $\mathbf{X}$, $\mathbf{Y}$ for states X and Y and $\underline{\mathbf{W}}_{X,Y}$ for $\mathbf{W}_{X,Y}$ it holds $\mathbf{X} = 0$ or $\mathbf{Y} = 0$ or $\underline{\mathbf{W}}_{X,Y} = 1$.

(c) For $\mu < 1$ in an equilibrium state the maximal equilibrium value $\underline{\mathbf{W}}_{X,Y}$ for $\mathbf{W}_{X,Y}$ is

$$\frac{\text{cs}(1,1)}{(1-\mu)+\text{cs}(1,1)} < 1$$

and the minimal equilibrium value $\mathbf{W}_{X,Y}$ is 0. For $\mu = 1$ the maximal value is 1 (when $\mathbf{X}$, $\mathbf{Y} > 0$ holds for the equilibrium values for the states X and Y) and the minimal value is 0 (which occurs when one of $\mathbf{X}$, $\mathbf{Y}$ is 0).

Corollary 1 is illustrated in the three examples shown in Box 14.2. Note that Table 14.2 summarizes these results.

Box 14.2 Different examples of Hebbian learning functions with variable separation with linear connection factor cc(W) = $1 - W$

Example 1 $c(V_1, V_2, W) = V_1V_2(1-W) + \mu W$

$$\text{cs}(V_1, V_2) = V_1V_2 \quad \text{cc}(W) = 1 - W$$

This is the example shown in Fig. 14.4

$$f_\mu(V_1, V_2) = \frac{\text{cs}(V_1, V_2)}{(1-\mu)+\text{cs}(V_1, V_2)} \tag{14.7}$$

Substitute cs(V_1, V_2) = V_1V_2 in (14.7) then $f_\mu(V_1, V_2) = \frac{V_1V_2}{(1-\mu)+V_1V_2}$

Maximal W is $W_{\max} = f_\mu(1,1) = \frac{1}{2-\mu}$, which for $\mu = 1$ is 1; minimal W is 0.

The equilibrium values shown in Fig. 14.4 can immediately be derived from this (recall $\mu = 0.8$):

Figure 14.4, top row $V_1 = 1$ $V_2 = 1$, then $f_\mu(1, 1) = \frac{1}{2-\mu}$ = 0.833333
Figure 14.4, middle row $V_1 = 0.6$, $V_2 = 0.6$, then
$f_\mu(0.6, 0.6) = 0.36/[(1 - 0.8) + 0.36] = 0.642857$
Figure 14.4, bottom row $V_1 = 0$, $V_2 = 0$, then $f_\mu(0, 0) = 0$

Example 2 $c(V_1, V_2, W) = \sqrt{V_1 V_2}(1 - W) + \mu W$

$$\begin{aligned} &cs(V_1, V_2) = \sqrt{V_1 V_2} \quad cc(W) = 1 - W \\ &f_\mu(V_1, V_2) = cs(V_1, V_2)/[(1 - \mu) + cs(V_1, V_2)] \end{aligned} \tag{14.8}$$

Substitute $cs(V_1, V_2) = \sqrt{V_1 V_2}$ in (14.8) to obtain

$$f_\mu(V_1, V_2) = \frac{\sqrt{V_1 V_2}}{(1 - \mu) + \sqrt{V_1 V_2}} \tag{14.9}$$

Maximal W is $W_{max} = f_\mu(1, 1) = \frac{1}{2-\mu}$, which for $\mu = 1$ is 1; minimal W is 0.

In a similar case as in Fig. 14.4, but now using this function, the following equilibrium values would be found

Top row $V_1 = 1$, $V_2 = 1$, then $f_\mu(1, 1) = \frac{1}{2-\mu}$ = 0.833333
Middle row $V_1 = 0.6$, $V_2 = 0.6$, then
$f_\mu(0.6, 0.6) = 0.6/[(1 - 0.8) + 0.6] = 0.75$
Bottom row $V_1 = 0$, $V_2 = 0$, then $f_\mu(0, 0) = 0$

Example 3 $c(V_1, V_2, W) = V_1 V_2(V_1 + V_2)(1 - W) + \mu W$

$$cs(V_1, V_2) = V_1 V_2(V_1 + V_2) \quad cc(W) = 1 - W$$

$$f_\mu(V_1, V_2) = \frac{cs(V_1, V_2)}{(1 - \mu) + cs(V_1, V_2)} \tag{14.10}$$

Substitute $cs(V_1, V_2) = V_1 V_2(V_1 + V_2)$ in (14.10) to obtain

$$f_\mu(V_1, V_2) = \frac{V_1 V_2(V_1 + V_2)}{(1 - \mu) + V_1 V_2(V_1 + V_2)} \tag{14.11}$$

Maximal W is $W_{max} = f_\mu(1, 1) = \frac{2}{(1-\mu)+2} = \frac{2}{3-\mu}$, which for $\mu = 1$ is 1; minimal W is 0.

In a similar case as in Fig. 14.4, but using this function the following equilibrium values would be found

Top row $V_1 = 1$, $V_2 = 1$, then $f_\mu(1, 1) = \frac{2}{3-\mu}$ = 0.909090

Middle row $V_1 = 0.6$, $V_2 = 0.6$, then $f_\mu(0.6, 0.6) = 0.36 * 1.2/[(1 - 0.8) + 0.36 * 1.2] = 0.632$

Bottom row $V_1 = 0$, $V_2 = 0$, then $f_\mu(0, 0) = 0$

14.6.2 *Hebbian Learning Functions with Variable Separation and Quadratic Connection Factor*

Next the specific class of Hebbian learning functions enabling variable separation with $\mathrm{cc}(W) = 1 - W^2$ is considered. In that case it holds

$$\mathrm{h}_{\mu}(W) = \frac{(1-\mu)W}{\mathrm{cc}(W)} = \frac{(1-\mu)W}{1-W^2} \tag{14.12}$$

and the inverse of h_{μ} can be determined algebraically as shown in Corollary 2. Inverting $\mathrm{h}_{\mu}(W)$ to get inverse $\mathrm{g}_{\mu}(W')$ now can be done as shown in Box 14.3:

Box 14.3 Inverting $\mathrm{h}_{\mu}(W)$ for quadratic connection factor $\mathrm{cc}(W) = 1 - W^2$

$$\begin{aligned} W' = \mathrm{h}_{\mu}(W) &= \frac{(1-\mu)W}{1-W^2} \\ (1-W^2)W' &= (1-\mu)W \\ -W' + (1-\mu)W + W^2W' &= 0 \end{aligned}$$

This is a quadratic equation in W:

$$W'W^2 + (1-\mu)W - W' = 0$$

As $W \geq 0$ the solution is

$$W = \frac{-(1-\mu) + \sqrt{(1-\mu)^2 + 4W'^2}}{2W'}$$

Therefore

$$\mathrm{g}_{\mu}(W') = \frac{-(1-\mu) + \sqrt{(1-\mu)^2 + 4W'^2}}{2W'}$$

So, from Box 14.3:

$$\mathrm{g}_{\mu}(W') = \frac{-(1-\mu) + \sqrt{(1-\mu)^2 + 4W'^2}}{2W'} \tag{14.13}$$

By substituting $W' = \mathrm{cs}(V_1, V_2)$ it follows

$$\mathrm{f}_{\boldsymbol{\mu}}(V_1, V_2) = \mathrm{g}_{\boldsymbol{\mu}}(\mathrm{cs}(V_1, V_2)) = \frac{-(1-\boldsymbol{\mu}) + \sqrt{(1-\boldsymbol{\mu})^2 + 4\mathrm{cs}(V_1, V_2)^2}}{2\mathrm{cs}(V_1, V_2)} \tag{14.14}$$

All this is summarised in the following:

Corollary 2 (cases for quadratic connection factor cc(*W*) = 1− *W*²)
Assume in a reified temporal-causal network the Hebbian learning combination function $\mathrm{c}(V_1, V_2, W)$ for $\mathbf{W}_{X,Y}$ with persistence parameter $\boldsymbol{\mu}$ enables variable separation by the two functions $\mathrm{cs}(V_1, V_2)$ monotonically increasing and $\mathrm{cc}(W)$ monotonically decreasing, where for the connection factor it holds $\mathrm{cc}(W) = 1 - W^2$. Then the following hold.

(a) When $\boldsymbol{\mu} < 1$ in an equilibrium state with equilibrium values **X**, **Y** for states *X* and *Y* and $\underline{\mathbf{W}}_{X,Y}$ for $\mathbf{W}_{X,Y}$ *it holds*

$$\underline{\mathbf{W}}_{X,Y} = \mathrm{f}_{\boldsymbol{\mu}}(\mathbf{X}, \mathbf{Y}) = \frac{-(1-\boldsymbol{\mu}) + \sqrt{(1-\boldsymbol{\mu})^2 + 4\mathrm{cs}(\mathbf{X}, \mathbf{Y})^2}}{2\mathrm{cs}(\mathbf{X}, \mathbf{Y})} < 1$$

(b) For $\boldsymbol{\mu} = 1$ in an equilibrium state with equilibrium values **X**, **Y** for states *X* and *Y* and $\underline{\mathbf{W}}_{X,Y}$ for $\mathbf{W}_{X,Y}$ it holds $\mathbf{X} = 0$ or $\mathbf{Y} = 0$ or $\underline{\mathbf{W}}_{X,Y} = 1$.

(c) For $\boldsymbol{\mu} < 1$ in an equilibrium state the maximal equilibrium value $\underline{\mathbf{W}}_{X,Y}$ for $\mathbf{W}_{X,Y}$ *is*

$$\frac{-(1-\boldsymbol{\mu}) + \sqrt{(1-\boldsymbol{\mu})^2 + 4\mathrm{cs}(1,1)^2}}{2\mathrm{cs}(1,1)} < 1$$

and the minimal equilibrium value $\underline{\mathbf{W}}_{X,Y}$ is 0. For $\boldsymbol{\mu} = 1$ the maximal value is 1 (when **X**, **Y** > 0 holds for the equilibrium values for the states *X* and *Y*) and the minimal value is 0 (which occurs when one of **X**, **Y** is 0).

Corollary 2 is illustrated in Example 4 in Box 14.4.

Box 14.4 Example of a Hebbian learning function with variable separation with quadratic connection factor $\mathrm{cc}(W) = 1 - W^2$
Example 4 $\mathrm{c}(V_1, V_2, W) = V_1V_2(V_1 + V_2)(1 - W^2) + \boldsymbol{\mu}W$

$$\mathrm{cs}(V_1, V_2) = V_1V_2(V_1 + V_2) \quad \mathrm{cc}(W) = 1 - W^2$$

$$\mathrm{f}_{\boldsymbol{\mu}}(V_1, V_2) = \frac{-(1-\boldsymbol{\mu}) + \sqrt{(1-\boldsymbol{\mu})^2 + 4\mathrm{cs}(V_1, V_2)^2}}{2\mathrm{cs}(V_1, V_2)} \tag{14.15}$$

Substitute cs(**X**, **Y**) = **XY** (**X** + **Y**)

$$f_{\mu}(\mathbf{X},\mathbf{Y}) = \frac{-(1-\mu)+\sqrt{(1-\mu)^2+4(\mathbf{XY}(\mathbf{X}+\mathbf{Y}))^2}}{2\mathbf{XY}(\mathbf{X}+\mathbf{Y})} \quad (14.16)$$

Maximal W is

$$W_{\max} = f_{\mu}(1,1) = \frac{-(1-\mu)+\sqrt{(1-\mu)^2+16}}{4}$$

which for $\mu = 1$ is 1; minimal W is 0. In a similar case as in Fig. 14.4, using this function the equilibrium values can be found by applying (18).

14.7 Discussion

In this chapter it was analysed how emerging behaviour of an adaptive network can be related to characteristics of reified network structure addressing adaptation principles. Parts of this chapter were adopted from (Treur 2018b). In particular this was addressed for an adaptive Mental Network based on Hebbian learning (Bi and Poo 2001; Gerstner and Kistler 2002; Hebb 1949; Keysers and Perrett 2004; Keysers and Gazzola 2014; Kuriscak et al. 2015; Zenke et al. 2017). The approach followed is based on network reification applied to Connectivity characteristics of a network expressed by connection weights; see Chap. 3. This makes that the base network is extended by reification states representing the adaptive connection weights. Applying the standard temporal-causal network structure characteristics, these reification states get their own combination functions assigned to define aggregation of the incoming impact. Such combination functions can be used to define certain types of Hebbian learning. Given these, relevant properties of the combination functions defining variants of the Hebbian adaptation principle have been identified together with their implied behaviour.

For different classes of Hebbian learning combination functions, emerging equilibrium values for the connection weight have been expressed as a function of the emerging equilibrium values of the connected states. The presented results do not concern results for just one type of network or function, as more often is found, but were formulated and proven at a more general level and therefore can be applied not just to specific networks but to classes of networks satisfying the identified relevant properties of reified network structure including the adaptation characteristics as specified by the Hebbian learning combination function.

References

Bi, G., Poo, M.: Synaptic modification by correlated activity: hebb's postulate revisited. Annu. Rev. Neurosci. **24**, 139–166 (2001)

Brauer, F., Nohel, J.A.: Qualitative Theory of Ordinary Differential Equations. Benjamin (1969)

Gerstner, W., Kistler, W.M.: Mathematical formulations of Hebbian learning. Biol. Cybern. **87**, 404–415 (2002)

Hebb, D.O.: The organization of behavior: a neuropsychological theory (1949)

Hirsch, M.W.: The dynamical systems approach to differential equations. Bull. (New Ser.) Am. Math. Soc. **11**, 1–64 (1984)

Keysers, C., Perrett, D.I.: Demystifying social cognition: a Hebbian perspective. Trends Cogn. Sci. **8**(11), 501–507 (2004)

Keysers, C., Gazzola, V.: Hebbian learning and predictive mirror neurons for actions, sensations and emotions. Philos. Trans. R. Soc. Lond. B Biol. Sci. **369**, 20130175 (2014)

Kuriscak, E., Marsalek, P., Stroffek, J., Toth, P.G.: Biological context of Hebb learning in artificial neural networks, a review. Neurocomputing **152**, 27–35 (2015)

Lotka, A.J.: Elements of Physical Biology. Williams and Wilkins Co. (1924), Dover Publications, 2nd ed. (1956)

Treur, J.: Network-Oriented Modeling: Addressing Complexity of Cognitive, Affective and Social Interactions. Springer Publishers (2016)

Treur, J.: Network reification as a unified approach to represent network adaptation principles within a network. In: Proceedings of the 7th International Conference on Theory and Practice of Natural Computing, TPNC'18. Lecture Notes in Computer Science, vol. 11324. pp. 344–358. Springer Publishers (2018a)

Treur, J.: Relating an adaptive network's structure to its emerging behaviour for Hebbian learning. In: Martín-Vide, C., Vega-Rodríguez, M.A., Fagan, D., O'Neill, M. (eds.) Theory and Practice of Natural Computing: 7th International Conference, TPNC 2018, Proceedings. Lecture Notes in Computer Science, vol. 11324. pp. 359–373, Springer Publishers (2018b)

Treur, J., Umair, M.: On rationality of decision models incorporating emotion-related valuing and Hebbian learning. In: Lu, B.-L., Zhang, L., Kwok, J. (eds.) Proceedings of the 18th International Conference on Neural Information Processing, ICONIP'11, Part III. Lecture Notes in Artificial Intelligence, vol. 7064, pp. 217–229. Springer Verlag (2011)

Zenke, F., Gerstner, W., Ganguli, S.: The temporal paradox of Hebbian learning and homeostatic plasticity. Neurobiology **43**, 166–176 (2017)

Part VII
Finalising

Chapter 15
Mathematical Details of Specific Difference and Differential Equations and Mathematical Analysis of Emerging Network Behaviour

Abstract In this chapter, additional mathematical details are presented for many of the chapters concerning the specific difference and differential equations, and mathematical analysis of emerging behaviour. For modeling and analysis of practical applications, insight into these details may not be necessary, but they may deepen insight from the mathematical and technical angle.

Keywords Difference equation · Differential equation · Network behaviour · Mathematical analysis

15.1 Introduction

As also discussed at the end of Chap. 1, in many chapters, especially from Chaps. 1 to 9, mathematical and procedural details were kept at a minimum to obtain optimal readability for a wide group of readers with diverse multidisciplinary backgrounds. As the Network-Oriented Modeling approach based on reified temporal-causal networks presented in this book abstracts from specific implementation details, making use of the dedicated software environment, modeling can be done without having to design procedural or algorithmic specifications. Moreover, a modeler does not even need to explicitly specify difference or differential equations to get a simulation done, as these are already taken care for by the software environment, based on the modeler's input in the form of the role matrices specifying the conceptual representation of the network model. Therefore, in Chaps. 1–9 all underlying specific procedural elements and difference or differential equations were usually not discussed, although the underlying universal difference and differential equation were briefly mentioned and discussed more extensively in Chap. 10. The only mathematical details that were addressed for design of a network model concern the combination functions used, most of which are already given in the combination function library. For analysis of the emerging behaviour of a network model also these combination functions are central, as the equilibrium equations are

J. Treur, *Network-Oriented Modeling for Adaptive Networks: Designing Higher-Order Adaptive Biological, Mental and Social Network Models*, Studies in Systems, Decision and Control 251, https://doi.org/10.1007/978-3-030-31445-3_15

based on them. Moreover, especially for Chaps. 11–14 in the current chapter (Sects. 15.6–15.9) details of the proofs are discussed that were omitted in those chapers.

However, for those readers who still want to see more mathematical details that are covered in the software environment, the current chapter presents these in more depth in different sections, as a kind of appendices to many of the chapters. These sections should be read in conjunction with the concerning chapter, since that chapter itself is not repeated here as a whole.

15.2 Two Different Formulations of Hebbian Learning Are Equivalent

In this section it is shown why the two forms of modeling adaptation (the hybrid form and the temporal-causal form) discussed in Chap. 1, Sects. 1.4.1 and 1.6.1 are mathematically equivalent. Recall from Chap. 1, Sect. 1.6.1 the Eqs. (1.1), (1.3) and (1.4), for network adaptation by Hebbian learning based on network reification, here renumbered to (15.1)–(15.3):

$$Y(t+\Delta t) = Y(t) + \boldsymbol{\eta}_Y[\mathbf{c}_Y(\boldsymbol{\omega}_{X_1,Y}X_1(t), \ldots, \boldsymbol{\omega}_{X_k,Y}X_k(t)) - Y(t)]\Delta t \tag{15.1}$$

$$\mathbf{hebb}_{\boldsymbol{\mu}}(V_1, V_2, W) = V_1 V_2(1 - W) + \boldsymbol{\mu} W \tag{15.2}$$

$$\boldsymbol{\mu} = 1 - \boldsymbol{\zeta}/\boldsymbol{\eta} \text{ or } \boldsymbol{\zeta} = (1 - \boldsymbol{\mu})\boldsymbol{\eta} \tag{15.3}$$

Moreover, in Fig. 1.2 in Chap. 1, Sect. 1.4.1 for the hybrid adaptation model the following equation is described:

$$\boldsymbol{\omega}_{X_i,Y}(t+\Delta t) = \boldsymbol{\omega}_{X_i,Y}(t) + (\boldsymbol{\eta} X_i(t)Y(t)(1 - \boldsymbol{\omega}_{X_i,Y}(t)) - \boldsymbol{\zeta}\boldsymbol{\omega}_{X_i,Y}(t))\Delta t \tag{15.4}$$

Based on relations (15.1), (15.2) and (15.3) it can be verified that the difference equation shown in (15.4) actually is mathematically equivalent to the standard Eq. (15.1) for reified temporal-causal networks using the above combination function (15.2) for the reification state. However, the latter formulation in (15.2) provides a more transparent and more unified format than (15.4). The equivalence can be found through rewriting of the mathematical formulas by elementary mathematical rules, starting from (15.4). In Box 15.1 for readers with less mathematical background the steps have been explained in some detail.

Box 15.1 Modeling the equation from the hybrid approach in (15.4) from Chap. 1, Sect. 1.4.1, Fig. 1.2 in the standard format for a temporal-causal network

$$\begin{aligned}
\boldsymbol{\omega}_{X_i,Y}(t+\Delta t) &= \boldsymbol{\omega}_{X_i,Y}(t) + (\boldsymbol{\eta} X_i(t)Y(t)(1-\boldsymbol{\omega}_{X_i,Y}(t)) - \boldsymbol{\zeta}\,\boldsymbol{\omega}_{X_i,Y}(t))\Delta t \\
&\quad \textit{from}(4) \\
&= \boldsymbol{\omega}_{X_i,Y}(t) + (\boldsymbol{\eta} X_i(t)Y(t)(1-\boldsymbol{\omega}_{X_i,Y}(t)) - (1-\boldsymbol{\mu})\boldsymbol{\eta}\,\boldsymbol{\omega}_{X_i,Y}(t))\Delta t \\
&\quad \textit{applying}(3) \\
&= \boldsymbol{\omega}_{X_i,Y}(t) + \boldsymbol{\eta}[X_i(t)Y(t)(1-\boldsymbol{\omega}_{X_i,Y}(t)) - (1-\boldsymbol{\mu})\,\boldsymbol{\omega}_{X_i,Y}(t)]\Delta t \\
&\quad (\textit{anti})\textit{distribution for}\ \boldsymbol{\eta} \\
&= \boldsymbol{\omega}_{X_i,Y}(t) + \boldsymbol{\eta}[X_i(t)Y(t)(1-\boldsymbol{\omega}_{X_i,Y}(t)) - \boldsymbol{\omega}_{X_i,Y}(t) + \boldsymbol{\mu} w_{X_i,Y}(t)]\Delta t \\
&\quad \textit{distribution for}\ \boldsymbol{\omega}_{X_i,Y}(t) \\
&= \boldsymbol{\omega}_{X_i,Y}(t) + \boldsymbol{\eta}[X_i(t)Y(t)(1-\boldsymbol{\omega}_{X_i,Y}(t)) + \boldsymbol{\mu} w_{X_i,Y}(t) - \boldsymbol{\omega}_{X_i,Y}(t)]\Delta t \\
&\quad \textit{commutation of}\ -\boldsymbol{\omega}_{X_i,Y}(t)\ \textit{and}\ \boldsymbol{\mu}\boldsymbol{\omega}_{X_i,Y}(t) \\
&= \boldsymbol{\omega}_{X_i,Y}(t) + \boldsymbol{\eta}[\mathbf{hebb}_{\boldsymbol{\mu}}(X_i(t), Y(t), \boldsymbol{\omega}_{X_i,Y}(t)) - \boldsymbol{\omega}_{X_i,Y}(t)]\Delta t \quad \textit{applying}(2)
\end{aligned}$$

As shown in Box 15.1, Eq. (15.4) for $\boldsymbol{\omega}_{X_i,Y}$ displayed in Fig. 1.2 in Chap. 1 can be rewritten into the following mathematically equivalent equation:

$$\boldsymbol{\omega}_{X_i,Y}(t+\Delta t) = w_{X_i,Y}(t) + \boldsymbol{\eta}[\mathbf{hebb}_{\boldsymbol{\mu}}(X_i(t), Y(t), \boldsymbol{\omega}_{X_i,Y}(t)) - \boldsymbol{\omega}_{X_i,Y}(t)]\Delta t$$

In terms of the reification state $\mathbf{W}_{X_i,Y}$ substituted for $\boldsymbol{\omega}_{X_i,Y}$ this is

$$\mathbf{W}_{X_i,Y}(t+\Delta t) = \mathbf{W}_{X_i,Y}(t) + \boldsymbol{\eta}[\mathbf{hebb}_{\boldsymbol{\mu}}(X_i(t), Y(t), \mathbf{W}_{X_i,Y}(t)) - \mathbf{W}_{X_i,Y}(t)]\Delta t$$

and this form is indeed exactly the standard equation form for a temporal-causal network applied to the reification state $\mathbf{W}_{X_i,Y}$, with $k = 3$, incoming impacts defined by the two upward (blue) arrows in Fig. 1.4 in Chap. 1, Sect. 1.4.2 and a connection to $\mathbf{W}_{X_i,Y}$ itself, and combination function $\mathbf{hebb}_{\boldsymbol{\mu}}(V_1, V_2, W)$ defined by (15.2) above.

15.3 Numerical Representation for an Example Reified Network Model

This section addresses the specific difference equations for the example reified network model described in Chap. 3, Sect. 3.7, as used in the software. From the specifications shown in Chap. 3, Sect. 3.7, Box 3.8, the difference equations are derived according to the format in (15.1) above, or (for the manager opinion state) according to the universal difference equation in Chap. 3, Sect. 3.5 (or in Chap. 10) as follows. Based on the role matrix specifications shown in Box 3.8 in Chap. 3, the difference equations for the three reification states are obtained as can be seen in Box 15.2.

Box 15.2 Difference equations for the reification states $\mathbf{H}_{\text{manageropinion}}$, $\mathbf{C}_{1,\text{manageropinion}}$, and $\mathbf{C}_{2,\text{manageropinion}}$

$$\mathbf{H}_{\text{manageropinion}}(t+\Delta t) = \mathbf{H}_{\text{manageropinion}}(t) + 0.5\,[\text{available time}(t) - \mathbf{H}_{\text{manageropinion}}(t)]\Delta t$$

$$\mathbf{C}_{1,\text{manageropinion}}(t+\Delta t) = \mathbf{C}_{1,\text{manageropinion}}(t)$$
$$+\,0.5[\mathbf{ssum}_{0.02}(0.01X_1(t), 0.01X_2(t), 0.01X_3(t), 0.01X_4(t), 0.01X_5(t), 0.01X_6(t), 0.01X_7(t),$$
$$-\,0.05\text{disappointment}(t)) - \mathbf{C}_{1,\text{manageropinion}}(t)]\Delta t$$

This can be rewritten into

$$\mathbf{C}_{1,\text{manageropinion}}(t+\Delta t) = \mathbf{C}_{1,\text{manageropinion}}(t) + 0.5[0.5X_1(t) + 0.5X_2(t) + 0.5X_3(t) + 0.5X_4(t)$$
$$+\,0.5X_5(t) + 0.5X_6(t) + 0.5X_7(t) - 2.5\text{disappointment}(t)$$
$$-\,\mathbf{C}_{1,\text{manageropinion}}(t)]\Delta t$$
$$\mathbf{C}_{2,\text{manageropinion}}(t+\Delta t) = \mathbf{C}_{2,\text{manageropinion}}(t)$$
$$+\,0.5\,[\text{disappointment}(t) - \mathbf{C}_{2,\text{manageropinion}}(t)]\Delta t$$

The difference equation for the base state manager opinion is given by the universal difference equation described in Chap. 3, Sect. 3.5 or in Chap. 10; see Box 15.3. According to formula (15.1), and the specifications in Chap. 3, Box 3.8, the base states available time and disappointment get the difference equations as shown in Box 15.3.

Box 15.3 Difference equations for the base states
Substituting the manager opinion state for Y in the universal difference equation and using the role matrix specifications in the row for X_8 in Chap. 3, Box 3.8 provides:

$$\text{manageropinion}(t+\Delta t) = \text{manageropinion}(t) + \mathbf{H}_{\text{manageropinion}}(t)$$
$$\left[\frac{\begin{array}{l}\mathbf{C}_{1,\text{manageropinion}}(t)\text{bcf}_1(\mathbf{W}_{X_1,\text{manageropinion}}(t)X_1(t), \ldots, \mathbf{W}_{X_7,\text{manageropinion}}(t)X_7(t))\\ \quad + \mathbf{C}_{2,\text{manageropinion}}(t)\text{bcf}_2(\mathbf{W}_{X_1,\text{manageropinion}}(t)X_1(t), \ldots, \mathbf{W}_{X_7,\text{manageropinion}}(t)X_7(t))\end{array}}{\mathbf{C}_{1,\text{manageropinion}}(t) + \mathbf{C}_{2,\text{manageropinion}}(t)} - \text{manageropinion}(t)\right]\Delta t$$

Using $\mathbf{W}_{X_i,\text{manageropinion}} = 1$ for all i (see the row for X_8 in **mcw** in Box 3.8), and $\text{bcf}_1(..) = \mathbf{ssum}_\lambda(..)$ and $\text{bcf}_2(..) = \mathbf{alogistic}_{\sigma,\tau}(..)$, this can be rewritten as

$$\text{manageropinion}(t+\Delta t) = \text{manageropinion}(t) + \mathbf{H}_{\text{manageropinion}}(t)$$
$$\left[\frac{\mathbf{C}_{1,\text{manageropinion}}(t)\,\mathbf{ssum}_7(X_1(t),\ldots,X_7(t)) + \mathbf{C}_{2,\text{manageropinion}}(t)\,\mathbf{alogistic}_{5,5.5}(X_1(t),\ldots,X_7(t))}{\mathbf{C}_{1,\text{manageropinion}}(t) + \mathbf{C}_{2,\text{manageropinion}}(t)}\right.$$
$$\left. - \text{manageropinion}(t)\right]\Delta t$$

According to temporal-causal format (1) above, and the role matrix specifications in Box 3.8, the base states available time and disappointment get the following difference equations:

$$\text{available time}(t+\Delta t) = \text{available time}(t) + 0.04[\mathbf{alogistic}_{18,0.2}(\text{available time}(t)) - \text{available time}(t)]\Delta t$$
$$\text{disappointment}(t+\Delta t) = \text{disappointment}(t) + 0.025[\mathbf{alogistic}_{18,0.2}(\text{disappointment}(t)) - \text{disappointment}(t)]\Delta t$$

Similarly, according to (1) and the role matrix specifications in Box 3.8 in Chap. 3, the group members X_i, i = 1, …, 7 get the following difference equations:

$$X_i(t+\Delta t) = X_i(t) + 0.005[\mathbf{ssum}_{\lambda_i}(\boldsymbol{\omega}_{X_1,X_i}X_1(t),\ldots,\boldsymbol{\omega}_{X_7,X_i}X_7(t)) - X_i(t)]\Delta t$$

where $\boldsymbol{\lambda}_i$ is the sum of the incoming weights $\boldsymbol{\omega}_{X_j,X_i}$ for X_i.

15.4 The Difference Equations for Combined Hebbian Learning and State-Connection Modulation

In Chap. 5, Sects. 5.3 and 5.4 an example reified network model was described in which the Hebbian learning adaptation principle is combined with the state-connection modulation adaptation principle. In the current section the underlying difference equations used in the implementation are shown in some detail. Recall that for the reification states of the connection weights the following combination functions were used. For *Hebbian learning* of a connection from state X_i to state X_j with connection weight reification state **W** the function described in (15.2) above where $\boldsymbol{\mu}$ is the persistence factor with 1 as full persistence. For *state-connection modulation* with control state cs_2 for connection weight reification state **W**:

$$\mathbf{scm}_{\boldsymbol{\alpha}}(V_1, V_2, W, V) = W + \boldsymbol{\alpha}\, VW(1-W) \tag{15.5}$$

where $\boldsymbol{\alpha}$ is the modulation parameter for **W** from cs_2, V is the single impact from cs_2, and W is the value of **W**; the V_1 and V_2 are auxiliary variables allowing to (partly) separate the arguments used in the two functions. For Hebbian learning separately the difference equation is:

$$\mathbf{W}(t+\Delta t) = \mathbf{W}(t) + \boldsymbol{\eta}_{\mathbf{W}}\left[\mathbf{c}_{\mathbf{W}}(X_i(t), X_j(t), \mathbf{W}(t)) - \mathbf{W}(t)\right]\Delta t \quad (15.6)$$

with

$$\mathbf{c}_{\mathbf{W}}(V_1, V_2, W) = \mathbf{hebb}_{\mu}(V_1, V_2, W) = V_1 V_2(1-W) + \mu\, W \quad (15.7)$$

For *state-connection modulation* with control state cs_2 for connection weight reification state **W** the difference equation is:

$$\mathbf{W}(t+\Delta t) = \mathbf{W}(t) + \boldsymbol{\eta}_{\mathbf{W}}\left[\mathbf{c}_{\mathbf{W}}(X_i(t), X_j(t), cs_2(t), \mathbf{W}(t)) - \mathbf{W}(t)\right]\Delta t \quad (15.8)$$

with

$$\mathbf{c}_{\mathbf{W}}(V_1, V_2, W, V) = \mathbf{scm}_{\alpha}(V_1, V_2, W, V) = W + \alpha V W(1-W) \quad (15.9)$$

Note that the first two auxiliary variables of $\mathbf{scm}_{\alpha}(V_1, V_2, W, V)$ are not used in the formula (15.9) for $\mathbf{scm}_{\alpha}(V_1, V_2, W, V)$. These variables are included to be able to combine this function with the Hebbian learning function while using the same sequence of variables. More specifically, this combination is done as follows. These two adaptive combination functions are used as a weighted average with γ_1 and γ_2 the combination function weights for $\mathbf{hebb}_{\mu}(V_1, V_2, W)$ and $\mathbf{scm}_{\alpha}(V_1, V_2, W, V)$, respectively, as follows:

$$\mathbf{W}(t+\Delta t) = \mathbf{W}(t) + \boldsymbol{\eta}_{\mathbf{W}}\left[\mathbf{c}_{\mathbf{W}}(X_i(t), X_j(t), \mathbf{W}(t), cs_2(t)) - \mathbf{W}(t)\right]\Delta t \quad (15.10)$$

with

$$\mathbf{c}_{\mathbf{W}}(V_1, V_2, W, V) = \gamma_1 \mathbf{hebb}_{\mu}(V_1, V_2, W) + \gamma_2 \mathbf{scm}_{\alpha}(V_1, V_2, W, V) \quad (15.11)$$

So, basically the difference equation for the reification state **W** for the weight of the connection from X_i to X_j is:

$$\begin{aligned}
\mathbf{W}(t+\Delta t) &= \mathbf{W}(t) + \boldsymbol{\eta}_{\mathbf{W}}[\gamma_1 \mathbf{hebb}_{\mu}(V_1, V_2, W) + \gamma_2 \mathbf{scm}_{\alpha}(V_1, V_2, W, V) - \mathbf{W}(t)]\Delta t \\
&= \mathbf{W}(t) + \boldsymbol{\eta}_{\mathbf{W}}[\gamma_1[V_1 V_2(1-W) + \mu W] + \gamma_2[W + \alpha V W(1-W)] - \mathbf{W}(t)]\Delta t \\
&= \mathbf{W}(t) + \boldsymbol{\eta}_{\mathbf{W}}[\gamma_1\left[X_i(t)X_j(t)(1-\mathbf{W}(t)) + \mu \mathbf{W}(t)\right] \\
&\quad + \gamma_2[\mathbf{W}(t) + \alpha cs_2(t)\mathbf{W}(t)(1-\mathbf{W}(t))] - \mathbf{W}(t)]\Delta t
\end{aligned}$$

However, also taking into account that the speed factor of **W** is adaptive and represented by a reification state **H**, the equation becomes:

$$\mathbf{W}(t+\Delta t) = \mathbf{W}(t) + \mathbf{H}(t)[\boldsymbol{\gamma}_1[X_i(t)X_j(t)(1-\mathbf{W}(t)) + \boldsymbol{\mu}\mathbf{W}(t)] + \boldsymbol{\gamma}_2[\mathbf{W}(t) + \boldsymbol{\alpha}\mathrm{cs}_2(t)\mathbf{W}(t)(1-\mathbf{W}(t))] - \mathbf{W}(t)]\Delta t$$

This can also be rewritten into a correct temporal-causal format (with speed factor by default 1), based on the universal difference equation as shown in Chap. 10:

$$\begin{aligned}\mathbf{W}(t+\Delta t) &= \mathbf{W}(t) + [\mathbf{H}(t)[\boldsymbol{\gamma}_1[X_i(t)X_j(t)(1-\mathbf{W}(t)) + \boldsymbol{\mu}\mathbf{W}(t)] \\ &\quad + \boldsymbol{\gamma}_2[\mathbf{W}(t) + \boldsymbol{\alpha}\mathrm{cs}_2(t)\mathbf{W}(t)(1-\mathbf{W}(t))]] \\ &\quad + (1-\mathbf{H}(t))\mathbf{W}(t) - \mathbf{W}(t)]\Delta t \\ &= \mathbf{W}(t) + [\mathbf{c}^*_{\mathbf{W}}(\mathbf{H}(t), X_i(t), X_j(t), \mathbf{W}(t), \mathrm{cs}_2(t)) - \mathbf{W}(t)]\Delta t\end{aligned}$$

where

$$\mathbf{c}^*_{\mathbf{W}}(H, V_1, V_2, W, V) = H[\boldsymbol{\gamma}_1[V_1V_2(1-W) + \boldsymbol{\mu}W] + \boldsymbol{\gamma}_2[W + \boldsymbol{\alpha}V\,W(1-W)]] + (1-H)W$$

15.5 Difference and Differential Equations for Multilevel Connection Weight Reification States

In Chap. 6, Sect. 6.3 the combination functions for the example model were described. In the current section, the difference equations used in implementation will be added. The base level and first reification level states are addressed in Box 15.4.

Box 15.4 Combination functions and difference equations for the base level and first reification level

Base level:

Base state X_i combination function and difference equation

The combination function for the base states X_i is basically the advanced logistic sum function $\mathbf{alogistic}_{\sigma,\tau}(..)$.

However, as the connection weights are reified at the first reification level, based on the universal combination function format, the following adaptive form for the combination function for the base states X_i is needed here:

$$\begin{aligned}\mathbf{c}^*_Y(W_1, \ldots, W_k, V_1, \ldots, V_k,) &= \mathbf{alogistic}_{\sigma,\tau}(W_1V_1, \ldots, W_kV_k) \\ &= \left[\frac{1}{1+\mathrm{e}^{-\sigma(W_1V_1+\ldots+W_kV_k-\tau)}} - \frac{1}{1+\mathrm{e}^{\sigma\tau}}\right](1+\mathrm{e}^{-\sigma\tau})\end{aligned}$$

Here W_i refers to connection weight reification state value $\mathbf{W}_{X_i,Y}(t)$ and V_i to state value $X_i(t)$.
This combination function defines the following difference equation for Y (see Chap. 4, Sect. 4.2, Table 4.1):

$$Y(t+\Delta t) = Y(t) + \boldsymbol{\eta}_Y[\mathbf{alogistic}_{\sigma,\tau}(\mathbf{W}_1(t)X_1(t), \ldots, \mathbf{W}_k(t)X_k(t)) - Y(t)]\Delta t$$

where

$$\mathbf{W}_i(t) = \mathbf{W}_{X_i,Y}(t)$$

First reification level:
Connection weight reification state $\mathbf{W}_{Y,Xi}$ combination function and difference equation
See Chap. 4, Sects. 4.3.2, and 4.2, or Chap. 3, Sect. 3.6.1, or (Treur 2016), Chap. 11, Sect. 11.7, the combination function **slhomo$_\alpha$(..)** for connection weight reification state $\mathbf{W}_{X_i,Y}$ is basically

$$\mathbf{slhomo}_\alpha(V_1, V_2, W) = W + \alpha W(1 - W)(\tau - |V_1 - V_2|)$$

where

- W refers to connection weight reification state value $\mathbf{W}_{X_i,Y}(t)$
- V_1 to $X_1(t)$ and V_2 to $X_2(t)$
- α is a homophily modulation factor for $\mathbf{W}_{X_i,Y}$
- τ is a homophily tipping point for $\mathbf{W}_{X_i,Y}$.

However, as the speed factor and tipping point are reified at the second reification level, based on the universal combination function format, the following adaptive form for the combination function for connection weight reification state $\mathbf{W}_{X_i,Y}$ is needed here:

$$\mathbf{c}^*_{\mathbf{W}_{X_i,Y}}(H, V_1, V_2, T, W) = H(W + \alpha W(1 - W)(T - |V_1 - V_2|)) + (1 - H)W$$

where

- H refers to the speed factor reification $\mathbf{H}_{\mathbf{W}_{X_i,Y}}(t)$ for $\mathbf{W}_{X_i,Y}$
- W to connection weight reification $\mathbf{W}_{X_i,Y}(t)$
- T to homophily tipping point reification state value $\mathbf{TP}_{\mathbf{W}_{X_i,Y}}(t)$ for $\mathbf{W}_{X_i,Y}$
- V_1 to $X_1(t)$ and V_2 to $X_2(t)$
- α is a homophily modulation factor.

This combination function (together with connection weights and speed factor 1) defines the following difference equation for connection weight reification state $\mathbf{W} = \mathbf{W}_{X_i,Y}$ (see Sect. 4.2, Table 4.1):

$$\mathbf{W}(t+\Delta t) = \mathbf{W}(t) + [\mathbf{H}(t)(\mathbf{W}(t) + \boldsymbol{\alpha}\,\mathbf{W}(t)(1-\mathbf{W}(t))(\mathbf{TP}(t)|X_i(t) - Y(t)|)) + (1-\mathbf{H}(t))\mathbf{W}(t) - \mathbf{W}]\Delta t$$

where

$$\mathbf{H}(t) = \mathbf{H}_{\mathbf{W}X_i,Y}(t) \quad \mathbf{TP}(t) = \mathbf{TP}_{\mathbf{W}X_i,Y}(t)$$

The reification states at the second reification level are addressed in Box 15.5 (homophily tipping point reification state), and Box 15.6 (connection weight speed factor reification state).

Box 15.5 Combination function and difference equation for the homophily tipping point reification state at the second reification level

Second reification level: tipping point reification state $\mathbf{TP}_{\mathbf{W}_{X_i,Y}}$ combination function and difference equation

The following combination function called *simple linear tipping point function* **sltip**$_{\mathbf{v},\boldsymbol{\alpha}}$**(..)** can be used for the second order reification state $\mathbf{TP}_{\mathbf{W}_{X_i,Y}}$ at the second reification level (upper, purple plane):

$$\mathbf{sltip}_{\mathbf{v},\boldsymbol{\alpha}}(W_1, \ldots, W_k, T) = T + \boldsymbol{\alpha}T(1-T)(\mathbf{v} - (W_1 + \ldots + W_k)/k)$$

where

- T refers to the homophily tipping point reification value $\mathbf{TP}_{\mathbf{W}_{X_i,Y}}(t)$ for $\mathbf{W}_{X_i,Y}$
- W_j to connection weight reification value $\mathbf{W}_{X_i,Y}(t)$
- $\boldsymbol{\alpha}$ is a modulation factor for the tipping point $\mathbf{TP}_{\mathbf{W}_{X_i,Y}}$
- $\mathbf{v}$ is a norm for Y for average connection weight $\mathbf{W}_{X_1,Y}$ to $\mathbf{W}_{X_k,Y}$

This function can be explained as follows. The norm parameter $\mathbf{v}$ indicates the preferred average level of the connection weights $\mathbf{W}_{X_i,Y}$ for person Y. The part $(\mathbf{v} - (W_1 + \ldots + W_k)/k)$ in the formula is positive when the current average connection weight $(W_1 + \ldots + W_k)/k$ is lower than this norm, and negative when it is higher that the norm. When T is not 0 or 1, in the first case, the combination function provides a value higher than T, which makes that the tipping point is increased, and as a consequence more connections are strengthened by the homophily adaptation, so the average connection weight will become more close to the norm $\mathbf{v}$. In the second case, the opposite takes place: the combination function provides a value lower than T, which makes that the tipping point is decreased, and as a consequence more connections are weakened by the homophily adaptation, so also now the average connection weight will become more close to the norm **v**. Together this makes that in principle (unless in the meantime other factors change) the average connection weight will approximate the norm **v**. The factor T $(1 - T)$ in the formula takes care that the values for T stay within the [0, 1] interval.

Together with connection weights and speed factor 1, this combination function defines the following difference equation for tipping point reification state $\mathbf{TP} = \mathbf{TP}_{\mathbf{W}_{X_i,Y}}$ (see Sect. 4.2, Table 4.2):

$$\mathbf{TP}(t+\Delta t) = \mathbf{TP}(t) + \boldsymbol{\eta}[[\mathbf{TP}(t) + \boldsymbol{\alpha}\mathbf{TP}(t)(1 - \mathbf{TP}(t))(\mathbf{v} - (\mathbf{W}_1(t) + \ldots + \mathbf{W}_k(t))/k)] - \mathbf{TP}(t)]\Delta t$$

where

$$\mathbf{W}_i(t) = \mathbf{W}_{X_i,Y}(t)$$

Box 15.6 Combination function and difference equation for the connection weight speed factor reification state at the second reification level

Second reification level:

speed factor reification state $\mathbf{H}_{\mathbf{W}Y,Xi}$ combination function and difference equation

For the adaptive connection adaptation speed factor the following combination function called *simple linear speed function* $\mathbf{slspeed}_{\mathbf{v},\boldsymbol{\alpha}}(..)$ can be considered making use of a similar mechanism using a norm for connection weights.

$$\mathbf{slspeed}_{\mathbf{v},\boldsymbol{\alpha}}(W_1, \ldots, W_k, , H) = H + \boldsymbol{\alpha} H(1 - H)(\mathbf{v} - (W_1 + \ldots + W_k)/k)$$

where

- H refers to $\mathbf{W}_{Y,X_i}$ speed factor reification value $\mathbf{H}_{\mathbf{W}_{X_i,Y}}(t)$
- W_j to connection weight reification value $\mathbf{W}_{X_i,Y}(t)$
- $\boldsymbol{\alpha}$ is a modulation factor for $\mathbf{H}_{\mathbf{W}_{X_i,Y}}$
- $\mathbf{v}$ is a norm for average of (incoming) connection weights for Y

This function can be explained as follows. Also here the norm parameter **v** indicates the preferred average level of the connection weights $\mathbf{W}_{X_i,Y}$ for person Y. The part $(\mathbf{v} - (W_1 + \ldots + W_k)/k)$ in the formula is positive when the current average connection weight $(W_1 + \ldots + W_k)/k$ is lower than this norm, and negative when it is higher that the norm. When H is not 0 or 1, in the first case, the combination function provides a value higher than H, which makes that the speed factor is increased, and the connections are changing faster by the homophily adaptation. In the second case, the combination function provides a value lower than H, which makes that the speed factor is decreased, and as a consequence the homophily adaptation speed is lower. The factor $H(1 - H)$ in the formula takes care that the values for H stay within the [0, 1] interval.

This combination function defines the following difference equation for speed factor reification state $\mathbf{H} = \mathbf{H}_{\mathbf{W}_{X_i,Y}}$:

$$\begin{aligned}\mathbf{H}(t+\Delta t) = \mathbf{H}(t) + \boldsymbol{\eta}[[\mathbf{H}(t) + \boldsymbol{\alpha}\mathbf{H}(t)(1 - \mathbf{H}(t))(\mathbf{v} \\ - (\mathbf{W}_1(t) + \ldots + \mathbf{W}_k(t))/k)] - \mathbf{H}(t)]\Delta t\end{aligned}$$

where

$$\mathbf{W}_i(t) = \mathbf{W}_{X_i,Y}(t)$$

15.6 Emerging Behaviour for Types of Aggregation and Types of Connectivity

This section presents a number of proofs that were left out from Chap. 11.

Proposition 6 Suppose a network with nonnegative connections has normalised scalar-free combination functions.

(a) If $X_1, \ldots, X_k$ are the states from which Y gets its incoming connections, and $X_1(t) = \ldots = X_k(t) = V$ for some common value V, then also $\mathbf{c}_Y(\boldsymbol{\omega}_{X_1,Y}X_1(t), \ldots, \boldsymbol{\omega}_{X_k,Y}X_k(t)) = V$.

(b) If, moreover, the combination functions are monotonic, and $X_1, \ldots, X_k$ are the states from which Y gets its incoming connections, and $V_1 \leq X_1(t), \ldots, X_k(t) \leq V_2$ for some values V_1 and V_2, then also $V_1 \leq \mathbf{c}_Y(\boldsymbol{\omega}_{X_1,Y}X_1(t), \ldots, \boldsymbol{\omega}_{X_k,Y}X_k(t)) \leq V_2$ and if $\boldsymbol{\eta}_Y\Delta t \leq 1$ and $V_1 \leq Y(t) \leq V_2$ then $V_1 \leq Y(t+\Delta t) \leq V_2$.

Proof

(a) This follows from

$$\begin{aligned}\mathbf{c}_Y(\boldsymbol{\omega}_{X_1,Y}X_1(t), \ldots, \boldsymbol{\omega}_{X_k,Y}X_k(t)) = \mathbf{c}_Y(\boldsymbol{\omega}_{X_1,Y}V, \ldots, \boldsymbol{\omega}_{X_k,Y}V) = V\mathbf{c}_Y(\boldsymbol{\omega}_{X_1,Y}, \ldots, \boldsymbol{\omega}_{X_k,Y}) \\ = V\end{aligned}$$

(b) This follows from $V_1 = V_1\mathbf{c}_Y(\boldsymbol{\omega}_{X_1,Y}, \ldots, \boldsymbol{\omega}_{X_k,Y}) = \mathbf{c}_Y(\boldsymbol{\omega}_{X_1,Y}V_1, \ldots, \boldsymbol{\omega}_{X_k,Y}V_1) \leq \mathbf{c}_Y(\boldsymbol{\omega}_{X_1,Y}X_1(t), \ldots, \boldsymbol{\omega}_{X_k,Y}X_k(t)) \leq \mathbf{c}_Y(\boldsymbol{\omega}_{X_1,Y}V_2, \ldots, \boldsymbol{\omega}_{X_k,Y}V_2) = V_2\mathbf{c}_Y(\boldsymbol{\omega}_{X_1,Y}, \ldots, \boldsymbol{\omega}_{X_k,Y}) = V_2$

The last part follows from

$$\begin{aligned} Y(t+\Delta t) &= Y(t) + \boldsymbol{\eta}_Y[\mathbf{c}_Y(\boldsymbol{\omega}_{X_1,Y}X_1(t),\ldots,\boldsymbol{\omega}_{X_k,Y}X_k(t)) - Y(t)]\Delta t \\ &= (1-\boldsymbol{\eta}_Y\Delta t)Y(t) + \boldsymbol{\eta}_Y\Delta t\mathbf{c}_Y(\boldsymbol{\omega}_{X_1,Y}X_1(t),\ldots,\boldsymbol{\omega}_{X_k,Y}X_k(t)) \end{aligned}$$

So $Y(t + \Delta t)$ is a weighted average with weights between 0 and 1 of $Y(t)$ and $\mathbf{c}_Y(\boldsymbol{\omega}_{X_1,Y}X_1(t), \ldots, \boldsymbol{\omega}_{X_k,Y}X_k(t))$ which both are in the interval $[V_1, V_2]$. Therefore $Y(t + \Delta t)$ itself also is in the interval $[V_1, V_2]$.

Theorem 1 (common state values provide equilibria) Suppose a network with nonnegative connections is based on normalised and scalar-free combination functions. Then the following hold.

(a) Whenever all states have the same value V, the network is in an equilibrium state.
(b) If for every state for its initial value V it holds $V_1 \leq V \leq V_2$, then for all t for every state Y it holds $V_1 \leq Y(t) \leq V_2$. In an achieved equilibrium for every state for its equilibrium value V it holds $V_1 \leq V \leq V_2$.

Proof

(a) It follows from Proposition 6(a) that the criterion of Lemma 1 is fulfilled.
(b) From Proposition 6(b) it follows by induction over the time steps that during a simulation for every state Y it holds $V_1 \leq Y(t) \leq V_2$ and therefore in a limit situation in an achieved equilibrium for every state for its equilibrium value V it holds $V_1 \leq V \leq V_2$.

■

Lemma 3 (Relating radical and max expressions) Suppose $a_1, \ldots, a_k$ are any nonnegative real numbers. Then

$$\lim_{n\to\infty} \sqrt[n]{a_1^n + \ldots + a_k^n} = \max(a_1, \ldots, a_k)$$

Proof First note that on the one hand

$$\max(a_1, \ldots, a_k) = \sqrt[n]{\max(a_1, \ldots, a_k)^n} \leq \sqrt[n]{a_1^n + \ldots + a_k^n}$$

and on the other hand

$$\begin{aligned} &\sqrt[n]{a_1^n + \ldots + a_k^n} \leq \sqrt[n]{\max(a_1, \ldots, a_k)^n + \ldots + \max(a_1, \ldots, a_k)^n} \\ &= \max(a_1, \ldots, a_k)\sqrt[n]{k} \end{aligned}$$

So

$$\max(a_1, \ldots, a_k) \leq \sqrt[n]{a_1^n + \ldots + a_k^n} \leq \max(a_1, \ldots, a_k)\sqrt[n]{k}$$

Now

$$\lim_{n \to \infty} \ln\left(\sqrt[n]{k}\right) = \lim_{n \to \infty} \ln(k)/n = 0$$

and therefore

$$\lim_{n \to \infty} \sqrt[n]{k} = \lim_{n \to \infty} e^{\ln\left(\sqrt[n]{k}\right)} = e^0 = 1.$$

This proves that for any nonnegative real numbers $a_1, \ldots, a_k$ it holds

$$\lim_{n \to \infty} \sqrt[n]{a_1^n + \ldots + a_k^n} = \max(a_1, \ldots, a_k)$$

■

Theorem 8 Let for each n the normalised Euclidean combination function $\mathbf{eucl}_{n,\lambda(n)}(V_1, \ldots, V_k)$ be given with scaling factor $\lambda(n)$, and let the normalised scaled maximum combination function $\mathbf{smax}_\lambda(V_1, \ldots, V_k)$ be given with scaling factor λ. Then for all $V_1, \ldots, V_k$ it holds

$$\lim_{n \to \infty} \mathbf{eucl}_{n,\lambda(n)}(V_1, \ldots, V_k) = \mathbf{smax}_\lambda(V_1, \ldots, V_k)$$

where

$$\lambda(n) = \omega_{X_1,Y}^n + \ldots + \omega_{X_k,Y}^n$$

and

$$\lambda = \max\left(\omega_{X_1,Y}, \ldots, \omega_{X_k,Y}\right)$$

Proof Recal the normalised formulas described in Table 11.4 in Chap. 11:

$$\mathbf{eucl}_{n,\lambda(n)}(V_1, .., V_k) = \sqrt[n]{\frac{V_1^n + \ldots + V_k^n}{\omega_{X_1,Y^n} + \ldots + \omega_{X_k,Y^n}}}$$

$$\mathbf{smax}_\lambda(V_1, \ldots, V_k) = \max(V_1, \ldots, V_k)/\max(\omega_{X_1,Y}, \ldots, \omega_{X_k,Y})$$

where

$$\lambda(n) = \omega_{X_1,Y^n} + \ldots + \omega_{X_k,Y^n}$$

and

$$\lambda = \max\left(\omega_{X_1,Y}, \ldots, \omega_{X_k,Y}\right)$$

Apply Lemma 3 to both $\omega_{X_1,Y}, \ldots, \omega_{X_k,Y}$ and $V_1, \ldots, V_k$ for $a_1, \ldots, a_k$ as follows

$$\begin{aligned}
\lim_{n\to\infty} \mathbf{eucl}_{n,\lambda(n)}(V_1, \ldots, V_k) &= \lim_{n\to\infty} \sqrt[n]{\frac{V_1^n + \ldots + V_k^n}{\lambda(n)}} \\
&= \frac{\lim_{n\to\infty} \sqrt[n]{V_1^n + \ldots + V_k^n}}{\lim_{n\to\infty} \sqrt[n]{\omega_{X_1,Y}^n + \ldots + {}_{X_k,Y}^n}} = \frac{\max(V_1, \ldots, V_k)}{\max\left(\omega_{X_1,Y}, \ldots, \omega_{X_k,Y}\right)} \\
&= \mathbf{smax}_\lambda(V_1, \ldots, V_k).
\end{aligned}$$

15.7 Using Strongly Connected Components to Explore Emerging Behaviour for a Class of Combination Functions for Any Type of Network Connectivity

This section presents a number of proofs that were left out from Chap. 12.

Proposition 1 Suppose the network is normalised.

(a) If the combination functions are scalar-free and X_1, …, X_k are the states from which Y gets its incoming connections, and $X_1(t)$ = … = $X_k(t)$ = V for some common value V, then also $\mathbf{c}_Y(\omega_{X_1,Y}X_1(t), \ldots, \omega_{X_k,Y}X_k(t)) = V$.

(b) If the combination functions are scalar-free and X_1, …, X_k are the states from which Y gets its incoming connections, and for U_1, …, U_k, V_1, …, V_k and $\alpha \geq 0$ it holds $V_i = \alpha\ U_i$, then $\mathbf{c}_Y(\omega_{X_1,Y}V_1, \ldots, \omega_{X_k,Y}V_k) = \alpha\mathbf{c}_Y(\omega_{X_1,Y}U_1, \ldots, \omega_{X_k,Y}U_k)$ If in this situation in two different simulations, state values $X_i(t)$ and $X_i'(t)$ are generated then $X_i'(t) = \alpha X_i(t) \Rightarrow X_i'(t+\Delta t) = \alpha X_i(t+\Delta t)$

(c) If the combination functions are additive and X_1, …, X_k are the states with outgoing connections to Y, then for values $U_1, \ldots, U_k$, $V_1, \ldots, V_k$ it holds

$$\begin{aligned}
\mathbf{c}_Y(\omega_{X_1,Y}(U_1+V_1), \ldots, \omega_{X_k,Y}(U_k+V_k)) &= \mathbf{c}_Y(\omega_{X_1,Y}U_1, \ldots, \omega_{X_k,Y}U_k) \\
&\quad + \mathbf{c}_Y(\omega_{X_1,Y}V_1, \ldots, \omega_{X_k,Y}V_k)
\end{aligned}$$

If in this situation in three different simulations, state values $X_i(t), X_i'(t)$ and $X_i''(t)$ are generated then

$$X_i''(t) = X_i(t) + X_i'(t) \Rightarrow X_i''(t+\Delta t) = X_i(t+\Delta t) + X_i'(t+\Delta t)$$

(d) If the combination functions are scalar-free and monotonically increasing, and $X_1, \ldots, X_k$ are the states with outgoing connections to Y, and $V_1 \leq X_1(t), \ldots, X_k(t) \leq V_2$ for some values V_1 and V_2, then also

$$V_1 \leq \mathbf{c}_Y(\boldsymbol{\omega}_{X_1,Y}X_1(t), \ldots, \boldsymbol{\omega}_{X_k,Y}X_k(t)) \leq V_2$$

and if $\boldsymbol{\eta}_Y \Delta t \leq 1$ and $V_1 \leq Y(t) \leq V_2$ then $V_1 \leq Y(t+\Delta t) \leq V_2$.

Proof

(a) This works as follows:

$$\begin{aligned}\mathbf{c}_Y(\boldsymbol{\omega}_{X_1,Y}X_1(t), \ldots, \boldsymbol{\omega}_{X_k,Y}X_k(t)) &= \mathbf{c}_Y(\boldsymbol{\omega}_{X_1,Y}V, \ldots, \boldsymbol{\omega}_{X_k,Y}V) = V\mathbf{c}_Y(\boldsymbol{\omega}_{X_1,Y}, \ldots, \boldsymbol{\omega}_{X_k,Y}) \\ &= V\end{aligned}$$

(b) can casily be verified
(c) can easily be verified.
(d) This follows from

$$\begin{aligned} V_1 &= V_1\mathbf{c}_Y(\boldsymbol{\omega}_{X_1,Y}, \ldots, \boldsymbol{\omega}_{X_k,Y}) = \mathbf{c}_Y(\boldsymbol{\omega}_{X_1,Y}V_1, \ldots, \boldsymbol{\omega}_{X_k,Y}V_1) \\ &\leq \mathbf{c}_Y(\boldsymbol{\omega}_{X_1,Y}X_1(t), \ldots, \boldsymbol{\omega}_{X_k,Y}X_k(t)) \leq \mathbf{c}_Y(\boldsymbol{\omega}_{X_1,Y}V_2, \ldots, \boldsymbol{\omega}_{X_k,Y}V_2) = V_2\mathbf{c}_Y(\boldsymbol{\omega}_{X_1,Y}, \ldots, \boldsymbol{\omega}_{X_k,Y}) = V_2\end{aligned}$$

and the second part from

$$\begin{aligned} Y(t+\Delta t) &= Y(t) + \boldsymbol{\eta}_Y[\mathbf{c}_Y(\boldsymbol{\omega}_{X_1,Y}X_1(t), \ldots, \boldsymbol{\omega}_{X_k,Y}X_k(t)) - Y(t)]\Delta t \\ &= \mathbf{c}_Y(\boldsymbol{\omega}_{X_1,Y}X_1(t), \ldots, \boldsymbol{\omega}_{X_k,Y}X_k(t))\boldsymbol{\eta}_Y\Delta t + Y(t)(1-\boldsymbol{\eta}_Y\Delta t) \\ &\leq V_2\boldsymbol{\eta}_Y\Delta t + V_2(1-\boldsymbol{\eta}_Y\Delta t) = V_2\end{aligned}$$

and similarly for V_1

$$\begin{aligned} Y(t+\Delta t) &= \mathbf{c}_Y(\boldsymbol{\omega}_{X_1,Y}X_1(t), \ldots, \boldsymbol{\omega}_{X_k,Y}X_k(t))\boldsymbol{\eta}_Y\Delta t + Y(t)(1-\boldsymbol{\eta}_Y\Delta t) \\ &\geq V_1\boldsymbol{\eta}_Y\Delta t + V_1(1-\boldsymbol{\eta}_Y\Delta t) = V_1\end{aligned}$$

■

Theorem 7 (equilibrium state values in relation to level 0 components in the linear case) Suppose the network N is normalised and the combination functions

are strictly monotonically increasing and linear. Assume that the states at level 0 that form a singleton component on their own are constant.
Then the following hold:

(a) For each state Y its equilibrium value is independent of the initial values of all states at some level $i > 0$. It is only dependent on the initial values for the states at level 0.
(b) More specifically, let $B_1, \ldots, B_p$ be the states in level 0 components. Then for each state Y its equilibrium value eq_Y is described by a linear function of the initial values $V_1, \ldots, V_p$ for $B_1, \ldots, B_p$, according to the following weighted average:

$$\mathrm{eq}_Y(V_1, \ldots, V_p) = d_{B_1,Y}V_1 + \ldots + d_{B_p,Y}V_p$$

Here the $d_{B_i,Y}$ are real numbers between 0 and 1 and the sum of them is 1:

$$d_{B_1,Y} + \ldots + d_{B_p,Y} = 1$$

(c) Each $d_{B_i,Y}$ is the equilibrium value for Y when the following initial values are used: $V_i = 1$ and all other initial values are 0:

$d_{Bi,Y} = eq_Y(0, \ldots, 0, 1, 0, \ldots, 0)$ with 1 as *ith* argument.

Proof From Proposition 1 it follows that he equilibrium value of Y is a linear function of the initial values of all states of N. Therefore the function is a linear combination of $e_i = \mathrm{eq}_Y(0, \ldots, 0, 1, 0, \ldots, 0)$ where only one state has initial value 1 and all other 0. An alternative, more theoretical linear algebra argument uses that the set of functions over time generated by the difference equations for different initial values forms an n-dimensional linear space with as basis the functions $d_i(t)$ generated for initial value 1 for state X_i and 0 for all other states. Therefore each generated function is a linear combination of such functions. By substituting $t = 0$ in them it is shown that the coefficients are the initial values and substituting t for an equilibrium shows that these initial values are the coefficients at that time point.

Now consider the different stratification levels. When all level 0 states have initial value 0, then by Theorem 5(a)(iii) they will have equilibrium value 0 as well. Then from Theorem 5(b)(ii) it follows that all states will have equilibrium value 0. In particular, this holds for cases that only one of the states at a level $i > 0$ have value 1 and all other states have initial value 0. This shows that from the linear combination the coefficient of these terms are 0. Therefore $\mathrm{eq}_Y(\ldots)$ is a function of $V_1, \ldots, V_p$ only.

■

Theorem 9 (equilibrium state values for components of level $i > 0$) Suppose the network is normalised, and consists of a strongly connected component plus a number of independent states $A_1, \ldots, A_p$ with outgoing connections to this strongly connected component. Then the following hold

(a) Suppose the combination functions are scalar-free and X_1, …, X_k are the states from which Y gets its incoming connections. If for U_1, …, U_k, V_1, …, V_k and $\alpha \geq 0$ it holds $V_i = \alpha\ U_i$ for all i, then $\mathbf{c}_Y(\boldsymbol{\omega}_{X_1,Y}V_1, \ldots, \boldsymbol{\omega}_{X_k,Y}V_k) = \boldsymbol{\alpha}\mathbf{c}_Y(\boldsymbol{\omega}_{X_1,Y}U_1, \ldots, \boldsymbol{\omega}_{X_k,Y}U_k)$

(b) Suppose the combination functions are additive and X_1, …, X_k are the states from which Y gets its incoming connections. Then if for values U_1, …, U_k, V_1, …, V_k, W_1, …, W_k it holds $W_i = U_i + V_i$ for all i, then

$$\mathbf{c}_Y(\boldsymbol{\omega}_{X_1,Y}W_1, \ldots, \boldsymbol{\omega}_{X_k,Y}W_k) = \mathbf{c}_Y(\boldsymbol{\omega}_{X_1,Y}U_1, \ldots, \boldsymbol{\omega}_{X_k,Y}U_k) + \mathbf{c}_Y(\boldsymbol{\omega}_{X_1,Y}V_1, \ldots, \boldsymbol{\omega}_{X_k,Y}V_k)$$

(c) Suppose all combination functions of the network N are linear. Then for given connection weights and speed factors, for each state Y the achieved equilibrium value for Y only depends on the equilibrium values V_1, …, V_p of states A_1, …, A_p; the function $\text{eq}_Y(V_1, \ldots, V_p)$ denotes this achieved equilibrium value for Y.

(d) Suppose the combination functions of the network N are linear. For the given connection weights and speed factors for each i let $d_{i,Y}$ be the achieved equilibrium value for state Y in a situation with equilibrium values $A_i = 1$ *and* $A_j = 0$ for all $j \neq i$, *i.e.*, $d_{i,Y} = eq_Y(0, \ldots, 0, 1, 0, \ldots, 0)$ with 1 as *ith* argument. Then in the general case for these given connection weights and speed factors, for each Y in the strongly connected component its equilibrium value is a linear, monotonically increasing, continuous and differentiable function $\text{eq}_Y(\ldots)$ of the equilibrium values V_1, …, V_p of A_1, …, A_p satisfying the following linear relation: $\text{eq}_Y(V_1, \ldots, V_p) = d_{1,Y}\ V_1 + \ldots + d_{p,Y}\ V_p$. Here the sum of the $d_{i,Y}$ *is 1:* $d_{1,Y} + \ldots + d_{p,Y} = 1$. In particular, the equilibrium values are independent of the initial values for all states Y different from A_1, …, A_p. If the combination functions of N are strictly increasing, then $d_{i,Y} > 0$ for all i, and $\text{eq}_Y(..)$ is also strictly increasing.

Proof (a) and (b) follow from Proposition 1
(c) From (a) and (b) it follows that the equilibrium value of Y is a linear function of the initial values of all states of N. Therefore the function is a linear combination of $e_i = \text{eq}_Y(0, \ldots, 0, 1, 0, \ldots, 0)$ where only one state has initial value 1 and all other 0. However, when all independent states have (constant) value 0, from Theorem 5(b) (ii) it follows that all states will have equilibrium value 0. In particular, this holds for cases that only one of the states that are not independent have initial value 1 and all other states have initial value 0. This shows that from the linear combination the coefficient e_i of these terms are 0. Therefore $\text{eq}_Y(\ldots)$ is a function of V_1, …, V_p only. From a) and b) it follows that $\text{eq}_Y(V_1, \ldots, V_p)$ is linear, as indicated above. Therefore

$$\begin{aligned}\text{eq}_Y(V_1, \ldots, V_p) &= \text{eq}_Y(V_1, 0, \ldots, 0) + \ldots + \text{eq}_Y(0, \ldots, 0, V_i, 0, .., 0) + \ldots + \text{eq}_Y(0, \ldots, V_p) \\ &= \text{eq}_Y(1, 0, \ldots, 0)V_1 + \ldots + \text{eq}_Y(0, \ldots, 0, 1, 0, .., 0)V_i + \ldots + \text{eq}_Y(0, \ldots, 1)V_p \\ &= d_{1,Y}V_1 + \ldots + d_{i,Y}V_i + \ldots + d_{p,Y}V_p\end{aligned}$$

■

15.8 Analysis of Emerging Behaviour for Classes of Homophily Functions

This section presents a number of proofs that were left out from Chap. 13.

Proposition 1 Suppose for any function $\mathrm{d}(\tau, D)$ it holds

$$\mathrm{d}(\tau, D) > 0 \text{ iff } D < \tau$$
$$\mathrm{d}(\tau, D) < 0 \text{ iff } D > \tau$$

Then the following hold:

(a) For any $\alpha > 0$ the function

$$\mathrm{c}(V_1, V_2, W) = W + \alpha W(1 - W)\mathrm{d}(\tau, |V_1 - V_2|)$$

satisfies the tipping point condition, but not strict.

(b) For any $\alpha > 0$ the function

$$\begin{aligned}\mathrm{c}'(V_1, V_2, W) = W &+ \alpha\,\mathrm{Pos}(\mathrm{d}(\tau, |V_1 - V_2|))\,(1 - W) \\ &- \alpha\,\mathrm{Pos}(-\mathrm{d}(\tau, |V_1 - V_2|))W\end{aligned}$$

satisfies the strict tipping point condition.

Proof

(a) The proof is mainly based on some algebraic rewriting.

Here as a first step it has to be proven that for any W with $0 < W < 1$ and all V_1, V_2 it holds

$$|V_1 - V_2| < \tau \Leftrightarrow \mathrm{c}(V_1, V_2, W) > W$$

This follows from

$$\begin{aligned}|V_1 - V_2| < \tau &\Leftrightarrow \mathrm{d}(\tau, |V_1 - V_2|) > 0 \Leftrightarrow \alpha W(1 - W)\mathrm{d}(\tau, |V_1 - V_2|) > 0 \\ &\Leftrightarrow \mathrm{c}(V_1, V_2, W) > W.\end{aligned}$$

Similarly the other two cases for $|V_1 - V_2| > \tau$ and $|V_1 - V_2| = \tau$ can be verified. From $\mathrm{c}(V_1, V_2, 0) = 0$ for all V_1, V_2 it follows that the strict tipping point requirement is not fulfilled.

(b) Also this proof is mainly based on some algebraic rewriting, thereby using Lemma 1.

First, $\mathrm{c}'(V_1, V_2, W)$ satisfies the tipping point condition; for any W with $0 < W < 1$ and all V_1, V_2 by Lemma 1 it holds:

$$\begin{aligned}|V_1 - V_2| < \tau &\Leftrightarrow \mathrm{d}(\tau, |V_1 - V_2|) > 0 \\ &\Leftrightarrow \mathrm{Pos}(\mathrm{d}(\tau, |V_1 - V_2|)) > 0 \,\text{and}\, \mathrm{Pos}(-\mathrm{d}(\tau, |V_1 - V_2|)) = 0\end{aligned}$$

Similarly the other two conditions.
It is strict because

$$|V_1 - V_2| < \tau \Rightarrow \mathrm{d}(\tau, |V_1 - V_2|) > 0 \Rightarrow \mathrm{c}'(V_1, V_2, 0) = \alpha\, \mathrm{Pos}(\mathrm{d}(\tau, |V_1 - V_2|) > 0$$

and

$$\begin{aligned}|V_1 - V_2| > \tau \Rightarrow \mathrm{d}(\tau, |V_1 - V_2|) < 0 \Rightarrow \mathrm{c}'(V_1, V_2, 1) \\ = 1 - \alpha\, \mathrm{Pos}(-\mathrm{d}(\tau, |V_1 - V_2|) < 1\end{aligned}$$

■

Proposition 2

(a) $\mathbf{log1hom}_{\tau,\alpha}(V_1, V_2, W)$ has tipping point τ, and is not strict
(b) $\mathbf{slog2hom}_{\tau,\alpha}(V_1, V_2, W)$ has tipping point τ, and is not strict
(c) $\mathbf{alog2hom}_{\tau,\alpha}(V_1, V_2, W)$ has a strict tipping point τ
(d) $\mathbf{exphomo}_{\tau,\sigma}(V_1, V_2, W)$ has a tipping point τ and is not strict

Proof

(a) This is based on some algebraic rewriting.

For $\mathbf{log1hom}_{\tau,\alpha}(V_1, V_2, W)$ suppose $0 < W < 1$, then for $D = |V_1 - V_2|$ it holds

$$\begin{aligned}D < \tau &\Leftrightarrow \mathrm{e}^{\sigma(D-\tau)} < 1 \Leftrightarrow (1 - W)\mathrm{e}^{\sigma(D-\tau)} < 1 - W \\ &\Leftrightarrow W + (1 - W)\mathrm{e}^{\sigma(D-\tau)} < W + 1 - W = 1 \\ &\Leftrightarrow \frac{W}{W + (1 - W)\mathrm{e}^{\sigma(D-\tau)}} > W\end{aligned}$$

Similarly the other conditions can be verified:

$$\begin{aligned}D = \tau &\Leftrightarrow \mathrm{e}^{\sigma(D-\tau)} = 1 \Leftrightarrow (1 - W)\mathrm{e}^{\sigma(D-\tau)} = 1 - W \\ &\Leftrightarrow W + (1 - W)\mathrm{e}^{\sigma(D-\tau)} = W + 1 - W = 1 \\ &\Leftrightarrow \frac{W}{W + (1 - W)\mathrm{e}^{\sigma(D-\tau)}} = W\end{aligned}$$

To verify that it has no strict tipping point τ: for any V_1, V_2 it holds

$$\mathbf{log1hom}_{\tau,\alpha}(V_1, V_2, 0) = 0$$

(b) This proof is based on Proposition 1. Applying Proposition 1(a) to $\mathbf{slog2hom}_{\tau,\alpha}(V_1, V_2, W)$ consider

$$d(\tau, D) = 0.5 - \frac{1}{1 + e^{-\sigma(D-\tau)}}$$

This function indeed satisfies the conditions of Proposition 1; therefore from Proposition 1(a) it follows that it has tipping point τ but not strict tipping point τ.

(c) Also this proof is based on Proposition 1. For $\mathbf{alog2hom}_{\tau,\alpha}(V_1, V_2, W)$ it follows from Proposition 1(b) with the same d(τ, D) as above that it has strict tipping point τ.

(d) This is based on some algebraic rewriting.
For W with $0 < W < 1$ it holds

$$\begin{aligned} D < \tau &\Leftrightarrow e^{\sigma(D-\tau)} < 1 \Leftrightarrow (1 - W)e^{\sigma(D-\tau)} < (1 - W) \\ &\Leftrightarrow 1 - (1 - W)e^{\sigma(D-\tau)} > 1 - (1 - W) = W \Leftrightarrow \mathbf{exphomo}_{\tau\sigma}(V_1, V_2, W) > W \end{aligned}$$

$$\begin{aligned} D = \tau &\Leftrightarrow e^{\sigma(D-\tau)} = 1 \Leftrightarrow (1 - W)e^{\sigma(D-\tau)} = (1 - W) \\ &\Leftrightarrow 1 - (1 - W)e^{\sigma(D-\tau)} = 1 - (1 - W) = W \Leftrightarrow \mathbf{exphomo}_{t,s}(V_1, V_2, W) = W \end{aligned}$$

$$\begin{aligned} D > \tau &\Leftrightarrow e^{\sigma(D-\tau)} > 1 \Leftrightarrow (1 - W)e^{\sigma(D-\tau)} > (1 - W) \\ &\Leftrightarrow 1 - (1 - W)e^{\sigma(D-\tau)} < 1 - (1 - W) = W \Leftrightarrow \mathbf{exphomo}_{\tau,\sigma}(V_1, V_2, W) < W \end{aligned}$$

This shows it has tipping point τ. It has no strict tipping point, as

$$\mathbf{exphomo}_{\tau,\sigma}(V_1, V_2, 1) = 1 \text{ for all } V_1, V_2.$$

The following proposition shows that weighted averages of functions with tipping point τ also have a tipping point τ, and the same for having a strict tipping point.

Proposition 3 A weighted average (with positive weights) of homophily combination functions with tipping point τ also has tipping point τ, and with strict tipping point τ, also has strict tipping point τ.

Proof This can be verified in a straightforward manner.

Suppose

$$c(V_1, V_2, W) = \gamma_1 c_1(V_1, V_2, W) + \ldots + \gamma_m c_m(V_1, V_2, W)$$

with $\gamma_1 + \ldots + \gamma_m = 1$. Suppose $0 < W < 1$. Then $|V_1 - V_2| < \tau \Rightarrow c_i(V_1, V_2, W) > W$ for all i, and therefore

$$c(V_1, V_2, W) > \gamma_1 W + \ldots + \gamma_m W = W$$

Similarly

$$|V_1 - V_2| = \tau \Rightarrow c(V_1, V_2, W) = W$$
$$|V_1 - V_2| > \tau \Rightarrow c(V_1, V_2, W) < W$$

Now suppose $c(V_1, V_2, W) > W$, then $|V_1 - V_2| = \tau$ or $|V_1 - V_2| > \tau$ cannot hold as they imply $c(V_1, V_2, W) = W$ or $c(V_1, V_2, W) < W$, therefore $|V_1 - V_2| < \tau$. The same for the other clauses. Moreover, suppose that the functions $c_i(V_1, V_2, W)$ all have strict tipping point τ. Then

If $|V_1 - V_2| < \tau$ then

$$c(V_1, V_2, 0) = \gamma_1 c_1(V_1, V_2, 0) + \ldots + \gamma_m c_m(V_1, V_2, 0) > 0$$

If $|V_1 - V_2| < \tau$ then

$$c(V_1, V_2, 1) = \gamma_1 c_1(V_1, V_2, 1) + \ldots + \gamma_m c_m(V_1, V_2, 1) < \gamma_1 + \ldots + \gamma_m = 1$$

Therefore also $c(V_1, V_2, W)$ has a strict tipping point.

■

Proposition 4

(a) When the homophily combination function $c(V_1, V_2, W)$ is symmetric, and initially the network is fully symmetric, then the network is continually fully symmetric.
(b) For every $n > 0$ a Euclidean combination function of **n**th degree is strictly monotonically increasing, scalar-free, and symmetric.

Proof

(a) This follows from the fact that in this case the difference equation for reification state $\mathbf{W}_{X,Y}$ for $\omega_{X,Y}$ is symmetric in X and Y.
(b) A Euclidean combination function is composed of strictly monotonic functions as each function $V_i \rightarrow V_i^n$ is monotonic for positive $\boldsymbol{n}$ and positive values V_i, and so are $W \rightarrow W/\lambda$ and $W \rightarrow W^{1/\boldsymbol{n}}$. From

$$\begin{aligned}
\mathbf{eucl}_{n,\lambda}(\alpha V_1, \ldots, \alpha V_k) &= \sqrt[n]{\frac{(\alpha V_1)^n + \ldots + (\alpha V_k)^n}{\lambda}} \\
&= \sqrt[n]{\frac{\alpha^n V_1^n + \ldots + \alpha^n V_k^n}{\lambda}} \\
&= \alpha \sqrt[n]{\frac{V_1^n + \ldots + V_k^n}{\lambda}} \\
&= \alpha\, \mathbf{eucl}_{n,\lambda}(V_1, \ldots, V_k)
\end{aligned}$$

it follows that it is scalar-free. The rest directly follows.

■

Lemma 3 Suppose the function $c(V_1, V_2, W)$ has tipping point τ for V_1 and V_2. Then

(i) The value 0 for $W_{X,Y}$ can only be reached from $\mathbf{W}_{X,Y}(t)$ with $0 < \mathbf{W}_{X,Y}(t) < 1$ if $|X(t) - Y(t)| > \tau$
(ii) The value 1 for $\mathbf{W}_{X,Y}$ can only be reached from $\mathbf{W}_{X,Y}(t)$ with $0 < \mathbf{W}_{X,Y}(t) < 1$ if $|X(t) - Y(t)| < \tau$.

Proof

(i) The proof is by contraposition. Suppose $0 < \mathbf{W}_{X,Y}(t) < 1$ holds and $|X(t) - Y(t)| > \tau$ does not hold. Then $|X(t) - Y(t)| \leq \tau$, and by Definition 1(a)(i) and (ii) it follows that $c(V_1, V_2, W) \geq W$, and therefore from the difference equation it follows that $\mathbf{W}_{X,Y}(t + \Delta t) \geq \mathbf{W}_{X,Y}(t)$ will not become lower and in particular will not reach 0.
(ii) is similar using Definition 1(a)(ii) and (iii).

Theorem 1 (Relations between equilibrium values for states and for connection weights) Suppose the function $c(V_1, V_2, W)$ has tipping point τ for V_1 and V_2 and an attracting equilibrium state is given with values $\underline{\mathbf{X}}$ for the states X and $\underline{\mathbf{W}}_{X,Y}$ for the connection weight reification states $\mathbf{W}_{X,Y}$. Then the following hold:

(a) If $|\underline{\mathbf{X}} \quad \underline{\mathbf{Y}}| < \tau$, then the equilibrium value $\underline{\mathbf{W}}_{X,Y}$ is 1; in particular this holds when $\underline{\mathbf{X}} = \underline{\mathbf{Y}}$. Therefore, if $\underline{\mathbf{W}}_{X,Y} < 1$, then $|\underline{\mathbf{X}} - \underline{\mathbf{Y}}| \geq \tau$, and, in particular, $\underline{\mathbf{X}} \neq \underline{\mathbf{Y}}$.
(b) If $|\underline{\mathbf{X}} - \underline{\mathbf{Y}}| > \tau$, then the equilibrium value $\underline{\mathbf{W}}_{X,Y}$ is 0. Therefore, if $\underline{\mathbf{W}}_{X,Y} > 0$, then $|\underline{\mathbf{X}} - \underline{\mathbf{Y}}| \leq \tau$.
(c) $0 < \underline{\mathbf{W}}_{X,Y} < 1$ implies $|\underline{\mathbf{X}} - \underline{\mathbf{Y}}| = \tau$.

Proof

(a) Suppose two states are given with equilibrium values $\underline{\mathbf{X}}$ and $\underline{\mathbf{Y}}$ with distance less than τ: $|\underline{\mathbf{X}} - \underline{\mathbf{Y}}| < \tau$. Given this, from the equilibrium equation $\mathbf{c}(\underline{\mathbf{X}}, \underline{\mathbf{Y}}, \underline{\mathbf{W}}_{X,Y}) = \underline{\mathbf{W}}_{X,Y}$, by Definition 1(ii) it follows that $0 < \underline{\mathbf{W}}_{X,Y} < 1$ cannot be true, and therefore $\underline{\mathbf{W}}_{X,Y} = 0$ or $\underline{\mathbf{W}}_{X,Y} = 1$. By Lemma 3(i) and the equilibrium being attracting it follows that $\underline{\mathbf{W}}_{X,Y} = 0$ can be excluded, so $\underline{\mathbf{W}}_{X,Y} = 1$.

The other statement, that if $\underline{\mathbf{W}}_{X,Y} < 1$, then $|\underline{\mathbf{X}} - \underline{\mathbf{Y}}| \geq \tau$, follows by logical contraposition.

(b) For $|\underline{\mathbf{X}} - \underline{\mathbf{Y}}| > \tau$ this is similar, using Definition 1(iii). The last statement follows from the contraposition of the previous one.
(c) This immediately follows from (a) and (b).

Lemma 4 Let a normalised network with nonnegative connections be given with combination functions that are monotonically increasing and scalar-free; then the following hold:

(a) (i) If for some state Y at time t for all nodes X with $\omega_{X,Y} > 0$ it holds $X(t) \leq Y(t)$, then $Y(t)$ is decreasing at t: $\mathbf{d}Y(t)/\mathbf{d}t \leq 0$.
(ii) If, moreover, the combination function is strictly increasing and a state X exists with $X(t) < Y(t)$ and $\omega_{X,Y} > 0$, then $Y(t)$ is strictly decreasing at t: $\mathbf{d}Y(t)(t)/\mathbf{d}t < 0$.

(b) (i) If for some state Y at time t for all nodes X with $\omega_{X,Y} > 0$ it holds $X(t) \geq Y(t)$, then $Y(t)$ is increasing at t: $\mathbf{d}Y(t)/\mathbf{d}t \geq 0$.
(ii) If, moreover, the combination function is strictly increasing and a state X exists with $X(t) > Y(t)$ and $\omega_{X,Y} > 0$, then $Y(t)$ is strictly increasing at t: $\mathbf{d}Y(t)(t)/\mathbf{d}t > 0$.

Proof The proofs for (a) and (b) are similar. Therefore only the proof for (a) is given.

(a) (i) This proof shows that lower values of the states with incoming connections can never increase the state value of state Y. More specifically, assume for all states X_i with (positive) outgoing connections to Y it holds $X_i(t) \leq Y(t)$. Therefore

$$\begin{aligned} c_Y(\omega_{X_1,Y} X_1(t), \ldots, \omega_{X_k,Y} X_k(t)) &\leq c_Y(\omega_{X_1,Y} Y(t), \ldots, w_{X_k,Y} Y(t)) \\ &= c_Y(\omega_{X_1,Y} \ldots, \omega_{X_k,Y}) Y(t) \\ &= Y(t) \end{aligned}$$

and by Lemma 2(i) and (iii) this implies $\mathbf{d}Y(t)/\mathbf{d}t \leq 0$.
(ii) This proof shows that a strictly lower values of one of the states with incoming connections will actually decrease the state value of state Y. More specifically, from being strictly monotonous it follows

$$c_Y(\omega_{X_1,Y} X_1(t), \ldots, \omega_{X_k,Y} X_k(t)) < c_Y(w_{X_1,Y} Y(t), \ldots, \omega_{X_k,Y} Y(t)) = Y(t)$$

and by Lemma 2(iii) this implies $\mathbf{d}Y(t)/\mathbf{d}t < 0$.

(b) This can be proven in a similar manner.

Theorem 2 (Equilibrium values $\underline{\mathbf{W}}_{X,Y}$ all 0 or 1) Suppose the network is weakly symmetric and normalised, and the combination functions for the social contagion for the base states are strictly monotonically increasing and scalar-free. Suppose that the combination functions $c(V_1, V_2, W)$ for the reification states for the connection weights have a tipping point τ. Then

(a) In an attracting equilibrium state for any states X, Y from $\underline{\mathbf{X}} \neq \underline{\mathbf{Y}}$ it follows $\underline{\mathbf{W}}_{X,Y} = 0$.
(b) In an attracting equilibrium state for any states X, Y with $\underline{\mathbf{X}} = \underline{\mathbf{Y}}$ it holds $\underline{\mathbf{W}}_{X,Y} = 0$ or $\underline{\mathbf{W}}_{X,Y} = 1$.
(c) If c(V_1, V_2, W) has a strict tipping point τ, then in an equilibrium state for any X, Y with $\underline{\mathbf{X}} = \underline{\mathbf{Y}}$ it holds $\underline{\mathbf{W}}_{X,Y} = 1$.

Proof

(a) The proof goes by reductio ad absurdum (reduce to absurdity); it shows that the opposite of the claimed statement cannot be true by deriving a contradiction from this opposite statement. So, suppose (oppositely) that in an attracting equilibrium state, states X and Y exist such that $\underline{\mathbf{X}} \neq \underline{\mathbf{Y}}$ and $\underline{\mathbf{W}}_{X,Y}$, $\underline{\mathbf{W}}_{Y,X} > 0$. Take a state X with this property with highest value $\underline{\mathbf{X}}$. Then for all states Z with $\underline{\mathbf{Z}} > \underline{\mathbf{X}}$ it holds $\underline{\mathbf{W}}_{X,Z} = \underline{\mathbf{W}}_{Z,X} = 0$. Therefore all states X_i with a nonzero (positive) outgoing connection weight to state X satisfy $\underline{\mathbf{X}}_i \leq \underline{\mathbf{X}}$. Moreover, one of these X_i is state Y with $\underline{\mathbf{X}} \neq \underline{\mathbf{Y}}$, so, as X has the highest value, it holds $\underline{\mathbf{Y}} < \underline{\mathbf{X}}$. Now apply Lemma 4(a)(ii) to this state X. It follows that $\mathbf{d}X(t)/\mathbf{d}t < 0$; therefore $X(t)$ cannot be not in equilibrium. This contradicts the premise that the network is in equilibrium. Therefore no nodes X and Y exist such that $\underline{\mathbf{X}} \neq \underline{\mathbf{Y}}$ and $\underline{\mathbf{W}}_{X,Y}$, $\underline{\mathbf{W}}_{Y,X} > 0$. This implies that $\underline{\mathbf{W}}_{X,Y} = 0$ and $\underline{\mathbf{W}}_{\omega Y,X} = 0$ for all nodes X and Y with $\underline{\mathbf{X}} \neq \underline{\mathbf{Y}}$.
(b) Also this proof goes by reductio ad absurdum (reduce to absurdity); also here it is shown that the opposite of the claimed statement cannot be true by deriving a contradiction from this opposite statement. So, suppose (oppositely) $\underline{\mathbf{X}} = \underline{\mathbf{Y}}$ and $0 < \mathbf{W}_{X,Y}(t) < 1$. Then by Definition 1(a)(i) from $X(t) = Y(t)$ it follows that c($X(t)$, $Y(t)$, $\mathbf{W}_{X,Y}(t)$) $> \mathbf{W}_{X,Y}(t)$. From this by Lemma 2(ii) it follows that $\mathbf{dW}_{X,Y}(t)/\mathbf{d}t > 0$: $\mathbf{W}_{X,Y}(t)$ is strictly increasing and is not in equilibrium. This contradicts the premise that the network is in equilibrium. Therefore in the equilibrium state when $\underline{\mathbf{X}} = \underline{\mathbf{Y}}$ it holds $\underline{\mathbf{W}}_{X,Y} = 0$ or $\underline{\mathbf{W}}_{X,Y} = 1$.
(c) From (b) it is already known that $\underline{\mathbf{W}}_{X,Y} = 0$ or $\underline{\mathbf{W}}_{X,Y} = 1$. The former option $\underline{\mathbf{W}}_{X,Y} = 0$ has to be excluded now. Also this goes by reductio ad absurdum (reduce to absurdity); also here it is shown that the option $\underline{\mathbf{W}}_{X,Y} = 0$ cannot be true by deriving a contradiction when this option is assumed. So, suppose $\underline{\mathbf{W}}_{X,Y} = 0$. If c($V_1$, V_2, W) is strict, and $|V_1 - V_2| < \tau$ then by Definition 1(b)(i) it holds c(V_1, V_2, 0) > 0, so by Lemma 1(ii) it follows that when $\underline{\mathbf{X}} = \underline{\mathbf{Y}}$, the value $\underline{\mathbf{W}}_{X,Y} = 0$ cannot be an equilibrium value, which contradicts the premise that the network is in equilibrium. Therefore in an equilibrium for any X, Y with $\underline{\mathbf{X}} = \underline{\mathbf{Y}}$ it holds $\underline{\mathbf{W}}_{X,Y} = 1$.

Theorem 3 (Partition and equilibrium values of nodes) Suppose the network is weakly symmetric and normalised, the combination functions for the social contagion for the base states are strictly monotonically increasing and scalar-free, and the combination functions for the reification states for the connection weights use tipping point τ and is strict and symmetric. Then in any attracting equilibrium state a partition of the set of states into disjoint subsets C_1, …, C_p occurs such that:

(i) For each C_i the equilibrium values for all the states in C_i are equal: $\underline{\mathbf{X}} = \underline{\mathbf{Y}}$ for all X, $Y \in C_i$.
(ii) Every C_i forms a fully connected network with weights 1: $\underline{\mathbf{W}}_{X,Y} = 1$ for all X, $Y \in C_i$.
(iii) Every two nodes in different C_i have connection weight 0: when $i \neq j$, then $X \in C_i$ and $Y \in C_j$ implies $\underline{\mathbf{W}}_{X,Y} = 0$.
(iv) Any two distinct equilibrium values of states $\underline{\mathbf{X}} \neq \underline{\mathbf{Y}}$ have distance $\geq \tau$. Therefore there are at most $p \leq 1 + 1/\tau$ communities C_i and equilibrium values $\underline{\mathbf{X}}$.

Proof Suppose in the equilibrium there are p distinct state values $V_1, \ldots, V_p$; then define the sets

$$C_i = \{\mathrm{X} | \underline{\mathbf{X}} = V_i\}$$

It can easily be verified in a straightforward manner that these sets fulfill what is claimed:

(i) By definition all state values in one C_i are equal.
(ii) From Theorem 2(c) it follows that all states with equal values have connections 1, therefore any C_i is fully connected.
(iii) This follows from Theorem 2(a).
(iv) Suppose for some X, Y it holds $| \underline{\mathbf{X}} - \underline{\mathbf{Y}} | < \tau$. Then by Definition 1(i) it follows $\mathrm{c}(V_1, V_2, 0) > 0$. Therefore 0 cannot be the equilibrium value $\underline{\mathbf{W}}_{X,Y}$; from Theorem 2(a) it follows that $\underline{\mathbf{X}} = \underline{\mathbf{Y}}$, and therefore X and Y are in one C_i. This implies that the state values in different C_i have distance $\geq \tau$.

■

Theorem 6 (Strongly connected components characterisation) Suppose the network is weakly symmetric, the combination functions for social contagion between the base nodes are strictly monotonically increasing, normalised and scalar-free, and the homophily combination functions for the connections weight reification states use tipping point τ and are strict and symmetric. Suppose at some time point t the following hold:

(i) Each strongly connected component C is fully connected and all states in C have a common state value.
(ii) All connections between states from different strongly connected components have weight 0 and the equilibrium values of these states have distance $> \tau$.

Then the network is in an equilibrium state.

Proof First in (a) it is proven that the state values are stationary; next, in b) it is proven that the connection weights are stationary. Having both stationary, the network is in equilibrium.

(a) Consider within any component C any state Y which has only nonzero incoming connections from states $X_1, \ldots, X_k$. Due to (ii) these necessarily belong to the same component C. As within C the state values are equal to one value V and each connection has weight 1 the following holds

$$\begin{aligned}\mathbf{aggimpact}_Y(t) &= \mathbf{c}_Y(X_1(t), \ldots, X_k(t)) \\ &= \mathbf{c}_Y(V, \ldots, V) \\ &= V\,\mathbf{c}_Y(1, \ldots, 1) \\ &= V \\ &= Y(t)\end{aligned}$$

and therefore by Lemma 2(i) it holds $dY(t)/dt = 0$, so $Y(t)$ is stationary.

(b) Next, it is proven that the connection weights are stationary. Consider the connection weight reification state $\mathbf{W}_{X,Y}$ for the connection from states X to Y in the same component C. Suppose as a perturbation from 1 it holds $\mathbf{W}_{X,Y}(t) < 1$. Given that $|X(t) - Y(t)| < \tau$, from Definition 1(i) it follows that $\mathbf{c}(X(t), Y(t), \mathbf{W}_{X,Y}(t)) > \mathbf{W}_{X,Y}(t)$, and therefore $\mathbf{d}\mathbf{W}_{X,Y}(t)/\mathbf{d}t > 0$, so it would move upward to 1. Therefore $\mathbf{W}_{X,Y}$ it is stationary. A similar argument for states X and Y in different components shows that $\mathbf{W}_{X,Y}$ would move downward to 0, and therefore is stationary.

As both the states and the connection weights have been proven stationary, it has been found that the network is in equilibrium.

■

15.9 Analysis of Emerging Behaviour for Classes of Hebbian Learning Functions

This section presents a number of proofs that were left out from Chap. 14.

Proposition 1 (functional relation for W) Suppose that $c(V_1, V_2, W)$ is a Hebbian learning function with persistence parameter μ.

(a) Suppose $\mu < 1$. Then the following hold:

   (i) The function $W \rightarrow c(V_1, V_2, W) - W$ on [0, 1] is strictly monotonically decreasing

   (ii) There is a unique function $f_\mu : [0, 1]x[0, 1] \rightarrow [0, 1]$ such for any V_1, V_2 it holds

$$c(V_1, V_2, f_\mu(V_1, V_2)) = f_\mu(V_1, V_2)$$

This function f_μ is a monotonically increasing function of V_1, V_2, and is implicitly defined by the above equation. Its maximal value is $f_\mu(1, 1)$ and minimum $f_\mu(0, 0) = 0$.

(b) Suppose $\mu = 1$. Then there is a unique function $f_1 : (0, 1]x(0, 1] \rightarrow [0, 1]$, such for any V_1, V_2 it holds

$$c(V_1, V_2, f_1(V_1, V_2)) = f_1(V_1, V_2)$$

This function f_1 is a constant function of V_1, V_2 with $f_1(V_1, V_2) = 1$ for all V_1, $V_2 > 0$ and is implicitly defined on (0, 1] x (0, 1] by the above equation.
If one of V_1, V_2 is 0, then any value of W satisfies the equation $c(V_1, V_2, W) = W$, so no unique function value for $f_1(V_1, V_2)$ can be defined then.

Proof

(a) Consider $\mu < 1$. Then by Definition 2(b) the function $W \rightarrow c(V_1, V_2, W) - \mu W$ is monotonically decreasing in W, and since $\mu - 1 < 0$ the function $W \rightarrow (\mu - 1)W$ is strictly monotonically decreasing in W. Therefore the sum of them is also strictly monotonically decreasing in W. Now this sum is

$$c(V_1, V_2, W) - \mu W + (\mu - 1)W = c(V_1, V_2, W) - W$$

So, the function $W \rightarrow c(V_1, V_2, W) - W$ is strictly monotonically decreasing in W; by Definition 2(d) it holds $c(V_1, V_2, 1) - 1 = \mu - 1 < 0$, and by Definition 2(c) $c(V_1, V_2, 0) - 0 \geq 0$. Therefore $c(V_1, V_2, W) - W$ has exactly 1 point with $c(V_1, V_2, W) - W = 0$; so for each V_1, V_2 the equation $c(V_1, V_2, W) - W = 0$ has exactly one solution W, indicated by $f_\mu(V_1, V_2)$; this provides a unique function f_μ: [0, 1] x [0, 1] $\rightarrow$ [0, 1] implicitly defined by $c(V_1, V_2, f_\mu(V_1, V_2)) = f_\mu(V_1, V_2)$. To prove that f_μ is monotonically increasing, the following. Suppose $V_1 \leq V_1'$ and $V_2 \leq V_2'$, then by monotonicity of $V_1, V_2 \rightarrow c(V_1, V_2, W)$ in Definition 2(a) it holds

$$0 = c(V_1, V_2, f_\mu(V_1, V_2)) - f_\mu(V_1, V_2) \leq c(V_1', V_2', f_\mu(V_1, V_2)) - f_\mu(V_1, V_2)$$

So

$$c(V_1', V_2', f_\mu(V_1, V_2)) - f_\mu(V_1, V_2) \geq 0$$

whereas

$$c(V_1', V_2', f_\mu(V_1', V_2')) - f_\mu(V_1', V_2') = 0$$

and therefore

$$c(V'_1, V'_2, f_\mu(V'_1, V'_2)) - f_\mu(V'_1, V'_2) \leq c(V'_1, V'_2, f_\mu(V_1, V_2)) - f_\mu(V_1, V_2)$$

By strict decreasing monotonicity of $W \rightarrow c(V_1, V_2, W) - W$ it follows that $f_\mu(V_1, V_2) > f_\mu(V'_1, V'_2)$ cannot hold, so $f_\mu(V_1, V_2) \leq f_\mu(V'_1, V'_2)$. This proves that f_μ is monotonically increasing. From this monotonicity of $f_\mu(..)$ it follows that $f_\mu(1, 1)$ is the maximal value and $f_\mu(0, 0)$ the minimal value. Now by Definition 1(d) it follows that $f_\mu(0, 0) = c(0, 0, f_\mu(0, 0)) = \mu\, f_\mu(0, 0)$ so $f_\mu(0, 0) = \mu\, f_\mu(0, 0)$, and as $\mu < 1$ this implies $f_\mu(0, 0) = 0$.

(b) Consider $\mu = 1$. When both V_1, V_2 are > 0, and $c(V_1, V_2, W) = W$, then $W = 1$, by Definition 1(d). This defines a function $f_1(V_1, V_2)$ of $V_1, V_2 \in (0, 1]$, this time $f_1(V_1, V_2) = 1$ for all $V_1, V_2 > 0$. When one of V_1, V_2 is 0 and $\mu = 1$, then also by Definition 1(d) always $c(V_1, V_2, W) = W$, so in this case multiple solutions for W are possible: every W is a solution, and therefore no unique function value for $f_1(V_1, V_2)$ can be defined then.

■

Proposition 2 (functional relation for *W* based on variable separation) Assume the Hebbian function $c(V_1, V_2, W)$ with persistence parameter μ enables variable separation by the two functions $cs(V_1, V_2)$ monotonically increasing and $cc(W)$ monotonically decreasing:

$$c(V_1, V_2, W) = cs(V_1, V_2)\, cc(W) + \mu\, W$$

Let $h_\mu(W)$ be the function defined for $W \in [0, 1)$ *by*

$$h_\mu(W) = \frac{(1-\mu)W}{cc(W)}$$

Then the following hold.

(a) When $\mu < 1$ the function $h_\mu(W)$ is strictly monotonically increasing, and has a strictly monotonically increasing inverse g_μ on the range $h_\mu([0, 1))$ of h_μ with $W = g_\mu(h_\mu(W))$ *for all* $W \in [0, 1)$.
(b) When $\mu < 1$ and $c(V_1, V_2, W) = W$, then $g_\mu(cs(V_1, V_2)) < 1$ and $W < 1$, and it holds

$$h_\mu(W) = cs(V_1, V_2)$$
$$W = g_\mu(cs(V_1, V_2))$$

So, in this case the function f_μ from Theorem 1 is the function composition $g_\mu \circ cs$ of cs followed by g_μ; it holds:

$$f_\mu(V_1, V_2) = g_\mu(cs(V_1, V_2))$$

(c) For $\mu = 1$ it holds $c(V_1, V_2, W) = W$ if and only if $V_1 = 0$ *or* $V_2 = 0$ or $W = 1$.
(d) For $\mu < 1$ the maximal value W with $c(V_1, V_2, W) = W$ is $g_\mu(cs(1, 1))$, and the minimal equilibrium value W is 0. For $\mu = 1$ the maximal value W *is* 1 (always when $V_1, V_2 > 0$ holds) and the minimal value is 0 (occurs when one of V_1, V_2 is 0).

Proof

(a) From $cc(W)$ monotonically decreasing in W it follows that $W \to 1/cc(W)$ is monotonically increasing on [0, 1). Moreover, the function W is strictly monotonically increasing; therefore for $\mu < 1$ the function $h_\mu(W) = (1 - \mu) W/cc(W)$ is strictly monotonically increasing. Therefore h_μ is injective and has an inverse function g_μ on the range of h_μ: a function g_μ with $g_\mu(h_\mu(W)) = W$ for all $W \in [0, 1)$.
(b) Suppose $\mu < 1$ and $c(V_1, V_2, W) = W$, then from Definition 2(d) it follows that $W = 1$ is excluded, since from both $c(V_1, V_2, W) = W$ and $c(V_1, V_2, W) = \mu W$ it would follow $\mu = 1$, which is not the case. Therefore $W < 1$, and the following hold

$$cs(V_1, V_2)cc(W) + \mu W = W$$
$$cs(V_1, V_2)cc(W) = (1 - \mu)W$$
$$cs(V_1, V_2) = (1 - \mu)W/cc(W) = h_\mu(W)$$

So, $h_\mu(W) = cs(V_1, V_2)$. Applying the inverse g_μ yields $W = g_\mu(h_\mu(W)) = g_\mu(cs(V_1, V_2))$.
Therefore in this case for the function f_μ from Theorem 1 it holds:

$$f_\mu(V_1, V_2) = W = g_\mu(cs(V_1, V_2)) < 1$$

so f_μ is the function composition g_μ o cs of cs(..) followed by g_μ.
(c) For $\mu = 1$ the equation $c(V_1, V_2, W) = W$ becomes $cs(V_1, V_2)\, cc(W) = 0$ and this is equivalent to $cs(V_1, V_2) = 0$ or $cc(W) = 0$. From the definition of separation of variables it follows that this is equivalent to $V_1 = 0$ or $V_2 = 0$ or $W = 1$.
(d) Suppose $\mu < 1$ and $c(V_1, V_2, W) = W$, then because cs(..) and g_μ are both monotonically increasing, the maximal W is $g_\mu(cs(1, 1))$, and the minimal W is $g_\mu(cs(0, 0))$. For $\mu = 1$ these values are 1 always when $V_1, V_2 > 0$, and any value in [0, 1] (including 0) when one of V_1, V_2 is 0. ■

Chapter 16
Using Network Reification for Adaptive Networks: Discussion

Abstract In this final chapter, the most important or most remarkable themes recurring at different places in this book are briefly summarized and reviewed. Subsequently the following themes are addressed: (1) How network reification can be used to model adaptive networks. (2) The formats in which conceptual representations of reified networks are expressed graphically using 3D pictures and role matrices. (3) The universal combination function, and the universal difference and differential equation as the basis for the numerical representation and implementation of reified networks. (4) Analysis of how emerging reified network behaviour relates to the reified network's structure. (5) The Network-Oriented design process based on reified networks. (6) The relation to longstanding themes in AI and beyond.

16.1 Adaptive Networks and Network Reification

This theme is a major theme throughout the whole book. It started in Chap. 1, Sects. 1.2 and 1.3 with an overview of many types of adaptive networks and domains in which adaptation of the network structure plays an important role. As adaptation principles can themselves also be adaptive, it is natural to consider adaptive networks of different orders. As described in Chap. 1, Sect. 1.3, it was found out that in the literature first-order adaptation occurs often, but second-order adaptation much less and third- or higher-order adaptation is almost absent. Modeling adaptive networks is usually done in a hybrid format: the network itself is described by some form of network modeling, but the adaptation principles are described using some form of procedural or algorithmic specification and programming to run the underlying difference equations. Thus a hybrid model consists of two components for two different types of models that interact with each other, as depicted in Chap. 1, Fig. 1.2.

In contrast to this hybrid approach, in Chap. 3 it was shown how network reification for temporal-causal networks can be used to model any adaptive network in a neat and declarative manner from a Network-Oriented Modeling perspective,

J. Treur, *Network-Oriented Modeling for Adaptive Networks: Designing Higher-Order Adaptive Biological, Mental and Social Network Models*, Studies in Systems, Decision and Control 251, https://doi.org/10.1007/978-3-030-31445-3_16

where also the adaptation principles are described in temporal-causal network format. Chapter 3 shows several examples of this for well-known first-order network adaptation principles, for example, shown in Figs. 3.4 to 3.10. In Chap. 4, second-order adaptive networks were addressed with an example (shown in Fig. 4.3) for plasticity and metaplasticity as considered in relatively recent empirical literature from the Cognitive Neuroscience domain.

Concerning the challenge of obtaining plausible adaptive network models of order >2, in Chaps. 7 and 8 three of such examples were presented. In Chap. 7 an example of such a reified adaptive network of order >2 was put forward that was inspired by literature on evolutionary processes (Fessler et al. 2005, 2015; Fleischman and Fessler 2011) as discussed in Chap. 1, Sect. 1.3. In Chap. 8, another example of a reified adaptive network of order >2 was described, this time inspired by Hofstadter (1979, 2007)'s ideas about Strange Loops; see also Sect. 16.6 below, and Fig. 16.3.

In the many examples in this book it was shown how a certain adaptation principle known from the literature can be formulated easily in reified temporal-causal network format. That this is possible, is not a coincidence, as in (Treur 2017) it was proven that any (state-determined) dynamical system as considered by Ashby (1960), Port and van Gelder (1995) can be modeled in temporal-causal network format. So, this gives confidence that in principle any hybrid adaptive network model can as well be (re)modelled as a reified temporal-causal network model.

It was shown by the many examples that the Network-Oriented Modeling approach based on network reification offers a huge potential for modeling quite complex adaptive network behaviour of any order. To execute such reified network models, it was shown in Chap. 9 that due to the unified approach for all levels of reification, quite compact software can be developed.

16.2 Conceptual Representations of Reified Networks: 3D Pictures and Role Matrices

Designing and presenting reified network models in a conceptually transparent manner comes with some minor challenges concerning structuring of pictures and data involved. For graphical conceptual representations of reified network structure pictures, this minor issue was resolved by using pictures that are literally transparent and are depicted in a 3D form with horizontal planes above each other for the base level and the reification levels, as shown in Fig. 16.1. Many pictures like this can be found in the book; e.g., Chap. 3, Figs. 3.4–3.10; Chap. 4, Fig. 4.3. This is in contrast to the flat 2D style of depicting networks, for example, in (Treur 2016) and many papers.

This 3D picture style has become a kind of standard now for these reified networks and can be found everywhere in the current book.

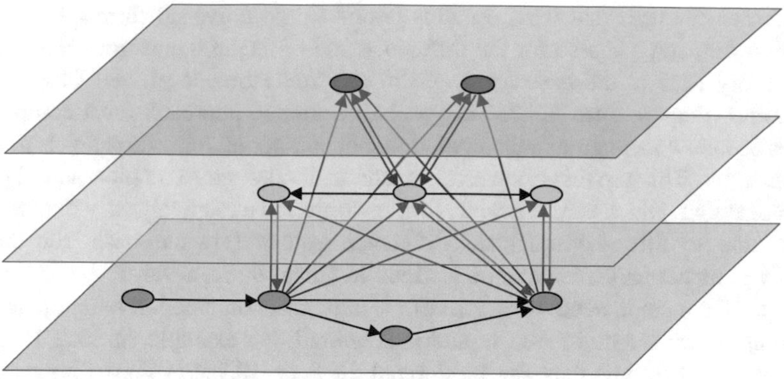

Fig. 16.1 The 3D style of graphical conceptual representations for reified networks

The other minor challenge of representing the relevant data was resolved by the choice for a specific table format to represent separately in a grouped form the different types or roles of these data (base connectivity, connection weights, speed factors, combination function weights, and combination function parameters) that define a specific reified network model: a separate role matrix for each of the roles played by these data. Also this format can be found at many places in the book (e.g., Chap. 2, Sect. 2.4, Box 2.1; Chap. 3, Sect. 3.7.2, Box 3.8; Chap. 4, Sect. 4.4.2, Box 4.1) and has now become a kind of standard. The modeling environment that was implemented makes use of this role matrix format as input (see Chap. 9, Sects. 9.2 and 9.4).

16.3 The Universal Combination Function and the Universal Difference and Differential Equation

Another remarkable theme that has occurred several times in the book concerns the universal combination function that was found for reified networks, and the universal difference and differential equation. For the non-reified case, in Chap. 2, Sect. 2.3.1 it was discussed how each state has its own combination function and basic difference and differential equation. These are specific for each state due to the specific values of the elements such as connection weights, speed factors, and combination function weights and parameters as specified in the role matrices. Given values for all these elements as specified in the role matrices, for each state its specific combination function and difference and differential equation can be derived in a standard manner as shown in Chap. 2, Sect. 2.4.2, Box 2.2

However, if a network is fully reified, there are not such static values for each base state, as they have become variables linked to their reification states.

The remarkable thing is that what remains is one single universal format for the combination function and also for the difference and differential equations that are used for every state at the base level and at any reification level; see Chap. 3, Sect. 3.5, and Chap. 4, Sect. 4.3.2. This enables a unified approach for a computational reified network engine implementation that handles all reification levels in a uniform manner. This provides powerful means to model very complex adaptive network behaviour, which can be adaptive of any order, in a relatively easy manner. And also due to this, the universal difference equation is the basis for the remarkably compact implementation described in Chap. 9, Sect. 9.4.

In Chap. 10 a more in depth treatment of the universal combination function and universal difference and differential equation is offered. For example, in Sect. 10.4, Box 10.4 it is shown how they can be derived. In Sect. 10.7 it is shown how the universal differential equation can be used for a compilation process that leads to the specification of a set of differential equations that can be run efficiently by any general purpose differential equation solver. This would be an alternative type of implementation, probably useful for large scale reified networks; this still has to be implemented and tested.

16.4 Analysis of How Emerging Reified Network Behaviour Relates to the Reified Network's Structure

The interesting theme how emerging network behaviour relates to network structure also occurs extensively in the book. The picture in Fig. 16.2 (also used in some other chapters) indicates the different relations for this theme.

The network structure is what is specified in the role matrices. The arrow at the bottom layer was addressed in more detail in Chap. 10, Sect. 10.6, Box 10.6 where it is shown how from the role matrices for the network structure characteristics the numerical representation in the form of the difference and differential equations can be derived that describes the network's dynamics. In Chap. 3, Sect. 3.7.4, and many examples in Sect. 3.6 it is shown how a reified network's emerging behaviour depends on the reified network's structure characteristics such as parameters of

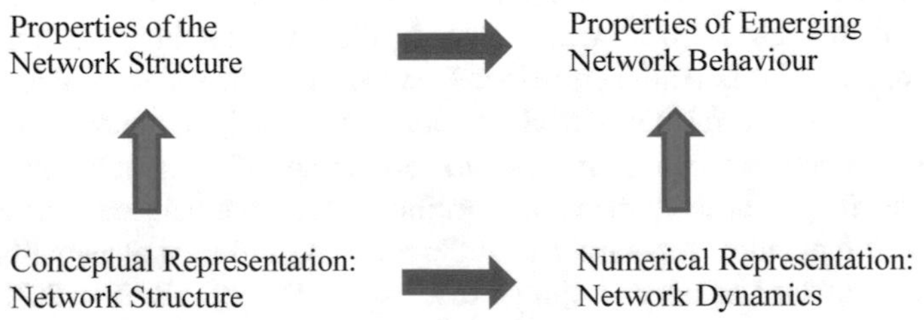

Fig. 16.2 Bottom layer: the conceptual representation defines the numerical representation. Top layer: properties of network structure entail properties of emerging network behaviour

combination functions. Similarly, in Chap. 5, Sect. 5.6; Chap. 6, Sect. 6.6, and Chap. 7, Sect. 7.5 this is shown for other example reified networks. All these deal with specific network models and relate to the bottom layer in Fig. 16.2.

However, in Chaps. 11 to 14 the step is made to the upper layer in Fig. 16.2. In these cases not specific network structures are considered, but classes of networks that satisfy certain network structure properties concerning connectivity and aggregation in terms of combination functions. In Chaps. 11 and 12 this is addressed for non-reified networks, leading to a number of results of the form that a certain set of network structure properties entails certain properties concerning the network's emerging behaviour, for example, that all states eventually get the same value. Similar analysis methods have been applied to reification states in Chaps. 13 and 14, in particular for adaptive bonding by homophily and for Hebbian learning, respectively. In this way in Chap. 13 results were found on how properties of reified network structure characteristics entail properties of community formation. In Chap. 14 results were found, for example, on how certain properties of combination functions for Hebbian learning entail the maximal final values of the connection weights.

Given that it is sometimes argued that emerging behaviour in general cannot be derived from the structure of a model, it might be felt as a surprise that this turns out not to be true for quite a number of cases, also nonlinear ones, as still many positive results have been found on the relation between network structure and network behaviour. All such results can be used for verification of network models. If in a simulation, behaviour is observed that contradicts one of these results, the network's implementation has to be inspected and improved.

16.5 The Network-Oriented Design Process Based on Reified Networks

The Network-Oriented Modeling approach for adaptive networks based on network reification supports the modeller in a number of ways. Design of a model takes place based on declarative building blocks concerning the network's structure such as the network's connectivity and the network's aggregation based on a choice of combination functions. As a network's behaviour is fully determined by the network's structure, also dynamics and (multi-order) adaptation of the designed network is specified in such declarative terms. The specification format based on role matrices that is used is declarative, implementation-independent and compact but detailed.

More specifically, the basic elements of the role matrices format are declarative: connection weights, combination functions, speed factors, and role matrices grouping them are declarative mathematical objects. Together these elements determine in a standard manner first-order difference or differential equations, which are declarative temporal specifications; for example, see Chap. 3, Sect. 3.5, and Chap. 4, Sect. 4.3.2; Chap. 9, Sect. 9.4; Chap. 10, Sect. 10.4. Indeed, a reified network model's behavior is fully determined by these declarative specifications, given some initial values. The modeling process is strongly supported by using such declarative building blocks: very complex (multi-order) adaptive patterns can be modeled easily, and in (temporal) declarative form. All procedural details are taken care of by the developed software environment, as described in Chap. 9, and hidden for the modeler. The generic setup of the combination function library makes it easy to add new combination functions that are automatically incorporated as soon as they have been added to the library; see Chap. 9, Sect. 9.3. The modeling approach also comes with simple means for mathematical analysis and background knowledge on the relation between a network's structure and the network's behaviour by which reified network models can be verified on correctness; see Chap. 3, Sects. 3.6 and 3.7.4; Chap. 5, Sect. 5.6; Chap. 6, Sect. 6.6; Chap. 7, Sect. 7.5, and Chaps. 11–14.

16.6 Relations to Longstanding Themes in AI and Beyond

The modeling approach presented in this book allows declarative modeling of dynamic and adaptive behaviour of multiple orders of adaptation. Traditionally declarative modeling approaches are a strong focus of AI. There are two longstanding themes in AI to which the work presented in this book contributes in particular: causal modeling (Kuipers 1984; Kuipers and Kassirer 1983; Pearl 2000) and metalevel architectures and metaprogramming (Bowen and Kowalski 1982; Demers and Malenfant 1995; Sterling and Beer 1989; Sterling and Shapiro 1996; Weyhrauch 1980). A main contribution to the causal modeling area is that this area is extended with dynamics and adaptivity of the causal modeling, not only addressing dynamics of the causal effects but also adaptive dynamics of the causal relations themselves. Without these extensions (multi-order) adaptive processes would be out of reach of causal modeling. A main contribution to the area of metalevel architectures and metaprogramming is that now network models are covered as well while traditionally the focus in this area is mainly on logical, functional and object-oriented modeling or programming approaches. The concepts of this area indeed show their value for Network-Oriented Modeling based on reified networks.

Also some areas outside AI are worth mentioning. One of them is the area of Cognitive (Neuro)Science and Philosophy of Mind (Kim 1996, 1998) where it is described how mental states function based on the causal networks they form; see Chap. 2, Sect. 2.2.1. The approach offered in this book applied in particular to Mental Networks can be considered as contributing to further formalisation and operationalisation of such perspectives. Also the more theoretical analysis of Hebbian learning (Hebb 1949) in Chap. 14 contributes to Cognitive (Neuro) Science.

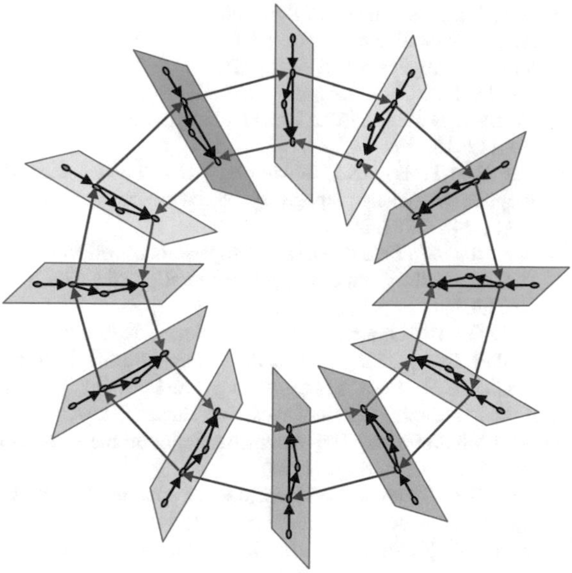

Fig. 16.3 Reified network model with cyclic reification level structure modeling Hofstadter (1979, 2007)'s Strange Loop

Another contribution to Philosophy of Mind and Cognitive (Neuro)Science is found in Chap. 8, where philosopher Hofstadter (1979, 2007)'s idea of Strange Loops as crucial element for human intelligence and consciousness is modeled by a three examples reified network with a cyclic reification level structure of the kind as summarized in Fig. 16.3 (for a bigger picture, see Chap. 8, Sect. 8.3, Fig. 8.5).

Another area with a long tradition worth mentioning is the area of Social Networks (e.g., Moreno 1934; Luce and Perry 1949; Leavitt 1951; Bott 1957; Banks and Carley 1996). The approach to Network-Oriented Modeling based on network reification provides a more principled approach to (multi-order) adaptive Social Networks than the often used hybrid approach; see Chap. 1, Sect. 1.4. Moreover, the more theoretical analyses of emerging behavioural patterns for social contagion in Chaps. 11 and 12 and for bonding based on homophily in Chap. 13 also particularly contribute to the area of Social Networks.

References

Ashby, W.R.: Design for a Brain. Chapman and Hall, London (second extended edition). First edition, 1952 (1960)

Banks, D.L., Carley, K.M.: Models for network evolution. J. Math. Sociol. **21**, 173–196 (1996)

Bott, E.: Family and Social Network: Roles, Norms and External Relationships in Ordinary Urban Families London, Tavistock Publications (1957)

Bowen, K.A., Kowalski, R.: Amalgamating language and meta-language in logic programming. In: Clark, K., Tarnlund, S. (eds.) Logic Programming, pp. 153–172. Academic Press, New York (1982)

Demers, F.N., Malenfant, J.: Reflection in logic, functional and object-oriented programming: a short comparative study. In: IJCAI'95 Workshop on Reflection and Meta-Level Architecture and their Application in AI, pp. 29–38 (1995)

Fessler, D.M.T., Clark, J.A., Clint, E.K.: Evolutionary psychology and evolutionary anthropology. In: Buss, D.M. (ed.) The Handbook of Evolutionary Psychology, pp. 1029–1046. Wiley and Sons (2015)

Fessler, D.M.T., Eng, S.J., Navarrete, C.D.: Elevated disgust sensitivity in the first trimester of pregnancy: evidence supporting the compensatory prophylaxis hypothesis. Evol. Hum. Behav. **26**(4), 344–351 (2005)

Fleischman, D.S., Fessler, D.M.T.: Progesterone's effects on the psychology of disease avoidance: support for the compensatory behavioral prophylaxis hypothesis. Horm. Behav. **59**(2), 271–275 (2011)

Hebb, D.O.: The Organization of Behavior: A Neuropsychological Theory (1949)

Hofstadter, D.R.: Gödel, Escher, Bach. Basic Books, New York (1979)

Hofstadter, D.R.: I Am a Strange Loop. Basic Books, New York (2007)

Kim, J.: Philosophy of Mind. Westview Press (1996)

Kim, J.: Mind in a Physical World: An Essay on the Mind-Body Problem and Mental Causation, MIT Press, Cambridge (1998)

Kuipers, B.J.: Commonsense reasoning about causality: Deriving behavior from structure, vol. 24. pp. 169–203, Artificial Intelligence (1984)

Kuipers, B.J., Kassirer, J.P.: How to discover a knowledge representation for causal reasoning by studying an expert physician. In: Proceedings of the Eighth International Joint Conference on Artificial Intelligence, IJCAI'83, pp. 49–56. William Kaufman, Los Altos, CA (1983)

Luce, R.D., Perry, A.: A method of matrix analysis of group structure. Psychometrika **14**, 95 (1949)

Leavitt, H.: Some effects of certain communication patterns on group performance. J. Abnorm. Soc. Psychol. **46**, 38 (1951)

Moreno, J.L.: Who Shall Survive? Nervous and Mental Disease. Publishing Company, Washington, DC (1934)

Pearl, J.: Causality. Cambridge University Press (2000)

Port, R.F., van Gelder, T.: Mind as motion: Explorations in the dynamics of cognition. MIT Press, Cambridge, MA (1995)

Sterling, L., Shapiro, E.: The Art of Prolog. MIT Press (Ch 17, pp. 319–356) (1996)

Sterling, L., Beer, R.: Metainterpreters for expert system construction. J. Log. Program. **6**, 163–178 (1989)

Treur, J.: Network-Oriented Modeling: Addressing Complexity of Cognitive, Affective and Social Interactions. Springer Publishers (2016)

Treur, J.: On the applicability of network-oriented modeling based on temporal-causal networks: why network models do not just model networks. J. Inf. Telecommun. **1**(1), 23–40 (2017)

Weyhrauch, R.W.: Prolegomena to a theory of mechanized formal reasoning. Artif. Intell. **13**, 133–170 (1980)

Printed in the United States
By Bookmasters